Asymptotic Multiple Scale Method in Time Domain

Asymptotic Multiple Scale Method in Time Domain

Multi-Degree-of-Freedom Stationary and Nonstationary Dynamics

Jan Awrejcewicz, Roman Starosta and
Grażyna Sypniewska-Kamińska

CRC Press is an imprint of the
Taylor & Francis Group, an **informa** business

First edition published 2022
by CRC Press
6000 Broken Sound Parkway NW, Suite 300, Boca Raton, FL 33487-2742

and by CRC Press
4 Park Square, Milton Park, Abingdon, Oxon, OX14 4RN

CRC Press is an imprint of Taylor & Francis Group, LLC

ISBN: 978-1-032-21941-7 (hbk)
ISBN: 978-1-032-21951-6 (pbk)
ISBN: 978-1-003-27070-6 (ebk)

DOI: 10.1201/9781003270706

Typeset in Nimbus
by KnowledgeWorks Global Ltd.

Access the Support Material: www.routledge.com/9781032219417

Contents

Preface

This book is addressed to the advanced-level audience including doctoral and graduate students, university and industry professors, researchers and scientists from various branches of engineering (mechanical, civil, electro-mechanical, etc.), as well as students and researchers from applied mathematics and physics. Its technical presentation level is appropriate to satisfy the mentioned audience consisting of scientists and engineers.

This condensed book offers challenging research, worked out by analytical approaches that explore the nonlinear features exhibited by various dynamic processes occurring in numerous branches of engineering and physics.

It has been observed over time that the various asymptotic methods (AMs), including multiple scale method (MSM), can be very attractive to engineers and applied nonlinear scientists in spite of significant progress in numerical techniques. The reason is motivated mainly by the intuitive development of ideas via experience of analytical study offered by MSM. Even in the case when we are interested in numerical solutions only an a priori AM/MSM analysis allows for choosing the most suitable analytical methods and sheds light on usually disordered and sometimes abundant numerically obtained data. We show and discuss efficiency of the used approaches by a study of numerous examples taken from mechanics and not necessarily limited to simple problems of 1 DoF systems.

It is shown through the book's material that MSM allows one not only to solve many problems of physics and technology (their application is not limited to problems with weak nonlinearity), but also to predict essential features of the analyzed nonlinear processes (vibrations). The latter may have either harmful effects from the point of view of applications or may give some benefits depending on needs and expectations.

Our material, though having its roots in development of nonlinear science, yields new questions and problems that science and technology have to face, and can be understood as a competition to the recently developed and widely used computational techniques based on the finite elements method (FEM), finite difference method (FDM), the Bubnov-Galerkin methods (BGM), etc. Our research experience shows that the MSM may not only predict archetypical features of nonlinear dynamic phenomena but also may compete with the results accuracy of the classical numerical methods.

Though the book deals with discrete mechanical systems in the form of coupled oscillators with a few degrees-of-freedom, the derived nonlinear governing ODEs are of the second-order and are transformed to their counterpart nondimensional forms. The latter can be obtained either by direct physical and mathematical modeling of numerous engineering problems where one may clearly separate the mass objects linked by massless stiffness and damping elements or via considering the continuous mass distribution of mechanical systems like rods, strings, beams, plates and shells governed by nonlinear PDEs.

Graduate and doctoral students can find (i) this material useful on the development of state-of-the-art multiple scale methods and their applications, focused on recent technological developments and the use of modern computational tools like *Mathematica*, (ii) introduction to the multiple scale method in time domain and (iii) numerous examples of vibrations of systems with a few degrees-of-freedoms, putting emphasis on detection and analysis of periodic orbits. The clear writing of the text implemented by numerous figures and computational results in graphic form allow the readers to understand complex phenomena such as nonlinear stationary and non-stationary processes, various resonance-jump phenomena (snap through), stability problems, etc. The book also includes novel contribution with regard to simplification of the problems via mathematical modeling, the employment of the limiting phase trajectories to quantify the nonlinear phenomena as well as discussion related to higher order asymptotic approximations (up to third-order ε^3).

As it has already been pointed out, the book material is not limited to studying only nonlinear ODEs and AEs (algebraic equations). The book is targeted also to researchers/engineers working in structural mechanics where the studied systems (rods, cables, beams, plates, shells and their interactions) are modeled by nonlinear partial differential equations (PDEs), i.e. they are of infinite dimension. However, the recent development of theoretical and computational approaches in many cases does not allow for reliable study of nonlinear PDEs, and usually the problem is truncated to studying of sets of nonlinear ODEs and AEs in the system time dependent coordinates. On the other hand, as it is well illustrated and documented in numerous papers and books, a strong reduction of the infinite dimensional problems to single-mode models (one second-order nonlinear ODE) are often quite poor idealizations of the actual studied problems governed by nonlinear PDEs.

Besides, in many cases the problems are reduced to a limited set of well-known archetypical oscillators and not necessarily are linked with the actual geometrical and mechanical properties of the studied continuous objects.

We think that our proposed methodology beyond the single-mode models may better approximate richness and variety of nonlinear interaction and complex behavior of structural members via two- and three-mode models (in fact, we consider 2 DoF and 3 DoF nonlinear problems).

All computations presented in the book are based on the *Mathematica* technical computing environment being widely available and extensively used by scientists and engineers. It allowed the authors to go beyond the traditional hand manipulation of analytical expressions and to use benefits of computer symbolic manipulations of the governing equations, simultaneously enriched by other computer-based methods employing computational and graphical visualization to achieve greater conceptual understanding. Clear equation manipulations yielded a concrete and tangible flavor and allows the reader to accommodate the offered flexible treatment of mathematical modeling and then the study of their own systems of nonlinear ODEs (and perhaps AEs).

In the first book chapter (section 1.1), literature review is presented. The emphasis is put on the historical background of the multiple scales method (MSM) and the recent results in this research field. The brief section 1.2 serves as an introduction to the well-known and widely used MSM.

Chapter 2 deals with nonlinear dynamics of the spring pendulum including resonant and non-resonant vibrations. The resonance curves are constructed and the stability analysis is carried out. Kinematically excited spring pendulum is analyzed in chapter 3. Three-time scales of MSM are employed and both parametric and primary resonances occurring either simultaneously or separately are investigated. Resonance response curves are presented with stability estimation of their branches.

The harmonically excited spring pendulum, whose longitudinal elasticity is of a cubic-type, is the subject of the analysis carried out in chapter 4. The problem is reduced to two first-order ODEs that are solved using the MSM. The initial value problem of the obtained modulation equations with regard to the amplitudes and phases is solved numerically. Comparison of the asymptotic (analytical) and numerical solutions exhibits their excellent coincidences. In spite of the steady state research, the chapter also offers analysis of the transient processes.

Chapter 5 is focused on dynamic behavior of a harmonically excited 3 DoF planar physical pendulum. The MSM combined with the asymptotic method yielded the modulated ODEs. Time histories and backbone resonance curves are obtained with emphasis put on stability of the periodic orbits. Three simultaneously occurred resonances are addressed as well as the energy transfer between vibration modes. The mathematical model of micromechanical gyroscope for measuring angular velocity consisting of two coupled nonlinear second-order ODEs is investigated in chapter 6. The employed MSM yields the approximate solutions to the initial value problem including both non-resonant and resonant dynamics. The derived analytical solutions are validated numerically.

Chapter 7 addresses problems of torsional vibrations (oscillations) of a two-disk rotating mechanical system. The governing equation is reduced to that describing the internal motion of the system, and then the latter is reduced to a first-order ODE with help of a complex representation of the phase variables. The two-time scales approach yielded the modulation equations. The fixed points are detected and studied. The amplitude-frequency curves are constructed including stability analysis. Stationary and nonstationary vibrations are discussed in detail, among others.

Dynamic behavior of oscillators with springs-in-series is investigated in chapter 8. In contrast to the previous mathematical models, the studied 1D oscillators are governed by differential and algebraic equations. The governing equations are solved either analytically or analytically and numerically. The non-resonant and resonant vibrations are analyzed based on the derived modulation equations with regard to the amplitude and phase. The obtained resonance curves are drawn in the form of dependences between the amplitude, modified phase and detuning parameter.

The nonlinear vibrations of rectangular micro-/nano-plates with an account of small-scale effects in a framework of the nonlocal elasticity theory are studied in chapter 9. Hamilton's principle yields the governing PDEs based on the Kirchhoff-Love hypotheses including the geometric nonlinearity introduced by the von Kármán model. The Bubnov-Galerkin approach reduces the problem to that of the second-order nonlinear ODEs. Then the MSM allowed one to derive the modulation equations, which serve for analysis of resonant and non-resonant vibrations. In particular, the so-called ambiguous and unambiguous back bone curves are detected allowing for prediction of the pull-in phenomena occurring during vibrations of micro/nanoplates.

The authors greatly appreciate the help of Mr. M. Kaźmierczak with the final manuscript preparation.

The financial support is acknowledged.and by the Polish National Science Center under the grant OPUS18 (UMO-2019/35/B/ST8/00980) and by the Ministry of Science and Higher Education in Poland, (0612/SBAD/3576).

Authors

Jan Awrejcewicz
Full professor
Faculty of Mechanical Engineering of the Lodz University of Technology
Member of the Polish Academy of Sciences
e-mail: jan.awrejcewicz@p.lodz.pl

Prof. Jan Awrejcewicz is Head of the Department of Automation, Biomechanics and Mechatronics at Lodz University of Technology and Head of the PhD School on "Mechanics". He is also recipient of Doctor Honoris Causa (Honorary Professor) of the Academy of Arts and Technology, Bielsko-Biala, Poland (2014), Czestochowa University of Technology, Poland (2014), Kielce University of Technology, Poland (2019), National Technical University "Kharkiv Polytechnic Institute", Ukraine (2019), Gdańsk University of Technology, Poland (2019), Prydniprowska State Akademy of Civil Engineering and Architekture, Dnipro, Ukraine (2021), Politehnica University Timisoara, Romania (2021), and Rector's Award for the best cited author from the Lodz University of Technology in 2021. His papers and research cover various disciplines of mechanics, material science, biomechanics, applied mathematics, automation, physics and computer-oriented sciences, with the main focus on nonlinear processes.

He authored/co-authored over 850 journal papers and refereed international conference papers and 53 monographs. He is editor-in-chief of 3 international journals and member of the editorial boards of 90 international journals (23 with IF) as well as editor of 33 books and 37 journal special issues. He also reviewed 45 monographs and textbooks and over 800 journal papers for about 140 journals, and he supervised 27 finalized and 6 ongoing PhD theses. He is a recipient of numerous scientific awards, including The Alexander von Humboldt Award for research and educational achievements (2010/2011 and 2016). For more information, please visit www.abm.p.lodz.pl

Roman Starosta is an Associate Professor of the Institute of Applied Mechanics at Poznan University of Technology (PUT), Poland. He received his MS in mechanical engineering from PUT at the Faculty of Mechanical Engineering and Management in 1993 and PhD in 2000 from the same university. He received his degree of habilitated doctor of technical sciences in 2012. He is the head of the Department of Technical Mechanics at PUT.

His scientific interests and research areas are dynamics of structures, fluid mechanics, asymptotic methods, and computational systems of algebra and symbolic transformations. He is the co-author of over 40 papers on nonlinear dynamics, energy harvesting, fluid-structure interactions, theory of elasticity, and teaching of mechanics.

Dr. Starosta's main professional activity has been education and research in engineering mechanics and fluid mechanics. He is the co-author of two textbooks on the subjects of computational mechanics and computer laboratory on fluid mechanics and biomechanics. He received several awards for being an outstanding teacher, including the medal from the Ministry of Education. He was a scholarship holder of the Foundation for Polish Science.

Dr. Starosta is a member of the main board of the Polish Society of Theoretical and Applied Mechanics. He was the chairman of several editions of the conference on Vibrations in Physical Systems.

Grażyna Sypniewska-Kamińska received her PhD in Technical Science with a specialty in Mechanics from the Poznan University of Technology in 1998. Her PhD thesis dealt with the dynamical interactions of porous media with the electromagnetic field. In 2019, she received the degree of Doctor of Science in Mechanical Engineering from the Poznan University of Technology. She is currently an Associate Professor at Poznan University of Technology. She is employed at the Institute of Applied Mechanics at the PUT since 1990. Her scientific and research interests concern nonlinear dynamics, asymptotic methods, computer methods in the area of applied mechanics of continuous and discrete systems, and inverse problems of heat conduction. She teaches mechanics, analytical mechanics, elasticity theory, mathematical physics as well as algorithmics, programming languages, and computer graphics at PUT. She is the co-author of an academic textbook in mathematical physics.

1 Introduction

This chapter provides motivation for writing this book. It is shown, based on the review of the state-of-the-art of the current development and application of the multiple scales method (MSM), that the latter allows the solving of many problems of vibrations of mechanical systems having a few degrees of freedom. In addition, in many cases also continuous systems dynamics governed by nonlinear PDEs can be reduced to a system of coupled oscillators, and it can be analyzed by MSM yielding reliable results. The chapter also presents benefits of the book over the existing books aiming at analytical treatment of nonlinear ODEs and PDEs. In addition, a short description of the main ideas of the MSM is given.

1.1 LITERATURE REVIEW

As the proposed book (monograph) deals with the asymptotic method (AM), particularly the multiple scale method (MSM), we begin our brief description of the state-of-the-art combination of these two approaches. Since the asymptotic methods have a long history in nonlinear science and there are dozens of papers devoted to that topic, we restrict our introductory background only to a few books devoted to the latter approach and methodology (be aware that sometimes the authors use the phrase "asymptotology" to point out the importance of this method as an independent branch of nonlinear science). However, in spite of the mentioned limitations, the reviewed books include long list of references, which can attract interest of the reader.

Roseau [182] used Fourier-Laplace transforms, operational calculus, special functions, and asymptotic methods to tackle problems borrowed from hydrodynamics and elasticity theory. Plane, cylindrical and spherical waves were considered. The book also addresses the methods of steepest descent, the asymptotic representation of Hankel's functions and Laplace transform.

De Bruijn [59] extended his pioneering study published in 1958 and devoted to asymptotic analysis through a rigorous process based on original worked examples. Book material covers the estimation of implicit functions and the roots of equations, various methods of estimating sums, saddle point method, and asymptotic behavior of solutions of differential equations.

Murray [148] proposed a useful introductory book to asymptotic analysis because of its capacity to circumvent theoretical and practical obstacles by addressing integrals on the real line and in the complex plane arising in different contexts and in solutions of differential equations. The book includes exercises and a short bibliography.

The monograph by Awrejcewicz et al. [18] reflects the multidisciplinary nature of the asymptotic approaches devoted to study problems occurring in mechanics, physics and applied mathematics. Both problems governed by nonlinear ODEs and

DOI: 10.1201/9781003270706-1

PDEs are addressed with emphasis on asymptotic and homogenization techniques, and Padé approximations.

Lakshmanan and Rajaseekar [115] addressed problems of nonlinear dynamics with emphasis on integrability, chaos and patterns. The book is well-suited for students, teachers and researchers in physics, mathematics and engineering needs, since it includes description of the fundamental concepts, theoretical background matched with experimental and numerical techniques, and supplemented by numerous examples.

Detailed and systematic treatment of asymptotic approaches in the theory of plates and shells, i.e. static and dynamic problems governed by nonlinear PDEs, is presented in the book by Awrejcewicz et al. [19]. The book presents basic principles of asymptotics and their applications, traditional and novel approaches to deal with regular and singular perturbations and the composite equations approach. It can be also useful as a handbook of methods of asymptotic integration.

Awrejcewicz and Krysko [21] revisited the classical asymptotic approaches matched with numerical simulations based on the integrated mathematical-analytical treatment of various objects and processes taken from interdisciplinary fields of mechanics, physics and applied mathematics. The book material provides comprehensive coverage of asymptotic approaches, regular and singular perturbations, 1D nonstationary nonlinear waves, PadÃl' approximations, oscillators with negative Duffing type stiffness, differential equations with discontinuities, as well as averaging of ribbed plates and chaos foresight.

Miller [144] employed the asymptotic expansion developed in the research context of wave propagation. It may serve as introduction (textbook) to asymptotic analysis. It deals with asymptotic evaluation of integrals and differential equations, locating zeros of Taylor polynomials of entire nonvanishing functions and counting integer lattice points in subsets of the plane with various geometrical properties of the boundary.

Awrejcewicz and Holicke [20] employed the asymptotic and perturbation techniques to develop the Melnikov-type methods applied to high dimensional systems governed by nonlinear ODEs and hence to predict the occurrence of nonsmoothed chaotic orbits. In particular, it emphasizes through a number of examples taken from mechanics that the developed approach may reveal prediction of both stick-slip and slip-slip chaotic orbits of mechanical system of coupled oscillators with dry Coulomb type friction.

Bouche et al. [43] solved the scattering problem at short wavelengths of electromagnetic fields diffracted by an object. It is illustrated how the asymptotic methods may provide closed form expansions for the diffracted fields.

Paulsen [166] employed theory of asymptotics to approximate limits, integrals, difference and differential equations. Both easy and difficult problems were solved.

Fikioris et al. [69] employed several asymptotic techniques useful to study research fields associated with electromagnetics and antennas. In particular, the idea of analytic continuation of functions of a single complex variable played a crucial role in the analysis. The application-oriented research was supplemented by material covering integration by parts and the Riemann-Lebesque lemma, the use of contour integration in conjunction with other methods, techniques related to Laplace's method and Watson's lemma, the asymptotic behavior of certain Fourier sine and cosine transforms, and the Poisson summation formula.

Andrianov et al. [10] presented the theoretical background of asymptotic approaches and their application in solving mechanical engineering-oriented problems of structural members, primarily plates (statics and dynamics) with mixed boundary conditions. The book offers the lesser-known approaches like the method of summation and construction of the asymptotically equivalent functions, the methods of small and large perturbation parameters, and the homotopy perturbations method. It may serve as the guiding source for solving problems related to nonlinear dynamics of rods and beams, and it explains bridges between the Adomian and homotopy perturbation approaches.

Chipot [51] investigated new issues relating to PDEs of asymptotic behavior of problems set in cylinders. The analysis included the existence and uniqueness of solution in unbounded domains, anisotropic singular perturbations, periodic behavior forced by periodic data, and anisotropic singular perturbations.

Andrianov et al. [11] proposed the construction of efficient and computationally oriented models of multi-parameter complex systems by applying asymptotic methods. Nonlocal and higher-order homogenized models are derived as well as local fields on the micro-level and the influence of so-called nonideal contact between the matrix and inclusions are modeled and analyzed. It includes investigation on perforated membranes, plates, shells, and the asymptotic modeling of imperfect nonlinear interfaces.

Wasow [230] reviewed and employed various asymptotic methods for ODEs. Regular and singular points, asymptotic power series, Jordan's canonical forms, and asymptotic power series were addressed.

Sung et al. [207] developed and employed the asymptotic modal analysis to predict and study the high-frequency vibroacoustic response of structural and acoustical systems. It is based on evaluation of the average modal response only at a center frequency of the bandwidth and does not sum the individual contributions from each mode to obtained final result.

It is demonstrated how the method can provide spatial information as well as the estimates of the accuracy of the solution for a particular number of modes. Both structural and acoustical systems subjected to random excitation are studied.

Mekhtiev [140] presented homogeneous solutions in static and dynamic problems of anisotropic theory of elasticity based on consideration of a hollow cylinder. Makhtiev developed a general theory for a transversally isotropic spherical shell including methods for constructing inhomogeneous and homogeneous solutions governing stress-strain behavior of the studied shells.

The so far described monographs are mainly concerned with rigorous asymptotic approaches, but there are numerous works employing the asymptotic approaches in the frame of the multiple scale method.

Rand and Holmes [177] studied a pair of weakly coupled van der Pol oscillators with emphasis paid to phase-locked periodic motions. The problem was reduced to the study of three algebraic equations.

Benedettini and Rega [40] employed a high-order perturbation analysis for the primary resonance while studying nonlinear dynamics of an elastic cable under planar excitation. Multivaluadness of the response curves were illustrated.

Luongo and Polone [130] carried out the multiple time-scale investigation for divergence-Hopf bifurcation of imperfect symmetric systems. The method was validated through analysis of a 2DoF rigid bar system (Augustin's model) and transversal flow.

Ghosh et al. [78] developed a multiple scale finite element method by combining the asymptotic homogenization theory with Voronoi cell finite element method for microstructural modeling and evaluation of stress and strains. Effects of size, shape, orientation, and distribution of heterogeneities on the local/global response were addressed.

Sanchez [186] implemented the MSM to study asymptotic solutions and normal forms for nonlinear oscillatory problems. A few Duffing type oscillators were employed to illustrate the power of the developed procedure.

Rega et al. [180] used two analytical approaches to construct asymptotic models for the nonlinear 3D responses of an elastic suspended shallow cable under harmonic excitation. Primary resonances of the in-plane and out-of-plane symmetric/antisymmetric modes were investigated based on the multiple scales method. Frequency-response curves were obtained and discussed.

Belhaq and Houssini [38] proposed a novel method for construction of an asymptotic expansion of the quasi-periodic solution based on the KBM (Krylov-Bogolubov-Mitropolskiy) method combined with the multiple scale method. The authors also considered features of chaos and its suppression based on 1DoF nonlinear oscillator parametrically and externally driven.

Leamy and Gottlieb [117] studied internal resonances in whirling strings with an account of longitudinal dynamics and material nonlinearities. Direct application of the multiple scales method on the three governing PDEs was addressed, as well as periodic, quasi-periodic, and aperiodic motions were detected and discussed.

Tuwankotta and Verhulst [215] studied resonances exhibited by 2DoF autonomous Hamiltonian systems based on the asymptotic analysis. The investigations included a mechanical system with symmetry, and the elastic pendulum and HÃl'non-Heiles family of Hamiltonians.

Belhaq and Lakrad [39] extended the classical multiple scales method by employing the Jacobian elliptic functions. The advantages of the proposed method were outlined.

Warmiński [223] analyzed a 1DoF nonlinear oscillator including van der Pol self-excitation term and parametric Mathieu term. Synchronization effects were demonstrated based on the averaging analytical approach.

Wirkus and Rand [236] used the averaging procedure to study the dynamics of two van der Pol oscillators coupled by delayed velocity. The simplified system of three-flow equations was obtained, and then stability of periodic motions of original system was investigated.

Belhaq et al. [36] found asymptotic solutions of damped nonlinear quasi-periodic Mathieu equations using a double multiple scales method. Explicit analytical approximations were reported and then compared with the numerical integration of the original governing ODE.

Lacarbonara et al. [112] employed two analytical methods, i.e. the full-basis Galerkin discretization approach and the direct treatment, both based on the multiple scales technique. Closed-form conditions for nonlinear orthogonality of the modes were discussed, among others.

Abdulle and Weinan [1] developed a numerical method called the finite difference heterogeneous MSM for solving multiscale parabolic systems. The novel method was supplemented by a few illustrative examples.

Janowicz [90] employed the MSM to study quantum-optical problems. The following problems were addressed: spontaneous emission, resonance fluorescence, and cavity quantum electrodynamics.

Luongo et al. [129] adopted the multiple scale method to study 1:1 resonant multiple Hopf bifurcations of discrete autonomous dynamical systems. They used fractional power expansion of a perturbation parameter, and they obtained m-order differential bifurcation equation in the complex amplitude of the unique critical vector. In order to illustrate the algorithm, mechanical systems subjected to aerodynamic forces triggering 1:1 resonant double Hopf bifurcations were presented.

Park and Liu [164] proposed a novel multiple scale method based on the bridging scale technique and illustrated its advantages in comparison to the classical multiple scale method. Fully nonlinear 1D problems were considered putting emphasis on coupling continuum and atomistic simulations.

Warmiński [224] studied vibrations of parametrically and self-excited 2DoF system with a Duffing nonlinearity. He detected, using the multiple scales method, synchronization phenomena near the principal resonances in the neighborhood of the first p1 and the second p_2 natural frequencies and near the combination resonance $(p_1 + p_2)/2$.

Abouhazim et al. [2] investigated three-period quasi-periodic oscillators in the vicinity of 2:2:1 resonance in a self-excited quasi-periodic Mathieu equation using the method of multiple scales. The efficiency of the method was validated.

Luongo and Di Egidio [128] employed the multiple scale method to a 1D continuous model to derive equations governing the system's asymptotic dynamics around a bifurcation point. Nonlinear, integro-differential equations of motion were derived and expanded up to cubic terms. Divergence, and Hopf and double zero bifurcations were revealed. The employed multiple scales analysis allowed one to study the three bifurcations and the relevant bifurcation equations were derived in their normal form.

Abouhazim et al. [3] investigated the damped cubic nonlinear quasiperiodic Mathieu equation in the vicinity of the principal 2:2:1 resonance. In particular, the effect of damping and nonlinearity on the resonant quasiperiodic motions were reported.

Srinil et al. [195] investigated resonant multimodal dynamics due to 2:1 internal resonances in the finite-amplitude free vibrations of horizontal/inclined cables. A second-order asymptotic analysis under planar 2:1 resonance was analyzed with a help of the multiple scales. Approximate horizontal/inclined cable models were validated numerically.

Kramer et al. [107] addressed the problems dealing with applications of MSM to three or more time scales.

Bakri et al. [34] studied persistence and bifurcations of Lyapunov manifolds by combination of averaging-normalization and numerical bifurcation methods. The developed theory was applied to the case of parametrically excited wave equations.

Kirillov and Verhulst [99] studied paradoxes of destabilization of a conservative or nonconservative system by small dissipation. Subsequent developments of the perturbation analysis of dissipation-reduced instabilities and application over this period, involving structural stability of matrices, Krein collision, and Hamilton-Hopf bifurcation were discussed.

Gottlieb and Cohen [79] studied the self-excited oscillations of a string on an elastic foundation under a nonlinear feed-forward force. The employed asymptotic multiple scale method yielded slowly varying evolution equations. The derived bifurcation structure included multiple regions of both stable and unstable coexisting periodic solutions defined by primary and secondary Hopf stability thresholds. They found existence of quasiperiodic combination-tone solutions and complex nonstationary solutions that emerged in a range of the asymptotically predicted unstable solutions.

Suchorsky et al. [205] employed two-variable expansion perturbation method to study oscillations of the van der Pol type systems with delayed feedback. The resulted amplitude-delay relation predicted two Hopf bifurcation curves, such that in the region between these curves, oscillations were quenched.

Warmiński [225] considered vibrations of nonlinear coupled parametrically and self-excited oscillators driven by an external harmonic force. Application of nonlinear normal modes theory yielded analytical solutions, which were validated by numerical simulations.

Zulli and Luongo [245] considered a 2DoF nonlinear system modeling dynamics of two towers exposed to turbulent window flow and linked by a nonlinear viscous device. Periodic and quasiperiodic solutions were revealed and studied by using a perturbation technique. In particular, effects of the viscous device on the dynamics of the structure were illustrated including mitigating the vibrations of the two independent towers.

Luongo and D'Annibale [127] investigated the combined effects of conservative and nonconservative loads on the mechanical behavior of an unshearable and inextensional visco-elastic beams. The post-critical analysis was carried out around a double-zero bifurcation using the multiple scale method. Destabilization region existence was detected.

Aginsky and Gottlieb [4] carried out the asymptotic investigation of the nonlinear fluid-structure interaction of an acoustically excited clamped panel immersed in an inviscid compressible fluid based on the multiple scale analysis. Both primary resonance and a 3:1 internal resonance between the panel fifth and ninth modes were studied. The coexisting bi-stable periodic solutions, and quasiperiodic response showing spatially periodic modal energy transfer for both panel and fluid were detected and studied.

Cacan et al. [48] developed an enriched multiple scales method and studied periodic solutions of the classical forced Duffing and van der Pol oscillators. Analytical investigations were validated by numerical simulations.

Settimi et al. [187] employed multiple scale asymptotics to carry out an external feedback control in a nonlinear continuum formulation of a noncontact AFM model. The investigation included controllable periodic dynamics and additional periodic and distinct quasiperiodic solution beyond the asymptotic stability thresholds.

Warmiński [226] investigated frequency locking in a nonlinear MEMS oscillator harmonically driven with an account of time delay. It was illustrated how interactions between self- and parametric excitation yielded either quasiperiodic or periodic responses. The amplitudes of the latter were found analytically by the multiple scale method in the second-order perturbation. The frequency locking zones and existence of the internal loop were controlled by selection of gains and time delay.

Mi and Gottlieb [141] derived and investigated nonlinear dynamics of a tethered sphere in uniform flow. The carried out asymptotic analysis combined with the multiple scales method revealed existence of quasiperiodic and nonstationary solutions.

Warmiński et al. [229] revisited nonlinear vibrations of a sagged cable with two in-plane and two out-of-plane vibration modes by the multiple time scale method. The nonlinear modal coupling as well multimodal responses were investigated.

Belhaq and Hamdi [37] considered energy harvesting from quasiperiodic vibrations of a van der Pol oscillator with time-varying delay amplitude coupled to an electromagnetic energy harvesting device. Double-step perturbation method was employed to estimate the amplitude of the quasiperiodic vibrations for the resonance and in the case of delay parametric resonance.

Jacobsen [89] employed the method of multiple scales to systems of ODEs and boundary layer problems of ODEs and PDEs. In particular, Schrödinger and Maxwell equations were studied.

Kalashnikov et al. [94] applied the multiscale approach to study nonlinear systems with noise. They gave hints to study systems ranging from lasers to nanostructures.

Mora and Gottlieb [146] analyzed vibrations of a parametrically excited microbeam-string affected by nonlinear damping. Both principal parametric resonance and 3:1 internal resonance were studied using the asymptotic multiple scales method. A bifurcation structure including the coexisting in-plane and out-of-phase solutions, Hopf bifurcations, and conditions for the loss of orbital stability combined with nonstationary quasiperiodic solutions and chaotic strange attractors were reported.

Kovaleva et al. [103] considered the liberational dynamics of the chains in paraffin crystals. The vicinity of the rotation phase transition and the nonlinear spectra of the system were studied.

Rusinek et al. [183] presented analytical solutions of a nonlinear 2DoF model of a human middle ear with shape memory element by the multiple time scales method. The parameters leading to nonperiodic or chaotic motion were also estimated.

Smirnov et al. [193] analyzed torsion oscillations in quasi-one-dimensional lattices with periodic potentials of the nearest neighbor interaction. Based on the asymptotic method, dispersion relations for the nonlinear normal modes for a wide range of oscillation amplitudes and wave numbers were reported. It was demonstrated how the localized oscillations (breathers) appeared near the long wavelength edge, while the short wavelength edge contained only dark solitons.

Wilbanks et al. [234] used the multiple scales technique to analyze two-scale command shaping feed-forward control method to reduce undesirable residual vibrations of traditional and nontraditional Duffing systems.

Pender et al. [168] studied how delayed information impacts queueing systems with a help of the asymptotic analysis. It was found that the time-varying arrival rate did not impact the critical delay unless the frequency of the time-varying arrival rate was twice that of the critical delay.

Elliot et al. [64] compared the direct normal form and MSM through frequency detuning by computing the backbone curves of 1DoF and 2DoF oscillators. It was shown that both of them gave identical results up to ε^2 order.

Kovaleva et al. [104] analyzed stationary and nonstationary oscillatory dynamics of the parametric pendulum by using limiting phase trajectory concept. A reduced order model was proposed allowing prediction existence of highly modulated regimes outside the traditionally considered range of initial conditions.

Guo and Rega [80] formulated solvability conditions while carrying out multiscale dynamic analysis of 1D structures with nonhomogeneous boundaries. The formulation allowed one to study four typical continuous structures, i.e. strings, cables, beams, and arches.

Kovaleva et al. [104] studied low- and high-amplitude oscillators of three nonlinear coupled pendula (trimer) beyond the quasilinear approximation. The reduction of the system dimension via the asymptotic procedure allowed for revealing energy exchange and nonstationary energy localization. The beating-like periodic and quasiperiodic recurrent energy exchange between the pendula was addressed.

Fronk and Leamy [72] revealed angle- and amplitude-dependent invariant waveforms, and plane-wave stability in 2D periodic media by higher-order multiple scales analysis. Simultaneous analysis of nonlinear shear lattices confirmed that inclusion of higher order terms in the injected waveforms significantly reduced the growth of higher harmonics. Implications for encryptions strategies and damage detection by using weakly nonlinear lattices were suggested.

Rand et al. [178] employed the perturbation technique combined with the multiple scale method to derive and analyze a simplified third-order model capturing the key features of the dynamics of microscale oscillators with thermos-optical feedback. In addition, the bifurcation diagram of the system was presented.

Clementi et al. [54] addressed the internal resonance of a 2DoF mechanical system with quadratic and cubic nonlinearities using the multiple scale method. A few unexpected or relevant features were highlighted, and hints to exploit the proposed results were proposed.

Guo and Rega [81] showed that the full-spectrum forced solutions at lower-order and the high-order coss-interactions with the structural modes were captured by the direct perturbation technique matched with the multiple scales method, but not by the discretized perturbation. Two different correction schemes were utilized to remove the occurred errors. The obtained results can be employed to analyze structures with initial curvature.

Warmiński [227] studied regular and chaotic vibrations of a nonlinear structure under the self-, parametric and external excitations. Approximate analytical solutions were derived based on the multiple scale method. The similarities and differences between the van der Pol and Rayleigh models were demonstrated for periodic, quasiperiodic and chaotic oscillations.

Warmiński et al. [228] analyzed nonlinear vibrations of an extensional beam with tip mass in slewing motion. The analytical solutions to the governing PDEs were found by the multiple scale method. Hardening and softening phenomena were addressed.

El-Dib [63] combined MSM with the parameter expansion to estimate stability of the Riemann-Liouville fractional derivative applied to the cubic delayed Duffing oscillator. The problem was reduced to study linear differential equation with the help of the algebraic one.

Lenci et al. [119] studied the internal resonances between the longitudinal and transversal vibrations of forces on Timoshenko beam with an axial end spring by means of the multiple scale method. The results reliability was discussed. Effects of jumping phenomena from hardening to softening by crossing the exact internal resonance value were illustrated.

In spite of the previously described direct application of the multiple scale method employed to dynamics of discrete (governed by ODEs) systems, the MSM (Multiple Scale Method) can be employed to analyze dynamics of the continuous mechanical systems like strings, beams, plates, and shells and their interplay governed, in general, by nonlinear PDEs. Though the latter systems have infinite dimension, there is a way to unveil the most important or true essence of the system's nonlinear features within two steps. First, the problem of infinite dimension is reduced to that of finite one, i.e. the governing nonlinear PDEs are reduced to finite number of nonlinear second-order ODEs by the numerous methods available (see Awrejcewicz et al. [25]). Second, the obtained ODEs or perhaps nonlinear ODEs and AEs, can be approximately solved by the asymptotic approaches and the multiple scales method. The latter reduced procedures from PDEs to ODEs may result in a new nonlinearities beyond the classical/typical ones, thus enriching the well-known gallery of the so called paradigmatic systems with a few degrees-of-freedom, usually 1DoF, 2DoF, and 3DoF systems. As it was widely illustrated, the paradigmatic systems may unveil existence of nonclassic and unusual nonlinear phenomena.

It is worthy to mention that the multiple scale method can also be directly employed to study PDEs governing dynamics (or statics) of structural members. For example, Caruntu and Oyervides [49] obtained the amplitude-frequency response of soft electrostatically actuated MEMS clamped circular plates to study the primary resonance. The MSM and ROM were used and compared, exhibiting the excellent agreement for amplitudes. The effects of voltage and damping parameters were also illustrated.

In what follows, we briefly address a few other books which may compete with our book.

Nayfeh and Mook [155] devoted a book to wide description of nonlinear oscillations as it is mentioned in its title. Description of the multiple scales method is addressed in section 2.33, 3.31, 4.33 and 8.1.2 to study conservative and nonconservative, parametrically excited 1DoF systems and traveling waves. The book oriented on engineers, physicists and applied mathematicians has gained wide acceptance in scientific community measured by the citations volume, but its scope is very large. It mainly covers comprehensive treatment of nonlinear vibrations from the point of view of perturbation theory partially including the multiple scale technique. *Our book is only target at asymptotic methods combined with multiple scale methods beyond the simple examples on 1DoF systems.*

Kervorkian and Cole [97] attempted to present various perturbation techniques to tackle the physical problems governed by ordinary and partial differential equations, integral equations, integro-differential equations, and difference equations. It is rather directed to applied mathematicians, since it provides more rigorous approach to perturbation problems. In particular, it addresses the singular perturbation problems including the layer-type problems and cumulative perturbation problems. The multiple scale method was described in chapter 3 including the Lindstedt's method and dealing with the second-order weakly nonlinear autonomous systems, second-order equations with variable coefficients, and second-order PDEs (section 4.4). Its 2nd edition (1985) entitled (Kervorkian and Cole [98]) was extended to cover a new material based on the previously published J.D. Cole's work (1968). Both books offered rather dense material with nontrivial and not enough clear mathematical concepts, and makes difficult to be used for engineers. *In contrast, our book describes only the multiple scale methods based on the clearly presented examples from mechanical engineering governed by nonlinear ODEs (the both mentioned books addressed mainly fluid-solid and fluid-fluid problems), examples of which can be easily repeated even by a newcomer via computer added symbolic computations.*

Nayfeh [150] unified various techniques of perturbations (asymptotic expansions) pointing out their similarities, advantages/disadvantages, as well as their limitations including a number of exercises. The method of multiple scales was described in chapter 6 including description of the method, application of the derivative expansion method, the two-variable expansion procedure, and its generalization. *However, all included examples were limited to consideration of 1DoF systems, in contrast to our book content.*

The next book of Nayfeh [151] stands as a comparison to the previously described book, and it was written in more elementary way to be accessible by advanced undergraduate and graduate students. Its content included novel material of asymptotic expansion and integrals and determination of the adjoins of homogeneous linear equations and the solvability conditions of linear nonhomogeneous problems. *In contrast, our book is aimed only at multiple scale methods devoted to studying engineering problems though it includes examples of various nonlinear and novel features revealed by systems with a few-degrees-of freedom.*

Bush [47] stands as an introduction to perturbation methods, addressing the problems of boundary layers and fluid flow but the multiple scale technique was limited to study only lubricated bearings. *In contrast, our book is only devoted to studying problems taken from solid mechanics with the use of only multiple scales method.*

Ramnath [176] is focused on application of MSM to aerospace problems. It bridges the gap between applied mathematics and engineering problems based on combining the asymptotic analysis, perturbation theory and multiple-scales method. *In contrast, results of our book, due to the introduction of non-dimensional governing equations, can be used in various branches of engineering, applied mathematics and physics. More complicated examples are provided with regard to the studied nonlinear ODEs offered by Ramnath.*

Pavliotis and Stuart [167] described a set of methods for simplification of a wide variety of problems mainly exhibiting two separated characteristic scales. The book material is based on the employment of averaging and homogenization methods to the problems governed by partial and ordinary differential equations, stochastic differential equations and Markov chains. It widely uses approximation of singularly perturbed linear equations. *In contrast, our book is only focused on the MSM and its applications to engineering problems while the mentioned previous book attempts to cover a wide-ranging setting in one volume. Also the use of the phrase multiscale methods do not necessarily address the problems covered by MSM.*

Finally, let us address the so-called reconstitution problem of solvability conditions and its impact on our book material.

In all the problems considered, there are so-called secular terms. They appear in the equations of the second and higher orders of approximation. These terms lead to the unbounded part of the solution of the differential equation. Therefore they must be zeroed to yield the so-called *solvability conditions*. The solvability conditions play an important role in the MSM. They are transformed using the real representation of the complex functions to *the modulation equations.*

The modulation equations describe the behavior of amplitudes and phases of the oscillations in slow time scales. These equations are derived for each approximation level, and hence they can be solved sequentially on the time scales τ_1, τ_2 etc. to obtain the relationship $a(\tau_1, \tau_2, ...)$ and $\psi(\tau_1, \tau_2, ...)$. However, such a method is inconvenient and not very readable. In this book, another approach to the reconstitution of modulation equations is applied. In essence, the solvability conditions for successive approximation levels are combined into a single equation for amplitudes and analogously for phases by returning to true-time τ, which is indeed the fastest time scale. Then these equations can be integrated to obtain amplitude and phase modulation at true time scale $\tau \equiv \tau_0$. The above procedure is called the *reconstitution method* [131, 152].

1.2 MULTIPLE SCALE METHOD

This section presents the main idea of the multiple scale method for solving the dynamical problems of motion of discrete systems. The multiple scale method, like other asymptotic approaches, is an analytical, approximate method. In general, it allows for a uniformly accurate asymptotic distribution over a much wider range of arguments than the classical small parameter method. Its main idea is to introduce different time scales to capture the phenomena specific to a given scale, e.g. slowly changing processes. This approach is described in detail in the numerous references mentioned in the previous section. Therefore, here we describe only briefly this approach while the detailed step-by-step procedures are presented in the subsequent chapters dealing with the case studies.

The outline of the method and its main steps used to study the motion of discrete systems are presented below. For the sake of simplicity, let's focus on a one-degree-of-freedom system. In general, the nonlinear equation of motion of such a system can be written as

$$D(u) = f(\tau), \tag{1.1}$$

where u is generalized coordinates describing the motion of the system, D is the nonlinear differential operator of second-order, and $f(\tau)$ is the known function of time describing external loading.

It is postulated that some elements of the differential operator D and the coefficients of the function f are of the order of the powers of a small parameter

$$D(u;\varepsilon,\varepsilon^2,...) = \Phi(f(\tau);\varepsilon,\varepsilon^2,...), \tag{1.2}$$

where Φ is a known function, whereas $0 < \varepsilon < 1$ is a so-called small perturbation parameter.

Then we assume the solution of equation (1.2) in the form of a power series expansion with respect to a small parameter

$$u(\tau;\varepsilon) = \sum_{k=0}^{k=\infty} \varepsilon^k u_k(\tau_0,\tau_1.\tau_2,...), \tag{1.3}$$

where $\tau_0 = \tau$, $\tau_1 = \varepsilon\tau$, $\tau_2 = \varepsilon^2\tau$ etc. are time scales treated in the further analysis as independent variables. In this approach the scale τ_{i+1} corresponds to slower processes than the scale τ_i. Substituting expansion (1.3) to equation (1.2) yields

$$D\left(\sum_{k=0}^{k=\infty} \varepsilon^k u_k(\tau_0,\tau_1.\tau_2,...);\varepsilon,\varepsilon^2,...\right) = \Phi(f(t);\varepsilon,\varepsilon^2,...). \tag{1.4}$$

Since unknown functions now depend on a number of independent variables, the derivatives of the operator D are understood as total derivatives, and they take the following form

$$\frac{d}{d\tau} = \sum_{k=0}^{\infty} \varepsilon^k \frac{\partial}{\partial\tau_k} = \frac{\partial}{\partial\tau_0} + \varepsilon\frac{\partial}{\partial\tau_1} + \varepsilon^2\frac{\partial}{\partial\tau_2} + ..., \tag{1.5}$$

$$\frac{d^2}{d\tau^2} = \frac{\partial^2}{\partial\tau_0^2} + 2\varepsilon\frac{\partial^2}{\partial\tau_0\partial\tau_1} + \varepsilon^2\left(\frac{\partial^2}{\partial\tau_1^2} + 2\frac{\partial^2}{\partial\tau_0\partial\tau_2}\right) + 2\varepsilon^3\frac{\partial^2}{\partial\tau_1\partial\tau_2}... \tag{1.6}$$

The definitions (1.5)-(1.6) are then entered into equation (1.4), which can be represented as

$$L(u_0) + \varepsilon L(u_1) + \varepsilon^2 L(u_2) + ... = F(u_0,u_1,u_2,...;\varepsilon,\varepsilon^2,...) + \Phi(f(t);\varepsilon,\varepsilon^2,...), \tag{1.7}$$

where L is a second-order linear partial derivative operator, and F stands for the difference of operators

$$F(u_0,u_1,u_2,...;\varepsilon,\varepsilon^2,...) = L(u_0)+\varepsilon L(u_1)+\varepsilon^2 L(u_2)+... \\ -D\left(\sum_{k=0}^{k=\infty}\varepsilon^k u_k(\tau_0,\tau_1,\tau_2,...);\varepsilon,\varepsilon^2,...\right). \tag{1.8}$$

After expanding the right side of equation (1.7) with respect to the powers of a small parameter ε, we get

$$L(u_0)+\varepsilon L(u_1)+\varepsilon^2 L(u_2)+... = \\ F_0(\tau_0,\tau_1,...)+\varepsilon F_1(u_0,\tau_0,\tau_1,...)+\varepsilon^2 F_2(u_0,u_1,\tau_0,\tau_1,...)+... \tag{1.9}$$

In the vibrations of discrete systems studied in this book, the differential operator has the form

$$L(u_i) = \frac{\partial u_i}{\partial \tau_0^2}+\omega^2 u_i, \tag{1.10}$$

where ω is the known parameter.

Due to the arbitrariness of ε, equation (1.9) is equivalent to the infinite recursive system of equations with partial derivatives

$$\begin{aligned} &\frac{\partial u_0}{\partial \tau_0^2}+\omega^2 u_0 = F_0(\tau_0,\tau_1,...),\\ &\frac{\partial u_1}{\partial \tau_0^2}+\omega^2 u_1 = F_1(u_0,\tau_0,\tau_1,...),\\ &\frac{\partial u_2}{\partial \tau_0^2}+\omega^2 u_2 = F_0(u_0,u_1,\tau_0,\tau_1,...),\\ &\cdots \end{aligned} \tag{1.11}$$

Awrejcewicz and Krysko [21] showed that the truncation of the power series (1.3) to the first m components

$$u(\tau;\varepsilon) = \sum_{k=0}^{k=m}\varepsilon^k u_k(\tau_0,\tau_1,\tau_2,...,\tau_m)+O(\varepsilon^{m+1}), \tag{1.12}$$

makes it uniformly useful for time of order $O(\varepsilon^{-(m+1)})$. Approximation of the solution by m components of the power series (1.3) using m time scales

$$u(\tau;\varepsilon) = \sum_{k=0}^{k=m}\varepsilon^k u_k(\tau_0,\tau_1,\tau_2,...,\tau_m), \tag{1.13}$$

makes the u_k functions to be determined from the recursive system of m differential equations

$$\begin{aligned}
&\frac{\partial u_0}{\partial \tau_0^2} + \omega^2 u_0 = F_0(\tau_0, \tau_1, ..., \tau_m),\\
&\frac{\partial u_1}{\partial \tau_0^2} + \omega^2 u_1 = F_1(u_0, \tau_0, \tau_1, ..., \tau_m),\\
&\frac{\partial u_2}{\partial \tau_0^2} + \omega^2 u_2 = F_0(u_0, u_1, \tau_0, \tau_1, ..., \tau_m),\\
&\cdots\\
&\frac{\partial u_m}{\partial \tau_0^2} + \omega^2 u_m = F_0(u_0, u_1, ..., u_{m-1}, \tau_0, \tau_1, ..., \tau_m).
\end{aligned} \tag{1.14}$$

The right hand side of the above equations may contain functions of various parameters of the system, time scales, lower-order solutions, and their derivatives for all time scales.

Due to the recursive nature of the system of differential equations to be solved, the so-called secular terms will appear in the equations of the second and higher-order approximation. Solutions to lower-order equations are periodic functions and play the role of nonhomogeneity in higher-order equations. So if the frequencies of these functions are equal to the eigenvalues of the differential operators of higher-order equations (which is inevitable), then the solutions will contain terms with the value of the vibration amplitude increasing indefinitely. These are secular terms that cause a fictitious resonance in the system. The occurrence of secular solutions is in contradiction with the experimentally observed periodic vibrations in nonlinear systems. For this reason, in higher-order equations, the terms leading to the secular solutions should be eliminated. The conditions for the elimination of secular terms are referred to as *the solvability conditions* to the problem.

The elimination of the secular terms ensures that the required condition is met

$$\frac{u_i}{u_{i-1}} < \infty, \tag{1.15}$$

which, in turn, is necessary to obtain an evenly usable solution u.

The solvability conditions are of particular importance in the presented method. After introducing the appropriate polar representation of the unknown complex functions, they lead to the amplitude and phase *modulation equations*. The asymptotic solutions obtained by the above method describe the behavior of the system away from resonance. Since the solution $u(\tau)$ has an analytical form, it can be used to provide qualitative information about the evolution of the system under study. In particular, it is possible to recognize the resonance conditions that may occur.

To study vibrations in the chosen resonance areas, the appropriate substitutions for frequencies must be made into equations (1.4). In addition, a small value called the *detuning parameter* is defined as the disturbance of the resonant frequency. This causes the modification of the solvability conditions and, consequently, allows one to determine the amplitude and phase modulations in the vicinity of resonance. This, in turn, allows for a detailed analysis of the behavior of the system under resonance conditions. Among other things, one can determine the amplitude-frequency characteristics, plot the resonance response curves, and test their stability. This procedure will be illustrated in the following chapters.

2 Spring Pendulum

This chapter presents the motion equations of the spring pendulum excited by harmonically changing forces using the Lagrange formalism. The equations are then transformed to their counterpart dimensionless form. Approximation of the trigonometric functions of unknown generalized coordinates by a few terms of the Taylor series and some assumptions concerning the order of the magnitude of parameters allow for replacing the primary nonlinear problem with a simpler one which is solved analytically, or partly analytically, through the employment of the multiple scales method. The variant of MSM with three variables describing the evolution of the system in time is used. The non-resonant vibrations and vibrations at double main-resonance are studied separately. The procedures serving for the construction of the approximate solutions for both cases are described in great detail. The accuracy of the solutions is assessed by the comparison with the solutions obtained numerically using procedure *NDSolve* of *Mathematica* software. Four algebraic equations that govern the steady-state are derived from the differential equations of modulation of the amplitudes and phases. The resonance response curves are determined by solving these algebraic equations numerically. The stability analysis in the Lyapunov sense indicates which one periodic steady state can be physically realized when the ambiguity of the resonance response occurs.

2.1 INTRODUCTION

A spring (elastic) pendulum serves as an archetypic system for extension of modeling of pendula-type systems, and it has a wide palette of applications in classical and quantum physics, and engineering though it requires rather complex use of the multiple scale method and asymptotic method to understand its dynamic behavior. In what follows, we give a brief review of the so far published results devoted to studying 2D spring pendulum.

Vitt and Gorelik [220] belong to the first researchers who studied the Fermi resonance and reduced the problem to consideration of the resonance 2:1 of a planar elastic pendulum.

Olsson [161] detected more parametric resonances exhibited by an elastic pendulum, while Lai [113] investigated the *recurrence phenomenon* of a string pendulum reducing the problem to the slow varying amplitude and phases, and comparing the experimental results with analytical predictions.

Breintenberg and Mueller [45] analyzed the elastic pendulum as the paradigm of a conservative, *autoparametric system* with an *internal resonance* by means of the slow-fluctuaction technique.

Cuerno et al. [57] investigated the chaotic dynamics of the elastic pendulum using the Poncaré map, the maximum Lyapunov exponent, the correlation function, and

DOI: 10.1201/9781003270706-2

the power spectrum. The paper was supplemented by analytical study and graphical representation concerning the applicability of the KAM theorem.

Anicin et al. [13] employed the classical *Ince-Strutt stability* chart to determining graphically the range of mass leading to instability effects while reducing the problem of the planar elastic pendulum to a linear approximation of its oscillations.

Bayly and Virgin [35] studied purely radial and swinging motion due to a 2:1 internal resonance of the elastic pendulum with 2 DoF. Based on the *Floquet theory*, bifurcation phenomena were analyzed analytically, numerically, and experimentally. In particular, characteristic multipliers were estimated experimentally in four-dimensional space and the *Karhunen-Loeve decomposition* concept was employed.

Davidovic et al. [58] detected the parabolas bounding the accessible domain in the x-y plane for small oscillation in the 2:1 resonance of an elastic pendulum.

Georgiou [77] illustrated how the planar elastic pendulum was yielded by a *singular perturbation* of the pendulum and how its dynamics was governed by global 2D *invariant manifolds* of motion. Poincaré sections of iso-energetic manifolds allowed to reveal that only motions near a *separatrix* emanating from the unstable region of the fast invariant manifold exhibited energy transfer.

Christensen [53] carried out improvements to the estimate of the mass necessary to induce the resonance between the longitudinal and transverse swing of the elastic pendulum. Several approximations were considered to simplify the problem.

Pokorny [171] derived the equations governing the motion of the three-dimensional heavy spring elastic pendulum. Stability conditions of vertical large amplitude oscillations were predicted analytically and validated numerically.

Cross [56] analyzed experimentally the energy transfer between the horizontal and vertical motion of an elastic pendulum.

Duka and Duka [62] focused on the effects of "peculiar" behavior of the spring pendulum including the parametric resonance and "pumping" swings. A new model of "pumping" swings was presented and the experimental validation of the studied parametric resonance was worked out.

Anurag et al. [14] investigated a planar elastic pendulum as an autonomous non-integrable Hamiltonian systems exhibiting autoparametric resonance corresponding to the 2:1 primary resonance. The combination of analytical and numerical approaches allowed for discovering that the order-chaos-order transition occurs due to the interaction between 2:1 resonances, i.e. one primary and another secondary.

Anurag et al. [15] studied numerically and analytically the quantum mechanics of planar elastic pendulum in libration. The *canonical perturbation theory* along with the Einstein-Brilloin-Keller quantization allowed one to find that the deviations in energy level statistics disappear by considering the states with the same symmetry. It was shown that the eight-point-out-of-time-correlator and the Ehrefest time can serve as the quantum correlator related to the second-largest Lyapunov exponent.

The work presented in this chapter is devoted to the study of spring pendulum, and extends our previous results obtained in references (Awrejcewicz et al. [31], Sypniewska et al. [208, 209]).

2.2 MATHEMATICAL MODEL

We are aimed at investigating the plane motion of a point mass m mounted on a spring-damper suspension. The scheme of the system is presented in Figure 2.1. The spring is assumed to be massless and having the nonlinearity of the cubic type. L_0 denotes the spring length in the non-stretched state. There are two purely viscous dampers in the system. The resistance force $\mathbf{R}$ depends on the point velocity $\mathbf{v}$ according to the relation $\mathbf{R} = -C_2\mathbf{v}$. The system is loaded by the torque, the magnitude of which changes harmonically accordingly to $M(t) = M_0 \cos(\Omega_2 t)$ and by the force $\mathbf{F}$ whose magnitude is also harmonic, i.e. $F(t) = F_0 \cos(\Omega_1 t)$. The total spring elongation $X(t)$ and the angle $\Phi(t)$ are assumed as the generalized coordinates.

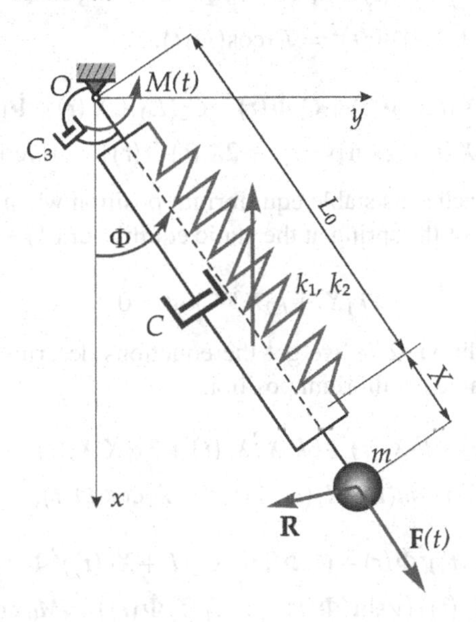

FIGURE 2.1 Spring pendulum.

The coordinates of the point denoted further as C in the reference frame Oxy are

$$X_C = (L_0 + X(t))\cos(\Phi(t)), \qquad Y_C = (L_0 + X(t))\sin(\Phi(t)). \tag{2.1}$$

The kinetic energy relative to the immovable reference frame is given by

$$T = \frac{1}{2}m\left(\dot{X}(t)^2 + (L_0 + X(t))^2\dot{\Phi}(t)^2\right). \tag{2.2}$$

The potential energy of the conservative forces can be written as

$$V = \frac{1}{2}k_1 X(t)^2 + \frac{1}{4}k_2 X(t)^4 - mg\cos(\Phi(t))(L_0 + X(t)), \tag{2.3}$$

where g denotes the gravitational acceleration.

All other forces acting on the pendulum are introduced into consideration as the generalized forces corresponding to the generalized coordinates, as follows

$$\begin{aligned} Q_X &= F_0 \cos(\Omega_1 t) - C_1 \dot{X}(t), \\ Q_\Phi &= M_0 \cos(\Omega_2 t) - C_3 \dot{\Phi}(t) - C_2 (L_0 + X(t))^2 \dot{\Phi}(t), \end{aligned} \tag{2.4}$$

where $C_1 = C + C_2$.

The Lagrange equations of the second kind yield the pendulum motion equations of the following form

$$\begin{aligned} & m\ddot{X}(t) + C_1 \dot{X}(t) + k_1 X(t) + k_2 X(t)^3 - mg\cos(\Phi(t)) - \\ & m(L_0 + X(t))\dot{\Phi}(t)^2 = F_0 \cos(\Omega_1 t), \end{aligned} \tag{2.5}$$

$$\begin{aligned} & m(L_0 + X(t))^2 \ddot{\Phi}(t) + C_3 \dot{\Phi}(t) + C_2 (L_0 + X(t))^2 \dot{\Phi}(t) + \\ & m(L_0 + X(t)) \left(g \sin(\Phi(t)) + 2\dot{X}(t)\dot{\Phi}(t) \right) = M_0 \cos(\Omega_2 t). \end{aligned} \tag{2.6}$$

The pendulum reaches its stable equilibrium position when $\Phi_e = 0$ and $X = X_e$, where the elongation of the spring at the static equilibrium X_e satisfies the following algebraic equation

$$k_1 X_e + k_2 X_e^3 - mg = 0 \tag{2.7}$$

Considering condition (2.7), we get the equations describing the motion of the system around the stable equilibrium position

$$\begin{aligned} & m\ddot{X}_1(t) + C_1 \dot{X}_1(t) + k_1 X_1(t) + 3k_2 X_e^2 X_1(t) + 3k_2 X_e X_1(t)^2 + k_2 X_1(t)^3 + \\ & mg(1 - \cos(\Phi(t))) - m(L + X_1(t))\dot{\Phi}(t)^2 = F_0 \cos(\Omega_1 t), \end{aligned} \tag{2.8}$$

$$\begin{aligned} & m(L + X_1(t))^2 \ddot{\Phi}(t) + C_3 \dot{\Phi}(t) + C_2 (L + X_1(t))^2 \dot{\Phi}(t) + \\ & m\, (L + X_1(t)) \left(g \sin(\Phi(t)) + 2\dot{X}_1(t)\dot{\Phi}(t) \right) = M_0 \cos(\Omega_2 t), \end{aligned} \tag{2.9}$$

where $X_1(t) = X(t) - X_e$, $L = L_0 + X_e$.

To transform equations (2.8)–(2.9) into the dimensionless form, the length of the spring at the stable equilibrium position $L = L_0 + X_e$ and the characteristic dimensionless time $\tau = t\omega_1$, where $\omega_1 = \sqrt{\frac{k_1}{m}}$ is the frequency of the simple harmonic oscillator of mass m and stiffness k_1, have been chosen as the reference quantities. The dimensionless motion equations of the system are as follows

$$\begin{aligned} & \ddot{\xi}(\tau) + c_1 \dot{\xi}(\tau) + \xi(\tau) + \alpha \left(3\, \xi_e^2\, \xi(\tau) + 3\xi_e \xi(\tau)^2 + \xi(\tau)^3 \right) - \\ & (1 + \xi(\tau))\, \dot{\varphi}(\tau)^2 + w^2\, (1 - \cos(\varphi(\tau))) = f_1 \cos(p_1 \tau), \end{aligned} \tag{2.10}$$

$$\begin{aligned} & (1 + \xi(\tau))^2 \ddot{\varphi}(\tau) + c_3 \dot{\varphi}(\tau) + c_2 (1 + \xi(\tau))^2 \dot{\varphi}(\tau) + \\ & 2(1 + \xi(\tau))\, \dot{\xi}(\tau)\, \dot{\varphi}(\tau) + w^2 (1 + \xi(\tau)) \sin(\varphi(\tau)) = f_2 \cos(p_2 \tau), \end{aligned} \tag{2.11}$$

where

$$\xi = \frac{X_1}{L}, \quad \tau = t\omega_1, \quad w = \frac{1}{\omega_1}\sqrt{\frac{g}{L}}, \quad \xi_e = \frac{X_e}{L}, \quad \alpha = \frac{k_2 L^2}{k_1}, \quad c_1 = \frac{C_1}{m\omega_1},$$

$$c_2 = \frac{C_2}{m\omega_1}, \quad c_3 = \frac{C_3}{L^2 m\omega_1}, \quad f_1 = \frac{F_0}{Lm\omega_1^2}, \quad f_2 = \frac{M_0}{L^2 m\omega_1^2}, \quad p_1 = \frac{\Omega_1}{\omega_1}, \quad p_2 = \frac{\Omega_2}{\omega_2}.$$

The functions $\xi(\tau)$ and $\varphi(\tau)$ are the dimensionless counterparts of the generalized coordinates $X_1(t)$, $\Phi(t)$. Equations (2.10)–(2.11) are supplemented with the initial conditions having the form

$$\xi(0) = \xi_0, \quad \dot{\xi}(0) = v_0, \quad \varphi(0) = \varphi_0, \quad \dot{\varphi}(0) = \omega_0, \tag{2.12}$$

where ξ_0, v_0, φ_0, ω_0 are known.

The dimensionless static elongation ξ_e of the spring satisfies the equilibrium condition in the form of the algebraic equation

$$\alpha\xi_e^3 + \xi_e - w^2 = 0. \tag{2.13}$$

2.3 SOLUTION METHOD

The approximate analytical solution to the initial value problem (2.10)–(2.12) is obtained using the multiple scales method (MSM). The trigonometric nonlinearities in equations (2.10)–(2.11) are approximated using the first three terms of Taylor's series expansion, i.e. we take

$$\sin\varphi \approx \varphi - \frac{\varphi^3}{3}, \qquad \cos\varphi \approx 1 - \frac{\varphi^2}{2}. \tag{2.14}$$

Following MSM, the system evolution in time is described using n variables of time nature. The number n usually takes a value of 2 or 3. Solving the problem concerning the pendulum with nonlinear spring, we have adopted three variables: τ_0, τ_1, τ_2. These variables are related to the dimensionless time τ in the following manner

$$\tau_i = \varepsilon^i \tau, \qquad i = 0, 1, 2. \tag{2.15}$$

The small parameter ε should satisfy a priori the inequalities $0 < \varepsilon \ll 1$.

From definition (2.15), it follows that the scale in which time τ and the time variable τ_0 are measured is the same. The units of the time variables τ_1 and τ_2 are much longer than the unit of the time variable τ_0, ε^{-1} and ε^{-2} times, respectively. Therefore the time variable τ_0 is usually called the fast time scale and the other time variables are regarded as the slow time scales.

In connection with the introduction of new time variables, the differential operators should be redefined. According to the chain rule, they take the form

$$\frac{d}{d\tau}=\sum_{k=0}^{2}\varepsilon^{k}\frac{\partial}{\partial\tau_k}=\frac{\partial}{\partial\tau_0}+\varepsilon\frac{\partial}{\partial\tau_1}+\varepsilon^2\frac{\partial}{\partial\tau_2}, \tag{2.16}$$

$$\frac{d^2}{d\tau^2}=\frac{\partial^2}{\partial\tau_0^2}+2\varepsilon\frac{\partial^2}{\partial\tau_0\partial\tau_1}+\varepsilon^2\left(\frac{\partial^2}{\partial\tau_1^2}+2\frac{\partial^2}{\partial\tau_0\partial\tau_2}\right)+2\varepsilon^3\frac{\partial^2}{\partial\tau_1\partial\tau_2}+O\left(\varepsilon^4\right). \tag{2.17}$$

Accordingly to MSM, we approximate the functions $\xi(\tau)$, $\varphi(\tau)$ by the following asymptotic expansions

$$\xi\left(\tau;\varepsilon\right)=\sum_{k=1}^{3}\varepsilon^{k}x_k\left(\tau_0,\ \tau_1,\ \tau_2\right)+O\left(\varepsilon^4\right), \tag{2.18}$$

$$\varphi\left(\tau;\varepsilon\right)=\sum_{k=1}^{3}\varepsilon^{k}\ \phi_k\left(\tau_0,\ \tau_1,\ \tau_2\right)+O\left(\varepsilon^4\right), \tag{2.19}$$

where the functions $x_k(\tau_0,\tau_1,\tau_2)$, $\phi_k(\tau_0,\tau_1,\tau_2)$, $k=1,...,3$ are sought.

Limiting the foregoing only to the case of weakly nonlinear spring pendulum, a few parameters describing the system and its loading are assumed to be small which can be expressed using the small parameter as follows

$$\alpha=\varepsilon^2\widehat{\alpha},\quad c_1=\varepsilon^2\widehat{c}_1,\quad c_2=\varepsilon^2\widehat{c}_2,\quad c_3=\varepsilon^2\widehat{c}_3,\quad f_1=\varepsilon^3\widehat{f}_1,\quad f_2=\varepsilon^3\widehat{f}_2. \tag{2.20}$$

The coefficients $\widehat{\alpha}$, $\widehat{c}_1$, $\widehat{c}_2$, $\widehat{c}_3$, $\widehat{f}_1$, $\widehat{f}_2$ can be understood as $O(1)$ when $\varepsilon\to 0$.

2.4 NON-RESONANT VIBRATION

The relations (2.14)–(2.20) substituted into equations of motion (2.10)–(2.11) yield the equations in which the small parameter appears in a few different powers. Omitting all terms of order $O\left(\varepsilon^4\right)$, we require that each of the equations be satisfied for any value of ε. After ordering the terms of the equations according to the powers of the small parameter, this requirement is realized by equating to zero all coefficients standing at the various powers of ε. Omitting the terms that are accompanied by ε in powers higher than three, one can obtain, the set of six differential equations with unknown functions $x_k\left(\tau_0,\ \tau_1,\ \tau_2\right)$, $\phi_k\left(\tau_0,\ \tau_1,\ \tau_2\right)$, $k=1,\ \ldots,\ 3$. We can organize the set of equations into three groups. To the first group belong the homogeneous equations that contain the terms standing at ε

$$\frac{\partial^2 x_1}{\partial\tau_0^2}+x_1=0, \tag{2.21}$$

$$\frac{\partial^2\phi_1}{\partial\tau_0^2}+w^2\phi_1=0. \tag{2.22}$$

Equations (2.21)–(2.22) are called the first-order approximation equations. The terms standing at ε^2 create the following equations of the second-order approximation

$$\frac{\partial^2 x_2}{\partial \tau_0^2} + x_2 = -2\,\frac{\partial x_1^2}{\partial \tau_0 \partial \tau_1} + \left(\frac{\partial \phi_1}{\partial \tau_0}\right)^2 - \frac{w^2}{2}\phi_1^2, \tag{2.23}$$

$$\frac{\partial^2 \phi_2}{\partial \tau_0^2} + w^2 \phi_2 = -2\frac{\partial^2 \phi_1}{\partial \tau_0 \partial \tau_1} - 2x_1 \frac{\partial^2 \phi_1}{\partial \tau_0^2} - 2\frac{\partial x_1}{\partial \tau_0}\frac{\partial \phi_1}{\partial \tau_0} - w^2 x_1 \phi_1. \tag{2.24}$$

The coefficients that are accompanied by ε^3 form the equations of the third-order approximation

$$\begin{aligned}\frac{\partial^2 x_3}{\partial \tau_0^2} + x_3 = {} & \hat{f}_1 \cos(p_1 \tau_0) - \hat{c}_1 \frac{\partial x_1}{\partial \tau_0} - 3\hat{\alpha}\,\xi_e^2\, x_1 - \frac{\partial^2 x_1}{\partial \tau_1^2} - \\ & 2\left(\frac{\partial^2 x_1}{\partial \tau_0 \partial \tau_2} + \frac{\partial^2 x_2}{\partial \tau_0 \partial \tau_1}\right) + x_1\left(\frac{\partial \phi_1}{\partial \tau_0}\right)^2 + 2\frac{\partial \phi_1}{\partial \tau_0}\left(\frac{\partial \phi_1}{\partial \tau_1} + \frac{\partial \phi_2}{\partial \tau_0}\right) - w^2 \phi_1 \phi_2,\end{aligned} \tag{2.25}$$

$$\begin{aligned}\frac{\partial^2 \phi_3}{\partial \tau_0^2} + w^2 \phi_3 = {} & \hat{f}_2 \cos(p_2 \tau_0) - (\hat{c}_2 + \hat{c}_3)\frac{\partial \phi_1}{\partial \tau_0} + \frac{1}{6} w^2 \phi_1^3 - w^2(x_2 \phi_1 + x_1 \phi_2) - \\ & \frac{\partial^2 \phi_1}{\partial \tau_1^2} - 2\frac{\partial^2 \phi_1}{\partial \tau_0 \partial \tau_2} - \frac{\partial^2 \phi_1}{\partial \tau_0^2}(x_1^2 + 2x_2) - 2\frac{\partial^2 \phi_2}{\partial \tau_0 \partial \tau_1} - 2x_1\left(2\frac{\partial^2 \phi_1}{\partial \tau_0 \partial \tau_1} + \frac{\partial^2 \phi_2}{\partial \tau_0^2}\right) - \\ & 2\frac{\partial x_1}{\partial \tau_0}\left(\frac{\partial \phi_1}{\partial \tau_1} + \frac{\partial \phi_2}{\partial \tau_0}\right) - 2\frac{\partial \phi_1}{\partial \tau_0}\left(\frac{\partial x_2}{\partial \tau_0} + \frac{\partial x_1}{\partial \tau_1}\right) - 2x_1 \frac{\partial x_1}{\partial \tau_0}\frac{\partial \phi_1}{\partial \tau_0}.\end{aligned} \tag{2.26}$$

The system of equations (2.21)–(2.26) is solved recursively. The general solutions of the equations of the first-order approximation containing unknown functions of τ_1 and τ_2 have the following form

$$x_1(\tau_0, \tau_1, \tau_2) = B_1(\tau_1, \tau_2)e^{i\tau_0} + \bar{B}_1(\tau_1, \tau_2)e^{-i\tau_0}, \tag{2.27}$$

$$\phi_1(\tau_0, \tau_1, \tau_2) = B_2(\tau_1, \tau_2)e^{iw\tau_0} + \bar{B}_2(\tau_1, \tau_2)e^{-iw\tau_0}, \tag{2.28}$$

where i denotes the imaginary unit, and $\bar{B}_j(\tau_1, \tau_2)$, $(j = 1,2)$, is the complex conjugate of the function $B_j(\tau_1, \tau_2)$, $j = 1,2)$.

It is worth noting that the differential operators of equations (2.21), (2.23), and (2.25) are the same. The *compatibility* of the operators also applies to equations (2.22), (2.24), and (2.26). Therefore starting with the second stage of the recursive solution procedure, the *secular terms* appear inevitably among the *inhomogeneous terms* on the left sides of equations (2.23)–(2.26). Substituting solutions (2.27)–(2.28) into equations (2.23)–(2.24) and detecting the secular terms one can get the following solvability conditions

$$\frac{\partial B_1}{\partial \tau_1} = 0, \quad \frac{\partial \bar{B}_1}{\partial \tau_1} = 0, \tag{2.29}$$

$$\frac{\partial B_2}{\partial \tau_1} = 0, \quad \frac{\partial \bar{B}_2}{\partial \tau_1} = 0. \tag{2.30}$$

From equations (2.29)–(2.30) follows that

$$\frac{\partial^2 B_1}{\partial \tau_2 \partial \tau_1} = 0, \qquad \frac{\partial^2 \bar{B}_1}{\partial \tau_2 \partial \tau_1} = 0, \tag{2.31}$$

$$\frac{\partial^2 B_2}{\partial \tau_2 \partial \tau_1} = 0, \qquad \frac{\partial^2 \bar{B}_2}{\partial \tau_2 \partial \tau_1} = 0. \tag{2.32}$$

Considering solvability conditions (2.29)–(2.32) one can solve the equations of the second-order approximation. Their particular solutions are as follows

$$x_2(\tau_0,\ \tau_1,\ \tau_2) = w^2 B_2 \bar{B}_2 + \frac{3w^2(e^{2iw\tau_0} B_2{}^2 + e^{-2iw\tau_0} \bar{B}_2^2)}{2(4w^2-1)}, \tag{2.33}$$

$$\begin{aligned}\phi_2(\tau_0,\tau_1,\tau_2) = &-\frac{w(w+2)e^{-i(w+1)\tau_0}(\bar{B}_1\bar{B}_2 + e^{2i(w+1)\tau_0} B_1 B_2)}{2w+1} + \\ &\frac{w(w-2)e^{-i(w-1)\tau_0}(B_1\bar{B}_2 + e^{2i(w-1)\tau_0}\bar{B}_1 B_2)}{2w-1}.\end{aligned} \tag{2.34}$$

Substituting solutions (2.27)–(2.28) and (2.33)–(2.34) into equations (2.25)–(2.26) and eliminating the secular terms one can obtain further solvability conditions in the form

$$B_1\left(\frac{6\left(w^2-1\right) w^2 B_2 \bar{B}_2}{4w^2-1} + 3\hat{\alpha}\, \xi_e^2 + i\, \hat{c}_1\right) + 2i\frac{\partial B_1}{\partial \tau_2} = 0, \tag{2.35}$$

$$\bar{B}_1\left(-\frac{6\left(w^2-1\right) w^2 B_2 \bar{B}_2}{4w^2-1} - 3\hat{\alpha}\, \xi_e^2 + i\, \hat{c}_1\right) + 2i\frac{\partial \bar{B}_1}{\partial \tau_2} = 0, \tag{2.36}$$

$$\begin{aligned}&2i\,\frac{\partial B_2}{\partial \tau_2} + i\,\left(\hat{c}_2 + \hat{c}_3\right) B_2 + \frac{6w\left(w^2-1\right)}{4\,w^2-1} B_1 \bar{B}_1 B_2 - \\ &\frac{w\left(8w^4 - 7w^2 - 1\right)}{8w^2-2} B_2^2 \bar{B}_2 = 0,\end{aligned} \tag{2.37}$$

$$\begin{aligned}&2i\frac{\partial \bar{B}_2}{\partial \tau_2} + i\,\left(\hat{c}_2 + \hat{c}_3\right) \bar{B}_2 - \frac{6w\left(w^2-1\right)}{4w^2-1} B_1 \bar{B}_1 \bar{B}_2 + \\ &\frac{w\left(8w^4 - 7w^2 - 1\right)}{8w^2-2} B_2 \bar{B}_2^2 = 0.\end{aligned} \tag{2.38}$$

After eliminating the secular terms from equations (2.25)–(2.26) enables obtaining the finite solutions. We get the following particular solutions

$$\begin{aligned}x_3\left(\tau_0,\ \tau_1,\ \tau_2\right) = &-3w\left(w+1\right)\frac{\left(e^{i(1+2w)\tau_0} B_1 B_2{}^2 + e^{-i(1+2w)\tau_0}\bar{B}_1\bar{B}_2^2\right)}{4\left(2w+1\right)} + \\ &3w\left(w-1\right)\frac{\left(e^{i(2w-1)\tau_0}\bar{B}_1 B_2{}^2 + e^{-i(2w-1)\tau_0} B_1 \bar{B}_2^2\right)}{4\left(2w-1\right)} + \frac{\hat{f}_1 \cos\left(p_1 \tau_0\right)}{1-p_1^2},\end{aligned} \tag{2.39}$$

$$\phi_3(\tau_0,\ \tau_1,\ \tau_2) = w\left(w^2+5w+6\right)\frac{\left(e^{\mathrm{i}(2+w)\tau_0}B_1^2B_2+e^{-\mathrm{i}(2+w)\tau_0}\bar{B}_1^2\bar{B}_2\right)}{4\,(2w+1)}+$$

$$w\left(w^2-5w+6\right)\frac{\left(e^{\mathrm{i}(w-2)\tau_0}\bar{B}_1^2B_2+e^{-\mathrm{i}(w-2)\tau_0}B_1^2\bar{B}_2\right)}{4\,(2w-1)}- \tag{2.40}$$

$$\left(49w^2-1\right)\frac{\left(e^{3\mathrm{i}w\tau_0}B_2^3+e^{-3\mathrm{i}w\tau_0}\bar{B}_2^3\right)}{48\,(4w^2-1)}+\frac{\hat{f}_2\cos(p_2\tau_0)}{w^2-p_2^2}.$$

The unknown complex-valued functions $B_1(\tau_1,\tau_2)$, $B_2(\tau_1,\tau_2)$ and their complex conjugates $\bar{B}_1(\tau_1,\tau_2)$, $\bar{B}_2(\tau_1,\tau_2)$ occur in the solutions (2.27)–(2.28), (2.33)–(2.34) and (2.39)–(2.40). The functions are restricted by the solvability conditions. All the functions do not depend on the variable τ_1. Taking into account this independence one can regard the solvability conditions (2.35)–(2.38) as the set of four ordinary differential equations of the first-order with respect to the functions $B_1(\tau_2)$, $B_2(\tau_2)$ and their complex conjugates. We present the unknown functions in the exponential form

$$B_1(\tau_2)=\frac{1}{2}b_1(\tau_2)\,e^{\mathrm{i}\psi_1(\tau_2)},\quad \bar{B}_1(\tau_2)=\frac{1}{2}b_1(\tau_2)\,e^{-\mathrm{i}\psi_1(\tau_2)}, \tag{2.41}$$

$$B_2(\tau_2)=\frac{1}{2}b_2(\tau_2)\,e^{\mathrm{i}\psi_2(\tau_2)},\quad \bar{B}_2(\tau_2)=\frac{1}{2}b_2(\tau_2)\,e^{-\mathrm{i}\psi_2(\tau_2)}, \tag{2.42}$$

where the functions $b_1(\tau_2)$, $b_2(\tau_2)$, $\psi_1(\tau_2)$, $\psi_2(\tau_2)$ are real-valued.

Inserting relationships (2.41)–(2.42) into differential equations (2.35)–(2.38), and then solving the latter with respect to the derivatives, we obtain

$$\frac{db_1}{d\tau_2}=-\frac{1}{2}\hat{c}_1b_1, \tag{2.43}$$

$$\frac{db_2}{d\tau_2}=-\frac{1}{2}(\hat{c}_2+\hat{c}_3)\,b_2, \tag{2.44}$$

$$\frac{d\psi_1}{d\tau_2}=\frac{3\left(w^2(w^2-1)b_2^2+2\left(4w^2-1\right)\xi_e^2\hat{\alpha}\right)}{4\,(4w^2-1)}, \tag{2.45}$$

$$\frac{d\psi_2}{d\tau_2}=\frac{w\left(w^2-1\right)\left(12b_1^2-\left(8w^2+1\right)b_2^2\right)}{16\,(4\,w^2-1)}. \tag{2.46}$$

Equations (2.43)–(2.46) govern the evolution of the functions $b_1(\tau_2)$, $b_2(\tau_2)$, $\psi_1(\tau_2)$, $\psi_2(\tau_2)$ in the slowest time variable. Conversion of equations (2.43)–(2.46) to the form in which only the primary time τ occurs is convenient. The conversion consists in the reversely using definition (2.16). So, multiplying the equations by ε^2, and then taking into account the relations (2.16) and (2.20), one can write

$$\frac{da_1}{d\tau}=-\frac{1}{2}c_1a_1(\tau), \tag{2.47}$$

$$\frac{da_2}{d\tau} = -\frac{1}{2}(c_2+c_3)\,a_2(\tau), \tag{2.48}$$

$$\frac{d\psi_1}{d\tau} = \frac{3w^2\left(w^2-1\right)a_2^2(\tau)}{16w^2-4} + \frac{3}{2}\xi_e^2\alpha, \tag{2.49}$$

$$\frac{d\psi_2}{d\tau} = \frac{w\left(w^2-1\right)\left(12a_1^2(\tau)-\left(8w^2+1\right)a_2^2(\tau)\right)}{16\left(4\,w^2-1\right)}, \tag{2.50}$$

where $a_i(\tau) = \varepsilon\, b_i(\tau),\quad i = 1,\, 2.$

According to equations (2.18)–(2.19), (2.27)–(2.28) and (2.41)–(2.42), the functions $a_1(\tau), a_2(\tau)$ represent the amplitudes of the pendulum vibration which are described by the solutions of the first-order approximation. In turn, the functions $\psi_1(\tau), \psi_2(\tau)$ are the phases of this vibration. Therefore, equations (2.47)–(2.50) are often called modulation equations.

The modulation equations are supplemented by the initial conditions

$$a_1(0) = a_{10},\quad a_2(0) = a_{20},\quad \psi_1(0) = \psi_{10},\quad \psi_2(0) = \psi_{20}, \tag{2.51}$$

where known quantities a_{10}, a_{20}, ψ_{10}, ψ_{20} are related to the initial values ξ_0, v_0, φ_0, ω_0. The exact solution of the initial value problem given by (2.47)–(2.50) and (2.51) has the following form

$$a_1(\tau) = a_{10}e^{-\frac{c_1\tau}{2}}, \tag{2.52}$$

$$a_2(\tau) = a_{20}e^{-\frac{1}{2}(c_2+c_3)\tau}, \tag{2.53}$$

$$\psi_1(\tau) = \frac{3}{2}\alpha\xi_e^2\tau - \frac{3a_{20}^2w^2\left(w^2-1\right)\left(e^{-(c_2+c_3)\tau}-1\right)}{4\left(4w^2-1\right)(c_2+c_3)} + \psi_{10}, \tag{2.54}$$

$$\begin{aligned}\psi_2(\tau) = {} & \frac{3a_{10}^2w\left(w^2-1\right)\left(1-e^{-c_1\tau}\right)}{4c_1\left(4w^2-1\right)} + \\ & \frac{a_{20}^2w\left(8w^4-7w^2-1\right)\left(e^{-(c_2+c_3)\tau}-1\right)}{16(c_2+c_3)\left(4w^2-1\right)} + \psi_{20}\,.\end{aligned} \tag{2.55}$$

Assembling the subsequent solutions in accordance with (2.18)–(2.19), but omitting all terms of order $O(4)$, we obtain the approximate solution of the considered problem

$$\begin{aligned}\xi(\tau) = {} & a_1(\tau)\cos(\tau+\psi_1(\tau)) + \\ & \frac{3w(w-1)\,a_1(\tau)a_2^2(\tau)}{32w-16}\cos((1-2w)\tau+\psi_1(\tau)-2\psi_2(\tau)) - \\ & \frac{3w(w+1)a_1(\tau)a_2^2(\tau)}{32w+16}\cos((1+2w)\tau+\psi_1(\tau)+2\psi_2(\tau)) + \\ & \frac{3w^2a_2^2(\tau)\cos(2w\tau+2\psi_2(\tau))}{16w^2-4} + \frac{w^2}{4}a_2^2(\tau) + \frac{f_1\cos(p_1\tau)}{1-p_1^2},\end{aligned} \tag{2.56}$$

$$
\begin{aligned}
\varphi(\tau) &= a_2(\tau)\cos(w\tau+\psi_2(\tau))+ \\
&\frac{w(w-2)a_1(\tau)a_2(\tau)}{4w-2}\cos((1-w)\tau+\psi_1(\tau)-\psi_2(\tau))+ \\
&\frac{w(w-3)(w-2)a_1^2(\tau)a_2(\tau)}{32w-16}\cos((2-w)\tau+2\psi_1(\tau)-\psi_2(\tau))+ \\
&\frac{w(w+2)(w+3)a_1^2(\tau)a_2(\tau)}{32w+16}\cos((2+w)\tau+2\psi_1(\tau)+\psi_2(\tau))- \\
&\frac{w(w+2)a_1(\tau)a_2(\tau)}{4w+2}\cos((1+w)\tau+\psi_1(\tau)+\psi_2(\tau))+ \\
&\frac{(1-49w^2)a_2^3(\tau)}{192(4w^2-1)}\cos(3w\tau+3\psi_2(\tau))+\frac{f_2\cos(p_2\tau)}{w^2-p_2^2},
\end{aligned}
\tag{2.57}
$$

where the functions $a_1(\tau)$, $a_2(\tau)$, $\psi_1(\tau)$, $\psi_2(\tau)$ are given by (2.52)–(2.55).

The first term of each of solutions (2.56) - (2.57) depict the solution to the first-order approximation. The expectation that these terms are dominant results from the assumption about the uniform validity of the asymptotic expansions (2.18)–(2.19).

Observe that the solution (2.56)–(2.57) fails when any of the denominators occurring in the formulae is equal to zero. Moreover, when any of the denominators is close to zero then the resonant vibration appears in the system. The approximate solution (2.56)–(2.57) based on the approach with three time scales allows one to recognize the following resonant cases:

(i) primary external resonances, when $p_1 \approx 1,\ p_2 \approx w$;
(ii) internal resonance, when $2w \approx 1$.

The resonant cases should be considered separately based on additionally formulated assumptions. Excluding the above-mentioned resonant cases, one can tell that the solution (2.56)–(2.57) describes the vibrational motion of the pendulum in non-resonant conditions.

To present and test the approximate solution that applies in the case of the non-resonant vibration, we assume the following values of the parameters which have an impact on the pendulum motion: $w = 0.13$, $\alpha = 0.05, c_1 = 0.01, c_2 = 0.01, c_3 = 0.01, f_2 = 0.001, f_1 = 0.001, p_1 = 1.65, p_2 = 0.89$. The initial values for the modulation problem (2.47)–(2.50) are: $a_{10} = 0.02$, $a_{20} = 0.05$, $\psi_{10} = 0$, $\psi_{20} = 0$.

The following initial values for the motion equation (2.10)–(2.11) remain in compliance with the above values of amplitudes and phases: $\xi_0 = 0.0193964$, $v_0 = -0.00009954005$, $\varphi_0 = 0.04876395$, $\omega_0 = -0.0005008056$.

The time histories of the generalized coordinates $\xi(\tau)$ and $\varphi(\tau)$ are presented in Figures 2.2–2.5. Figures 2.2 and 2.3 concern the initial stage of movement in which the transient vibration is realized. In each of these two graphs, two curves are depicted, namely: the approximate analytical solution obtained using MSM drawn as the solid black line, and the numerical solution obtained using the standard *NDSolve* of the *Mathematica* software. One can observe high compliance between both solutions.

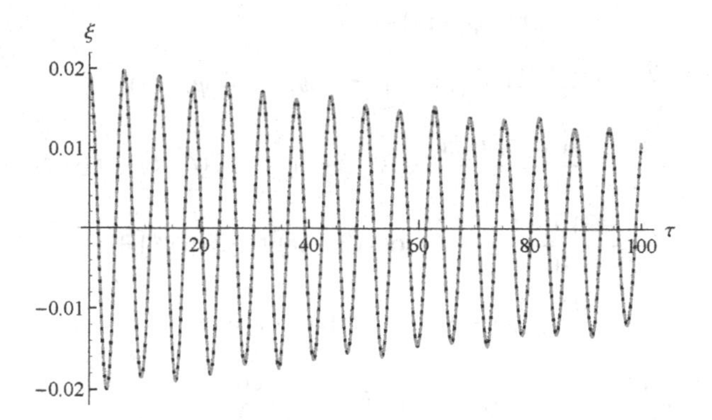

FIGURE 2.2 The transient longitudinal vibration $\xi(\tau)$ estimated numerically (dotted line) and analytically using MSM (solid line).

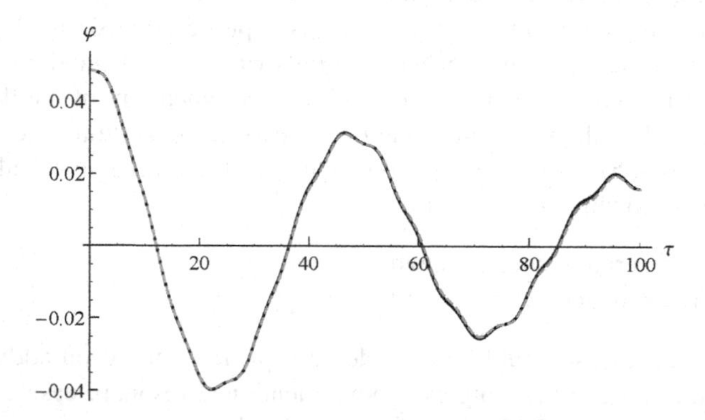

FIGURE 2.3 The transient swing vibration $\varphi(\tau)$ estimated numerically (dotted line) and analytically using MSM (solid line).

In Figures 2.4 and 2.5, the stationary pendulum oscillations are presented which occur after the transient stage. The same presentation method has been applied as previously. So, we can observe the comparison of the solution obtained in the analytical manner (MSM) and the solution determined by means of the *NDSolve* procedure.

The accuracy evaluation of the approximate solutions is estimated using the following measure

$$\delta_i = \sqrt{\frac{1}{\tau_e - \tau_s} \int_{\tau_s}^{\tau_e} \left(H_i\left(\xi_a(\tau), \varphi_a(\tau)\right) - 0\right)^2 d\tau}, \qquad i = 1,2, \tag{2.58}$$

where $H_1\left(\xi_a(\tau), \varphi_a(\tau)\right)$ and $H_2\left(\xi_a(\tau), \varphi_a(\tau)\right)$ stand for the differential operators, i.e. the left sides of the motion equations (2.10)–(2.11), $\xi_a(\tau), \varphi_a(\tau)$ are the

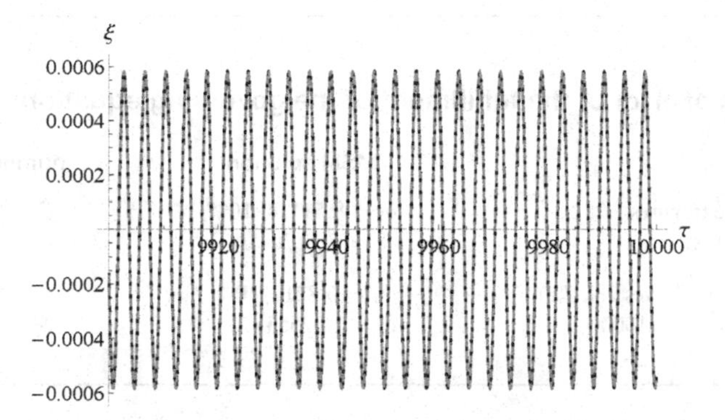

FIGURE 2.4 The stationary longitudinal vibration $\xi(\tau)$ estimated numerically (dotted line) and analytically using MSM (solid line).

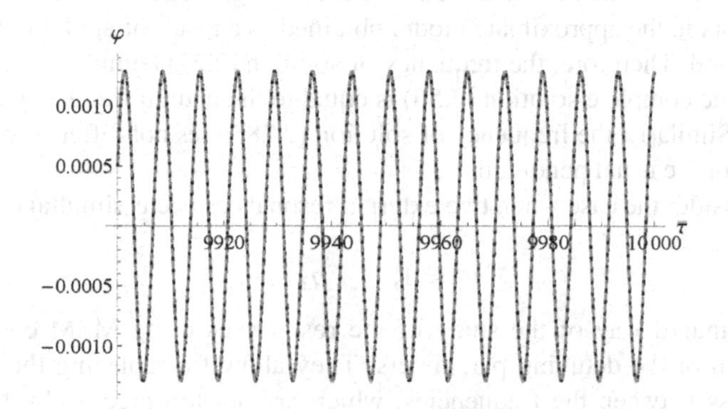

FIGURE 2.5 The stationary swing vibration $\varphi(\tau)$ estimated numerically (dotted line) and analytically using MSM (solid line).

approximate solutions obtained using the MSM as well as numerically, and τ_s and τ_e denote the chosen instants. The measure proposed evaluates the error of the fulfillment of the governing equations. The functions $\xi_a(\tau), \varphi_a(\tau)$, irrespective of the way of their obtainment, satisfy the motion equations only approximately.

The values of error (2.58) for the analytical and numerical approximate solutions discussed above are collected in Table 2.1.

It is worth noting that the truncation error resulting from approximation (2.14) reduces additionally the accuracy of the solutions obtained using MSM in contrast to the numerical solutions which are free of the influence of that approximation.

TABLE 2.1
The values of error of the fulfillment of the governing equations

	MSM solution	numerical solution
transient stage (Figures 2.2, 2.3) $\tau_s = 0,\ \tau_e = 100$	$\delta_1 = 8.51114 \cdot 10^{-6}$ $\delta_2 = 3.1170 \cdot 10^{-5}$	$\delta_1 = 1.9235 \cdot 10^{-4}$ $\delta_2 = 1.7791 \cdot 10^{-6}$
stationary stage (Figures 2.4, 2.5) $\tau_s = 9900,\ \tau_e = 10000$	$\delta_1 = 6.8212 \cdot 10^{-6}$ $\delta_2 = 1.62555 \cdot 10^{-5}$	$\delta_1 = 1.4505 \cdot 10^{-7}$ $\delta_2 = 2.6979 \cdot 10^{-7}$

2.5 RESONANT VIBRATION AT SIMULTANEOUSLY OCCURRING EXTERNAL RESONANCES

The assumption about the weak character of the nonlinearities makes that all couplings occurring in the motion equations have no significant impact on the differential operators in the approximate model obtained as a result of applying the multiple scales method. Therefore, the frequency of solution (2.27) being the dominant component of the complete solution (2.56) is equal to the natural frequency of the linear oscillator. Similarly, the frequency of solution (2.28) does not differ from the natural frequency of the usual pendulum.

We consider the case when two external resonances occur simultaneously, that is when

$$p_1 \approx 1, \qquad p_2 \approx w. \tag{2.59}$$

The standard way of the study of the resonances using MSM consists in the introduction of the detuning parameters. They allow for replacing the relations of the nearness between the frequencies, which are not too precise, by the relations of the equality. Introducing the detuning parameters σ_1 and σ_2, we can write the resonance conditions (2.59) as follows

$$p_1 = 1 + \sigma_1, \qquad p_2 = w + \sigma_2, \tag{2.60}$$

where σ_1 and σ_2 as dimensionless quantities are any, but rather small, numbers. We assume that they are of order $O\left(\varepsilon^2\right)$, so

$$\sigma_1 = \varepsilon^2 \hat{\sigma}_1, \quad \sigma_2 = \varepsilon^2 \hat{\sigma}_2. \tag{2.61}$$

The coefficients $\hat{\sigma}_1$, $\hat{\sigma}_2$ are understood as $O(1)$ when $\varepsilon^2 \to 0$. Inserting all previously made assumptions, i.e. (2.11)–(2.15) and also (2.60)–(2.61) into equations of motion (2.10)–(2.11), we get the equations the terms of which contain the small parameter ε. Arranging them with respect to the powers of the small parameter ε and then requiring that each of the equations should be fulfilled for any value ε, leads, at omitting all terms accompanied by ε in powers higher than three, to the set of six differential equations

$$\frac{\partial^2 x_1}{\partial \tau_0^2} + x_1 = 0, \tag{2.62}$$

$$\frac{\partial^2 \phi_1}{\partial \tau_0^2} + w^2 \phi_1 = 0, \tag{2.63}$$

$$\frac{\partial^2 x_2}{\partial \tau_0^2} + x_2 = -2\,\frac{\partial x_1^2}{\partial \tau_0 \partial \tau_1} + \left(\frac{\partial \phi_1}{\partial \tau_0}\right)^2 - \frac{w^2}{2}\phi_1^2, \tag{2.64}$$

$$\frac{\partial^2 \phi_2}{\partial \tau_0^2} + w^2 \phi_2 = -2\frac{\partial^2 \phi_1}{\partial \tau_0 \partial \tau_1} - 2x_1\frac{\partial^2 \phi_1}{\partial \tau_0^2} - 2\frac{\partial x_1}{\partial \tau_0}\frac{\partial \phi_1}{\partial \tau_0} - w^2 x_1 \phi_1, \tag{2.65}$$

$$\begin{aligned} \frac{\partial^2 x_3}{\partial \tau_0^2} + x_3 = {} & \hat{f}_1 \cos\left(\left(1+\varepsilon^2\hat{\sigma}_1\right)\tau_0\right) - \hat{c}_1\frac{\partial x_1}{\partial \tau_0} - \\ & 3\hat{\alpha}\,\xi_e^2\, x_1 - \frac{\partial^2 x_1}{\partial \tau_1^2} - 2\left(\frac{\partial^2 x_1}{\partial \tau_0 \partial \tau_2} + \frac{\partial^2 x_2}{\partial \tau_0 \partial \tau_1}\right) + \\ & x_1\left(\frac{\partial \phi_1}{\partial \tau_0}\right)^2 + 2\frac{\partial \phi_1}{\partial \tau_0}\left(\frac{\partial \phi_1}{\partial \tau_1} + \frac{\partial \phi_2}{\partial \tau_0}\right) - w^2\phi_1\phi_2, \end{aligned} \tag{2.66}$$

$$\begin{aligned} \frac{\partial^2 \phi_3}{\partial \tau_0^2} + w^2\phi_3 = {} & \hat{f}_2 \cos\left(\left(w+\varepsilon^2\hat{\sigma}_2\right)\tau_0\right) - \left(\hat{c}_2+\hat{c}_3\right)\frac{\partial \phi_1}{\partial \tau_0} + \\ & \frac{1}{6}w^2\phi_1^3 - w^2\left(x_2\phi_1 + x_1\phi_2\right) - \frac{\partial^2 \phi_1}{\partial \tau_1^2} - 2\frac{\partial^2 \phi_1}{\partial \tau_0 \partial \tau_2} - \\ & 2\frac{\partial^2 \phi_2}{\partial \tau_0 \partial \tau_1} - \frac{\partial^2 \phi_1}{\partial \tau_0^2}\left(x_1^2 + 2x_2\right) - 2x_1\left(2\frac{\partial^2 \phi_1}{\partial \tau_0 \partial \tau_1} + \frac{\partial^2 \phi_2}{\partial \tau_0^2}\right) - \\ & 2\frac{\partial x_1}{\partial \tau_0}\left(\frac{\partial \phi_1}{\partial \tau_1} + \frac{\partial \phi_2}{\partial \tau_0}\right) - 2\frac{\partial \phi_1}{\partial \tau_0}\left(\frac{\partial x_2}{\partial \tau_0} + \frac{\partial x_1}{\partial \tau_1}\right) - 2x_1\frac{\partial x_1}{\partial \tau_0}\frac{\partial \phi_1}{\partial \tau_0}. \end{aligned} \tag{2.67}$$

In differential equations (2.62)–(2.67), the unknows are the functions x_1, x_2, x_3 and ϕ_1, ϕ_2, ϕ_3 that are the subsequent terms of asymptotic expansions (2.18) and (2.19), respectively. The system (2.62)–(2.67) is solved recursively starting from equations of the first-order approximation (2.62) and (2.63) are mutually independent and homogeneous. Their general solutions are

$$x_1(\tau_0,\, \tau_1,\, \tau_2) = B_1(\tau_1,\, \tau_2)\, e^{i\tau_0} + \bar{B}_1(\tau_1,\, \tau_2)\, e^{-i\tau_0}, \tag{2.68}$$

$$\phi_1(\tau_0,\, \tau_1,\, \tau_2) = B_2(\tau_1,\, \tau_2)\, e^{iw\tau_0} + \bar{B}_2(\tau_1,\, \tau_2)\, e^{-iw\tau_0}, \tag{2.69}$$

where the functions $B_1(\tau_1, \tau_2)$, $B_2(\tau_1, \tau_2)$ and their complex conjugates $\bar{B}_1(\tau_1, \tau_2)$, $\bar{B}_2(\tau_1, \tau_2)$ are unknown.

Substituting solutions (2.68)–(2.69) into equations of the second-order approximation (2.64)–(2.65) generates the following solvability conditions

$$\frac{\partial B_1}{\partial \tau_1} = 0, \quad \frac{\partial \bar{B}_1}{\partial \tau_1} = 0, \tag{2.70}$$

$$\frac{\partial B_2}{\partial \tau_1} = 0, \quad \frac{\partial \bar{B}_2}{\partial \tau_1} = 0. \tag{2.71}$$

The following conclusions result from the above solvability conditions

$$\frac{\partial^2 B_1}{\partial \tau_2 \partial \tau_1} = 0, \quad \frac{\partial^2 \bar{B}_1}{\partial \tau_2 \partial \tau_1} = 0, \tag{2.72}$$

$$\frac{\partial^2 B_2}{\partial \tau_2 \partial \tau_1} = 0, \quad \frac{\partial^2 \bar{B}_2}{\partial \tau_2 \partial \tau_1} = 0. \tag{2.73}$$

The particular solutions of equations (2.64)–(2.65) obtained after eliminating the secular terms and consequent relations (2.72)–(2.73) are

$$x_2(\tau_0, \tau_1, \tau_2) = w^2 B_2 \bar{B}_2 + \frac{3w^2\left(e^{2iw\tau_0} B_2{}^2 + e^{-2iw\tau_0} \bar{B}_2^2\right)}{2(4w^2-1)}, \tag{2.74}$$

$$\phi_2(\tau_0, \tau_1, \tau_2) = -\frac{w(w+2)e^{-i(w+1)\tau_0}\left(\bar{B}_1\bar{B}_2 + e^{2i(w+1)\tau_0} B_1 B_2\right)}{2w+1} + \frac{w(w-2)e^{-i(w-1)\tau_0}\left(B_1\bar{B}_2 + e^{2i(w-1)\tau_0}\bar{B}_1 B_2\right)}{2w-1}. \tag{2.75}$$

Substituting solutions (2.68)–(2.69) and (2.74)–(2.75) into equations of the third-order approximation (2.66)–(2.67) causes that some next secular terms occur. Due to the expected bounded character of the oscillations, the secular terms must be eliminated from the equations, which leads to the following solvability conditions

$$-B_1\left(\frac{6(w^2-1)w^2 B_2\bar{B}_2}{4w^2-1} + 3\hat{\alpha}\xi_e^2 + i\,\hat{c}_1\right) - 2i\frac{\partial B_1}{\partial \tau_2} + \frac{1}{2}i\hat{f}_1 e^{i\varepsilon^2\hat{\sigma}_1\tau_0} = 0, \tag{2.76}$$

$$\bar{B}_1\left(-\frac{6(w^2-1)w^2 B_2\bar{B}_2}{4w^2-1} - 3\hat{\alpha}\xi_e^2 + i\hat{c}_1\right) + 2i\frac{\partial \bar{B}_1}{\partial \tau_2} + \frac{1}{2}i\hat{f}_1 e^{-i\varepsilon^2\hat{\sigma}_1\tau_0} = 0, \tag{2.77}$$

$$-2iw\frac{\partial B_2}{\partial \tau_2} - \frac{6w^2(w^2-1)B_1\bar{B}_1 B_2}{4w^2-1} + \frac{w^2(8w^4-7w^2-1)B_2^2\bar{B}_2}{2(4w^2-1)} - iw(\hat{c}_2+\hat{c}_3)B_2 + \frac{1}{2}i\hat{f}_2 e^{i\varepsilon^2\hat{\sigma}_2\tau_0} = 0, \tag{2.78}$$

$$2iw\frac{\partial \bar{B}_2}{\partial \tau_2} - \frac{6w^2(w^2-1)B_1\bar{B}_1\bar{B}_2}{4w^2-1} + \frac{w^2(8w^4-7w^2-1)B_2\bar{B}_2^2}{2(4w^2-1)} + iw(\hat{c}_2+\hat{c}_3)\bar{B}_2 + \frac{1}{2}i\hat{f}_2 e^{-i\varepsilon^2\hat{\sigma}_2\tau_0} = 0. \tag{2.79}$$

Removing all the secular terms from equations of the third-order approximation and taking relations (2.72)–(2.73) into account allows for obtaining the finite solution. The particular solutions to equations (2.67)–(2.68) have the form

$$x_3(\tau_0,\tau_1,\tau_2) = -3w(w+1)\frac{\left(e^{\mathrm{i}(1+2w)\tau_0}B_1{B_2}^2 + e^{-\mathrm{i}(1+2w)\tau_0}\bar{B}_1\bar{B}_2^2\right)}{4(2w+1)} + 3w(w-1)\frac{\left(e^{\mathrm{i}(2w-1)\tau_0}\bar{B}_1{B_2}^2 + e^{-\mathrm{i}(2w-1)\tau_0}B_1\bar{B}_2^2\right)}{4(2w-1)}, \tag{2.80}$$

$$\phi_3(\tau_0,\tau_1,\tau_2) = w\left(w^2+5w+6\right)\frac{\left(e^{\mathrm{i}(2+w)\tau_0}B_1^2B_2 + e^{-\mathrm{i}(2+w)\tau_0}\bar{B}_1^2\bar{B}_2\right)}{4(2w+1)} + w\left(w^2-5w+6\right)\frac{\left(e^{\mathrm{i}(w-2)\tau_0}\bar{B}_1^2B_2 + e^{-\mathrm{i}(w-2)\tau_0}B_1^2\bar{B}_2\right)}{4(2w-1)} - \left(49w^2-1\right)\frac{\left(e^{3\mathrm{i}w\tau_0}B_2^3 + e^{-3\mathrm{i}w\tau_0}\bar{B}_2^3\right)}{48\left(4w^2-1\right)}. \tag{2.81}$$

The unknown functions $B_1(\tau_1,\ \tau_2)$, $B_2(\tau_1,\ \tau_2)$ and their complex conjugates $\bar{B}_1(\tau_1,\ \tau_2)$, $\bar{B}_2(\tau_1,\ \tau_2)$ are restricted by the solvability conditions. Equations (2.70)–(2.71) show that all the functions do not depend on the time variable τ_1. Therefore, the exponential representation of these functions in form (2.41)–(2.42) remains useful. Inserting relationships (2.41)–(2.42) into differential equations (2.76)–(2.79), and then solving the latter with respect to the derivatives, we get the set of four ordinary differential equations of the first-order

$$\frac{db_1}{d\tau_2} = -\frac{1}{2}\hat{c}_1 b_1 + \frac{1}{2}\hat{f}_1\sin(\hat{\sigma}_1\tau_2 - \psi_1), \tag{2.82}$$

$$\frac{db_2}{d\tau_2} = -\frac{1}{2}\left(\hat{c}_2+\hat{c}_3\right)b_2 + \frac{1}{2w}\hat{f}_2\sin(\hat{\sigma}_2\tau_2 - \psi_2), \tag{2.83}$$

$$\frac{d\psi_1}{d\tau_2} = \frac{3w^2(w^2-1)b_2^2}{4\left(4w^2-1\right)} + \frac{3}{2}\xi_e^2\widehat{\alpha} - \frac{\hat{f}_1\cos(\hat{\sigma}_1\tau_2-\psi_1)}{2\,b_1}, \tag{2.84}$$

$$\frac{d\psi_2}{d\tau_2} = \frac{w\left(w^2-1\right)\left(12b_1^2 - \left(8w^2+1\right)b_2^2\right)}{16\left(4\,w^2-1\right)} - \frac{\hat{f}_2\cos(\hat{\sigma}_2\tau_2-\psi_2)}{2\,w\,b_2}. \tag{2.85}$$

Multiplying equations (2.82)–(2.83) by ε^3, equations (2.84)–(2.85) by ε^2, and then taking into account relationships (2.15), (2.16), (2.20) and (2.61) yield the differential equations formulated in the primal time variable τ, of the following form

$$\frac{da_1}{d\tau} = -\frac{1}{2}c_1a_1(\tau) + \frac{1}{2}f_1\sin(\sigma_1\tau - \psi_1(\tau)), \tag{2.86}$$

$$\frac{da_2}{d\tau} = -\frac{1}{2}\left(c_2+c_3\right)a_2(\tau) + \frac{1}{2\,w}f_2\sin(\sigma_2\tau - \psi_2(\tau)), \tag{2.87}$$

$$\frac{d\psi_1}{d\tau} = \frac{3w^2(w^2-1)a_2^2(\tau)}{4(4w^2-1)} + \frac{3}{2}\xi_e^2\alpha - \frac{f_1 \cos(\sigma_1\tau - \psi_1(\tau))}{2\,a_1(\tau)}, \tag{2.88}$$

$$\frac{d\psi_2}{d\tau} = \frac{w(w^2-1)\left(12a_1^2(\tau) - (8w^2+1)\,a_2^2(\tau)\right)}{16(4\,w^2-1)} - \frac{f_2 \cos(\sigma_2\tau - \psi_2(\tau))}{2\,w\,a_2(\tau)}, \tag{2.89}$$

where $a_i(\tau) = \varepsilon\, b_i(\tau)$, $i = 1,\ 2$.

It is worth underlying that both equations (2.43)–(2.46) as well as (2.47)–(2.50), despite the formal differences, describe the modulation of the vibration amplitudes and phases in the slowest time scale. Indeed, taking into account definition (2.16) and assuming, accordingly to (2.70)–(2.71), that the amplitudes and phases are functions only of the slowest time scale τ_2, one can write

$$\frac{da_i}{d\tau} = \varepsilon^2\frac{da_i}{d\tau_2} = \varepsilon^3\frac{db_i}{d\tau_2}, \qquad \frac{d\psi_i}{d\tau} = \varepsilon^2\frac{d\psi_i}{d\tau_2} \quad \text{for} \quad i = 1,2. \tag{2.90}$$

The initial conditions

$$a_1(0) = a_{10}, \quad a_2(0) = a_{20}, \quad \psi_1(0) = \psi_{10}, \quad \psi_2(0) = \psi_{20}, \tag{2.91}$$

where the known quantities a_{10}, a_{20}, ψ_{10}, ψ_{20} are agreed with the initial values ξ_0, v_0, φ_0, ω_0, complete the formulation of the initial-value problem which is necessary to find the amplitudes and phases. In contrast to the case of the non-resonant vibration, the current initial-value problem cannot be solved analytically.

Inserting solutions (2.68)–(2.69), (2.74)–(2.75) and (2.80)–(2.81) into asymptotic expansions (2.18)–(2.19), and skipping the terms of order $O(4)$, we obtain the approximate solution describing the problem of the spring pendulum vibration at doubled external resonance

$$\begin{aligned}
\xi(\tau) &= a_1(\tau)\cos(\tau + \psi_1(\tau)) + \\
&\frac{3w(w-1)\ a_1(\tau)\,a_2^2(\tau)}{32w-16}\cos((1-2w)\,\tau + \psi_1(\tau) - 2\psi_2(\tau)) - \\
&\frac{3w(w+1)\,a_1(\tau)\,a_2^2(\tau)}{32w+16}\cos((1+2w)\,\tau + \psi_1(\tau) + 2\psi_2(\tau)) + \\
&\frac{3w^2a_2^2(\tau)\cos(2w\tau + 2\psi_2(\tau))}{16w^2-4} + \frac{w^2}{4}a_2^2(\tau),
\end{aligned} \tag{2.92}$$

$$\begin{aligned}
\varphi(\tau) &= a_2(\tau)\cos(w\tau + \psi_2(\tau)) + \\
&\frac{w(w-2)\,a_1(\tau)\,a_2(\tau)}{4w-2}\cos((1-w)\,\tau + \psi_1(\tau) - \psi_2(\tau)) + \\
&\frac{w(w-3)\,(w-2)\,a_1^2(\tau)\,a_2(\tau)}{32w-16}\cos((2-w)\,\tau + 2\psi_1(\tau) - \psi_2(\tau)) + \\
&\frac{w(w+2)\,(w+3)\,a_1^2(\tau)\,a_2(\tau)}{32w+16}\cos((2+w)\,\tau + 2\psi_1(\tau) + \psi_2(\tau)) - \\
&\frac{w(w+2)\,a_1(\tau)\,a_2(\tau)}{4w+2}\cos((1+w)\,\tau + \psi_1(\tau) + \psi_2(\tau)) + \\
&\frac{(1-49w^2)\,a_2^3(\tau)}{192(4w^2-1)}\cos(3w\tau + 3\psi_2(\tau)),
\end{aligned} \tag{2.93}$$

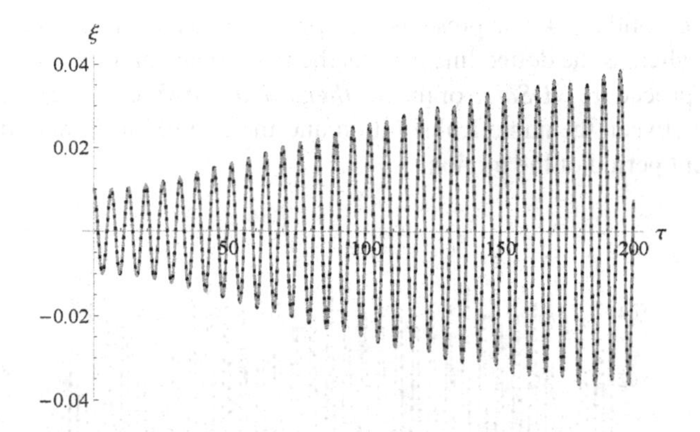

FIGURE 2.6 The transient longitudinal vibration $\xi(\tau)$ in doubled external resonance estimated numerically (dotted line) and analytically using MSM (solid line).

where the functions $a_1(\tau)$, $a_2(\tau)$, $\psi_1(\tau)$, $\psi_2(\tau)$ are numerically obtained solutions of the modulation equations (2.86)–(2.89).

The values of the parameters assumed in the simulation concerning the case of the doubled external resonance are as follows: $w = 0.13$, $\alpha = 0.05$, $c_1 = 0.006$, $c_2 = 0.002$, $c_3 = 0.002$, $f_2 = 0.0005$, $f_1 = 0.0005$, $\sigma_1 = 0.01$, $\sigma_2 = -0.001$.

The initial values for the modulation problem (2.86)–(2.89) are: $a_{10} = 0.01$, $a_{20} = 0.05$, $\psi_{10} = 0$, $\psi_{20} = 0$.

The following initial values for the motion equation (2.10)–(2.11) remain in compliance with the above values of the amplitudes and phases $\xi_0 = 0.0099767$, $v_0 = -0.00002991$, $\varphi_0 = 0.05002699$, $\omega_0 = -0.0001001346$.

The time course of the generalized coordinates $\xi(\tau)$ and $\varphi(\tau)$ is presented in Figures 2.6–2.9. Figures 2.6–2.7 concern the transient stage of the vibration. As

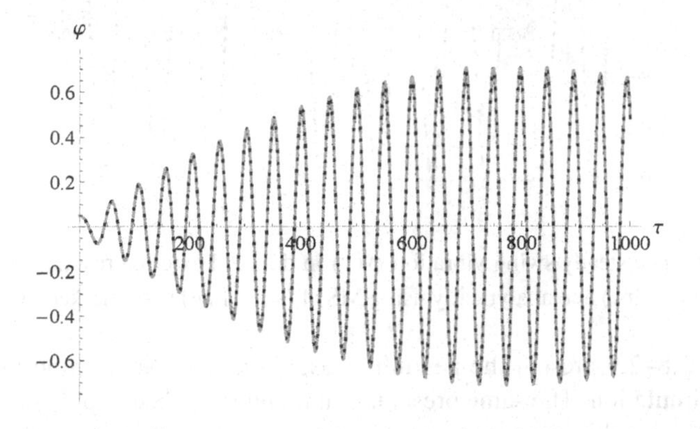

FIGURE 2.7 The transient swing vibration $\varphi(\tau)$ in doubled external resonance estimated numerically (dotted line) and analytically using MSM (solid line).

previously, the solid black line presents the approximate analytical solution obtained using MSM whereas the dotted line depicts the numerical solution obtained utilizing the standard procedure *NDSolve* of the *Mathematica* software. One can observe high compliance between the numerical solution and the approximate analytical solution at the transient period of the motion.

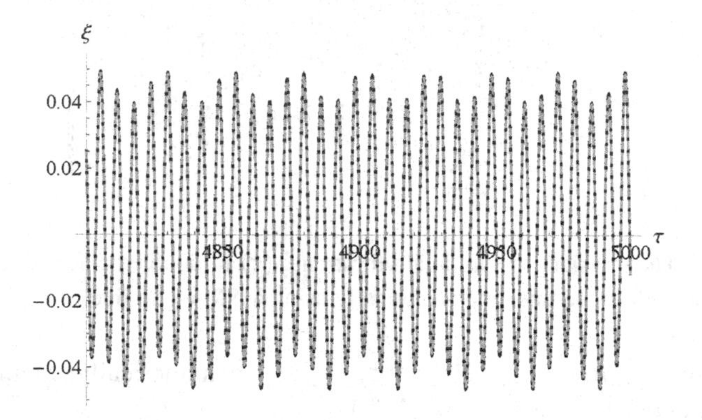

FIGURE 2.8 The quasi-periodic longitudinal vibration $\xi(\tau)$ in doubled external resonance estimated numerically (dotted line) and analytically using MSM (solid line).

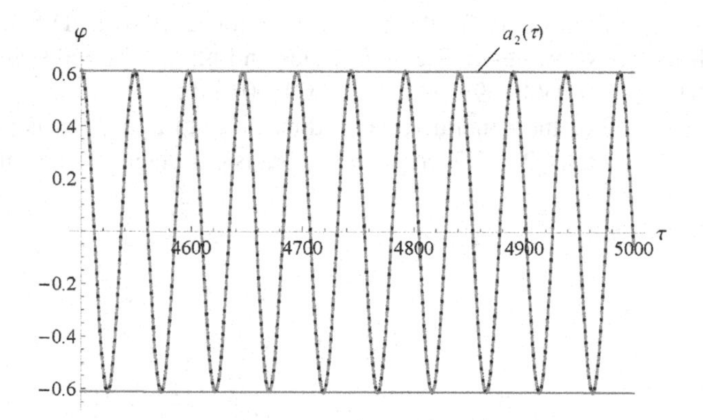

FIGURE 2.9 The steady swing vibration $\varphi(\tau)$ in doubled external resonance estimated numerically (dotted line) and analytically using MSM (solid line) with marked amplitude $a_2(\tau)$.

Figures 2.8–2.9 present the pendulum oscillations on some time interval at the end of the simulation. The same presentation manner has been applied as previously. One may observe the comparison of the solution obtained analytically using MSM and the solution determined using the *NDSolve* procedure. The spring pendulum oscillations in the longitudinal direction do not set in the simulation duration. The vibration in the transversal motion fixes oneself after the transient stage. The functions $a_2(\tau)$ and $-a_2(\tau)$ representing the amplitudes whose graphs are drawn in Figure 2.9 by the solid gray lines are almost constant. Both graphs pass with high accuracy through all maxima and the minima, respectively, of the plots of the fast-changing oscillations which confirms the stationary nature of the oscillations. Despite the different course of the transversal and longitudinal vibration in this phase of motion, the solution obtained using MSM shows the high agreement with the numerical solution of the motion equations in the primal form.

The values of error calculated accordingly to formula (2.58) for the analytical and numerical approximate solutions presented in Figures 2.6–2.9 are collected in Table 2.2.

TABLE 2.2

The values of error of the fulfillment of the governing equations for the resonant vibration

	MSM solution	numerical solution
transient stage (Figures 2.6, 2.7) $\tau_s = 0, \ \tau_e = 500$	$\delta_1 = 7.9824 \cdot 10^{-5}$ $\delta_2 = 7.6137 \cdot 10^{-5}$	$\delta_1 = 5.2192 \cdot 10^{-5}$ $\delta_2 = 9.4442 \cdot 10^{-8}$
stationary stage (Figures 2.8, 2.9) $\tau_s = 4500, \ \tau_e = 5000$	$\delta_1 = 6.0736 \cdot 10^{-5}$ $\delta_2 = 9.6197 \cdot 10^{-5}$	$\delta_1 = 6.9400 \cdot 10^{-7}$ $\delta_2 = 1.5851 \cdot 10^{-7}$

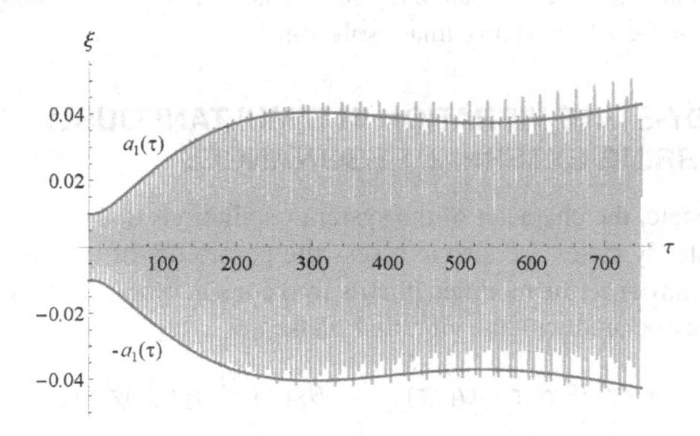

FIGURE 2.10 The transient longitudinal vibration graph $\xi(\tau)$ and the amplitudes $a_1(\tau)$ and $-a_1(\tau)$.

The sense of the curves representing the modulation of the amplitudes is explained in Figures 2.10–2.11 in which the same graphs of the amplitudes $a_1(\tau)$ and $-a_1(\tau)$ are depicted by the dark gray lines. The approximate solution determined by

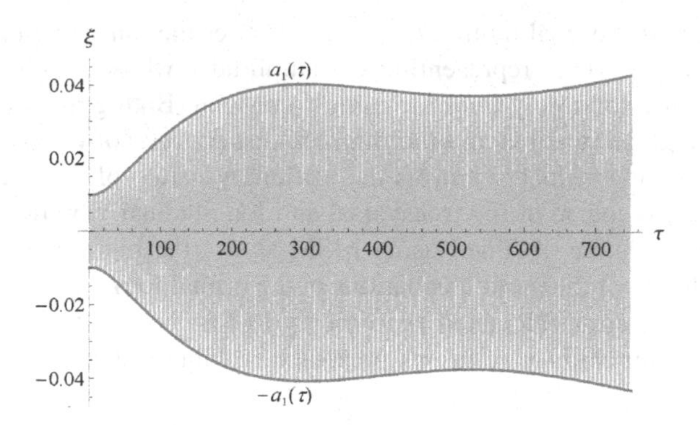

FIGURE 2.11 The first term of the approximate solution $\xi_1(\tau)$ and the amplitudes $a_1(\tau)$ and $-a_1(\tau)$.

(2.92) is presented in Figure 2.10, whereas in Figure 2.11 only the first term of this solution, i.e. $\xi_1(\tau) = a_1(\tau)\cos(\tau + \psi_1(\tau))$, is shown. Both plots of the amplitudes in time pass strictly through the maxima and the minima of the graph of the first term of the approximate solution. Treating them as the lines tangent to the maxima and the minima of the graph of the whole approximate solution is justified only in an approximate sense. The quality and accuracy of such an interpretation are strictly connected to the matter of the uniform validity of the asymptotic expansions (2.18)–(2.19). Each subsequent term of the uniformly valid expansion is small with respect to the previous one. It means that the second and the following terms of the expansion disturb in a small measure, the tangency relationship between the amplitude graph and the graph of the whole approximate solution.

2.6 STEADY-STATE VIBRATION IN SIMULTANEOUSLY OCCURRING EXTERNAL RESONANCES

In the steady-state, the character of the system oscillations does not change in time. To study the steady states and also to assess the possibility of their occurrence, it is convenient to introduce the modified phases into consideration. The modified phases θ_1 and θ_2 are associated with the vibration phases ψ_1 , ψ_2 in the following way

$$\theta_1(\tau) = \sigma_1\tau - \psi_1(\tau), \qquad \theta_2(\tau) = \sigma_2\tau - \psi_2(\tau). \tag{2.94}$$

After inserting the modified phases into equations (2.86)–(2.89), we obtain the following system of four differential equations

$$\frac{da_1}{d\tau} = -\frac{1}{2}c_1 a_1(\tau) + \frac{1}{2}f_1 \sin(\theta_1(\tau)), \tag{2.95}$$

$$\frac{da_2}{d\tau} = -\frac{1}{2}(c_2 + c_3)a_2(\tau) + \frac{1}{2\,w}f_2 \sin(\theta_2(\tau)), \tag{2.96}$$

$$\sigma_1 - \frac{d\theta_1}{d\tau} = \frac{3w^2(w^2-1)a_2^2(\tau)}{4(4w^2-1)} + \frac{3}{2}\xi_e^2\alpha - \frac{f_1\cos(\theta_1(\tau))}{2\,a_1(\tau)}, \tag{2.97}$$

$$\sigma_2 - \frac{d\theta_2}{d\tau} = \frac{w\,(w^2-1)\left(12a_1^2(\tau) - \left(8w^2+1\right)a_2^2(\tau)\right)}{16\,(4\,w^2-1)} - \frac{f_2\cos(\theta_2(\tau))}{2\,w\,a_2(\tau)}. \tag{2.98}$$

The system (2.95)–(2.98) with unknown functions $a_i(\tau)$ and $\theta_i(\tau)$, $i = 1,\ 2$, is an autonomous dynamical system. According to the assumptions about the non-stationary vibration, we postulate that the time derivatives of the amplitudes and the modified phases become equal to zero. The amplitudes a_1, a_2 and modified phases θ_1, θ_2 are therefore constant quantities. The steady-state vibration is governed by the following set of four equations

$$-c_1a_1 + f_1\sin(\theta_1) = 0, \tag{2.99}$$

$$-(c_2+c_3)\,a_2 + \frac{f_2}{w}\sin(\theta_2) = 0, \tag{2.100}$$

$$\frac{3w^2(w^2-1)a_2^2}{2\,(4w^2-1)} + 3\xi_e^2\alpha - \frac{f_1\cos(\theta_1)}{a_1} - 2\,\sigma_1 = 0, \tag{2.101}$$

$$\frac{w\,(w^2-1)\left(12a_1^2 - \left(8w^2+1\right)a_2^2\right)}{8\,(4\,w^2-1)} - \frac{f_2\cos(\theta_2)}{w\,a_2} - 2\,\sigma_2 = 0. \tag{2.102}$$

The values satisfying the system (2.99)–(2.102) represent the fixed amplitudes and modified phases of the spring pendulum vibration in the case of *two main resonances that occur simultaneously*.

There are algebraic and also trigonometric nonlinearities in the system. The modified phases can be eliminated from equations (2.99)–(2.102) using the trigonometric identities which allows one to express the amplitude-frequency dependencies as a set of two algebraic equations. We have

$$\left(\frac{c_1a_1}{f_1}\right)^2 + \frac{a_1^2\left(3\,w^4a_2^2 - 3\,w^2a_2^2 - 6\,\alpha\,\xi_e^2 + 24w^2\alpha\,\xi_e^2 + 4\sigma_1 - 16w^2\sigma_1\right)^2}{4f_1^2(-1+4w^2)^2} - 1 = 0, \tag{2.103}$$

$$\frac{w^2a_2^2\left(12a_1^2\left(w-w^3\right) + a_2^2\left(8w^5 - 7w^3 - w\right) + 16\sigma_2\left(4w^2-1\right)\right)^2}{64\,f_2^2(4w^2-1)^2} + \frac{(c_2+c_3)^2w^2a_2^2}{f_2^2} - 1 = 0. \tag{2.104}$$

Each of the polynomials occurring on the left sides of equations (2.103)–(2.104) is of degree six, hence the number of the solutions of the system, i.e. the pairs $(a_1,\ a_2)$, equals thirty-six. Not all of these solutions are real. Let $G_1(a_1,a_2)$ denotes the left side of equation (2.103), and $G_2(a_1,a_2)$ be the left side of equation (2.104).

Introducing these functions allows one to write the system (2.103)–(2.104) in the following compact form

$$G_1(a_1, a_2) = 0, \tag{2.105}$$

$$G_2(a_1, a_2) = 0. \tag{2.106}$$

The polynomials $G_1(a_1, a_2)$ and $G_2(a_1, a_2)$ are the even functions of their arguments which makes that one can limit looking for the solution of the system (2.105)–(2.106) only to the region in which $a_1 > 0$ and $a_2 > 0$.

The solutions of system (2.103)–(2.104) can be interpreted using the graphs proposed by Starosta et al. [201]. The graphs are drawn on a plane. The Cartesian coordinates defined on the plane represent the values of the amplitudes of the longitudinal and the swing vibration. The solution to each of equations (2.105)–(2.106) is represented by points forming a certain plane curve. The cross points of these curves are the solution to the algebraic system. In Figure 2.12, is shown how the detuning parameters affect the number of the solutions of the system (2.103)–(2.104). For the exemplification, the following values of the parameters are assumed: $w = 0.13$, $\alpha = 0.05$, $c_1 = 0.0005$, $c_2 = 0.0002$, $c_3 = 0.0002$, $f_2 = 0.0004$, $f_1 = 0.0004$.

While studying the resonance, the responcse curves play a very important role in showing the relationship between the magnitude of the vibration amplitude and the frequency of the forcing force. The resonance response (back bone) curves shown in Figure 2.13 present the dependence of the amplitudes a_1 and a_2 on the detuning parameter σ_2. The value of the other detuning parameter σ_1 is assumed to be -0.005. In turn, the relationship between the amplitudes a_1 and a_2 and the detuning parameter σ_1 is illustrated in Figure 2.14. In this case, there is assumed $\sigma_2 = -0.005$. The results are obtained for the following values of the system parameters: $w = 0.13$, $\alpha = 0.03$, $c_1 = 0.002$, $c_2 = 0.0021$, $c_3 = 0.001$, $f_2 = 0.0003$, $f_1 = 0.0003$.

The points forming the resonance response curves are determined numerically using the standard procedure *NSolve* of the *Mathematica*.

The shape of resonant curves strongly depends on the values of the system parameters, among others on the damping coefficients. For example, the resonance response curves presented in Figures 2.15–2.16 have been obtained when we changed only the value of the damping coefficient c_2 leaving the values of the other parameters. The value of c_2 corresponding to the graphs is equal to 0.0023.

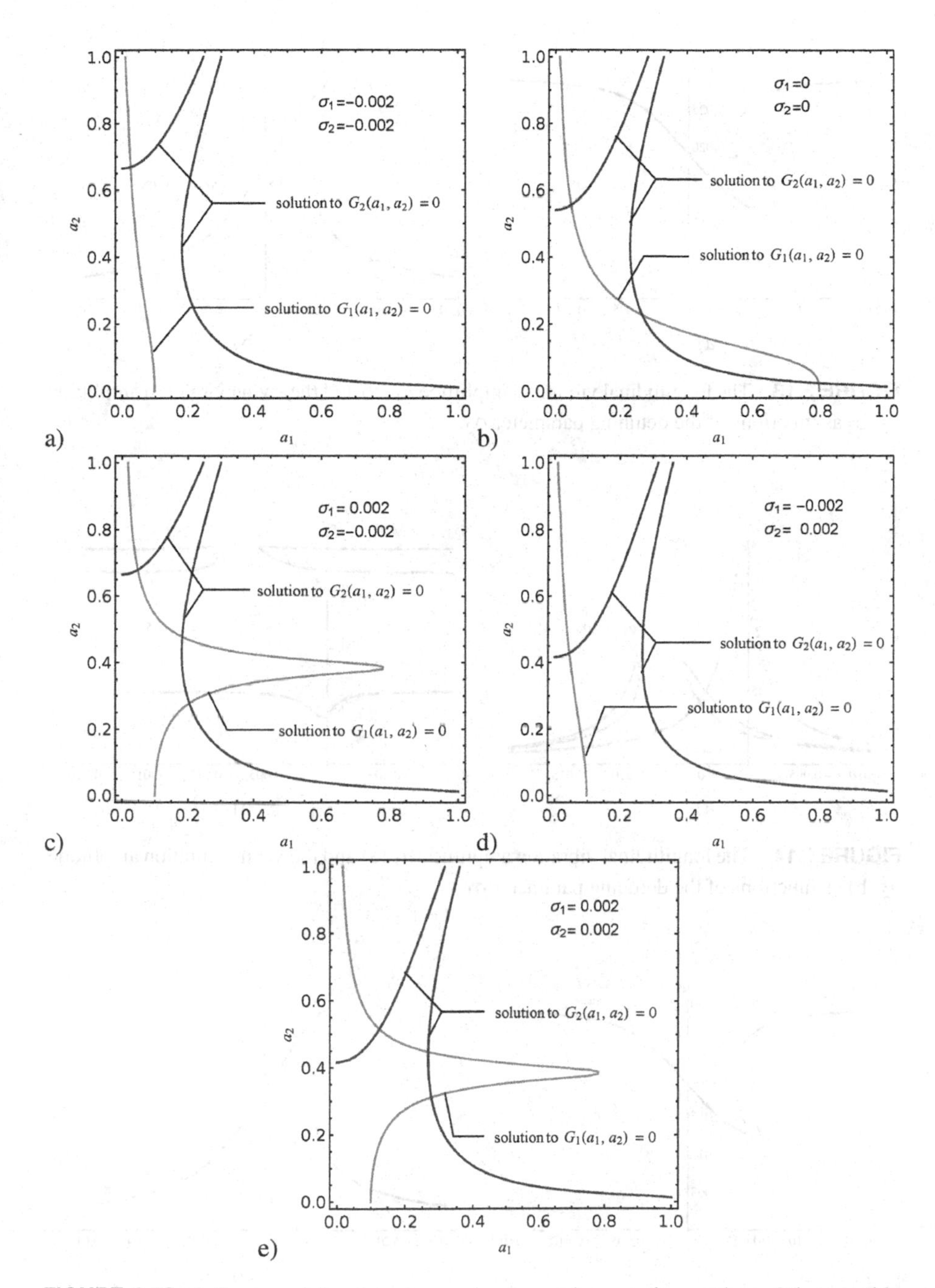

FIGURE 2.12 Influence of the detuning parameters values on the number of the possible steady states: a) $\sigma_1 = -0.002$, $\sigma_2 = -0.002$; b) $\sigma_1 = 0$, $\sigma_2 = 0$; c) $\sigma_1 = 0.002$, $\sigma_2 = -0.002$; d) $\sigma_1 = -0.002$, $\sigma_2 = 0.002$; e) $\sigma_1 = 0.002$, $\sigma_2 = 0.002$.

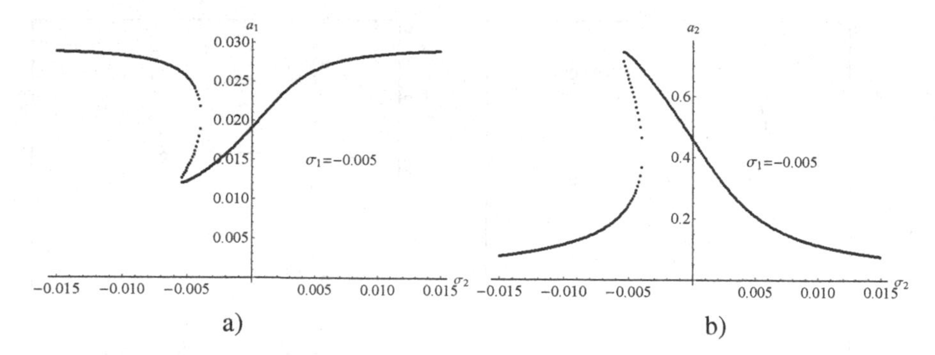

FIGURE 2.13 The longitudinal vibration amplitude a_1 (a) and the swing vibration amplitude a_2 (b) as functions of the detuning parameter σ_2.

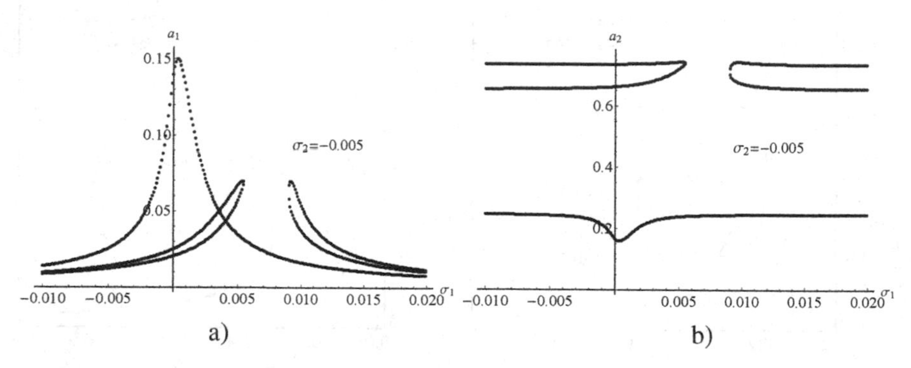

FIGURE 2.14 The longitudinal vibration amplitude a_1 (a) and the swing vibration amplitude a_2 (b) as functions of the detuning parameter σ_1.

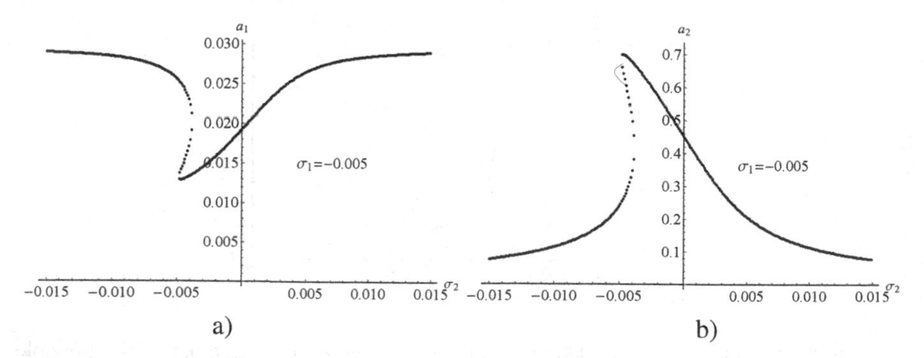

FIGURE 2.15 The longitudinal vibration amplitude a_1 (a) and the swing vibration amplitude a_2 (b) as functions of the detuning parameter σ_2 for $c_2 = 0.0023$.

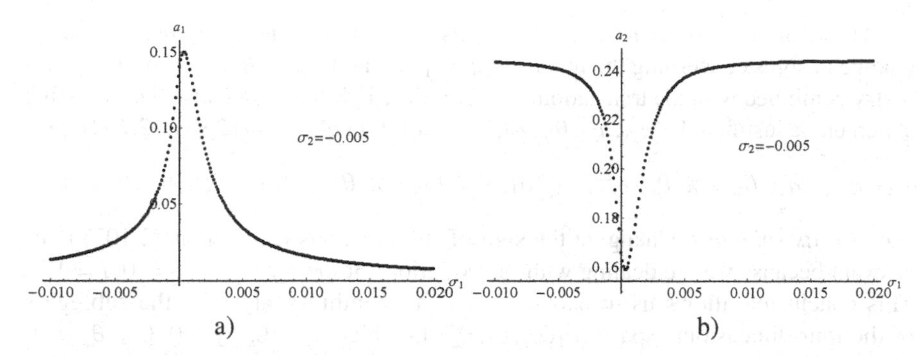

FIGURE 2.16 The longitudinal vibration amplitude a_1 (a) and the swing vibration amplitude a_2 (b) as functions of the detuning parameter σ_1 for $c_2 = 0.0023$.

2.7 STABILITY ANALYSIS

One of the features of nonlinear systems is the possibility of the occurrence of several periodic vibrations with different amplitude values at a given resonance frequency. There are, therefore, several possible stationary resonant states though not all of them are stable. The analysis of the stability of the steady-state oscillations stands for an important aspect of the resonance study. Each periodic solution, whose amplitude was determined by the methods described in the previous section, should be examined for stability.

The periodic steady solutions are, in fact, the fixed points of the modulation equations (2.95)–(2.98). The fixed points of this dynamical system are solutions of equations (2.99)–(2.102). Being interested only in the determination of the vibration amplitudes, we carried out the procedure leading to the elimination of the modified phases in the previous chapter. This approach cannot be applied here, because the stability study requires the confirmation both the amplitudes and the modified phases do not change with time.

Let $g_1(a_1,\theta_1)$, $g_2(a_2,\theta_2)$, $g_3(a_1,a_2,\theta_1)$ and $g_4(a_1,a_2,\theta_2)$ denote the functions standing on the left sides of the subsequent equations (2.99)–(2.102). The functions $g_1(a_1,\theta_1)$ and $g_2(a_2,\theta_2)$ are odd and periodic, so they change their signs under the following transformations

$$\begin{aligned} g_1(-a_1,\theta_1+\pi) &= -g_1(a_1,\theta_1), \\ g_2(-a_2,\theta_2+\pi) &= -g_2(a_2,\theta_2). \end{aligned} \tag{2.107}$$

In turn, the functions $g_3(a_1,a_2,\theta_1)$ and $g_4(a_1,a_2,\theta_2)$ are invariant under the following transformation

$$\begin{aligned} g_3(-a_1,-a_2,\theta_1+\pi) &= g_3(a_1,a_2,\ \theta_1), \\ g_3(-a_1,a_2,\theta_1+\pi) &= g_3(a_1,a_2,\ \theta_1), \\ g_4(-a_1,-a_2,\theta_2+\pi) &= g_4(a_1,a_2,\theta_2). \\ g_4(a_1,-a_2,\theta_2+\pi) &= g_4(a_1,a_2,\theta_2). \end{aligned} \tag{2.108}$$

The invariance properties of the functions $g_3(a_1,a_2,\theta_1)$ and $g_4(a_1,a_2,\theta_2)$ and the change in signs concerning the functions $g_1(a_1,\theta_1)$ and $g_2(a_2,\theta_2)$ at the mirror symmetry combined with the translation of the modified phases by π cause the following statement is justified. If $(a_{1s},a_{2s},\theta_{1s},\theta_{2s})$ *is a solution of system (2.99)–(2.102), then*

$$(-a_{1s},-a_{2s},\theta_{1s}+\pi,\theta_{2s}+\pi),\ \ (-a_{1s},a_{2s},\theta_{1s}+\pi,\theta_{2s}),\ \ (a_{1s},-a_{2s},\theta_{1s},\theta_{2s}+\pi)$$

are also its solutions. Changing the sign of the functions in relations (2.107) is irrelevant because we are dealing with the equations of the form $g_i(\ldots)=0$, $i=1,2$. This conclusion allows us to narrow down the stability analysis to the subregion of the four-dimensional space $(a_1,a_2,\theta_1,\theta_2)$ in which $a_1>0$, $a_2>0$, $0<\theta_1<\pi$, $0<\theta_2<\pi$.

Let us introduce new variables

$$t_1=\tan\left(\frac{\theta_1}{2}\right),\qquad t_2=\tan\left(\frac{\theta_2}{2}\right).\tag{2.109}$$

The following substitutions

$$\sin\theta_1=\frac{2t_1}{1+t_1^2},\qquad \cos\theta_1=\frac{1-t_1^2}{1+t_1^2},\tag{2.110}$$

$$\sin\theta_2=\frac{2t_2}{1+t_2^2},\qquad \cos\theta_2=\frac{1-t_2^2}{1+t_2^2}\tag{2.111}$$

allow one to avoid the difficulties concerning the ambiguity of the sign when the relationship between the sine and the cosine are involved, and the Pythagorean identity is used. Using the substitutions we obtain the system of the following four algebraic equations

$$c_1a_1=\frac{2f_1t_1}{1+t_1^2},\tag{2.112}$$

$$(c_2+c_3)\,a_2=\frac{2f_2t_2}{w\left(1+t_2^2\right)},\tag{2.113}$$

$$4\,\sigma_1-6\,\xi_e^2\alpha=\frac{3w^2(w^2-1)a_2^2}{4w^2-1}-\frac{2f_1\left(1-t_1^2\right)}{\left(1+t_1^2\right)\,a_1},\tag{2.114}$$

$$\sigma_2=\frac{w\left(w^2-1\right)\left(12a_1^2-\left(8w^2+1\right)a_2^2\right)}{16(4\,w^2-1)}-\frac{f_2\left(1-t_2^2\right)}{2\,w\left(1+t_2^2\right)\,a_2}.\tag{2.115}$$

The system of equations (2.112)–(2.115) is solved numerically using the standard procedure *NSolve* of the *Mathematica*. The solutions of the system for different values of the detuning parameters can be presented in the form of the resonance response curves. Now, also the relationships between the modified phases and the detuning parameters are regarded as the resonance response curves. The curves obtained for the same values of parameters as in the case which is presented in Figures 2.13–2.14 are shown in Figures 2.17–2.18. Making use of the symmetry properties, there are plotted only the single branch of each of these curves.

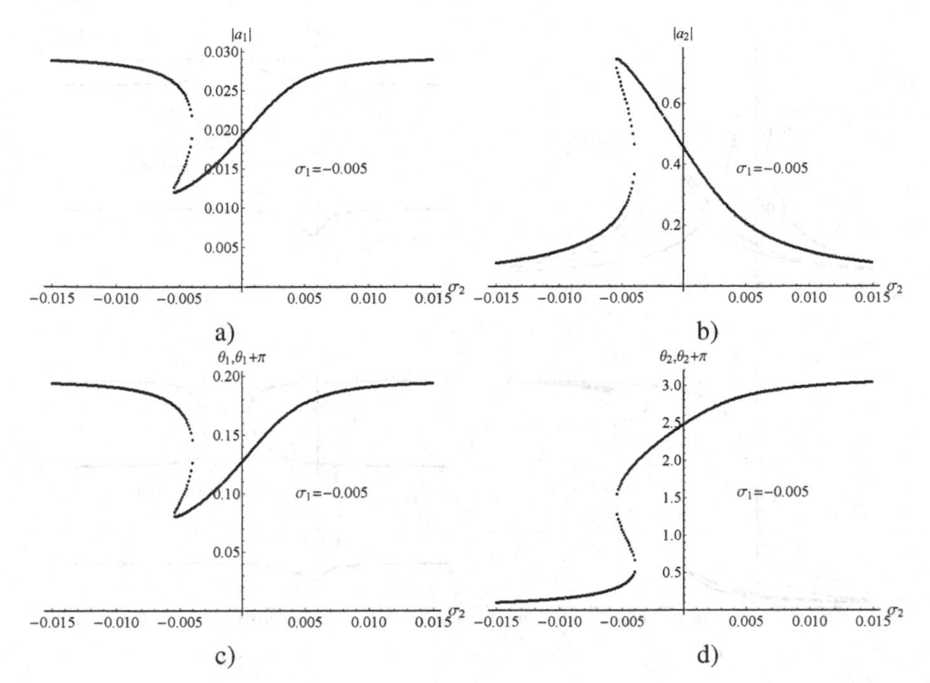

FIGURE 2.17 The amplitudes and modified phases a) $|a_1|$, b) $|a_2|$, c) θ_1, d) θ_2 versus the detuning parameter σ_2 for the following fixed values: $w=0.13$, $\alpha=0.03$, $c_1=0.002$, $c_2=0.0021$, $c_3=0.001$, $f_2=0.0003$, $f_1=0.0003$.

Let $(a_{1s}, a_{2s}, \theta_{1s}, \theta_{2s})$ be any fixed point of the autonomous dynamical system (2.95)–(2.98). That is tantamount to stating that the values $a_{1s}, a_{2s}, \theta_{1s}, \theta_{2s}$ satisfy the steady-state equations (2.99)–(2.102). To examine the stability of this point in the sense of Lyapunov, we focus on the non-stationary solutions of the system (2.95)–(2.98) that are close to the steady-state $(a_{1s}, a_{2s}, \theta_{1s}, \theta_{2s})$. Introducing the functions $\tilde{a}_1(\tau), \tilde{a}_2(\tau), \tilde{\theta}_1(\tau), \tilde{\theta}_2(\tau)$ that can be treated as small perturbations, one can write these non-stationary solutions in the form

$$\begin{aligned} a_1(\tau) &= a_{1s} + \tilde{a}_1(\tau), \quad a_2(\tau) = a_{2s} + \tilde{a}_2(\tau), \\ \theta_1(\tau) &= \theta_{1s} + \tilde{\theta}_1(\tau), \quad \theta_2(\tau) = \theta_{2s} + \tilde{\theta}_2(\tau). \end{aligned} \tag{2.116}$$

Substituting of the functions (2.116) into equations (2.95)–(2.98) gives

$$\frac{d\tilde{a}_1}{d\tau} = -\frac{1}{2}c_1\left(a_{1s} + \tilde{a}_1(\tau)\right) + \frac{f_1}{2}\sin\left(\theta_{1s} + \tilde{\theta}_1(\tau)\right), \tag{2.117}$$

$$\frac{d\tilde{a}_2}{d\tau} = -\frac{1}{2}(c_2+c_3)\left(a_{2s} + \tilde{a}_2(\tau)\right) + \frac{f_2}{2\,w}\sin\left(\theta_{2s} + \tilde{\theta}_2(\tau)\right), \tag{2.118}$$

$$\frac{d\tilde{\theta}_1}{d\tau} = \sigma_1 - \frac{3w^2(w^2-1)(a_{2s}+\tilde{a}_2(\tau))^2}{4(4w^2-1)} - \frac{3}{2}\xi_e^2\alpha + \frac{f_1\cos\left(\theta_{1s}+\tilde{\theta}_1(\tau)\right)}{2(a_{1s}+\tilde{a}_1(\tau))}, \tag{2.119}$$

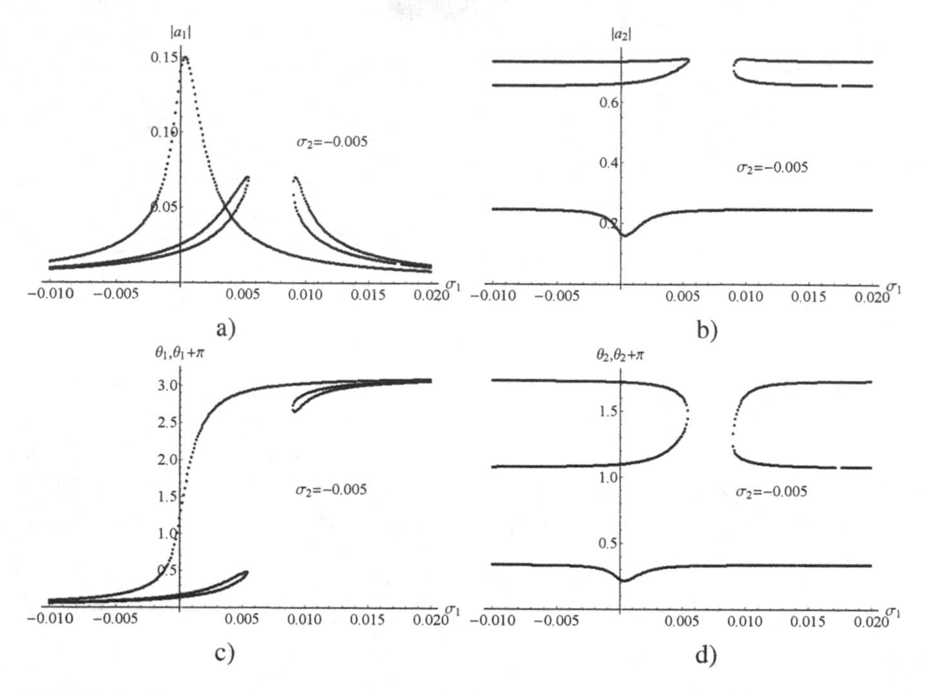

FIGURE 2.18 The amplitudes and modified phases a) $|a_1|$, b) $|a_2|$, c) θ_1, d) θ_2 versus the detuning parameter σ_1 for the following fixed values: $w=0.13,\ \alpha=0.03,\ c_1=0.002,\ c_2=0.0021,\ c_3=0.001, f_2=0.0003,\ f_1=0.0003$.

$$\frac{d\tilde{\theta}_2}{d\tau}=\sigma_2-\frac{w\left(w^2-1\right)\left(12(a_{1s}+\tilde{a}_1(\tau))^2-\left(8w^2+1\right)(a_{2s}+\tilde{a}_2(\tau))^2\right)}{16\left(4\,w^2-1\right)}+$$
$$\frac{f_2\cos\left(\theta_{2s}+\tilde{\theta}_2(\tau)\right)}{2w\,(a_{2s}+\tilde{a}_2(\tau))}. \tag{2.120}$$

The functions occurring on the right sides of equations (2.117)–(2.120) are expanded in their Taylor series around the fixed point $(a_{1s},a_{2s},\theta_{1s},\theta_{2s})$. Taking into account only the terms to order $n=1$, we get the following linearized form of equations (2.117)–(2.120)

$$\frac{d\tilde{a}_1}{d\tau}=-\frac{1}{2}c_1\,(a_{1s}+\tilde{a}_1(\tau))+\frac{f_1}{2}\left(\sin\theta_{1s}+\cos\theta_{1s}\tilde{\theta}_1(\tau)\right), \tag{2.121}$$

$$\frac{d\tilde{a}_2}{d\tau}=-\frac{1}{2}(c_2+c_3)\,(a_{2s}+\tilde{a}_2(\tau))+\frac{f_2}{2\,w}\left(\sin\theta_{2s}+\cos\theta_{2s}\tilde{\theta}_2(\tau)\right), \tag{2.122}$$

$$\frac{d\tilde{\theta}_1}{d\tau}=\sigma_1-\frac{3}{2}\xi_e^2\alpha-\frac{3w^2(w^2-1)}{4\,(4w^2-1)}a_{2s}\,(a_{2s}+2\tilde{a}_2(\tau))+$$
$$\frac{f_1}{2a_{1s}^2}\left(\cos\theta_{1s}\,(\tilde{a}_1(\tau)-a_{1s})+a_{1s}\sin\theta_{1s}\tilde{\theta}_1(\tau)\right), \tag{2.123}$$

$$\frac{d\tilde{\theta}_2}{d\tau} = \sigma_2 - \frac{3w\left(w^2-1\right)}{4\left(4\,w^2-1\right)} a_{1s}\left(a_{1s}+2\tilde{a}_1\left(\tau\right)\right) - \frac{w\left(w^2-1\right)\left(8w^2+1\right)}{16\left(4\,w^2-1\right)} a_{2s}\left(a_{2s}+2\tilde{a}_2\left(\tau\right)\right) + \frac{f_2}{2wa_{2s}^2}\left(\cos\theta_{2s}\left(\tilde{a}_2\left(\tau\right)-a_{2s}\right)+a_{2s}\sin\theta_{2s}\tilde{\theta}_2\left(\tau\right)\right). \tag{2.124}$$

Taking into account that the fixed point $(a_{1s}, a_{2s}, \theta_{1s}, \theta_{2s})$ satisfy the steady-state equations (2.99)–(2.102), we can simplify equations (2.121)–(2.124) to the form

$$\frac{d\tilde{a}_1}{d\tau} = -\frac{c_1}{2}\tilde{a}_1\left(\tau\right) + \frac{f_1}{2}\cos\theta_{1s}\tilde{\theta}_1\left(\tau\right), \tag{2.125}$$

$$\frac{d\tilde{a}_2}{d\tau} = -\frac{1}{2}\left(c_2+c_3\right)\tilde{a}_2\left(\tau\right) + \frac{f_2}{2\,w}\cos\theta_{2s}\tilde{\theta}_2\left(\tau\right), \tag{2.126}$$

$$\frac{d\tilde{\theta}_1}{d\tau} = \frac{3w^2(w^2-1)a_{2s}\tilde{a}_2\left(\tau\right)}{2\left(4w^2-1\right)} - \frac{f_1}{2\,a_{1s}^2}\left(\cos\theta_{1s}\tilde{a}_1\left(\tau\right)+a_{1s}\sin\theta_{1s}\tilde{\theta}_1\left(\tau\right)\right), \tag{2.127}$$

$$\frac{d\tilde{\theta}_2}{d\tau} = \frac{w\left(w^2-1\right)\left(\left(8w^2+1\right)a_{2s}\tilde{a}_2\left(\tau\right)-12\,a_{1s}\tilde{a}_1\left(\tau\right)\right)}{8\left(4\,w^2-1\right)} - \frac{f_2}{2wa_{2s}^2}\left(\cos\theta_{2s}\tilde{a}_2\left(\tau\right)+a_{2s}\sin\theta_{2s}\tilde{\theta}_2\left(\tau\right)\right). \tag{2.128}$$

The dynamical system (2.125)–(2.128) is linear with respect to the perturbations $\tilde{a}_1\left(\tau\right), \tilde{a}_2\left(\tau\right), \tilde{\theta}_1\left(\tau\right), \tilde{\theta}_2\left(\tau\right)$. Thus, the applied procedure makes that the issue of studying the stability of stationary resonant states has been replaced with the much simpler but approximate problem of the stability of the linear dynamic system.

Equations (2.125)–(2.128) are homogeneous, and all their coefficients are constant. The general solution of the system can be presented in the form of the linear combination of the exponential functions

$$\begin{aligned} \tilde{a}_i\left(\tau\right) &= D_1^{(i)}e^{\lambda_1\tau} + D_2^{(i)}e^{\lambda_2\tau} + D_3^{(i)}e^{\lambda_3\tau} + D_4^{(i)}e^{\lambda_4\tau}, \quad i=1,2, \\ \tilde{\theta}_i\left(\tau\right) &= D_1^{(i+2)}e^{\lambda_1\tau} + D_2^{(i+2)}e^{\lambda_2\tau} + D_3^{(i+2)}e^{\lambda_3\tau} + D_4^{(i+2)}e^{\lambda_4\tau}, \quad i=1,2, \end{aligned} \tag{2.129}$$

where λ_j, $j=1,..,4$, are the roots of the characteristic equation

$$|\mathbf{A} - \lambda\,\mathbf{I}\,| = 0. \tag{2.130}$$

The symbol **I** denotes the identity matrix, and **A**

$$\mathbf{A}=\begin{bmatrix} -\frac{c_1}{2} & 0 & \frac{f_1\cos\theta_{1s}}{2} & 0 \\ 0 & -\frac{c_2+c_3}{2} & 0 & \frac{f_2\cos\theta_{2s}}{2w} \\ -\frac{f_1\cos\theta_{1s}}{2a_{1s}^2} & -\frac{3w^2(w^2-1)a_{2s}}{2(4w^2-1)} & -\frac{f_1 sin\theta_{1s}}{2a_{1s}} & 0 \\ -\frac{3w^2(w^2-1)a_{1s}}{2(4w^2-1)} & \frac{w(w^2-1)(8w^2+1)a_{2s}}{8(4w^2-1)}-\frac{f_2\cos\theta_{2s}}{2wa_{2s}^2} & 0 & -\frac{f_2 sin\theta_{2s}}{2wa_{2s}} \end{bmatrix} \quad (2.131)$$

is the characteristic matrix of the differential equations (2.125)–(2.128). The roots λ_j , $j=1,..,4,$ of the characteristic equation are the eigenvalues of the matrix **A**. If the real parts of all eigenvalues of the matrix **A** are negative, then the fixed point $(a_{1s}, a_{2s}, \theta_{1s}, \theta_{2s})$ equivalent to the steady-state solution is asymptotically stable in the sense of Lyapunov, of course only in the sense of the linear approximation.

Making use of the procedure described above the stability of the resonance response curves shown in Figures 2.17–2.18 was evaluated. The eigenvalues of matrix A are determined using the standard procedure *Eigenvalues* of the *Mathematica* software.

In Figures 2.19–2.20, the resonance response curves are depicted by the gray points, whereas the black dots present the fixed points that are asymptotically stable

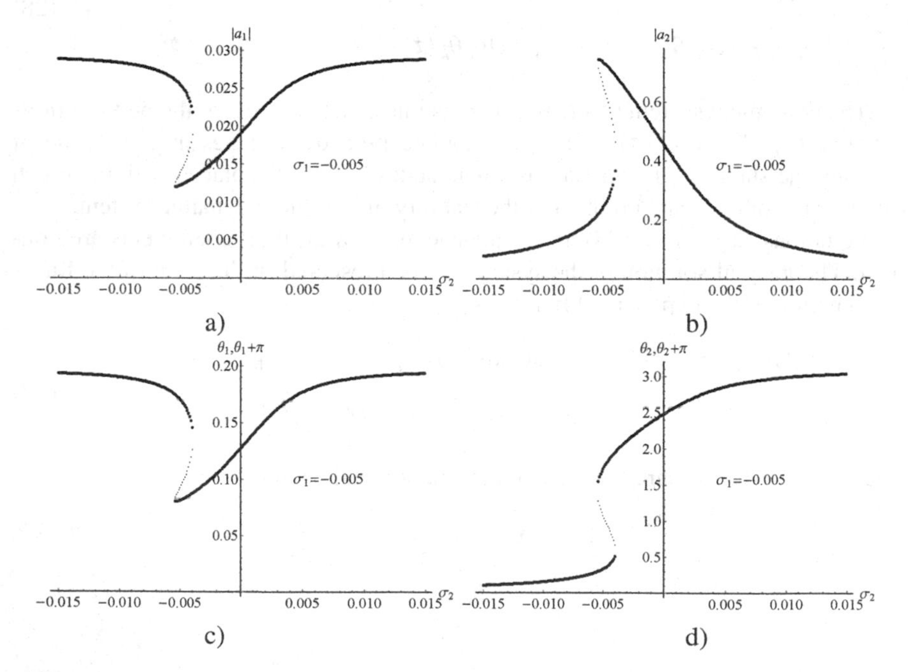

FIGURE 2.19 The resonance response curves versus the detuning parameter σ_2 : a) $|a_1|$, b) $|a_2|$, c) θ_1, d) θ_2. Stable branches are drawn with larger dot size.

in the sense of Lyapunov. The values of the parameters are the same as previously, i.e. we take $w = 0.13$, $\alpha = 0.03$, $c_1 = 0.002$, $c_2 = 0.0021$, $c_3 = 0.001$, $f_2 = 0.0003$, $f_1 = 0.0003$.

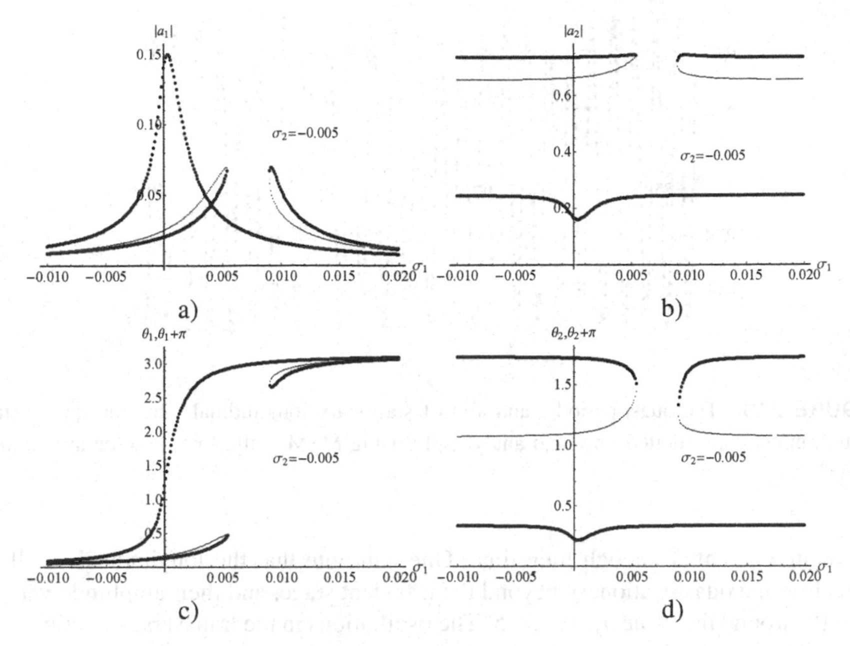

FIGURE 2.20 The resonance response curves versus the detuning parameter σ_1: a) $|a_1|$, b) $|a_2|$, c) θ_1, d) θ_2. Stable branches are drawn with larger dot size.

From Figures 2.19–2.20 one can read that for $\sigma_1 = -0.005$ and $\sigma_2 = -0.005$ the values of the stable amplitudes are as follows

$$a_{1s} \approx 0.012 \quad \text{or} \quad a_{1s} \approx 0.025,$$
$$a_{2s} \approx 0.239 \quad \text{or} \quad a_{2s} \approx 0.734.$$

Assuming the same values of parameters as in the considered case, i.e. $w = 0.13$, $\alpha = 0.03$, $c_1 = 0.002$, $c_2 = 0.0021$, $c_3 = 0.001$, $f_2 = 0.0003$, $f_1 = 0.0003$, $\sigma_1 = -0.005, \sigma_2 = -0.005$, and additionally putting the following initial values $a_{10} = 0.04$, $a_{20} = 0.1$, $\psi_{10} = 0$, $\psi_{20} = 0$, the equations of motion (2.10)–(2.11) were solved employing the method of multiple scales as well numerically. The following initial values for the motion equation remain in compliance with the initial values of the amplitudes and phases: $\xi_0 = 0.0399092$, $v_0 = -0.00003972$, $\varphi_0 = 0.100214$, $\omega_0 = -0.0015554$.

The time courses of the generalized coordinates $\xi(\tau)$ and $\varphi(\tau)$ are presented in Figures 2.21–2.22. The solid black line presents the approximate analytical solution obtained using MSM and by the dotted line is depicted the numerical solution calculated with the use of the standard procedure *NDSolve* of the *Mathematica* software.

The dark gray lines plotted in Figures 2.21–2.22 present the graphs of the amplitudes $a_1(\tau)$ and $a_2(\tau)$, respectively. In the case of the steady-state, both amplitudes

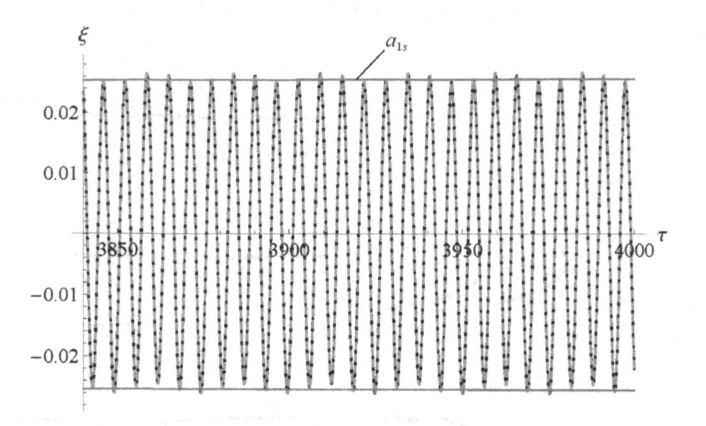

FIGURE 2.21 The quasi-periodic and almost stationary longitudinal vibration $\xi(\tau)$ estimated numerically (dotted line) and analytically using MSM (solid line) and the amplitude $a_{1s}(\tau)$.

fix themselves after enough long time. One can note that the longitudinal oscillations become quasi-stationary beyond the transient stage, and their amplitude varies slightly around the value $a_{1s} \approx 0.025$. The oscillations in the transverse direction become stationary at the end of the simulation, and their amplitude is almost equal to the value $a_{2s} \approx 0.239$.

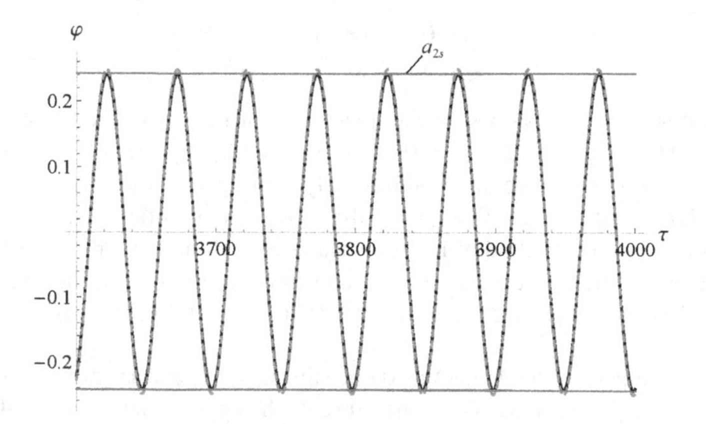

FIGURE 2.22 The stationary swing vibration $\varphi(\tau)$ estimated numerically (dotted line) and analytically using MSM (solid line) and the amplitude $a_{2s}(\tau)$.

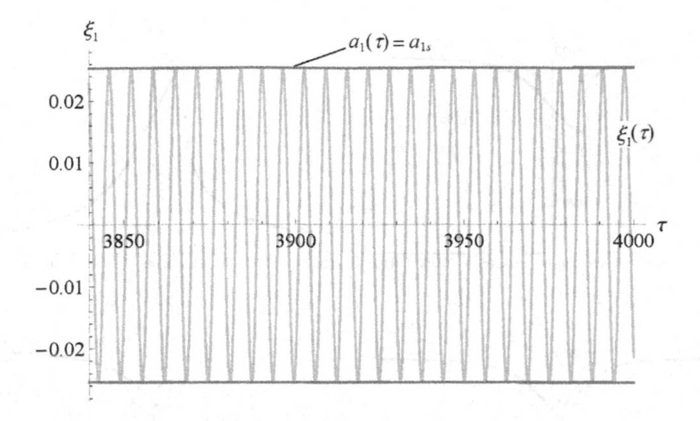

FIGURE 2.23 The amplitude $a_1(\tau)$ and the first term $\xi_1(\tau)$ of the approximate solution.

The graphs of the first terms of the approximate solutions obtained employing the multiple scales method, i.e. $\xi_1(\tau) = a_1(\tau)\cos(\tau + \psi_1(\tau))$ and $\varphi_1(\tau) = a_2(\tau)\cos(w\tau + \psi_2(\tau))$ together with the plots of amplitudes $a_1(\tau)$, $-a_1(\tau)$ and $a_2(\tau)$, $-a_2(\tau)$ are depicted in Figures 2.23–2.24. One can note that each of the amplitude graphs passes exactly through the maxima or the minima of the diagram presenting the time course of the first term of the approximate solution.

The shape of the resonance response curves depends strongly on the values of the parameters occurring in the modulation equations. A certain sight on the matter is provided by the curves graphs shown in Figures 2.25–2.28. They show, how the values of the amplitudes of the harmonic forces affect the resonant response. To focus

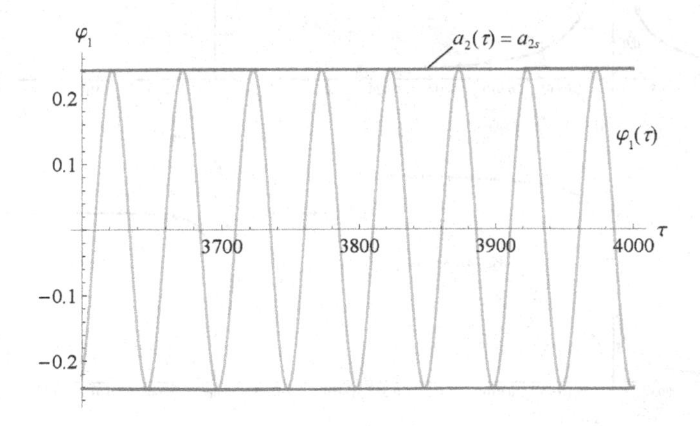

FIGURE 2.24 The amplitude $a_2(\tau)$ and the first term $\xi_1(\tau)$ of the approximate solution.

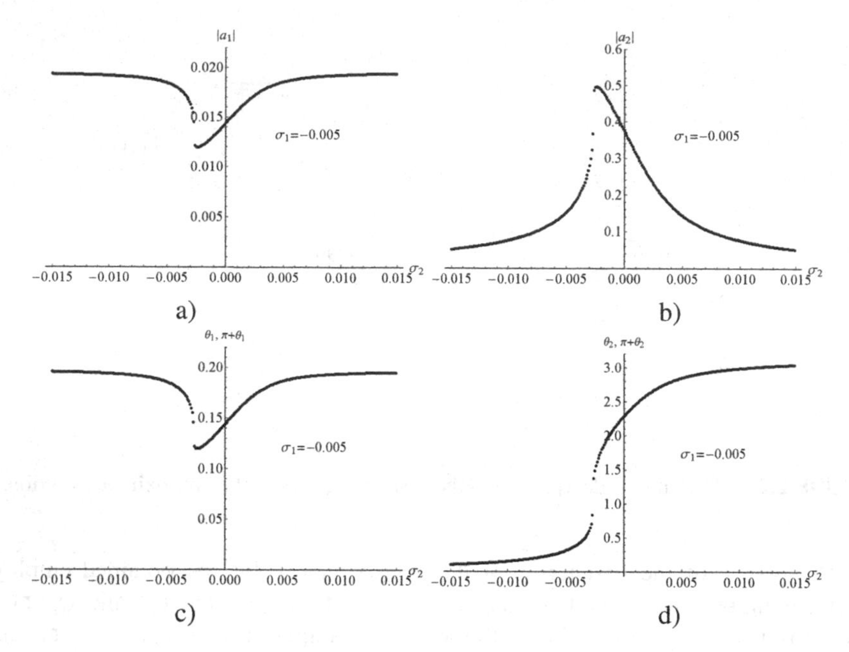

FIGURE 2.25 The unambiguous resonance response curves versus the detuning parameter σ_2 for $f_1 = f_2 = 0.0002$: a) $|a_1|$, b) $|a_2|$, c) θ_1, d) θ_2.

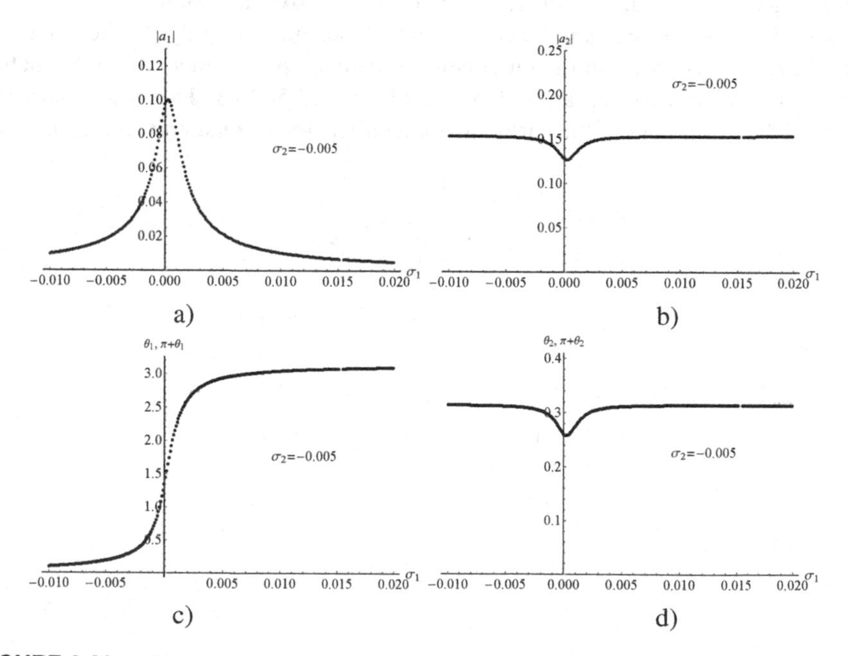

FIGURE 2.26 The unambiguous resonance response curves versus the detuning parameter σ_1 for $f_1 = f_2 = 0.0002$: a) $|a_1|$, b) $|a_2|$, c) θ_1, d) θ_2.

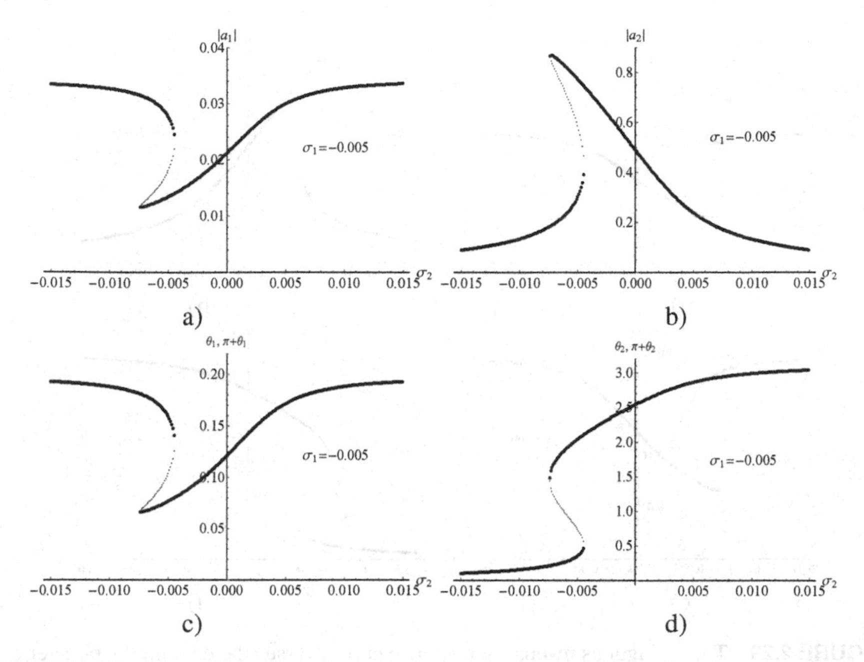

FIGURE 2.27 The ambiguous resonance response curves versus the detuning parameter σ_2 for $f_1 = f_2 = 0.00035$: a) $|a_1|$, b) $|a_2|$, c) θ_1, d) θ_2. Stable branches are drawn with larger dot size.

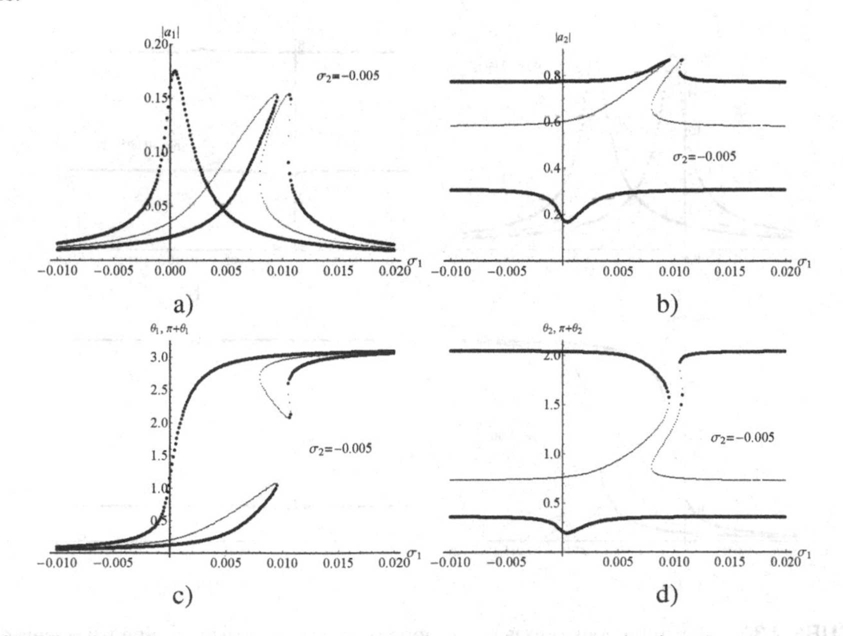

FIGURE 2.28 The ambiguous resonance response curves versus the detuning parameter σ_1 for $f_1 = f_2 = 0.00035$: a) $|a_1|$, b) $|a_2|$, c) θ_1, d) θ_2. Stable branches are drawn with larger dot size.

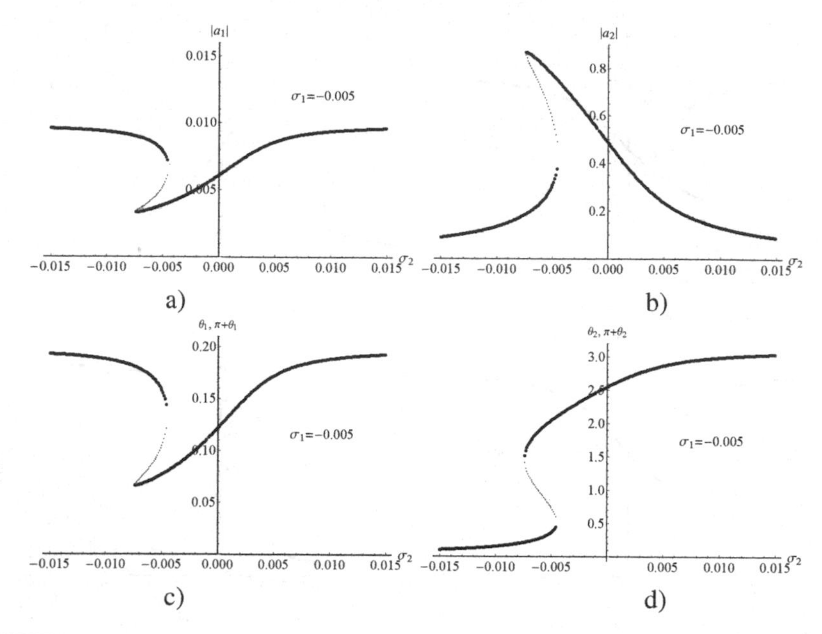

FIGURE 2.29 The ambiguous resonance response curves versus the detuning parameter σ_2 for $f_1 = 0.0001$, $f_2 = 0.00035$: a) $|a_1|$, b) $|a_2|$, c) θ_1 , d) θ_2. Stable branches are drawn with larger dot size.

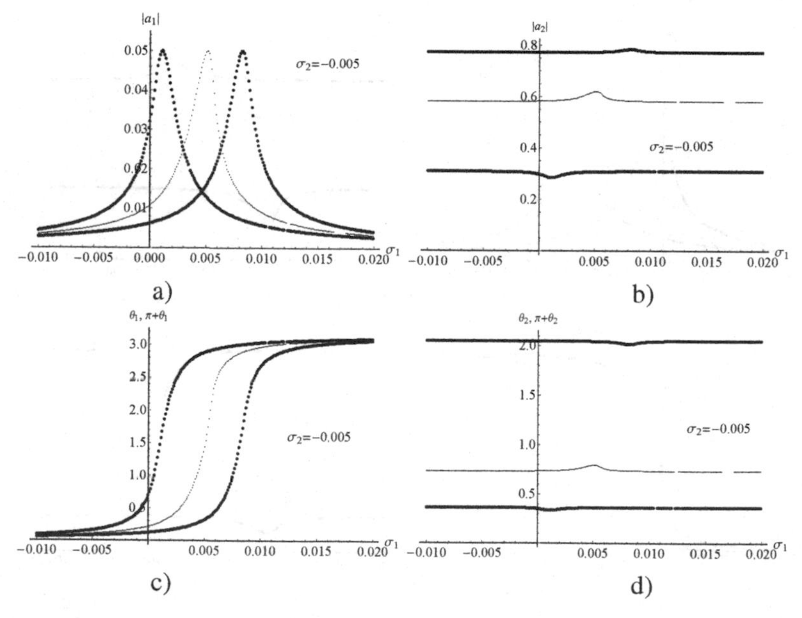

FIGURE 2.30 The ambiguous resonance response curves versus the detuning parameter σ_1 for $f_1 0.0001$, $f_2 = 0.00035$: a) $|a_1|$, b) $|a_2|$, c) θ_1, d) θ_2. Stable branches are drawn with larger dot size.

on the impact of only these two parameters, the others have been set as constant. The following values are assumed: $w = 0.13$, $\alpha = 0.03$, $c_1 = 0.002$, $c_2 = 0.0021$, $c_3 = 0.001$. The values of the amplitudes f_1 and f_2 are listed in the captions for the figures. The stable fixed points corresponding to the stable resonant steady states are plotted in dark gray.

2.8 CLOSING REMARKS

A few aspects of the dynamics of the spring pendulum with the elastic force govern by the cubic law and the viscous damping have been studied. The case when the magnitudes of the axial force and the torque forcing the pendulum change harmonically has been investigated. The initial value problem derived using the Lagrange equations of the second kind has been transformed to its counterpart dimensionless form.

The weak type of the nonlinearities assumed in the model causes that the dynamical and elastic couplings occurring in the system become also weak. Thus, the resonant frequencies of the spring pendulum are close to the natural frequencies of the corresponding linearized system.

The approximate solution to the initial value problem concerning the nonresonant vibration has been obtained employing the multiple scales method. The problem of modulation of amplitudes and phases which is an integral part of the MSM procedure was solved analytically. The accuracy of the approximate solution obtained using MSM was assessed by the absolute error measuring the satisfaction of the solutions of the governing equations.

The analytical form of the approximate solution obtained by MSM yields the frequencies in the neighborhood of which the resonances occur. The resonant oscillations have been considered separately and the solutions concerning them have also been derived, employing MSM but with additionally formulated assumptions regarding the frequencies of the forcing. Two simultaneously existing main resonances have been investigated. The more complicated form of the equations governing the modulation of the amplitudes and phases in resonance shows that a direct application of numerical methods to solve them becomes inevitable. This is why the asymptotic solution to the initial problem regarding the resonant case is estimated in a semi-analytical way. The accuracy of the solutions obtained in this way was estimated by applying an error which is a measure of satisfying the differential equations of the model adequate to the resonance. The solutions have also been compared with the numerical ones. Both tests indicate that if the appropriate assumptions on which the model is based are satisfied, the solutions can be treated as reliable and validated. Attention has been paid to the interpretation of the amplitudes being the solutions to the modulation problem.

The system of four algebraic equations, the solutions which depict the periodic steady-states have been derived from the modulation equations. The symmetry and invariance properties of the left sides of the algebraic equations imply certain dependencies of the mutual symmetry between the roots. It allows one to make the representative analysis of the resonance response curves in several regions of the four-dimensional space of the amplitudes and phases. In turn, it is also possible to eliminate the modified phases from the algebraic equations and construct a separate formulation concerning only the amplitudes. The formulation has the form of a system of two algebraic equations while each of them is of the sixth order. Stable branches of the resonance response curves including the branches of the curves depicting the dependencies between the phases and the detuning parameters have been determined based on the linear theory of the small perturbations.

3 Kinematically Excited Spring Pendulum

In this chapter, the problem of kinematic excitation is studied based on a spring pendulum suspended at the moving point. It is assumed that this point moves along a given curve and its motion is fully defined. Additionally, the pendulum is loaded by the external time-dependent force and moment, whereas damping effects are of viscous nature. The governing equations are derived using Lagrange's formalism and then transformed into the dimensionless form. Some necessary assumptions about the mathematical model have been made to enable it to be tested with the multiple scale method. The system's dynamics is studied in the vicinity of the stable equilibrium position, so the trigonometric functions of the swing angle are developed into the Taylor series up to the third-order. The MSM was used with three-time scales that enable the study of the behavior of the pendulum both in and out of resonance. Parametric and primary resonances appearing either simultaneously or separately are examined. The calculation procedure is described in detail. The asymptotic solution is then compared with the numeric one. The comparison indicates excellent consistency between analytical and numerical results. Steady-state analysis of the modulation equations allows obtaining resonance response curves, the stability of which is investigated in the Lyapunov sense and presented graphically.

3.1 INTRODUCTION

Kinematic excitation appears in many problems of applied mechanics and is crucial in some technical problems. It can be caused by contact forces in some devices, mechanisms, vehicles, or engineering structures. In addition, machine supports may be exposed to kinematic excitation due to ground vibrations caused by road or rail traffic, seismic vibrations, or the action of other machines.

There are many examples of problems with kinematic excitation in the literature. Pielorz [170] described dynamic tests of low structures subjected to kinematic excitation caused by transverse waves.

Szolc [210] investigated the serious interactions between a trolley of a modern passenger car and the track caused by kinematic and parametric excitation from the track and kinematic excitation caused by polygonization of the wheel rim. The dynamic behavior of self-propelled devices as a result of kinematic excitation caused by bumpy roads is the subject of the work carried out by Axinti and Axinti [33]. Such problems also occur at the nanoscale. The possibility of the existence of undamped kinematic excitation in one-dimensional molecular crystals was investigated by Kozmidis-Luburić et al. [106].

DOI: 10.1201/9781003270706-3

Weibel et al. [232], Tondl and Nabergoj [212], Zhu et al. [244], Starosta and Awrejcewicz [197] studied the behavior of various types of pendulums as a good and intuitive example of a nonlinear system.

Amer and Bek [7] investigated the chaotic response of a harmonically excited spring pendulum moving along a circular path. It is a nonlinear system with two degrees of freedom and serves as a good example for several engineering applications such as roll ship motion.

An inverted pendulum driven by two fibres and affected by a gravitation force was studied by Polach et al. [172]. It was shown that the mass of the fibres influences strongly the pendulum vibration. However, the paper did not offer experimental verification of the constructed model.

Fritzkowski and Kamiński [71] considered 2D motion of a hanging rope with kinematic excitation. Discrete pendulum-type model was employed to study nonlinear dynamics of the mechanical system with a help of the maximum Lyapunov exponent and via fast Fourier transform, and bifurcation diagrams. Both regular and chaotic dynamics were reported.

Amer et al. [8] studied the response of a nonlinear multi-degrees of freedom system represented by an elastic pendulum moving in an elliptic path. The MSM allowed to derive the analytical solutions up to the third approximation order. All possible resonances were detected and the modulation equations were solved numerically. In addition, stability problem of periodic solutions was addressed.

Some kind of pendulum is also tested in this chapter. It is a spring pendulum with a suspension point moving along a designated path. Apart from the kinematic excitation and assumed external forces, the inertial coupling can lead to auto-parametric vibration excitation. The two-degrees of freedom system explored in this chapter is a good example for developing and testing analytical methods because of its simplicity and intuitive predictability. In this section, we focus on detecting resonance conditions, and then a selected case of resonances occurring simultaneously is examined. The applied multiple scale method allows one to find the number of all possible amplitudes in a steady-state for selected parameters and to predict their values. The amplitude-frequency response curves are presented in the case of two resonances occurring simultaneously.

The calculations were made with the use of computer algebra and the symbolic manipulation system *Mathematica*. Most of the operations are performed automatically using special procedures created by the authors.

It should be noted that similar problem has earlier been considered by the authors (Awrejcewicz and Starosta [26], Starosta et al. [199, 201]).

3.2 THE PHYSICAL AND MATHEMATICAL MODEL

We consider the mathematical spring pendulum of mass m with kinematic excitation caused by the motion of the suspension point. Its motion is fully defined by the following parametric equations

$$x = R_x \cos(\Omega_x t), \qquad y = R_y \sin(\Omega_y t), \tag{3.1}$$

where R_x, R_y, Ω_x and Ω_y are known.

This means that the suspension point moves harmonically in two perpendicular directions, so its path is Lissajou's curve.

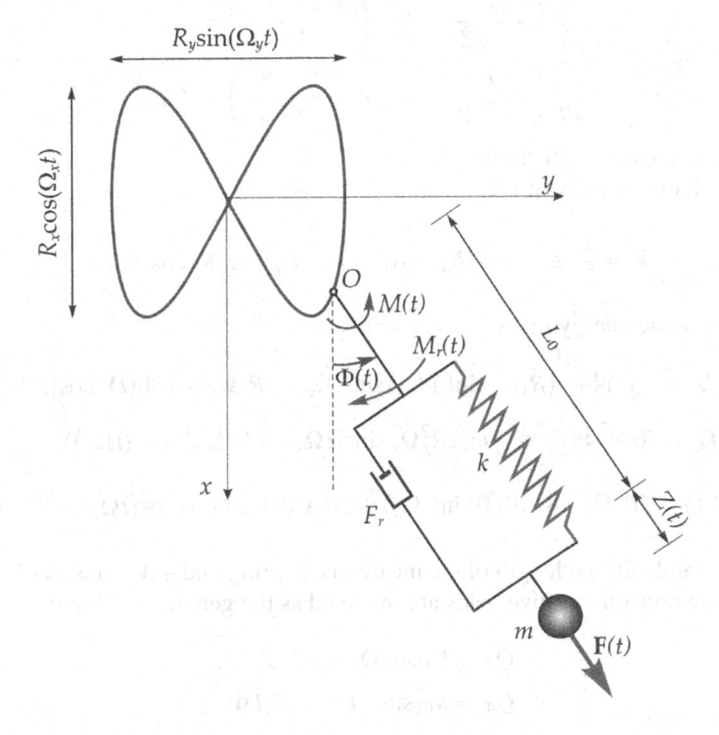

FIGURE 3.1 Spring pendulum with kinematic excitation.

The studied system is presented in Figure 3.1. We assume that the motion is planar in the vertical plane OXY. The spring elongation Z and the angle Φ between the vertical line and the axis of the pendulum are assumed as the generalized coordinates.

Further model assumptions are as follows:

- the whole mass m of the system is concentrated at the material point,
- the spring is linear with the elastic constant k,
- the pendulum is loaded by the by external torque $M(t) = M_0 \sin(\Omega_2 t)$ and the external force $F(t) = F_0 \sin(\Omega_1 t)$, whose direction crosses the suspension point O,
- resistance of motion is described by the viscous model, so the longitudinal vibrations are damped by the force $F_r = C_1 \dot{Z}(t)$ acting on the mass m, and the swing oscillations are damped by the moment $M_r = C_2 L \dot{\Phi}(t)$, where C_1 and C_2 are damping coefficients.

The equations of motion are derived using the Lagrange equations of the second kind

$$\frac{d}{dt}\left(\frac{\partial (T-V)}{\partial \dot{Z}}\right)-\left(\frac{\partial (T-V)}{\partial Z}\right)=Q_Z,$$
$$\frac{d}{dt}\left(\frac{\partial (T-V)}{\partial \dot{\Phi}}\right)-\left(\frac{\partial (T-V)}{\partial \Phi}\right)=Q_\Phi, \tag{3.2}$$

where Q_Z and Q_Φ are generalized forces.

The potential energy of the system is

$$V=\frac{1}{2}kZ^2-mg\left(R_x\cos\left(\Omega_x t\right)+\left(L_0+Z\right)R_y\cos\left(\Phi\right)\right), \tag{3.3}$$

while the kinetic energy reads

$$\begin{aligned} T&=m(L_0+Z(t))\dot{\Phi}(t)(R_x\Omega_x\sin(\Phi(t))\sin(t\Omega_x)+R_y\Omega_y\cos(\Phi(t))\cos(t\Omega_y))\\ &+\frac{1}{2}m(L_0+Z(t))^2\dot{\Phi}(t)^2+\frac{1}{4}m(2R_x^2\Omega_x^2\sin^2(t\Omega_x)+2R_y^2\Omega_y^2\cos^2(t\Omega_y))\\ &+\frac{1}{2}m\dot{Z}(t)(-2R_x\Omega_x\cos(\Phi(t))\sin(t\Omega_x)+2R_y\Omega_y\sin(\Phi(t))\cos(t\Omega_y)+\dot{Z}(t)), \end{aligned} \tag{3.4}$$

where L_0 stands for the length of the unstretched spring and g denotes Earth's acceleration. The nonconservative loads are modeled as the generalized forces

$$\begin{aligned} Q_Z&=F_0\sin\left(\Omega_1 t\right)-C_1\dot{Z},\\ Q_\Phi&=M_0\sin\left(\Omega_2 t\right)-C_2L\dot{\Phi}, \end{aligned} \tag{3.5}$$

where $L=L_0+\delta_{eq}$ is the length of the pendulum at its static equilibrium position and $\delta_{eq}=\frac{mg}{k}$ is the spring elongation at this position.

Introducing (3.3)–(3.5) into (3.2) we get the equations of motion

$$\begin{aligned} &m\ddot{Z}(t)+C_1\dot{Z}(t)-m\cos\left(\Phi(t)\right)\left(g+R_x\Omega_x^2\cos\left(t\Omega_x\right)\right)+Z(t)\left(k-m\dot{\Phi}(t)^2\right)-\\ &L_0m\dot{\Phi}(t)^2-mR_y\Omega_y^2\sin\left(\Phi(t)\right)\sin\left(t\Omega_y\right)=F_0\sin\left(t\Omega_1\right), \end{aligned} \tag{3.6}$$

$$\begin{aligned} &\left(L_0+Z(t)\right)\ddot{\Phi}(t)+C_2L\dot{\Phi}(t)+m\left(L_0+Z(t)\right)\left(\sin\left(\Phi(t)\right)\left(g+R_x\Omega_x^2\cos\left(t\Omega_x\right)\right)\right.\\ &\left.+R_y\Omega_y^2\left(-\cos\left(\Phi(t)\right)\right)\sin\left(t\Omega_y\right)+2\dot{\Phi}(t)\dot{Z}(t)\right)=M_0\left(\sin\left(t\Omega_2\right)\right) \end{aligned} \tag{3.7}$$

Equations (3.6)–(3.7) are supplemented by the initial conditions for generalized coordinates and their first derivatives of the following form

$$Z(0)=Z_0,\quad \dot{Z}(0)=V_0,\quad \Phi(0)=\Phi_0,\quad \dot{\Phi}(0)=\omega_0, \tag{3.8}$$

where $Z_0, V_0, \Phi_0, \omega_0$ are known.

It is convenient to transform the initial value problem (3.6)–(3.8) to the dimensionless form. Let us introduce the new dimensionless generalized coordinates

$z(\tau) = \frac{Z(t)-\delta_{eq}}{L}$ and $\varphi(\tau)$ as functions of the dimensionless time $\tau = \omega_1 t$. The dimensionless frequencies are defined as follows

$$w = \frac{\omega_2}{\omega_1}, \quad p_1 = \frac{\Omega_1}{\omega_1}, \quad p_2 = \frac{\Omega_2}{\omega_1}, \quad p_x = \frac{\Omega_x}{\omega_1}, \quad p_y = \frac{\Omega_y}{\omega_1},$$

where $\omega_1 = \sqrt{k/m}$, $\omega_2 = \sqrt{g/L}$.

The other parameters read

$$r_x = \frac{R_x}{L}, \quad r_y = \frac{R_y}{L}, \quad c_1 = \frac{C_1}{m\omega_1}, \quad c_2 = \frac{C_2}{mL\omega_1}, \quad f_1 = \frac{F_0}{mL\omega_1^2}, \quad f_2 = \frac{M_0}{mL^2\omega_1^2}.$$

Introducing the defined above definitions into equations (3.6)–(3.7) leads to the dimensionless governing equations in the form

$$\begin{aligned}&\ddot{z}(\tau)+c_1\dot{z}(\tau)-p_x^2 r_x\cos(\varphi(\tau))\cos(\tau p_x)-p_y^2 r_y\sin(\varphi(\tau))\sin(\tau p_y)-\\&\dot{\varphi}(\tau)^2-w^2\cos(\varphi(\tau))+w^2-z(\tau)\,\dot{\varphi}(\tau)^2+z(\tau)=f_1\sin(p_1\tau),\end{aligned} \tag{3.9}$$

$$\begin{aligned}&(z(\tau)+1)^2\ddot{\varphi}(\tau)+c_2\dot{\varphi}(\tau)+(z(\tau)+1)\sin(\varphi(\tau))\left(p_x^2 r_x\cos(\tau p_x)+w^2\right)-\\&p_y^2 r_y(z(\tau)+1)\cos(\varphi(\tau))\sin(\tau p_y)+2(z(\tau)+1)\,\dot{\varphi}(\tau)\,\dot{z}(\tau)=f_2\sin(p_2\tau),\end{aligned} \tag{3.10}$$

with the initial conditions

$$z(0)=z_0, \quad \dot{z}(0)=v_0, \quad \varphi(0)=\varphi_0, \quad \dot{\varphi}(0)=\omega_0, \tag{3.11}$$

where $z_0 = \frac{Z_0-\delta_{eq}}{L}$, $v_0 = \frac{V_0}{L\omega_1}$.

Definitions of dimensionless forces allow us to estimate their size. These quantities can be written as follows

$$f_1 = \frac{F_0}{mL\omega_1^2} = \frac{F_0}{mL\frac{k}{m}} = \frac{F_0}{kL}, \quad f_2 = \frac{M_0}{mL^2\omega_1^2} = \frac{M_0}{mL^2\frac{k}{m}} = \frac{M_0}{(kL)L}. \tag{3.12}$$

In the denominator of the above expressions, there is a product kL that is equal to the elastic force when the spring elongates up to 100%. In most engineering applications, such elongation could need very high force, so it allows to assume that the dimensionless forces f_1 and f_2 are far less than unity

$$f_1 \ll 1, \qquad f_2 \ll 1.$$

Moreover, we assume small values of the damping coefficients that meet the engineering practice in most practical cases.

3.3 ASYMPTOTIC SOLUTION

The multiple scales technique is applied to solve the initial value problem (3.9)–(3.11). The asymptotic approach demand to expand the trigonometric functions in power series. The three first terms of Taylor's series approximation are adopted, so we take

$$\sin\varphi = \varphi - \frac{\varphi^3}{6}, \quad \cos\varphi = 1 - \frac{\varphi^2}{2}. \tag{3.13}$$

The expansions (3.13) transform the governing equations (3.9)–(3.10) to the form which is suitable for the application of MSM

$$\begin{aligned} &\ddot{z}(\tau) + c_1\dot{z}(\tau) - p_x^2 r_x \left(1 - \frac{\varphi(\tau)^2}{2}\right)\cos(\tau p_x) - \\ &p_y^2 r_y \left(\varphi(\tau) - \frac{\varphi(\tau)^3}{6}\right)\sin(\tau p_y) - \dot{\varphi}(\tau)^2 - \left(w^2\left(1 - \frac{\varphi(\tau)^2}{2}\right)\right) + \\ &w^2 - z(\tau)\,\dot{\varphi}(\tau)^2 + z(\tau) = f_1 \sin(p_1\tau), \end{aligned} \tag{3.14}$$

$$\begin{aligned} &(z(\tau)+1)^2\ddot{\varphi}(\tau) + c_2\dot{\varphi}(\tau) + \\ &\left(\varphi(\tau) - \frac{\varphi(\tau)^3}{6}\right)(z(\tau)+1)\left(p_x^2 r_x \cos(\tau p_x) + w^2\right) - \\ &p_y^2 r_y \left(1 - \frac{\varphi(\tau)^2}{2}\right)(z(\tau)+1)\sin(\tau p_y) + \\ &2(z(\tau)+1)\,\dot{\varphi}(\tau)\,\dot{z}(\tau) = f_2 \sin(p_2\tau). \end{aligned} \tag{3.15}$$

According to the MSM, the time scales should be introduced instead of the time τ. We adopt three time scales

$$\tau_0 = \tau, \qquad \tau_1 = \varepsilon\tau, \qquad \tau_2 = \varepsilon^2\tau, \tag{3.16}$$

where $0 < \varepsilon \ll 1$ is the dimensionless small parameter.

The definitions (3.16) indicate τ_0 as the fast time scale while τ_1 and τ_2 are slower and slower time scales.

Similar to the former chapter the generalized coordinates are looked for in the form of asymptotic expansions

$$z(\tau;\varepsilon) = \sum_{k=1}^{3} \varepsilon^k x_k(\tau_0, \tau_1, \tau_2) + O(\varepsilon^4), \tag{3.17}$$

$$\varphi(\tau;\varepsilon) = \sum_{k=1}^{3} \varepsilon^k \phi_k(\tau_0, \tau_1, \tau_2) + O(\varepsilon^4). \tag{3.18}$$

The new unknown functions $x_k(\tau_0,\ \tau_1,\ \tau_2)$ and $\phi_k(\tau_0,\ \tau_1,\ \tau_2)$ are dependent on three variables, so their time derivatives must be derived as follows

$$\frac{d}{d\tau} = \sum_{k=0}^{2} \varepsilon^k \frac{\partial}{\partial \tau_k} = \frac{\partial}{\partial \tau_0} + \varepsilon\frac{\partial}{\partial \tau_1} + \varepsilon^2\frac{\partial}{\partial \tau_2}, \tag{3.19}$$

$$\frac{d^2}{d\tau^2} = \frac{\partial^2}{\partial \tau_0^2} + 2\varepsilon \frac{\partial^2}{\partial \tau_0 \partial \tau_1} + \varepsilon^2 \left(\frac{\partial^2}{\partial \tau_1^2} + 2\frac{\partial^2}{\partial \tau_0 \partial \tau_2} \right) + \\ 2\varepsilon^3 \frac{\partial^2}{\partial \tau_1 \partial \tau_2} \ldots + O\left(\varepsilon^4\right). \tag{3.20}$$

In the further study, we admit weak damping effects and small values of the external and kinematic excitations. Therefore, we assume that the appropriate parameters are of the order of some powers of the small parameter:

$$c_1 = \varepsilon^2 \tilde{c}_1, \quad c_2 = \varepsilon^2 \tilde{c}_2, \quad f_1 = \varepsilon^3 \tilde{f}_1, \quad f_2 = \varepsilon^3 \tilde{f}_2, \quad r_x = \varepsilon^2 \tilde{r}_x, \quad r_y = \varepsilon^2 \tilde{r}_y. \tag{3.21}$$

On the other hand, the parameters $\tilde{c}_1, \tilde{c}_2, \tilde{f}_1, \tilde{f}_2, \tilde{r}_x, \tilde{r}_y$ are of the order of 1.

Afterward, the relations (3.21) are substituted into the governing equations (3.14)–(3.15). So, taking into consideration the asymptotic expansions (3.17)–(3.18) and definitions of the time derivative yield the equations where the small parameter ε appears in various powers. In the further analysis, we omit all terms of order $O\left(\varepsilon^4\right)$. Afterward, the equations are ordered by collecting the components that are multiplied by ε of the same power. The equation should be satisfied for any value of the small parameter, which is fulfilled when the coefficient standing at the various powers of ε are equal to zero. This leads to the set of the six equations with partial derivatives. These equations can be grouped as the approximation of first, second, and third-order according to the power of ε.

(i) Equations of order "1" (coefficient at ε^1)

$$\frac{\partial^2 x_1}{\partial \tau_0^2} + x_1 = 0, \tag{3.22}$$

$$\frac{\partial^2 \phi_1}{\partial \tau_0^2} + w^2 \phi_1 = 0, \tag{3.23}$$

(ii) Equations of order "2" (coefficient at ε^2)

$$\frac{\partial^2 x_2}{\partial \tau_0^2} + x_2 = -2\frac{\partial^2 x_1}{\partial \tau_0 \partial \tau_1} + \left(\frac{\partial \phi_1}{\partial \tau_0} \right)^2 + \\ p_x^2 \tilde{r}_x \cos\left(p_x \tau_0\right) - \frac{1}{2} w^2 \phi_1^2, \tag{3.24}$$

$$\frac{\partial^2 \phi_2}{\partial \tau_0^2} + w^2 \phi_2 = -2\frac{\partial x_1}{\partial \tau_0}\frac{\partial \phi_1}{\partial \tau_0} - 2x_1 \frac{\partial^2 \phi_1}{\partial \tau_0^2} - \\ 2\frac{\partial^2 \phi_1}{\partial \tau_0 \partial \tau_1} + p_y^2 \tilde{r}_y \sin\left(p_y \tau_0\right) - w^2 x_1 \phi_1, \tag{3.25}$$

(iii) Equations of order "3" (coefficient at ε^3)

$$\frac{\partial^2 x_3}{\partial \tau_0^2}+x_3=\tilde{f}_1\sin(p_1\tau_0)+p_y^2\tilde{r}_y\phi_1\sin(p_y\tau_0)-w^2\phi_1\phi_2-\frac{\partial^2 x_1}{\partial \tau_1^2}-\tilde{c}_1\frac{\partial x_1}{\partial \tau_0}$$
$$+2\frac{\partial \phi_1}{\partial \tau_0}\frac{\partial \phi_1}{\partial \tau_1}+x_1\left(\frac{\partial \phi_1}{\partial \tau_0}\right)^2+2\frac{\partial \phi_1}{\partial \tau_0}\frac{\partial \phi_2}{\partial \tau_1}-2\frac{\partial^2 x_1}{\partial \tau_0\partial \tau_2}-2\frac{\partial^2 x_2}{\partial \tau_0\partial \tau_1}, \tag{3.26}$$

$$\frac{\partial^2 \phi_3}{\partial \tau_0^2}+w^2\phi_3=\tilde{f}_2\sin(p_2\tau_0)+p_y^2\tilde{r}_yx_1\sin(p_y\tau_0)-p_x^2\tilde{r}_x\phi_1\cos(p_y\tau_0)-$$
$$w^2\phi_1x_2-w^2\phi_2x_1+\frac{1}{6}w^2\phi_1^3-\frac{\partial^2 \phi_1}{\partial \tau_1^2}-2\frac{\partial \phi_1}{\partial \tau_1}\frac{\partial x_1}{\partial \tau_0}-\tilde{c}_2\frac{\partial \phi_1}{\partial \tau_0}-2\frac{\partial \phi_1}{\partial \tau_0}\frac{\partial x_1}{\partial \tau_1}-$$
$$2x_1\frac{\partial \phi_1}{\partial \tau_0}\frac{\partial x_1}{\partial \tau_0}-2\frac{\partial \phi_1}{\partial \tau_0}\frac{\partial x_2}{\partial \tau_0}-2\frac{\partial \phi_2}{\partial \tau_0}\frac{\partial x_1}{\partial \tau_0}-2\frac{\partial^2 \phi_1}{\partial \tau_0\partial \tau_2}-4x_1\frac{\partial^2 \phi_1}{\partial \tau_0\partial \tau_1}- \tag{3.27}$$
$$2\frac{\partial^2 \phi_2}{\partial \tau_0\partial \tau_1}-x_1^2\frac{\partial^2 \phi_1}{\partial \tau_0^2}-2x_2\frac{\partial^2 \phi_1}{\partial \tau_0^2}-2x_1\frac{\partial^2 \phi_2}{\partial \tau_0^2}.$$

The original initial governing equations are substituted by the set of equations (3.22)–(3.27), which can be solved recursively, i.e. the solution of the lower order approximation is introduced to the equations of the higher-order approximation.

The general solutions to equations (3.22)–(3.23) are

$$x_1(\tau_0,\tau_1,\tau_2)=B_1(\tau_1,\tau_2)e^{\mathrm{i}\tau_0}+\bar{B}_1(\tau_1,\tau_2)e^{-\mathrm{i}\tau_0}, \tag{3.28}$$

$$\phi_1(\tau_0,\tau_1,\tau_2)=B_2(\tau_1,\tau_2)e^{\mathrm{i}w\tau_0}+\bar{B}_2(\tau_1,\tau_2)e^{-\mathrm{i}w\tau_0}, \tag{3.29}$$

where $B_i(\tau_1,\tau_2)$ are unknown functions of the slow time scales, $\bar{B}_i(\tau_1,\tau_2)$ are their complex conjugates ($\mathrm{i}=\sqrt{-1}$) for $i=1,2$.

When introducing the solutions (3.28)–(3.29) into equations (3.24)–(3.25) the secular terms emerge, that yield an infinite solution. Elimination of the secular terms demand the satisfaction of the equations

$$\frac{\partial B_1}{\partial \tau_1}=0,\qquad \frac{\partial \bar{B}_1}{\partial \tau_1}=0, \tag{3.30}$$

$$\frac{\partial B_2}{\partial \tau_1}=0,\qquad \frac{\partial \bar{B}_2}{\partial \tau_1}=0, \tag{3.31}$$

which implies

$$\frac{\partial^2 B_1}{\partial \tau_2\partial \tau_1}=0,\qquad \frac{\partial^2 \bar{B}_1}{\partial \tau_2\partial \tau_1}=0, \tag{3.32}$$

$$\frac{\partial^2 B_2}{\partial \tau_2\partial \tau_1}=0,\qquad \frac{\partial^2 \bar{B}_2}{\partial \tau_2\partial \tau_1}=0. \tag{3.33}$$

Taking into account the solutions (3.28), (3.29) and conditions (3.30)–(3.33), the solution of the second-order equations (3.24)–(3.25) is as follows

$$x_2(\tau_0,\tau_1,\tau_2)=w^2B_2\bar{B}_2+\frac{3w^2B_2^2e^{2\mathrm{i}w\tau_0}}{4(4w^2-1)}+\frac{p_x^2\tilde{r}_xe^{\mathrm{i}p_x\tau_0}}{2(1-p_x^2)}+CC, \tag{3.34}$$

$$\phi_2(\tau_0,\tau_1,\tau_2)=\frac{(w-2)\,w\bar{B}_1B_2e^{\mathrm{i}(w-1)\tau_0}}{2w-1}-\frac{(w+2)\,wB_1B_2e^{\mathrm{i}(w+1)\tau_0}}{2w+1}+$$
$$\frac{\mathrm{i}p_y^2\tilde{r}_ye^{\mathrm{i}p_y\tau_0}}{2\left(p_y^2-w\right)}+CC, \tag{3.35}$$

where CC stands for the complex conjugates of the preceding terms.

The solutions (3.28), (3.29) and (3.34), (3.35) are then introduced into equations (3.26), (3.27), which cause occurrence of the secular terms, that leads to the solvability conditions

$$\frac{6\left(w^2-1\right)w^2B_1B_2\bar{B}_2}{4w^2-1}+\mathrm{i}B_1\tilde{c}_1+2\mathrm{i}\frac{\partial B_1}{\partial \tau_2}=0, \tag{3.36}$$

$$-\frac{6\left(w^2-1\right)w^2\bar{B}_1B_2\bar{B}_2}{4w^2-1}+\mathrm{i}\bar{B}_1\tilde{c}_1+2\mathrm{i}\frac{\partial \bar{B}_1}{\partial \tau_2}=0, \tag{3.37}$$

$$-\frac{6w\left(w^2-1\right)}{4w^2-1}B_1\bar{B}_1B_2+\frac{w\left(8w^4-7w^2-1\right)}{8w^2-2}B_2^2\bar{B}_2-\mathrm{i}\tilde{c}_2B_2-2\mathrm{i}\frac{\partial B_2}{\partial \tau_2}=0, \tag{3.38}$$

$$-\frac{6w\left(w^2-1\right)}{4w^2-1}B_1\bar{B}_1\bar{B}_2+\frac{w\left(8w^4-7w^2-1\right)}{8w^2-2}B_2\bar{B}_2^2+\mathrm{i}\tilde{c}_2\bar{B}_2+2\mathrm{i}\frac{\partial \bar{B}_2}{\partial \tau_2}=0. \tag{3.39}$$

Taking into account the solvability conditions (3.36)–(3.39), we can solve equations (3.26), (3.27). We get

$$x_3(\tau_0,\tau_1,\tau_2)=-\frac{3B_1B_2^2w(w+1)\,e^{\mathrm{i}\tau_0(1+2w)}}{8w+4}+\frac{3\bar{B}_1B_2^2(w-1)\,we^{\mathrm{i}\tau_0(2w-1)}}{8w-4}$$
$$+\frac{\mathrm{i}B_2p_y^3\tilde{r}_y(p_y+2w)\,e^{\mathrm{i}\tau_0(p_y+w)}}{2(p_y-w)(p_y+w-1)(p_y+w)(p_y+w+1)} \tag{3.40}$$
$$+\frac{\mathrm{i}\bar{B}_2p_y^3\tilde{r}_y(p_y-2w)\,e^{\mathrm{i}\tau_0(p_y-w)}}{2(p_y-w-1)(p_y-w)(p_y-w+1)(p_y+w)}+\frac{\mathrm{i}\tilde{f}_1e^{\mathrm{i}p_1\tau_0}}{2\left(p_1^2-1\right)}+CC,$$

$$\phi_3(\tau_0,\tau_1,\tau_2)=\frac{B_2\bar{B}_1^2w\left(w^2-5w+6\right)e^{\mathrm{i}\tau_0(w-2)}}{8w-4}+$$
$$\frac{B_1^2B_2w\left(w^2+5w+6\right)e^{\mathrm{i}\tau_0(w+2)}}{8w+4}+$$
$$\frac{B_2p_x\tilde{r}_x\left(2wp_x+p_x^2+w^2-1\right)e^{\mathrm{i}\tau_0(p_x+w)}}{2(p_x-1)(p_x+1)(p_x+2w)}-$$
$$\frac{\mathrm{i}B_1p_y^3(p_y+2)\,\tilde{r}_ye^{\mathrm{i}\tau_0(p_y+1)}}{2(p_y-w)(p_y-w+1)(p_y+w)(p_y+w+1)}-\frac{B_2^3\left(49w^2-1\right)e^{3\mathrm{i}\tau_0w}}{48\left(4w^2-1\right)}+ \tag{3.41}$$
$$\frac{\mathrm{i}B_1(p_y-2)\,p_y^3\tilde{r}_ye^{-\mathrm{i}\tau_0(p_y-1)}}{2(p_y-w-1)(p_y-w)(p_y+w-1)(p_y+w)}+$$
$$\frac{B_2p_x\tilde{r}_x\left(-2wp_x+p_x^2+w^2-1\right)e^{-\mathrm{i}\tau_0(p_x-w)}}{2(p_x-1)(p_x+1)(p_x-2w)}+\frac{\mathrm{i}\tilde{f}_2e^{\mathrm{i}p_2\tau_0}}{2\left(p_2^2-w^2\right)}+CC.$$

The unknown complex functions $B_1, \bar{B}_1, B_2$ and $\bar{B}_2$, which appear in the solutions (3.28), (3.29), (3.34), (3.35), (3.40) and (3.41), are dependent on only the slower time scale τ_2, which results from (3.30), (3.31), and can be determined from the solvability conditions (3.36)–(3.39). In further analysis the complex amplitudes B_i and $\bar{B}_i$ are expressed in the following exponential form

$$\begin{aligned} &B_i(\tau_2) = \frac{1}{2}\tilde{a}_i(\tau_2)\, e^{i\psi_i(\tau_2)}, \quad \bar{B}_i(\tau_2) = \frac{1}{2}\tilde{a}_i(\tau_2)\, e^{-i\psi_i(\tau_2)}, \\ &a_i = \varepsilon\tilde{a}_i, \quad i = 1,2, \end{aligned} \tag{3.42}$$

where the functions $a_1(\tau_2), a_2(\tau_2), \psi_1(\tau_2), \psi_2(\tau_2)$ are real-valued.

Inserting substitutions (3.42) into solvability conditions (3.36)–(3.39) and then solving them for the derivatives leads to the equations

$$\frac{d\tilde{a}_1}{d\tau_2} = -\frac{1}{2}\tilde{c}_1\tilde{a}_1, \tag{3.43}$$

$$\frac{d\psi_1}{d\tau_2} = \frac{3\left(w^2(w^2-1)\tilde{a}_2^2\right)}{4\,(4w^2-1)}, \tag{3.44}$$

$$\frac{d\tilde{a}_2}{d\tau_2} = -\frac{1}{2}\tilde{c}_2\tilde{a}_2, \tag{3.45}$$

$$\frac{d\psi_2}{d\tau_2} = \frac{w\,(w^2-1)\left(12\tilde{a}_1^2 - (8w^2+1)\,\tilde{a}_2^2\right)}{16\,(4w^2-1)}. \tag{3.46}$$

The above equations can be transformed to the real time τ by multiplying them by ε^2 and using the formula for the derivative (3.19), and then by substituting the original denotations according to (3.21). The mentioned procedure yields the following first-order amplitude-phase ODEs:

$$\frac{da_1}{d\tau} = -\frac{1}{2}c_1 a_1, \tag{3.47}$$

$$\frac{d\psi_1}{d\tau} = \frac{3\left(w^2(w^2-1)a_2^2\right)}{4\,(4w^2-1)}, \tag{3.48}$$

$$\frac{da_2}{d\tau} = -\frac{1}{2}c_2 a_2, \tag{3.49}$$

$$\frac{d\psi_2}{d\tau} = \frac{w\,(w^2-1)\left(12a_1^2 - (8w^2+1)\,a_2^2\right)}{16\,(4w^2-1)}. \tag{3.50}$$

The functions $a_1(\tau)$, $a_2(\tau)$ and $\psi_1(\tau)$, $\psi_2(\tau)$ represent the amplitudes and phases of the solution in the first-order approximation. Therefore, equations (3.47)–(3.50) describe the modulation of these quantities in time.

The set (3.47)–(3.50) can be solved analytically for the initial conditions

$$a_1(0) = a_{10}, \quad a_2(0) = a_{20}, \quad \psi_1(0) = \psi_{10}, \quad \psi_2(0) = \psi_{20}, \tag{3.51}$$

where a_{10}, a_{20}, ψ_{10}, ψ_{20} are known, and they are related to the initial values (3.11).

The exact solution to the initial value problem (3.47)–(3.51) is as follows

$$a_1(\tau) = a_{10}e^{-\frac{c_1\tau}{2}}, \tag{3.52}$$

$$\psi_1(\tau) = \frac{3a_{20}^2 w^2 (w^2-1)(1-e^{-c_2\tau})}{4(4w^2-1)c_2} + \psi_{10}, \tag{3.53}$$

$$a_2(\tau) = a_{20}e^{-\frac{c_2\tau}{2}}, \tag{3.54}$$

$$\begin{aligned}\psi_2(\tau) = {} & \frac{3a_{10}^2 w (w^2-1)(1-e^{-c_1\tau})}{4c_1(4w^2-1)} - \\ & \frac{a_{20}^2 w (w^2-1)(1+8w^2)(1-e^{-c_2\tau})}{16c_2(4w^2-1)} + \psi_{20}.\end{aligned} \tag{3.55}$$

Now we can write the asymptotic solution of the original problem (3.9)–(3.11). The solutions of the first, second, and third-order approximation (3.28), (3.29), (3.34), (3.35) and (3.40), (3.41) are substituted into the expansion (3.17), (3.18) taking into consideration formulae (3.19), (3.20) and expressing the complex functions B_i, $\bar{B}_i$ by the polar form (3.42). Using the original notation (according to (3.21)) the solution has the following form

$$\begin{aligned} z(\tau) = {} & a_1\cos(\eta_1) - \frac{a_2 p_y^3 r_y (p_y-2w)\sin(\tau p_y - \eta_2)}{2(p_y-w-1)(p_y-w)(p_y-w+1)(p_y+w)} - \\ & \frac{a_2 p_y^3 r_y (p_y+2w)\sin(\eta_2+\tau p_y)}{2(p_y-w)(p_y+w-1)(p_y+w)(p_y+w+1)} + \frac{3a_2^2 w^2\cos(2\eta_2)}{16w^2-4} + \\ & \frac{1}{4}a_2^2 w^2 + \frac{3a_1 a_2^2 (w-1) w\cos(-\eta_1+2\eta_2)}{32w-16} - \\ & \frac{3a_1 a_2^2 w (w+1)\cos(\eta_1+2\eta_2)}{32w+16} + \frac{f_1\sin(p_1\tau)}{1-p_1^2} + \frac{2p_x^2 r_x\cos(\tau p_x)}{2-2p_x^2}, \end{aligned} \tag{3.56}$$

$$\begin{aligned} \varphi(\tau) = {} & a_2\cos(\eta_2) + \frac{a_2 p_x r_x (2wp_x + p_x^2 + w^2 - 1)\cos(\eta_2+\tau p_x)}{2(p_x-1)(p_x+1)(p_x+2w)} + \\ & \frac{a_2 p_x r_x \left((p_x-w)^2-1\right)\cos(\tau p_x - \eta_2)}{2(p_x^2-1)(p_x-2w)} + \\ & \frac{a_1 (p_y-2) p_y^3 r_y \sin(\tau p_y - \eta_1)}{2(p_y-w-1)(p_y-w)(p_y+w-1)(p_y+w)} + \end{aligned} \tag{3.57}$$

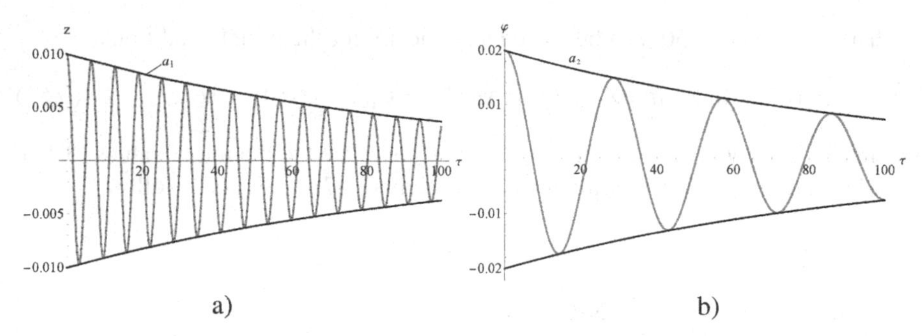

FIGURE 3.2 Time history of the free vibration in the longitudinal and transverse directions: a) $z(\tau)$; b) $\varphi(\tau)$ estimated analytically (MSM) and numerically (dotted line).

$$\frac{a_1 p_y^3 (p_y+2) r_y \sin(\eta_1+\tau p_y)}{2(p_y-w)(p_y-w+1)(p_y+w)(p_y+w+1)} + \frac{a_2^3(1-49w^2)\cos(3\eta_2)}{192(4w^2-1)} +$$
$$\frac{a_1^2 a_2 (w-3)(w-2) w \cos(2\eta_1-\eta_2)}{32w-16} + \frac{a_1^2 a_2 w (w+2)(w+3)\cos(2\eta_1+\eta_2)}{32w+16} +$$
$$\frac{a_1 a_2 (w-2) w \cos(\eta_1-\eta_2)}{4w-2} - \frac{a_1 a_2 w (w+2)\cos(\eta_1+\eta_2)}{4w+2} -$$
$$\frac{f_2 \sin(p_2\tau)}{p_2^2-w^2} - \frac{p_y^2 r_y \sin(\tau p_y)}{p_y^2-w^2},$$

where the relations $\eta_1 = \tau + \psi_1(\tau)$ and $\eta_2 = w\tau + \psi_2(\tau)$ are used to shorten the notation.

The quantities a_1, a_2, ψ_1 and ψ_2 describe respectively amplitudes and phases of the first approximations in the asymptotic solution (the first component on the right in (3.56) and (3.57)). Since excitations appear at the higher-order asymptotic solutions, the functions $a_1(\tau)$ and $a_2(\tau)$ describe amplitude changing of the free oscillations. This is illustrated in Figure 3.2 for the following fixed parameters: $w = 0.22$, $f_1 = 0$, $f_2 = 0, p_1 = 0.8, p_2 = 2.7, p_x = 0.72, p_y = 1.43, r_x = 0, r_y = 0, c_1 = 0.02, c_2 = 0.02, z_0 = 0.01, \varphi_0 = 0.02, v_0 = 0, \omega_0 = 0$.

Apart from the asymptotic solution, the numerical one is also shown in the graphs. Finding solutions requires a coincidence of the initial conditions (3.11) and (3.51). This is guaranteed by substituting $z_0 = 0.01, \varphi_0 = 0.02, v_0 = 0, \omega_0 = 0$ and $\tau = 0$ into (3.56) and (3.57), and by solving the algebraic equations with regard to $a_{10}, \psi_{10}, a_{20}, \psi_{20}$. The determined values follow: $a_{10} = 0.01001, a_{20} = 0.01998, \psi_{10} = -0.01005, \psi_{20} = -0.04656$.

The graphs presented in Figure 3.2 indicate high compatibility of the time history obtained by MSM and numerically. The MSM gives a very good solutions approximation also for oscillations excited kinematically and by the external forces. The comparison of the time histories for forced vibrations is presented in Figures 3.3 and 3.4. The time histories are reported for the following fixed parameters: $w = 0.22$, $f_1 = 0.005$, $f_2 = 0.005$, $p_1 = 0.8$, $p_2 = 2.7$, $p_x = 0.72$, $p_y = 1.43$, $r_x = 0.003$, $r_y = 0.005$, $c_1 = 0.002$, $c_2 = 0.002$, $z_0 = 0.001$, $\varphi_0 = 0.002$, $v_0 = 0$, $\omega_0 = 0$.

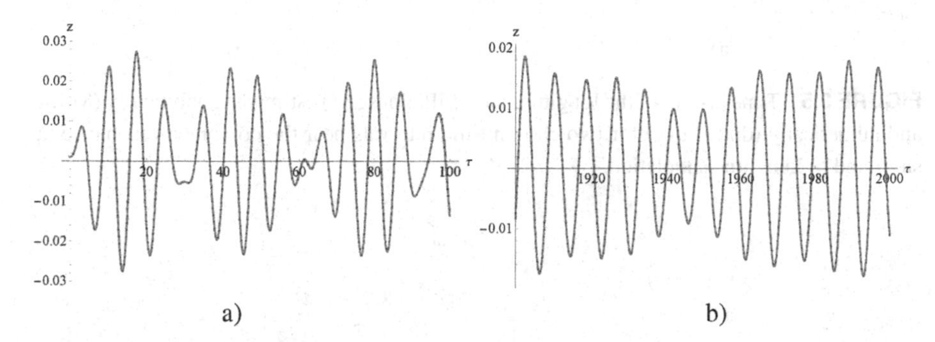

FIGURE 3.3 Time history of the longitudinal oscillations $z(\tau)$ for two chosen time intervals: for a short (a) and long-time (b) simulation estimated analytically (MSM) and numerically (dotted line).

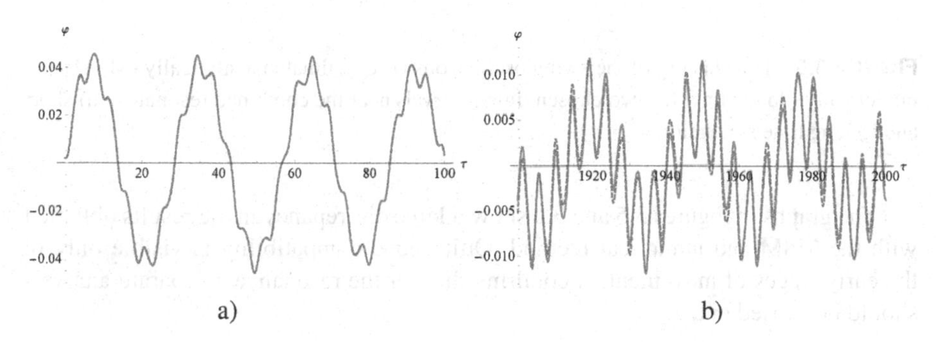

FIGURE 3.4 Time history of the swing oscillations $\varphi(\tau)$ for two chosen time intervals: for a short (a) and long-time (b) simulation estimated analytically (MSM) and numerically (dotted line).

The graphs presented in Figures 3.3 and 3.4 indicate that good compatibility of the MSM and the numerical solution is achieved also for the loaded system. We observe such a situation when the vibrations take place away from resonance. When the system is close to any resonance the asymptotic solutions (3.56) and (3.57) do not describe the motion correctly. For example, the comparison of the MSM and numerical solutions near the resonance ($p_y \approx 1 - w$) is shown in Figures 3.5 and 3.6. The data was the same as above, except for $p_y = 0.785$.

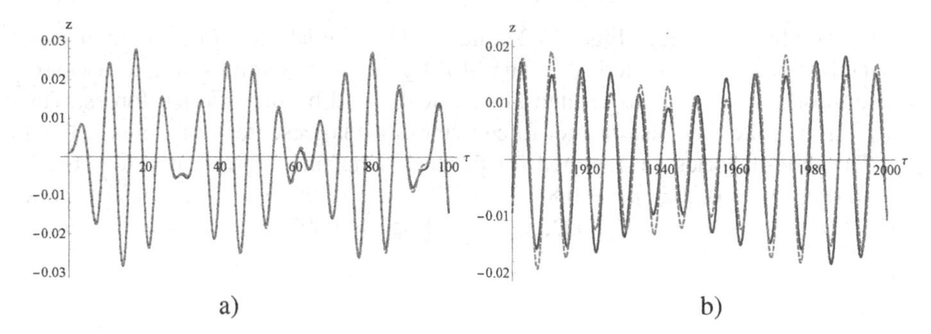

FIGURE 3.5 Time history of the longitudinal oscillations $z(\tau)$ estimated analytically (MSM) and numerically (dotted line) for two chosen time intervals near the combined resonance: a) short and b) long time simulation.

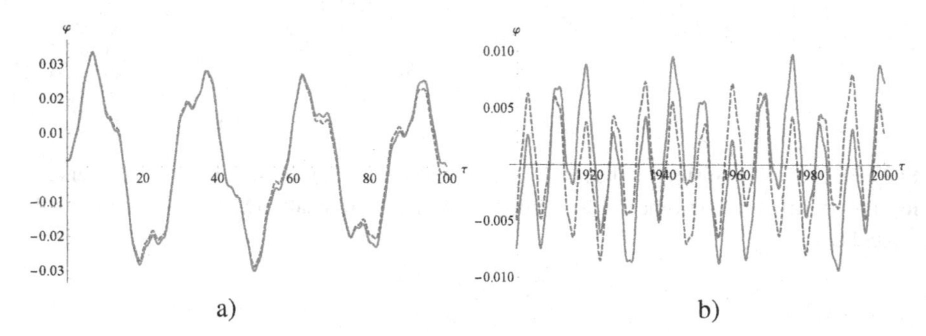

FIGURE 3.6 Time history of the swing oscillations $\varphi(\tau)$ estimated analytically (MSM) and numerically (dotted line) for two chosen time intervals near the combined resonance: a) short and b) long time simulation.

The graphs in Figures 3.5 and 3.6 show a large discrepancy in the results obtained with the MSM and numerical method. Quite good compatibility is visible only in the early stages of movement. It confirms that for the resonance a separate analysis should be carried out.

3.4 RESONANCE OSCILLATIONS

As one may guess, the asymptotic solution (3.56)–(3.57) is not valid in the resonance conditions due to some of their components which tend to infinity for certain values of parameters. This phenomenon appears due to zeroing some denominators. Hence, the conditions under which the denominators approach zero are precisely the conditions of resonance occurrence.

The analytical form of the asymptotic solutions (3.56)–(3.57) allows one to recognize all relevant cases of resonance that can appear in the system. In solutions of even higher orders of approximation, new resonances of higher order and combined resonances may appear, but from a practical point of view, they are less important, as generally, even slight damping suppresses them effectively. This is because higher-order solutions stand for a minimum correction compared to lower-order solutions in which all the factors affecting the motion have already appeared.

All resonances till the third-order can be recognized in this way. They can be classified as follows:

- main (primary) external resonances when $p_1 \approx 1$, $p_2 \approx w$;
- resonance of the spring caused by kinematic excitation when $p_x \approx 1$;
- resonance of the pendulum caused by kinematic excitation when $p_y \approx w$, $p_x \approx 2w$;
- internal resonance when $1 \approx 2w$;
- combined resonances when $p_y \approx \pm(1-w)$, $p_y \approx \pm(1+w)$.

If the natural frequencies satisfy the above resonance conditions, the system behavior could be very complex. This may happen in particular when two or more resonances appear simultaneously.

In order to obtain the MSM solution of the dynamic problem near the resonance, the parameters at which the given resonance occurs should be inserted into the equations of motion. There are eight different kinds of resonance in the tested system. Since the physical system discussed in this chapter has two degrees of freedom, the multiple scale method allows analyzing a situation in which at the same time there are up to two resonant conditions. This makes it possible to test the simultaneous occurrence of 28 different pairs of resonances. Our analysis will be restricted to the following case study.

Let us examine parametric and primary resonances appearing simultaneously, i.e. $p_x \approx 1, p_2 \approx w$. To study the latter resonances, we introduce the new so-called detuning parameters σ_1 and σ_2 as a measure of the distance from the pure resonance:

$$p_x = 1 + \sigma_1 = 1 + \varepsilon^2 \tilde{\sigma}_1 \qquad \text{and} \qquad p_2 = w + \sigma_2 = w + \varepsilon^2 \tilde{\sigma}_2. \tag{3.58}$$

The appearance of a small parameter ε in (3.58) indicates the assumed small deviation of values p_x and p_2 from the pure resonance.

The introduction into equations (3.14)–(3.15) all the previously made assumptions (3.16)–(3.21) and (3.58) causes that the small parameter ε appears in various powers. Therefore we can perform a similar procedure as in the previous section, which allows to obtain the set of six differential equations with partial derivatives of the following form

(i) Equations of order "1" (coefficient at ε^1)

$$\frac{\partial^2 x_1}{\partial \tau_0^2} + x_1 = 0, \tag{3.59}$$

$$\frac{\partial^2 \phi_1}{\partial \tau_0^2} + w^2 \phi_1 = 0, \tag{3.60}$$

(ii) Equations of order "2" (coefficient at ε^2)

$$\frac{\partial^2 x_2}{\partial \tau_0^2} + x_2 = -2\frac{\partial^2 x_1}{\partial \tau_0 \partial \tau_1} + \left(\frac{\partial \phi_1}{\partial \tau_0}\right)^2 + p_x^2 \tilde{r}_x \cos\left(\tau_0\left(1+\varepsilon^2\tilde{\sigma}_1\right)\right) - \frac{1}{2}w^2\phi_1^2, \tag{3.61}$$

$$\frac{\partial^2 \phi_2}{\partial \tau_0^2} + w^2\phi_2 = -2\frac{\partial x_1}{\partial \tau_0}\frac{\partial \phi_1}{\partial \tau_0} - 2x_1\frac{\partial^2 \phi_1}{\partial \tau_0^2} - 2\frac{\partial^2 \phi_1}{\partial \tau_0 \partial \tau_1} + p_y^2 \tilde{r}_y \sin\left(p_y \tau_0\right) - w^2 x_1 \phi_1, \tag{3.62}$$

(iii) Equations of order "3" (coefficient at ε^3)

$$\frac{\partial^2 x_3}{\partial \tau_0^2} + x_3 = \tilde{f}_1 \sin\left(p_1 \tau_0\right) + p_y^2 \tilde{r}_y \phi_1 \sin\left(p_y \tau_0\right) - w^2\phi_1\phi_2 - \frac{\partial^2 x_1}{\partial \tau_1^2} - \tilde{c}_1\frac{\partial x_1}{\partial \tau_0} + 2\frac{\partial \phi_1}{\partial \tau_0}\frac{\partial \phi_1}{\partial \tau_1} + x_1\left(\frac{\partial \phi_1}{\partial \tau_0}\right)^2 + 2\frac{\partial \phi_1}{\partial \tau_0}\frac{\partial \phi_2}{\partial \tau_1} - 2\frac{\partial^2 x_1}{\partial \tau_0 \partial \tau_2} - 2\frac{\partial^2 x_2}{\partial \tau_0 \partial \tau_1}, \tag{3.63}$$

$$\frac{\partial^2 \phi_3}{\partial \tau_0^2} + w^2\phi_3 = \tilde{f}_2 \sin\left(\tau_0\left(w+\varepsilon^2\tilde{\sigma}_2\right)\right) + p_y^2 \tilde{r}_y x_1 \sin\left(p_y \tau_0\right) - p_x^2 \tilde{r}_x \phi_1 \cos\left(\tau_0\left(1+\varepsilon^2\tilde{\sigma}_1\right)\right) - w^2\phi_1 x_2 - w^2\phi_2 x_1 + \frac{1}{6}w^2\phi_1^3 - \frac{\partial^2 \phi_1}{\partial \tau_1^2} - 2\frac{\partial \phi_1}{\partial \tau_1}\frac{\partial x_1}{\partial \tau_0} - \tilde{c}_2\frac{\partial \phi_1}{\partial \tau_0} - 2\frac{\partial \phi_1}{\partial \tau_0}\frac{\partial x_1}{\partial \tau_1} - 2x_1\frac{\partial \phi_1}{\partial \tau_0}\frac{\partial x_1}{\partial \tau_0} - 2\frac{\partial \phi_1}{\partial \tau_0}\frac{\partial x_2}{\partial \tau_0} - 2\frac{\partial \phi_2}{\partial \tau_0}\frac{\partial x_1}{\partial \tau_0} - 2\frac{\partial^2 \phi_1}{\partial \tau_0 \partial \tau_2} - 4x_1\frac{\partial^2 \phi_1}{\partial \tau_0 \partial \tau_1} - 2\frac{\partial^2 \phi_2}{\partial \tau_0 \partial \tau_1} - x_1^2\frac{\partial^2 \phi_1}{\partial \tau_0^2} - 2x_2\frac{\partial^2 \phi_1}{\partial \tau_0^2} - 2x_1\frac{\partial^2 \phi_2}{\partial \tau_0^2}. \tag{3.64}$$

Following the previous section, system of equations (3.59)–(3.64) is solved recursively.

The solutions to equations (3.59)–(3.60) are:

$$x_1\left(\tau_0,\tau_1,\tau_2\right) = B_1\left(\tau_1,\tau_2\right)e^{\mathrm{i}\tau_0} + \bar{B}_1\left(\tau_1,\tau_2\right)e^{-\mathrm{i}\tau_0}, \tag{3.65}$$

$$\phi_1\left(\tau_0,\tau_1,\tau_2\right) = B_2\left(\tau_1,\tau_2\right)e^{\mathrm{i}w\tau_0} + \bar{B}_2\left(\tau_1,\tau_2\right)e^{-\mathrm{i}w\tau_0}, \tag{3.66}$$

where $B_i\left(\tau_1,\tau_2\right)$ and $\bar{B}_i\left(\tau_1,\tau_2\right)$ are unknown complex functions and their complex conjugates.

When substituting solutions (3.65) and (3.66) into (3.61)–(3.62), the secular terms appear that should be eliminated. This leads to the following conditions

$$\frac{1}{2}p_x^2\tilde{r}_x e^{\mathrm{i}\tilde{\sigma}_1\tau_0\varepsilon^2} - 2\mathrm{i}\frac{\partial B_1}{\partial \tau_1} = 0, \qquad \frac{1}{2}p_x^2\tilde{r}_x e^{-\mathrm{i}\tilde{\sigma}_1\tau_0\varepsilon^2} + 2\mathrm{i}\frac{\partial \bar{B}_1}{\partial \tau_1} = 0, \tag{3.67}$$

$$\frac{\partial B_2}{\partial \tau_1} = 0, \qquad \frac{\partial \bar{B}_2}{\partial \tau_1} = 0. \tag{3.68}$$

The conditions (3.67), (3.68) imply

$$\frac{\partial^2 B_1}{\partial \tau_2 \partial \tau_1} = 0, \qquad \frac{\partial^2 \bar{B}_1}{\partial \tau_2 \partial \tau_1} = 0, \tag{3.69}$$

$$\frac{\partial^2 B_2}{\partial \tau_2 \partial \tau_1} = 0, \qquad \frac{\partial^2 \bar{B}_2}{\partial \tau_2 \partial \tau_1} = 0. \tag{3.70}$$

After eliminating secular terms, the solution of equations (3.61)–(3.62) are

$$x_2(\tau_0, \tau_1, \tau_2) = w^2 B_2 \bar{B}_2 + \frac{3w^2 B_2^2 e^{2iw\tau_0}}{4(4w^2-1)} + CC, \tag{3.71}$$

$$\phi_2(\tau_0, \tau_1, \tau_2) = \frac{(w-2) w\bar{B}_1 B_2 e^{i(w-1)\tau_0}}{2w-1} - \frac{(w+2) wB_1 B_2 e^{i(w+1)\tau_0}}{2w+1} + \frac{ip_y^2 \tilde{r}_y e^{ip_y\tau_0}}{2(p_y^2 - w)} + CC. \tag{3.72}$$

Next, the solutions (3.65)–(3.66) and (3.71)–(3.72) are substituted into equations (3.63)–(3.64) and the new secular terms appear. Since the motion of the pendulum is bounded, the terms that cause an unlimited increase of the generalized coordinates must be eliminated. This leads to the following solvability conditions

$$\frac{6(w^2-1) w^2 B_1 B_2 \bar{B}_2}{4w^2-1} + iB_1 \tilde{c}_1 + 2i\frac{\partial B_1}{\partial \tau_2} = 0, \tag{3.73}$$

$$-\frac{6(w^2-1) w^2 \bar{B}_1 B_2 \bar{B}_2}{4w^2-1} + i\bar{B}_1 \tilde{c}_1 + 2i\frac{\partial \bar{B}_1}{\partial \tau_2} = 0, \tag{3.74}$$

$$-\frac{6w(w^2-1)}{4w^2-1} B_1 \bar{B}_1 B_2 + \frac{w(8w^4 - 7w^2 - 1)}{8w^2-2} B_2^2 \bar{B}_2 - i\tilde{c}_2 B_2 - \frac{i}{2} e_0^{i\varepsilon^2 \tau \tilde{\sigma}_2} \tilde{f}_2 - 2i\frac{\partial B_2}{\partial \tau_2} = 0, \tag{3.75}$$

$$-\frac{6w(w^2-1)}{4w^2-1} B_1 \bar{B}_1 \bar{B}_2 + \frac{w(8w^4 - 7w^2 - 1)}{8w^2-2} B_2 \bar{B}_2^2 + i\tilde{c}_2 \bar{B}_2 + \frac{i}{2} e_0^{-i\varepsilon^2 \tau \tilde{\sigma}_2} \tilde{f}_2 + 2i\frac{\partial \bar{B}_2}{\partial \tau_2} = 0. \tag{3.76}$$

An account of conditions (3.73)–(3.76) in equations (3.63)–(3.64) eliminates all secular terms, and the following solution is found

$$x_3(\tau_0, \tau_1, \tau_2) = -\frac{3B_1 B_2^2 w(w+1) e^{i\tau_0(1+2w)}}{8w+4} + \frac{3\bar{B}_1 B_2^2 (w-1) we^{i\tau_0(2w-1)}}{8w-4} + \frac{iB_2 p_y^3 \tilde{r}_y (p_y + 2w) e^{i\tau_0(p_y+w)}}{2(p_y - w)(p_y + w - 1)(p_y + w)(p_y + w + 1)} + \frac{i\bar{B}_2 p_y^3 \tilde{r}_y (p_y - 2w) e^{i\tau_0(p_y-w)}}{2(p_y - w - 1)(p_y - w)(p_y - w + 1)(p_y + w)} + \frac{i\tilde{f}_1 e^{ip_1\tau_0}}{2(p_1^2 - 1)} + CC, \tag{3.77}$$

$$\phi_3(\tau_0,\tau_1,\tau_2) = \frac{B_2\bar{B}_1^2 w\left(w^2-5w+6\right)e^{\mathrm{i}\tau_0(w-2)}}{8w-4} + \frac{B_1^2 B_2 w\left(w^2+5w+6\right)e^{\mathrm{i}\tau_0(w+2)}}{8w+4} + \frac{B_2 p_x^2 \tilde{r}_x (1+w)(w-1)^2 e^{\mathrm{i}\tau_0\left(w-1-\varepsilon^2\tilde{\sigma}_1\right)}}{2\left(1+\varepsilon^2\tilde{\sigma}_1\right)\left(2w-1-\varepsilon^2\tilde{\sigma}_1\right)(2w-1)}$$
$$\frac{\mathrm{i}B_1 p_y^3 (p_y+2)\tilde{r}_y e^{\mathrm{i}\tau_0(p_y+1)}}{2(p_y-w)(p_y-w+1)(p_y+w)(p_y+w+1)} - \frac{B_2^3\left(49w^2-1\right)e^{3\mathrm{i}\tau_0 w}}{48\left(4w^2-1\right)} + \tag{3.78}$$
$$\frac{\mathrm{i}B_1(p_y-2)p_y^3\tilde{r}_y e^{-\mathrm{i}\tau_0(p_y-1)}}{2(p_y-w-1)(p_y-w)(p_y+w-1)(p_y+w)} + \frac{B_2 p_x \tilde{r}_x (w-1)(w+1)^2 e^{\mathrm{i}\tau_0\left(w+1+\varepsilon^2\tilde{\sigma}_1\right)}}{2\left(1+\varepsilon^2\tilde{\sigma}_1\right)\left(2w+1+\varepsilon^2\tilde{\sigma}_1\right)(2w+1)} + CC.$$

The solutions (3.65), (3.66), (3.71), (3.72) and (3.77), (3.78) of subsequent orders of approximation can be collected to the final form according to (3.17), (3.18). Before doing that, let us express the complex functions $B_1(\tau_1,\tau_2)$ and $B_2(\tau_1,\tau_2)$ in the following counterpart form

$$\begin{aligned} B_1(\tau_1,\tau_2) &= \frac{1}{2}\tilde{a}_1(\tau_1,\tau_2)e^{\mathrm{i}\psi_1(\tau_1,\tau_2)}, \quad \bar{B}_1(\tau_1,\tau_2) = \frac{1}{2}\tilde{a}_1(\tau_1,\tau_2)e^{-\mathrm{i}\psi_1(\tau_1,\tau_2)}, \\ B_2(\tau_1,\tau_2) &= \frac{1}{2}\tilde{a}_2(\tau_1,\tau_2)e^{\mathrm{i}\psi_2(\tau_1,\tau_2)}, \quad \bar{B}_2(\tau_1,\tau_2) = \frac{1}{2}\tilde{a}_2(\tau_1,\tau_2)e^{-\mathrm{i}\psi_2(\tau_1,\tau_2)}, \end{aligned} \tag{3.79}$$

where $\tilde{a}_i(\tau_1,\tau_2) = \varepsilon a_i(\tau_1,\tau_2)$ and $\psi_i(\tau_1,\tau_2)$ are real-valued.

Introduction of definitions (3.79) to the solvability conditions (3.67), (3.68) and (3.73)–(3.76) yields the set of eight first-order ODEs

$$\frac{d\tilde{a}_1}{d\tau_1} = \frac{1}{2}p_x^2\tilde{r}_x \sin\left(\tilde{\sigma}_1\tau_0\varepsilon^2 - \psi_{11}\right), \tag{3.80}$$

$$\frac{d\psi_1}{d\tau_1} = -\frac{p_x^2\tilde{r}_x\cos\left(\tilde{\sigma}_1\tau_0\varepsilon^2-\psi_{11}\right)}{2\tilde{a}_1}, \tag{3.81}$$

$$\frac{d\tilde{a}_2}{d\tau_1} = 0, \tag{3.82}$$

$$\frac{d\psi_2}{d\tau_1} = 0, \tag{3.83}$$

$$\frac{d\tilde{a}_1}{d\tau_2} = -\frac{1}{2}\tilde{a}_1\tilde{c}_1, \tag{3.84}$$

$$\frac{d\psi_1}{d\tau_2} = \frac{3\tilde{a}_2^2 w^2\left(w^2-1\right)}{16w^2-4}, \tag{3.85}$$

$$\frac{d\tilde{a}_2}{d\tau_2} = -\frac{\tilde{a}_2\tilde{c}_2 w + \tilde{f}_2\cos\left(\tilde{\sigma}_2\tau_2-\psi_{21}\right)}{2w}, \tag{3.86}$$

$$\frac{d\psi_2}{d\tau_2} = -\frac{\tilde{f}_2\sin\left(\tilde{\sigma}_2\tau_2-\psi_{21}\right)}{2\tilde{a}_2 w} - \frac{w\left(w^2-1\right)\left(\tilde{a}_2^2\left(8w^2+1\right)-12\tilde{a}_1^2\right)}{64w^2-16}. \tag{3.87}$$

Multiplying equations (3.81), (3.83) by ε, equations (3.80), (3.82), (3.85), (3.87) by ε^2, and equations (3.84), (3.86) by ε^3, and then making use of the relations (3.16), (3.19)–(3.21), and (3.58) we can write the modulation equations in time τ with regard to the original notations. Namely, we have

$$\frac{da_1}{d\tau} = -\frac{1}{2}c_1 a_1 + \frac{1}{2}p_x^2 r_x \sin(\sigma_1 \tau - \psi_1), \tag{3.88}$$

$$\frac{d\psi_1}{d\tau} = \frac{3\left(w^2(w^2-1)a_2^2\right)}{4(4w^2-1)} - \frac{p_x^2 r_x \cos(\sigma_1 \tau - \psi_1)}{2a_1}, \tag{3.89}$$

$$\frac{da_2}{d\tau} = -\frac{1}{2}c_2 a_2 - \frac{f_2 \cos(\sigma_2 \tau - \psi_2)}{2w}, \tag{3.90}$$

$$\frac{d\psi_2}{d\tau} = -\frac{f_2 \sin(\sigma_2 \tau - \psi_2)}{2a_2 w} + \frac{3a_1^2 w(w^2-1)}{4(4w^2-1)} + \frac{a_2^2 w(-8w^4+7w^2+1)}{16(4w^2-1)}, \tag{3.91}$$

and the initial conditions follow

$$a_1(0) = a_{10}, \quad a_2(0) = a_{20}, \quad \psi_1(0) = \psi_{10}, \quad \psi_2(0) = \psi_{20}. \tag{3.92}$$

The known quantities a_{01}, a_{02}, ψ_{01} and ψ_{02} are related to the initial values z_0, v_0, φ_0 and ω_0. However, due to nonlinearity, the above initial value problem (3.88)–(3.92) cannot be solved analytically, in contrast to the problem considered in the previous section for non-resonant vibrations.

Introducing all the subsequent solutions (3.65), (3.66), (3.71), (3.72), (3.77), (3.78) into the assumed form of solution (3.17), (3.18), using polar representation (3.79) of the complex functions $B_1(\tau_1, \tau_2)$ and $B_2(\tau_1, \tau_2)$ and returning to the original notations, the following approximate analytical solution to the problem (3.9)–(3.11) in the case of two resonances are obtained

$$\begin{aligned} z(\tau) = {} & a_1\cos(\eta_1) + \frac{a_2 p_y^3 r_y (p_y - 2w)\sin(-\tau p_y + \eta_2)}{2(p_y - w - 1)(p_y - w)(p_y - w + 1)(p_y + w)} - \\ & \frac{a_2 p_y^3 r_y (p_y + 2w)\sin(\tau p_y + \eta_2)}{2(p_y - w)(p_y + w - 1)(p_y + w)(p_y + w + 1)} + \frac{3a_2^2 w^2 \cos(2\eta_2)}{16w^2 - 4} + \\ & \frac{1}{4}a_2^2 w^2 + \frac{3a_1 a_2^2 (w-1) w\cos(\eta_1 - 2\eta_2)}{32w - 16} - \\ & \frac{3a_1 a_2^2 w(w+1)\cos(\eta_1 + 2\eta_2)}{32w + 16} + \frac{f_1 \sin(p_1 \tau)}{1 - p_1^2}, \end{aligned} \tag{3.93}$$

$$\varphi(\tau) = a_2\cos(\eta_2) + \frac{a_2(w-1)^2(w+1)p_x^2 r_x \cos(\tau(\sigma_1+1)-\eta_2)}{2(\sigma_1+1)(2w-1)(-\sigma_1+2w-1)} - \frac{a_2(w-1)(w+1)^2 p_x^2 r_x \cos(\tau(\sigma_1+1)+\eta_2)}{2(\sigma_1+1)(2w+1)(\sigma_1+2w+1)} + \frac{a_1 p_y^3(p_y+2) r_y \sin(\tau p_y+\eta_1)}{2(p_y-w)(p_y-w+1)(p_y+w)(p_y+w+1)} - \frac{a_1(p_y-2)p_y^3 r_y \sin(\eta_1-\tau p_y)}{2(p_y-w-1)(p_y-w)(p_y+w-1)(p_y+w)} + \frac{a_2^3(1-49w^2)\cos(3\eta_2)}{192(4w^2-1)} + \frac{a_1^2 a_2(w-3)(w-2)w\cos(2\eta_1-\eta_2)}{32w-16} + \frac{a_1^2 a_2 w(w+2)(w+3)\cos(2\eta_1+\eta_2)}{32w+16} + \frac{a_1 a_2(w-2)w\cos(\eta_1-\eta_2)}{4w-2} - \frac{a_1 a_2 w(w+2)\cos(\eta_1+\eta_2)}{4w+2} - \frac{p_y^2 r_y \sin(\tau p_y)}{p_y^2-w^2}, \tag{3.94}$$

where $\eta_1 = \tau + \psi_1(\tau)$, $\eta_2 = w\tau + \psi_2(\tau)$.

Functions $a_1(\tau)$, $a_2(\tau)$, $\psi_1(\tau)$ and $\psi_2(\tau)$ describe time-dependent amplitudes and phases of the solution of the first-order approximations (3.65), (3.66). With the help of (3.79), the solution takes the form

$$z_1(\tau) = a_1\cos(\tau+\psi_1(\tau)) \quad \text{and} \quad \varphi_1(\tau) = a_2\cos(w\tau+\psi_2(\tau)).$$

Time histories of functions $a_1(\tau)$ and $a_2(\tau)$ are presented in Figure 3.7. The graphs are obtained based on formulas (3.88)–(3.92) by the numerical integration. Moreover, the solution of the first-order approximation is shown in Figure 3.7, confirms the physical meaning of the amplitude modulations $a_1(\tau)$, $a_2(\tau)$.

The following fixed parameters are used while constructing Figures 3.7, 3.8, and 3.9: $\sigma_1 = -0.01, \sigma_2 = -0.012, w = 0.89,\ r_x = 0.004, r_y = 0.012, c_1 = 0.004,\ c_2 = 0.003,\ f_1 = 0.001, f_2 = 0.002, p_1 = 4.3, p_2 = \sigma_2 + w, p_x = \sigma_1 + 1, p_y = 2.7$.

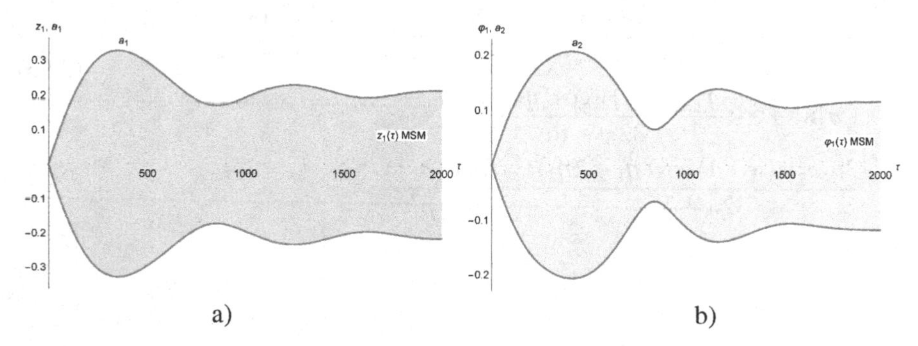

FIGURE 3.7 Time history of the longitudinal z_1 (a) and swing φ_1 (b) vibration and their amplitudes for first-order approximation estimated analytically (MSM).

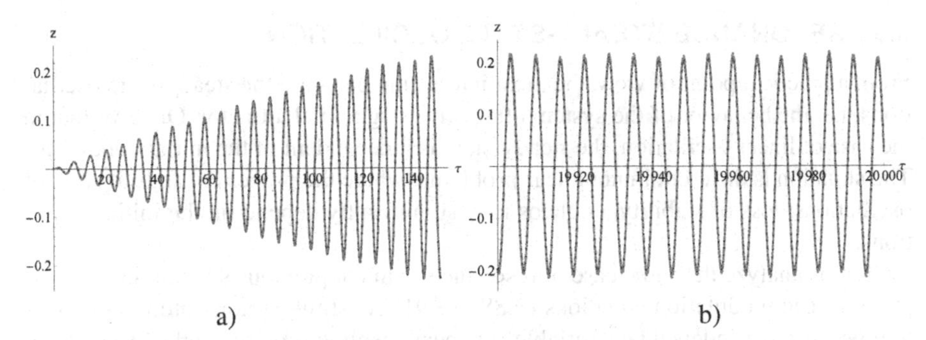

FIGURE 3.8 Time history of the longitudinal vibration for a short (a) and long-time (b) simulation estimated analytically (MSM) and numerically (dotted line).

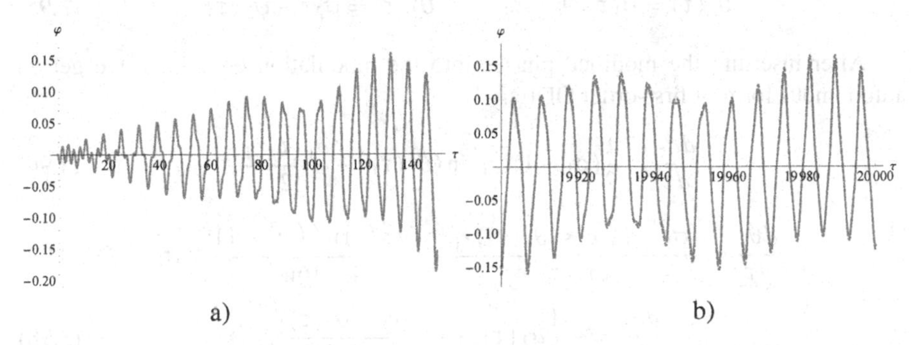

FIGURE 3.9 Time history of the swing vibration for a short (a) and long-time (b) simulation estimated analytically (MSM) and numerically (dotted line).

The initial conditions for the modulation problem (3.88)–(3.92) are assumed in the following form

$$a_{10} = 0.001, \quad a_{20} = 0.001, \quad \psi_{10} = 0, \psi_{20} = 0.$$

The initial conditions of the generalized coordinates (3.11) associated with the above data are as follows:

$$z_0 = 0.001, \quad v_0 = -0.0002868, \quad \varphi_0 = 0.001, \quad \omega_0 = -0.03744.$$

The comparison of the time histories of generalized coordinates obtained by MSM and by numerical simulation is presented in Figures 3.8 and 3.9. The time interval shown includes transient oscillations in the initial motion state and its steady-state achieved after a long time.

One can enjoy the high order accuracy between results obtained by the asymptotic and numerical methods. This validates the correctness of the analytical results, as well as the powerful use of the MSM.

3.5 RESONANCE STEADY-STATE OSCILLATION

From a practical point of view, the most interesting parts are the steady-state oscillations, i.e. the behavior of the system after achieving a final attractor. Observe that in the case of linear oscillation, the steady-state is independent of the initial conditions. The situation is different in nonlinear problems, where the process of stabilization of oscillations, loss of stability, or chaos may significantly depend on the initial conditions.

Let us analyze the same case of resonance as in the previous section, i.e. $p_x \approx 1$, $p_2 \approx w$. The modulation equations (3.88)–(3.91) constitute a non-autonomous system because the independent variable τ appears explicitly on the right. It is convenient to introduce to the system (3.88)–(3.91) the following new modified phases

$$\theta_1(\tau) = \sigma_1\tau - \psi_1(\tau), \qquad \theta_2(\tau) = \sigma_2\tau - \psi_2(\tau). \tag{3.95}$$

After inserting the modified phases into the modulation equations, we get an autonomous form of first-order ODEs:

$$\frac{da_1}{d\tau} = \frac{1}{2}(\sigma_1+1)^2 r_x \sin(\theta_1(\tau)) - \frac{a_1(\tau)c_1}{2}, \tag{3.96}$$

$$\frac{d\theta_1}{d\tau} = \frac{(\sigma_1+1)^2\cos(\theta_1(\tau))r_x}{2a_1(\tau)} + \frac{3a_2^2(\tau)w^2(w^2-1)}{4-16w^2} + \sigma_1, \tag{3.97}$$

$$\frac{da_2}{d\tau} = -\frac{1}{2}a_2(\tau)c_2 - \frac{f_2\cos(\theta_2(\tau))}{2w}, \tag{3.98}$$

$$\begin{aligned}\frac{d\theta_2}{d\tau} = {} & \frac{f_2\sin(\theta_2(\tau))}{2a_2(\tau)w} + \frac{3a_1^2(\tau)w(w^2-1)}{4-16w^2} + \\ & \frac{a_2^2(\tau)w(8w^4-7w^2-1)}{64w^2-16} + \sigma_2.\end{aligned} \tag{3.99}$$

The unknown functions in (3.96)–(3.99) are $a_1(\tau)$, $a_2(\tau)$, $\theta_1(\tau)$ and $\theta_2(\tau)$. The above form of the equations allows studying the steady-state motion of the pendulum, which corresponds to zeroing time derivatives $\frac{da_i}{d\tau}$ and $\frac{d\theta_i}{d\tau}$ for $i = 1,2$. Introduction of this condition to the modulation equations (3.96)–(3.99) yields the set of nonlinear algebraic equations with unknown amplitudes a_1, a_2 and modified phases θ_1, θ_2 of the following form

$$(\sigma_1+1)^2\sin(\theta_1)r_x - a_1c_1 = 0, \tag{3.100}$$

$$\frac{(\sigma_1+1)^2\cos(\theta_1)r_x}{2a_1} + \frac{3a_2^2w^2(w^2-1)}{4-16w^2} + \sigma_1 = 0, \tag{3.101}$$

$$-a_2c_2 - \frac{f_2\cos(\theta_2)}{w} = 0, \tag{3.102}$$

$$\frac{f_2\sin(\theta_2)}{2a_2w} + \frac{3a_1^2w(w^2-1)}{4-16w^2} + \frac{a_2^2w(8w^4-7w^2-1)}{64w^2-16} + \sigma_2 = 0. \tag{3.103}$$

Equations (3.100)–(3.103) describe the limiting behavior of a system with damping and in the case of two resonances occurring simultaneously.

By eliminating the trigonometric functions from (3.100)–(3.103), one can obtain the relationships between amplitudes and frequencies in the following form:

- for the kinematic resonance

$$a_1^2c_1^2 + a_1^2\left(\frac{3a_2^2w^2\left(w^2-1\right)}{2-8w^2} + \sigma_1\right)^2 + r_x^2(\sigma_1+1)^4 = 0, \tag{3.104}$$

- for the external resonance

$$\frac{1}{4}a_2^2c_2^2 + a_2^2\left(\frac{3a_1^2w\left(w^2-1\right)}{4-16w^2} + \frac{a_2^2w\left(8w^4-7w^2-1\right)}{64w^2-16} + \sigma_2\right)^2 = \frac{f_2^2}{4w^2}, \tag{3.105}$$

where a_1 and a_2 are amplitudes of the longitudinal and swing oscillations, respectively.

Case study 1: Single resonances

The analysis presented above applies to the presence of two resonances at the same time. It is also possible to investigate the occurrence of these resonances separately. In order to do it in a simple way, one may select the chosen resonance, i.e. $p_x \approx 1$ or $p_2 \approx w$ and then employ them to equations (3.14) and (3.15).

Proceeding similarly to section 3.4, for the case of kinematic resonance $p_x \approx 1$, the following equations for steady-state amplitudes and phases are obtained

$$(\sigma_1+1)^2\sin(\theta_1)\,r_x - a_1c_1 = 0, \tag{3.106}$$

$$\frac{(\sigma_1+1)^2\cos(\theta_1)\,r_x}{2a_1} + \frac{3a_2^2w^2\left(w^2-1\right)}{4-16w^2} + \sigma_1 = 0, \tag{3.107}$$

$$a_2c_2 = 0, \tag{3.108}$$

$$\frac{3a_1^2a_2w\left(w^2-1\right)}{1-4w^2} + \frac{a_2^3w\left(8w^4-7w^2-1\right)}{4\left(4w^2-1\right)} = 0. \tag{3.109}$$

Equation (3.108) shows that for the studied case, the amplitude of the swing motion tends to zero, while the amplitude-frequency relationship for the amplitude of elastic vibrations determined by the elimination of trigonometric functions from equations (3.106) and (3.107) has the following form

$$a_1^2\left(c_1^2+4\sigma_1^2\right) = (\sigma_1+1)^4r_x^2. \tag{3.110}$$

Figure 3.10 presents diagrams showing the dependence of the amplitude a_1 on the parameter σ_1, determined based on equation (3.110) for different values of r_x and for different damping coefficients c_1.

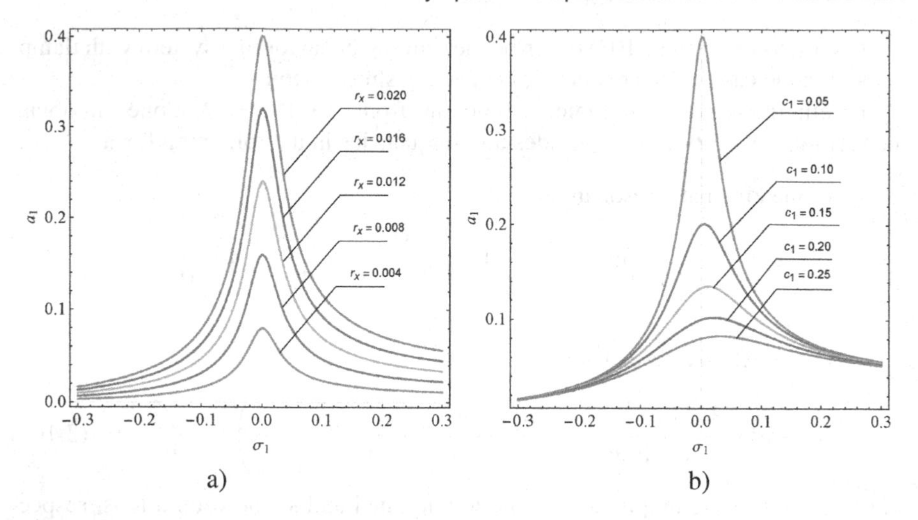

FIGURE 3.10 Amplitude a_1 vs. σ_1 for $c_1 = 0.05$ and various r_x (a) and amplitude a_1 vs. σ_1 for $r_x = 0.02$ and various c_1 (b).

By examining the external resonance case $p_2 \approx w$ and proceeding similarly to the previous case, one can obtain the steady-state vibration equations for amplitudes and phases in the following form

$$a_1 c_1 = 0, \tag{3.111}$$

$$\frac{3a_2^2 a_1 w^2 \left(w^2 - 1\right)}{4 - 16w^2} = 0, \tag{3.112}$$

$$-a_2 c_2 - \frac{f_2 \cos\left(\theta_2\right)}{w} = 0, \tag{3.113}$$

$$\frac{f_2 \sin\left(\theta_2\right)}{2a_2 w} + \frac{3a_1^2 w \left(w^2 - 1\right)}{4 - 16w^2} + \frac{a_2^2 w \left(8w^4 - 7w^2 - 1\right)}{64w^2 - 16} + \sigma_2 = 0. \tag{3.114}$$

Equation (3.111) shows the disappearance of elastic oscillations. The relations (3.113) and (3.114) make it possible to obtain the dependence of the amplitude a_2 on the parameter σ_2, similarly as before, by eliminating the trigonometric functions, which yields the following algebraic formula

$$\frac{1}{4} a_2^2 c_2^2 + a_2^2 \left(\frac{a_2^2 w \left(8w^4 - 7w^2 - 1\right)}{64w^2 - 16} + \sigma_2 \right)^2 = \frac{f_2^2}{4w^2}. \tag{3.115}$$

Let us analyze some examples of the behavior of the system in external resonance condition. Figure 3.11 shows the plotted families of curves for different values of the amplitude of the external load f_2 and for different values of the damping coefficient c_2.

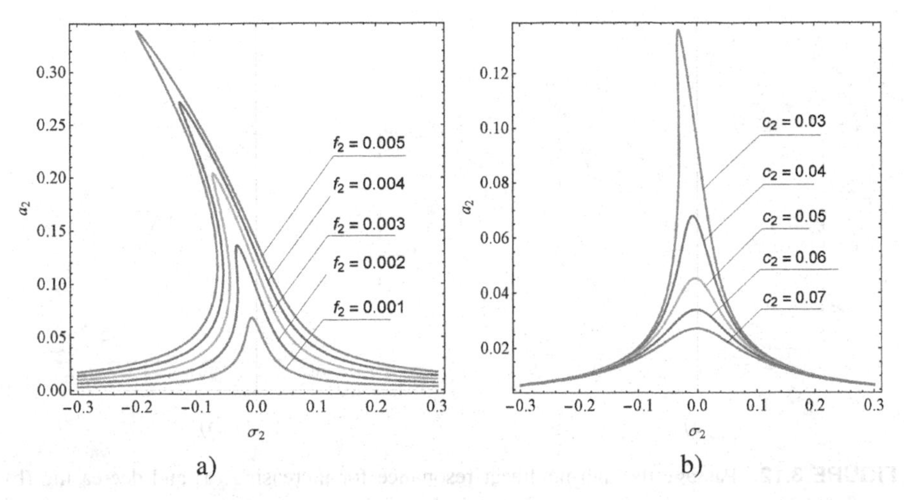

FIGURE 3.11 Amplitude a_2 vs. $\sigma_2$for $c_2 = 0.03$ and $w = 0.49$ for various f_2 (a) and amplitude a_2 vs. σ_2 for $f_2 = 0.002$ and $w = 0.49$ for various c_2 (b).

The analysis of the graphs presented in Figure 3.10 indicates that the amplitude a_1 increases monotonically with the increase of r_x, while it decreases with the increase of the damping coefficient c_1. A similar tendency is shown by the resonance curves presented in Figure 3.11. They show the monotonic dependence of the amplitude a_2 on the amplitude of the external load f_2 and on the damping coefficient c_2. The curves in Figure 3.11 describe a nonlinear resonance. The slope of the resonance backbone curve is tailored in such a way that for some values of σ_2 three solutions are possible. It indicates that for the same parameters, oscillations of different amplitude may occur. This is related to the so-called jump phenomenon when the system passes through nonlinear resonance. It is the case, for example, for the resonance curve corresponding to $f_2 = 0.004$ presented in Figure 3.11a. The magnitude of the vibration amplitude depends in this case on the history of the movement.

Let us follow the slow passage through the resonance for this case. Figure 3.12 presents the path along which the vibration amplitude changes as the parameter σ_2 increases or decreases. The bold gray curve (Figure 3.12a) shows the changes in the amplitude when passing through the resonance for the quantity σ_2 increasing from negative to positive value, i.e. from $p_2 < w$ to $p_2 > w$. Finally, Figure 3.12b shows the resonance path in the reverse direction, i.e. from $p_2 > w$ to $p_2 < w$.

The vertical lines in the presented figures correspond to a sudden change in the amplitude of the oscillations. It follows that for certain values of σ_2 the oscillation amplitude depends on the previous state of the system. The passage of a single degree of freedom system through nonlinear resonance is discussed in detail in the paper [199].

It is also possible to discuss solutions in resonance for different values of the parameter w. It should be noted that the change in the dimensionless frequency w

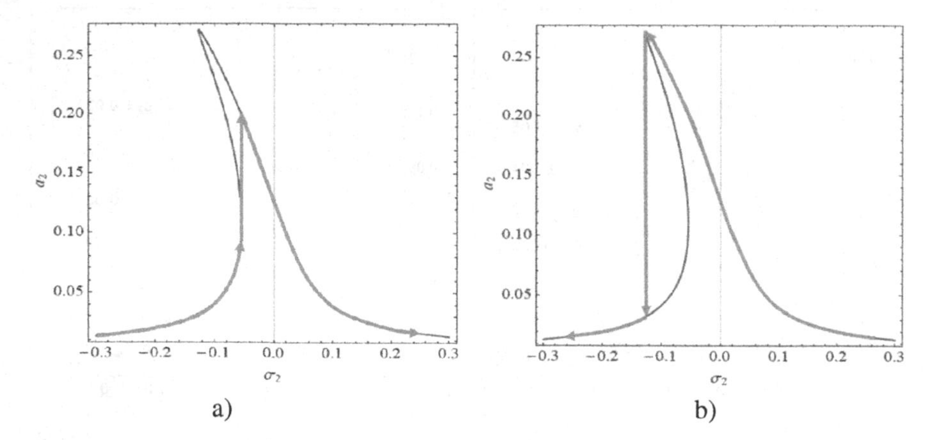

FIGURE 3.12 Passage through nonlinear resonance for increasing (a) and decreasing (b) parameter σ_2.

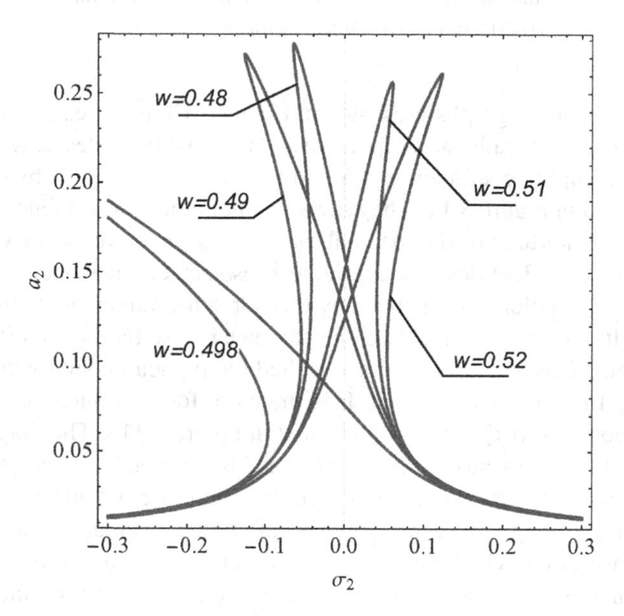

FIGURE 3.13 Amplitude a_2 vs. detuning parameter σ_2 for various nondimensional frequency w.

interplays with the mechanical properties of the studied system, as it depends on its parameters.

Figure 3.13 shows the effect of the dimensionless frequency w on the slope of the backbone curve for swing vibrations at the main resonance. The slope of the resonance curve changes with the change in parameter w. In particular, the qualitative change in the nature of the resonance response occurs for the system where $w = 0.5$.

For $w < 0.5$ and $w > 1$, the effect of the "hard" characteristic is observed, i.e. the slope of the resonance curve to the left. However, for $0.5 < w < 1$, the resonance characteristic changes to be "soft", i.e. the slope of the curves bends to the right. For the case $w = 1$, the nonlinear effects disappear and the resonance curve is not bent. It is found that the sign of the coefficient standing at the highest power of a_2 in equation (3.115)

$$coeff = \frac{w\left(8w^4 - 7w^2 - 1\right)}{64w^2 - 16} \tag{3.116}$$

is responsible for the type of characteristic. The behavior of this coefficient vs. w is reported in Figure 3.14.

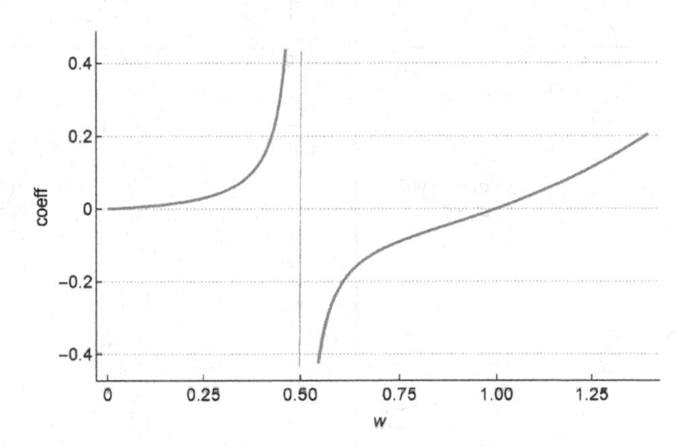

FIGURE 3.14 The dependence of the coefficient at the highest power od a_2 in equation (3.115) on w.

For $w < 0.5$, the slope of the curve is the greater, the larger is the absolute value of the coefficient (3.116). For $w = 0.5$, the coefficient (3.116) changes its sign, while for $w < 0.4$, the absolute value of this quantity is very small, which results in a resonance characteristic close to the linear counterpart.

Case study 2: Simultaneous resonances

In the analyzed case, when two resonances occur simultaneously, the expressions (3.104) and (3.105) should be treated as a system of nonlinear algebraic equations with unknowns a_1 and a_2. Each of equations (3.104), (3.105) is a sixth-degree polynomial, hence the number of solutions of the system, i.e. pairs (a_1, a_2), may achieve thirty six. However, as expected only some of them are real and hence realized physically. All theoretically possible steady states can be illustrated on a certain plane in which the Cartesian coordinates correspond to the amplitudes a_1 and a_2 of the longitudinal and swing oscillations, respectively.

In Figure 3.15, the black line stands for the geometric location of the roots of equation (3.104), while the gray line presents the solutions of equation (3.105). The points of intersection of both curves define a solution of the system of equations

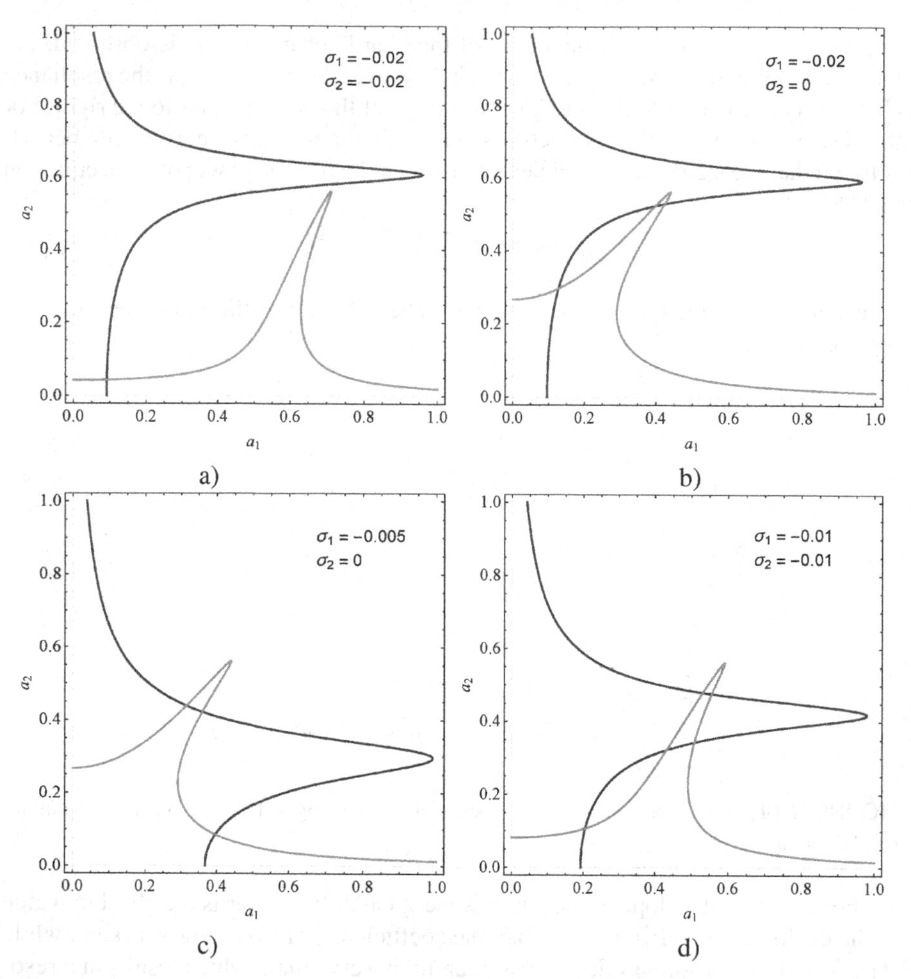

FIGURE 3.15 Graphical presentation of the solutions of the system (3.104)–(3.105) for different values of σ_1 and σ_2: a) $\sigma_1 = -0.02$, $\sigma_2 = -0.02$; b) $\sigma_1 = -0.02$, $\sigma_2 = 0$; c) $\sigma_1 = -0.005$, $\sigma_2 = 0$; d) $\sigma_1 = -0.01$, $\sigma_2 = -0.01$. The intersection points of the curves identify the steady-state amplitudes of oscillations.

(3.104)–(3.105) and indicate the values of the amplitudes of longitudinal and swing oscillations of the pendulum in its steady-state. However, since the solutions determined in this way may be stable or unstable, it opens the door to another problem investigation.

The solutions presented in Figure 3.15 are found for the following fixed parameters: $w = 0.89, r_x = 0.004$, $f_2 = 0.0015$, $c_1 = 0.004$, $c_2 = 0.003$.

The number of possible solutions to the system (3.104)–(3.105) depends on the values of the parameters and the distance from the pure resonance. The smallest number of steady amplitudes is 1, and the largest equals 5, which is easy to see in the exemplary drawings presented in Figure 3.15.

Resonance response curves play an important role in the study of oscillations which allows one to estimate the magnitude of the oscillations amplitude determined in the resonance conditions. The resonance curves presented in Figure 3.16 show the dependence of the amplitudes a_1 and a_2 as a function of the detuning parameter σ_1. In turn, Figure 3.17 presents the dependence of a_1 and a_2 on the detuning parameter σ_2. The graphs presented in Figures 3.16 and 3.17 are reported for the same data as above. The longitudinal vibration amplitude a_1 (a) and the swing vibration amplitude a_2 (b) vs. functions of the detuning parameter σ_2 for $\sigma_1 = -0.01$, are examined.

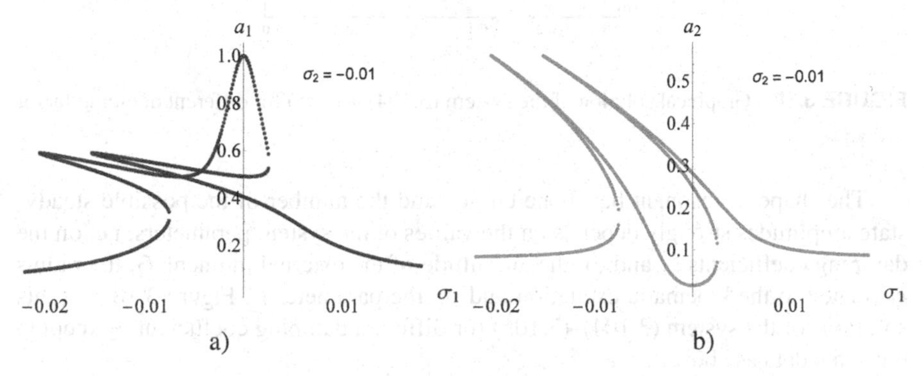

FIGURE 3.16 The longitudinal vibration amplitude a_1 (a) and the swing vibration amplitude a_2 (b) vs. the detuning parameter σ_1 for $\sigma_2 = -0.01$.

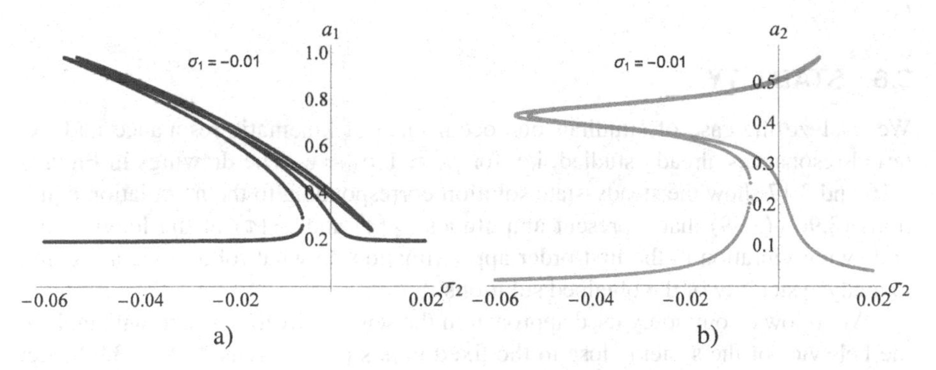

FIGURE 3.17 The longitudinal vibration amplitude a_1 (a) and the swing vibration amplitude a_2 (b) vs. functions of the detuning parameter σ_2 for $\sigma_1 = -0.01$.

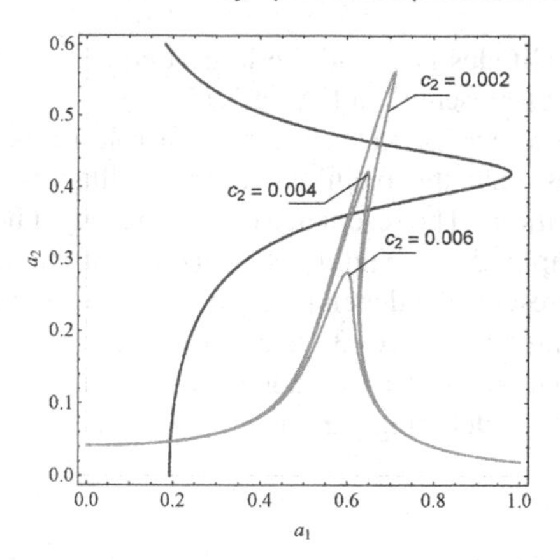

FIGURE 3.18 Graphical solution of the system (3.104)–(3.105) for different damping factor c_2.

The shape of resonant backbone curves and the number of the possible steady-state amplitudes strongly depends on the values of the system parameters, i.e. on the damping coefficients c_1 and c_2, the amplitude of the external moment f_2, the radius r_x related to the kinematic excitation and on the parameter w. Figure 3.18 presents solutions of the system (3.104)–(3.105) for different damping coefficient c_2 keeping the same data as above.

As it can be seen from Figure 3.18, sufficiently low damping increases the number of possible steady-state vibration amplitudes. The number of solutions of the system (3.104)–(3.105) also depends on changes in other parameters like f_2, c_1 and r_x.

3.6 STABILITY

We analyze the case of simultaneous occurrence of kinematic resonance and external resonances already studied, i.e. for $p_x \approx 1$ $p_2 \approx w$. The drawings in Figures 3.16 and 3.17 show the steady-state solution corresponding to the modulation equations (3.96)–(3.99) that represent amplitudes $a_1(\tau)$ and $a_2(\tau)$ of the longitudinal and swing vibration in the first-order approximation. In what follows we are going to analyze stability of the obtained solutions.

We follow a commonly used approach in the sense of Lyapunov and will analyze the behavior of the system close to the fixed points of equations (3.96)–(3.99). Let us write the non-stationary solution close to the stationary one in the following form

$$\begin{aligned} a_1(\tau) &= a_{1s} + \tilde{a}_1(\tau), \quad a_2(\tau) = a_{2s} + \tilde{a}_2(\tau), \\ \theta_1(\tau) &= \theta_{1s} + \tilde{\theta}_1(\tau), \quad \theta_2(\tau) = \theta_{2s} + \tilde{\theta}_2(\tau), \end{aligned} \tag{3.117}$$

where a_{1s}, a_{2s}, θ_{1s}, θ_{2s} are fixed points of the system of equations (3.96)–(3.99), i.e. they satisfy equations (3.100)–(3.103), and $\tilde{a}_1(\tau)$, $\tilde{a}_2(\tau)$, $\tilde{\theta}_1(\tau)$ and $\tilde{\theta}_2(\tau)$ are small disturbances of these quantities. Substituting the functions (3.117) into equations (3.96)–(3.99) the following first-order ODEs with regard to perturbations are obtained

$$\frac{d\tilde{a}_1}{d\tau} = \frac{1}{2}(\sigma_1+1)^2 r_x \sin\left(\theta_{1s}+\tilde{\theta}_1(\tau)\right) - \frac{(a_{1s}+\tilde{a}_1(\tau))c_1}{2}, \tag{3.118}$$

$$\frac{d\tilde{\theta}_1}{d\tau} = \frac{(\sigma_1+1)^2\cos\left(\theta_{1s}+\tilde{\theta}_1(\tau)\right)r_x}{2(a_{1s}+\tilde{a}_1(\tau))} + \frac{3(a_{2s}+\tilde{a}_2(\tau))^2 w^2(w^2-1)}{4-16w^2} + \sigma_1, \tag{3.119}$$

$$\frac{d\tilde{a}_2}{d\tau} = -\frac{1}{2}(a_{2s}+\tilde{a}_2(\tau))c_2 - \frac{f_2\cos\left(\theta_{2s}+\tilde{\theta}_2(\tau)\right)}{2w}, \tag{3.120}$$

$$\begin{aligned}\frac{d\tilde{\theta}_2}{d\tau} = {} & \frac{f_2\sin\left(\theta_{2s}+\tilde{\theta}_2(\tau)\right)}{2(a_{2s}+\tilde{a}_2(\tau))w} + \frac{3(a_{1s}+\tilde{a}_1(\tau))^2(\tau)w(w^2-1)}{4-16w^2} + \\ & \frac{(a_{1s}+\tilde{a}_1(\tau))^2 w(8w^4-7w^2-1)}{64w^2-16} + \sigma_2.\end{aligned} \tag{3.121}$$

Next, the right-hand sides of equations (3.118)–(3.121) are expanded into the power series around fixed steady-state solutions a_{1s}, a_{2s}, θ_{1s}, θ_{2s}. Considering only the terms of the first-order approximations, the following linearized system of equations for the disturbances $\tilde{a}_1(\tau)$, $\tilde{a}_2(\tau)$, $\tilde{\theta}_1(\tau)$ and $\tilde{\theta}_2(\tau)$ is obtained

$$\begin{aligned}\frac{d\tilde{a}_1}{d\tau} = {} & -\frac{a_{1s}c_1}{2} - \frac{1}{2}c_1\tilde{a}_1(\tau) + \frac{1}{2}\cos(\theta_{1s})\tilde{\theta}_1(\tau)(\sigma_1+1)^2 r_x + \\ & \frac{1}{2}\sin(\theta_{1s})(\sigma_1+1)^2 r_x,\end{aligned} \tag{3.122}$$

$$\begin{aligned}\frac{d\tilde{\theta}_1}{d\tau} = \sigma_1 - {} & \frac{\tilde{a}_1(\tau)\cos(\theta_{1s})(\sigma_1+1)^2 r_x}{2a_{1s}^2} - \\ & \frac{\sin(\theta_{1s})\tilde{\theta}_1(\tau)(\sigma_1+1)^2 r_x}{2a_{1s}} + \frac{\cos(\theta_{1s})(\sigma_1+1)^2 r_x}{2a_{1s}} + \\ & \frac{3a_{2s}^2(w^2-1)w^2}{4-16w^2} + \frac{3a_{2s}(w^2-1)w^2\tilde{a}_2(\tau)}{2-8w^2},\end{aligned} \tag{3.123}$$

$$\frac{d\tilde{a}_2}{d\tau} = -\frac{a_{2s}c_2}{2} - \frac{1}{2}c_2\tilde{a}_2(\tau) + \frac{f_2\sin(\theta_{2s})\tilde{\theta}_2(\tau)}{2w} - \frac{f_2\cos(\theta_{2s})}{2w}, \tag{3.124}$$

$$\begin{aligned}\frac{d\tilde{\theta}_2}{d\tau} = {} & \sigma_2 + \frac{3a_{1s}^2 w(w^2-1)}{4-16w^2} + \frac{3a_{1s}w(w^2-1)\tilde{a}_1(\tau)}{2-8w^2} - \frac{f_2\tilde{a}_2(\tau)\sin(\theta_{2s})}{2a_{2s}^2 w} \\ & + \frac{a_{2s}^2 w(8w^4-7w^2-1)}{64w^2-16} + \frac{a_{2s}w(8w^4-7w^2-1)\tilde{a}_2(\tau)}{32w^2-8} + \\ & \frac{f_2\cos(\theta_{2s})\tilde{a}_2(\tau)}{2a_{2s}w} + \frac{f_2\sin(\theta_{2s})}{2a_{2s}w}.\end{aligned} \tag{3.125}$$

Since the values a_{1s}, a_{2s}, θ_{1s}, θ_{2s} satisfy equations (3.100)–(3.103), the perturbation linear ODEs take the following form

$$\frac{d\tilde{a}_1}{d\tau} = -\frac{1}{2}c_1\tilde{a}_1(\tau) + \frac{1}{2}\cos(\theta_{1s})\tilde{\theta}_1(\tau)(\sigma_1+1)^2 r_x, \tag{3.126}$$

$$\frac{d\tilde{\theta}_1}{d\tau} = -\frac{\tilde{a}_1(\tau)\cos(\theta_{1s})(\sigma_1+1)^2 r_x}{2a_{1s}^2} - \frac{\sin(\theta_{1s})\tilde{\theta}_1(\tau)(\sigma_1+1)^2 r_x}{2a_{1s}} + \frac{3a_{2s}(w^2-1)w^2\tilde{a}_2(\tau)}{2-8w^2}, \tag{3.127}$$

$$\frac{d\tilde{a}_2}{d\tau} = -\frac{1}{2}c_2\tilde{a}_2(\tau) + \frac{f_2\sin(\theta_{2s})\tilde{\theta}_2(\tau)}{2w}, \tag{3.128}$$

$$\frac{d\tilde{\theta}_2}{d\tau} = \frac{3a_{1s}w(w^2-1)\tilde{a}_1(\tau)}{2-8w^2} - \frac{f_2\tilde{a}_2(\tau)\sin(\theta_{2s})}{2a_{2s}^2 w} + \frac{a_{2s}w(8w^4-7w^2-1)\tilde{a}_2(\tau)}{32w^2-8} + \frac{f_2\cos(\theta_{2s})\tilde{\theta}_2(\tau)}{2a_{2s}w}. \tag{3.129}$$

Now, the stability of the solutions of the system of ODEs (3.126)–(3.129) can be analyzed with the standard procedure regarding the linear first-order differential equations.

Since the system (3.126)–(3.129) is homogeneous with constant coefficients, its general solution has the form of a superposition of the exponent functions

$$\tilde{a}_i = \sum_{k=1}^{4} D_{ik}e^{\lambda_k\tau}, \quad \tilde{\theta}_i = \sum_{k=1}^{4} E_{ik}e^{\lambda_k\tau}, \quad i=1,2, \tag{3.130}$$

where D_{ik} and E_{ik} are constants dependent on the initial conditions, and λ_k are the roots of the characteristic equation

$$|\mathbf{A}-\lambda\mathbf{I}| = 0. \tag{3.131}$$

Here $\mathbf{I}$ is the identity matrix while $\mathbf{A}$ is the characteristic matrix of the system (3.126)–(3.129).

The roots λ_k of the characteristic polynomial (3.131) are eigenvalues of the matrix $\mathbf{A}$ having the following explicit form

$$\mathbf{A} = \begin{bmatrix} -\frac{c_1}{2} & 0 & \frac{p_x^2 r_x\cos\theta_{1s}}{2} & 0 \\ 0 & -\frac{c_2}{2} & 0 & \frac{f_2\sin\theta_{2s}}{2w} \\ -\frac{p_x^2 r_x\cos\theta_{1s}}{2a_{1s}^2} & \frac{3a_{2s}w^2(w^2-1)}{2-8w^2} & -\frac{p_x^2 r_x\sin\theta_{1s}}{2a_{1s}} & 0 \\ \frac{3a_{1s}w(w^2-1)}{2-8w^2} & \frac{a_{2s}w(8w^4-7w^2-1)}{32w^2-8} - \frac{f_2\sin\theta_{2s}}{2a_{2s}^2 w} & 0 & \frac{f_2\cos\theta_{2s}}{2a_{2s}w} \end{bmatrix} \tag{3.132}$$

where $p_x = (1-\sigma_1)$.

The solution form (3.130) implies that the perturbations $\tilde{a}_1(\tau)$, $\tilde{a}_2(\tau)$, $\tilde{\theta}_1(\tau)$ and $\tilde{\theta}_2(\tau)$ tend to zero in time when all values of λ_k are negative. If this condition is satisfied, the steady-state solution $(a_{1s}, a_{2s}, \theta_{1s}, \theta_{2s})$ is asymptotically stable in the sense of Lyapunov.

Now, we proceed to consider the resonance response curves yielded by equations (3.100)–(3.103) and test their stability using the previously employed technique. The results presented in Figures 3.19–3.22 are calculated for the same data as in Figures 3.15–3.17, i.e. $w = 0.89, r_x = 0.004, f_2 = 0.0015, c_1 = 0.004, c_2 = 0.003$. The black color in these figures denotes the stable branch of the resonance curves.

The analysis of the stability of the fixed points of the modulation equations (3.96)–(3.99) shows that for example taking $\sigma_1 = -0.01$ and $\sigma_2 = -0.01$ the amplitudes are stabilized on the values $a_{1s} = 0.20496$, $a_{2s} = 0.10666$, which can also be read from Figure 3.15d. The time history of the amplitude modulation curves for that case is presented in Figure 3.23.

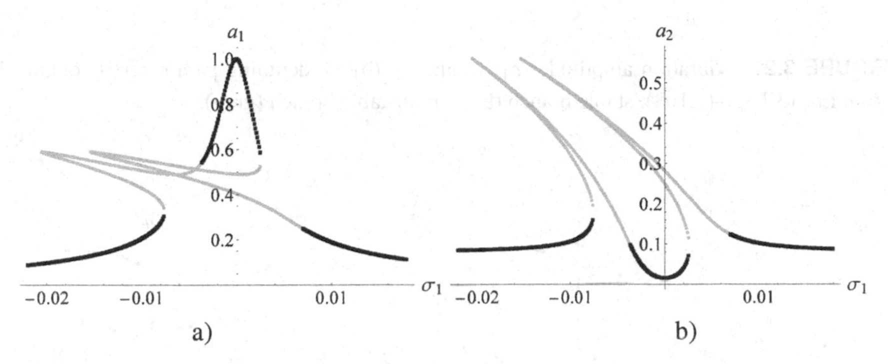

FIGURE 3.19 The vibration amplitudes a_1 (a) and a_2 (b) vs. detuning parameter σ_1 obtained from Eqs (3.100)–(3.103): stable branch (black); unstable branch (gray).

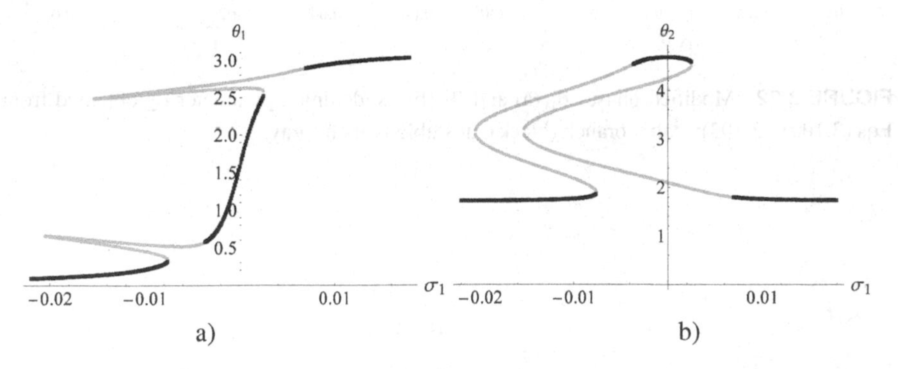

FIGURE 3.20 The modified phases θ_1 (a) and θ_2 (b) vs. detuning parameter σ_1 obtained from Eqs (3.100)–(3.103): stable branch (black); unstable branch (gray).

Note that the graphs in Figures 3.19–3.22 indicate some excitation frequencies for which there are no steady vibrations at all. This is the case, for example, for $\sigma_1 = -0.006$ and $\sigma_2 = -0.01$. The time history of the modulation curves for this case is shown in Figure 3.24.

The lack of stable solutions for some parameters under resonance conditions is the result of intensive energy exchange between the system and the environment.

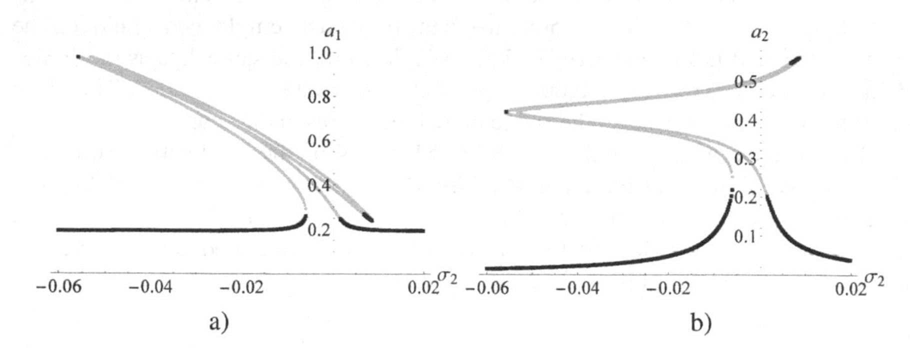

FIGURE 3.21 Vibration amplitudes a_1 (a) and a_2 (b) vs. detuning parameter σ_2 obtained from Eqs (3.100)–(3.103): stable branch (black); unstable branch (gray).

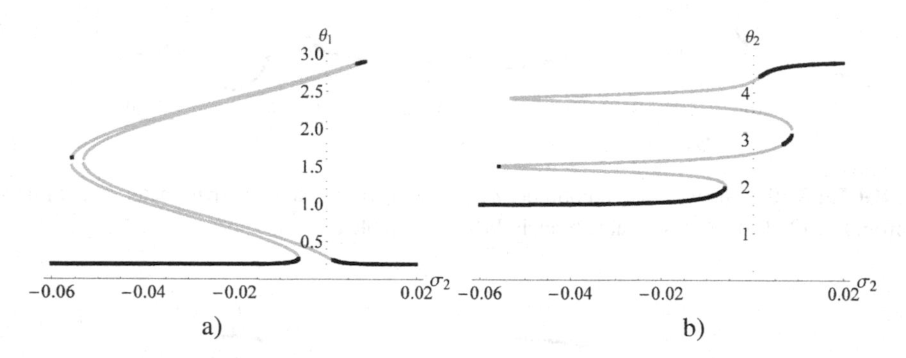

FIGURE 3.22 Modified phases θ_1 (a) and θ_2 (b) vs. detuning parameter σ_2 obtained from Eqs (3.100)–(3.103): stable branch (black); unstable branch (gray).

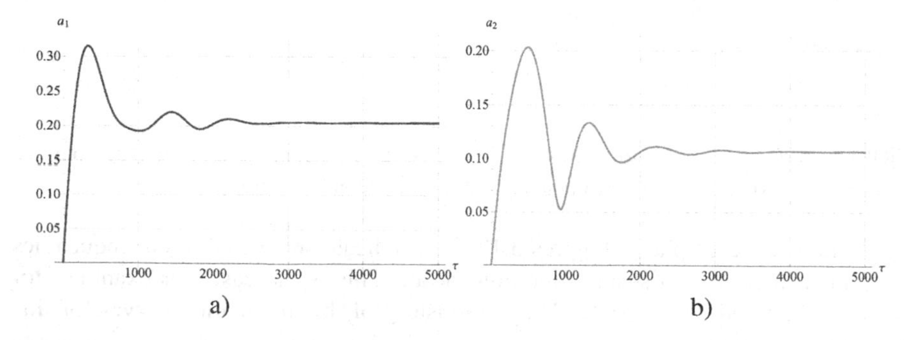

FIGURE 3.23 Time histories of the amplitudes a_1 (a) and a_2 (b) modulation–case of stable oscillations.

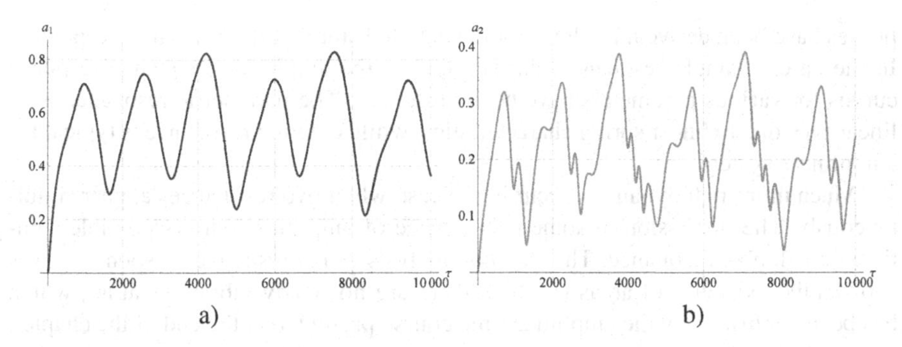

FIGURE 3.24 Time histories of the amplitudes a_1 (a) and a_2 (b) modulation–case of non-stable oscillations.

3.7 CLOSING REMARKS

The nonlinear two degrees of freedom system has been examined. The subject of the asymptotic analysis was a spring pendulum suspended in the moving point. Swing motion and couplings between generalized coordinates are the sources of nonlinearity in the mathematical model. The spring has been assumed as linear. The equation of motion has been derived using Lagrange's equations of the second type. Afterward, some assumptions concerning the smallness of some parameters were made, and the governing equations have been transformed to the dimensionless form. The dimensionless initial value problem was subjected to a detailed asymptotic analysis using MSM with three-time scales. In this way, the solution up to the third-order of approximation has been achieved.

At first, the non-resonant vibration was analyzed. In this case, the derived modulation equations of amplitudes and phases have been solved analytically, therefore an approximate but fully analytical solution has been obtained. It was shown that the solution is very consistent with the result of numerical analysis, which confirms the correctness of the applied approximate analytical approach. Moreover, the analytical form of the solution allows recognizing all resonance conditions till the third-order approximation. Thus, eight types of resonances were identified.

A separate study was performed for resonance conditions. A detailed analysis of external and parametric resonances has been provided, and the asymptotic solution up to third-order approximation has been achieved. The modulation equations obtained as a result of reducing the secular terms have in this case too complicated form to solve them analytically, so the numerical approach is necessary at this stage. The comparison of the asymptotic solution to the numerical ones shows very good agreement between results obtained in these two ways.

The steady-state motion when two resonances occur simultaneously was examined in detail, and a separate analysis was carried out for single parametric and external resonances. For this purpose, the modified phases were introduced to the modulation equations which transform them into the autonomous form. From these equations, the amplitude-frequency relationships in which there is neither time nor phases have been derived. It allows for the detailed study of the resonance responses. In the case of single resonance, the families of the amplitude-frequency response curves for various parameters have been presented. The parametric resonance is of linear type due to linear spring characteristics, while external resonance is typical for a nonlinear system.

Much more rich dynamics occur in the case when two resonances appear simultaneously. The discussion of some phase space of amplitudes shows possible solutions in a double resonance. The stability analysis of the resonance response curves showed the existence of areas in which there are no steady vibrations at all, which has been confirmed by the amplitude time course presented at the end of the chapter.

4 Spring Pendulum Revisited

This chapter is again devoted to the study of a spring pendulum (see chapter 2) but with emphasis put on another method of analysis. The governing equations have been transformed to the complex form according to the method proposed by Manevitch and Musenko [136]. This allows one to decrease the order of the differential equations. In order to employ the multi-scale method, the selected parameters are assumed to be small, however the approach used does not require the assumption of the smallness of the parameter responsible for the nonlinearity of the spring.

The elimination of terms leading to unbounded solutions yields the solvability conditions, which in turn allow the derivation of the equations with regard to amplitude and phase modulation. These dependencies are used for qualitative and quantitative analysis of motion in both steady and unsteady states. The behavior of the system is presented in several ways. There are presented time histories, resonance response curves, and phase trajectories in the domain of amplitudes and phases. The stability of the derived periodic orbits in the sense of Lyapunov is also addressed.

4.1 INTRODUCTION

In engineering practice, steady-state vibrations are the most common. However, in areas of strong resonance and with a small damping factor, the transient vibrations can last for a relatively long time. Under specific conditions non-steady vibrations may be even a permanent feature of the system. In such systems, an intensive energy exchange takes place between subsystems or between the system and the environment. This problem is widely discussed by Sado [184], Shivamoggi [189] and Manevitch and Manevitch [135]. Problems of this type occurring in nonlinear systems are usually investigated using numerical methods due to the occurred mathematical difficulties. In recent years, there has been a lot of interest among scientists in approximate analytical methods in the study of engineering problems (see Awrejcewicz and Krysko [21]). Particularly interesting seems to be the concept for studying both steady and transient states, proposed by Manevitch and Musenko [136].

Though pendulums are relatively simple systems, yet they can be used to simulate the dynamics of various types of engineering devices and machine parts. The currently studied system, in comparison to that investigated in chapter 2, differs in the model of external interactions and damping effects. Pendulum-type mechanical systems with nonlinear and parametric interactions exhibit rich nonlinear behavior, so understanding and predicting them plays an important role both from a theory and application point of view. Couplings occurring in equations of motion may cause autoparametric excitation and the related intensive energy exchange between vibration modes. This issue is addressed by Starosvetsky and Gendelman [203]. Energy transfer in systems with many degrees of freedom is a well-known phenomenon that has been studied by many authors including Sado and Gajos [185] and Starosta et al.

DOI: 10.1201/9781003270706-4

[200]. A key role in the analysis of vibrations of nonlinear systems is the prediction and determination of thresholds (critical values of parameters) at which there is a qualitative change in dynamics of the system. Such a critical value of the parameter responsible for the nonlinearity of the spring is determined in this chapter.

4.2 THE PHYSICAL AND MATHEMATICAL MODEL

The spring pendulum studied in the chapter consists of the concentrated mass suspended on the massless nonlinear spring. The pendulum is excited by the external time-dependent forces $\vec{F_1}$ and $\vec{F_2}$ of magnitudes $F_1(t) = F_1\cos(\Omega_1 t)$ and $F_2(t) = F_2\cos(\Omega_2 t)$ which act along the pendulum and in transversal direction, respectively (see Figure 4.1). The damping effects are modeled as viscous forces which are present both in longitudinal and swing motion of the pendulum. Since the motion of the pendulum is assumed to be planar and takes place in the vertical plane, the system has two degrees of freedom (2 DoF).

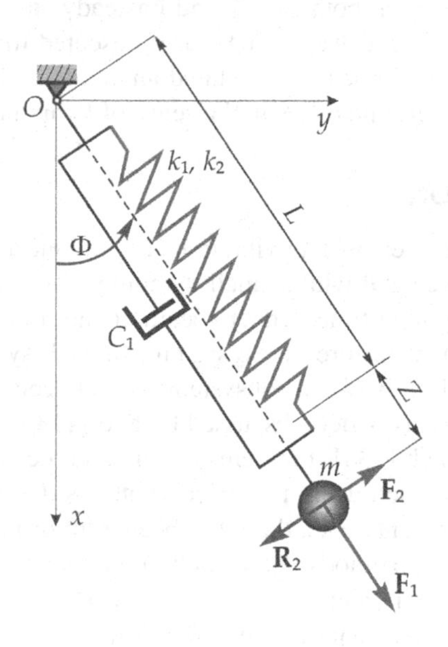

FIGURE 4.1 The planar spring pendulum.

The equations of motion are derived using Lagrange's formalism. The Lagrangian of the system has the form

$$L = g\,m(L_0 + Z(t))\cos(\Phi(t)) - \frac{1}{2}k_1 Z(t)^2 - \frac{1}{4}k_2 Z(t)^4 + \frac{1}{2}m\left((L_0 + Z(t))^2\dot{\Phi}(t)^2 + \dot{Z}(t)^2\right) \tag{4.1}$$

where m is mass of the pendulum, L_0 stands for the length of the unstretched spring, k_1 and k_2 are spring elastic constants and g is the Earth's acceleration. The time-dependent spring elongation $Z(t)$ and the angle $\Phi(t)$ are assumed as the generalized coordinates.

The pendulum reaches the stable equilibrium position at $\Phi \equiv \Phi_e = 0$ and $Z \equiv Z_e$ where Z_e is the static elongation of the spring which satisfies the following equilibrium equation

$$k_1 Z_e + k_2 Z_e^3 = mg. \tag{4.2}$$

The generalized forces include external excitations $\vec{F}_1$ and $\vec{F}_2$ and damping forces $\vec{R}_1$ and $\vec{R}_2$ of magnitudes $C_1 \dot{Z}(t)$ and $C_2 \dot{\Phi}(t)$, where C_1 and C_2 are damping coefficients. We have

$$Q_Z = F_1 \sin(\Omega_1 t) - C_1 \dot{Z}(t), \tag{4.3}$$

$$Q_\Phi = F_2 (L_0 + Z(t)) \sin(\Omega_2 t) - C_2 (L_0 + Z(t)) \dot{\Phi}(t). \tag{4.4}$$

Introducing Lagrangian (4.1) and the generalized forces (4.3)–(4.4) into Lagrange's equations of the second kind, the following equations of motion are obtained

$$\begin{aligned} &m \ddot{Z}(t) + C_1 \dot{Z}(t) + k_1 Z(t) + k_2 Z(t)^3 = \\ &F_1 \cos(t\, \Omega_1) + g\, m \cos(\Phi(t)) + m (L_0 + Z(t)) \dot{\Phi}(t)^2, \end{aligned} \tag{4.5}$$

$$\begin{aligned} &(L_0 + Z(t)) \left(m (L_0 + Z(t)) \ddot{\Phi}(t) + \dot{\Phi}(t) \left(C_2 + 2m\, \dot{Z}(t) \right) + \right. \\ &\left. + g\, m \sin(\Phi(t)) - F_2 \cos(t\, \Omega_2) \right) = 0. \end{aligned} \tag{4.6}$$

The initial-value problem includes the equations of motion (4.5)–(4.6) and initial conditions for generalized coordinates and their first derivatives, i.e. we have

$$Z(0) = Z_0, \quad \dot{Z}(0) = V_0, \quad \Phi(0) = \Phi_0, \quad \dot{\Phi}(0) = \omega_0 \tag{4.7}$$

where Z_0, V_0, Φ_0, ω_0 are known.

It is convenient to transform the initial value problem (4.5)–(4.7) to its counterpart dimensionless form. For this purpose, we define the following dimensionless frequencies

$$w = \frac{\omega_2}{\omega_1}, \qquad p_1 = \frac{\Omega_1}{\omega_1}, \qquad p_2 = \frac{\Omega_2}{\omega_1},$$

where $\omega_1 = \sqrt{k_1/m}$, $\omega_2 = \sqrt{g/L}$ are frequencies of the free vibrations of the longitudinal and swing motion. $L = L_0 + Z_e$ is the length of the pendulum at the static equilibrium position which is considered as a characteristic dimension, and we take

$$c_1 = \frac{C_1}{m\omega_1}, \quad c_2 = \frac{C_2}{mL^2\omega_1}, \quad f_1 = \frac{F_1}{mL\omega_1^2}, \quad f_2 = \frac{F_2}{mL^2\omega_1^2}, \quad \alpha = \frac{k_2 L^2}{m\omega_1^2}.$$

The new dimensionless generalized coordinates $z(\tau) = \frac{Z(t) - Z_e}{L}$ and $\varphi(\tau)$ are functions of the dimensionless time $\tau = t\, \omega_1$. Introducing all the above definitions, the governing equations take the following form

$$\begin{aligned} &\ddot{z}(\tau) + c_1 \dot{z}(\tau) + z(\tau) + 3\alpha z_e z(\tau)^2 + 3\alpha z_e^2 z(\tau) + \alpha z(\tau)^3 + \\ &w^2 (1 - \cos(\varphi(\tau))) - (z(\tau) + 1) \dot{\varphi}(\tau)^2 = f_1 \cos(p_1 \tau), \end{aligned} \tag{4.8}$$

$$(z(\tau)+1)\left((z(\tau)+1)\ddot{\varphi}(\tau)+\dot{\varphi}(\tau)(c_2+2\dot{z}(\tau))+w^2\sin(\varphi(\tau))\right)= \\ (z(\tau)+1)f_2\cos(p_2\tau). \tag{4.9}$$

Observe that the dimensionless static elongation $z_e = Z_e/L$ of the spring must satisfy the algebraic equilibrium equation

$$z_e + \alpha z_e^3 = w^2. \tag{4.10}$$

Equation (4.9) has, among other, one trivial solution $z(\tau) = -1$ which is omitted in further analysis, because it has no physical meaning.

Vibrations of the pendulum are investigated in the vicinity of the stable equilibrium position, so the trigonometric functions can be approximated by their power expansions of the form

$$\sin\varphi = \varphi - \frac{\varphi^3}{6}, \qquad \cos\varphi = 1 - \frac{\varphi^2}{2}, \tag{4.11}$$

which limits the angle φ about to $\pi/6$ with the precision of 4 significant digits.

Taking the above considerations into account, the equations of motion take the form

$$\ddot{z}(\tau)+c_1\dot{z}(\tau)+z(\tau)+\alpha z(\tau)^3+3\alpha z_e z(\tau)^2+3\alpha z_e^2 z(\tau)+ \\ \frac{1}{2}w^2\varphi(\tau)^2-(z(\tau)+1)\dot{\varphi}(\tau)^2 = f_1\cos(p_1\tau), \tag{4.12}$$

$$(z(\tau)+1)\ddot{\varphi}(\tau)+\dot{\varphi}(\tau)(c_2+2\dot{z}(\tau))+ \\ w^2\left(\varphi(\tau)-\frac{\varphi(\tau)^3}{6}\right) = f_2\cos(p_2\tau), \tag{4.13}$$

with the initial conditions

$$z(0)=z_0, \quad \dot{z}(0)=v_0, \quad \varphi(0)=\varphi_0, \quad \dot{\varphi}(0)=\beta_0. \tag{4.14}$$

Therefore, the initial value problem (4.12)–(4.14) containing coupled nonlinear differential equations of the second-order and the algebraic equation (4.10) is further investigated.

4.3 COMPLEX REPRESENTATION

Let us introduce the phase space coordinates $v(\tau)$ and $\beta(\tau)$ and rewrite the initial value problem (4.12)–(4.14) in the form of four differential equations of the first-order

$$\dot{z}(\tau) = v(\tau), \tag{4.15}$$

$$\dot{\varphi}(\tau) = \beta(\tau), \tag{4.16}$$

$$\dot{v}(\tau)+c_1v(\tau)+z(\tau)+\alpha z(\tau)^3+3\alpha z_e z(\tau)^2+3\alpha z_e^2 z(\tau)+\frac{1}{2}w^2\varphi(\tau)^2- \\ (z(\tau)+1)\beta(\tau)^2 = f_1\cos(p_1\tau), \tag{4.17}$$

$$(z(\tau)+1)\dot{\beta}(\tau)+\beta(\tau)(c_2+2v(\tau))+w^2\left(\varphi(\tau)-\frac{\varphi(\tau)^3}{6}\right)=f_2\cos(p_2\tau), \quad (4.18)$$

with the initial conditions

$$z(0)=z_0, \quad v(0)=v_0, \quad \varphi(0)=\varphi_0, \quad \beta(0)=\beta_0. \quad (4.19)$$

Then, following the idea proposed by Mavevitch and Musenko [136], the new complex-valued functions

$$\Psi_z(\tau)=v(\tau)+\mathrm{i}z(\tau), \qquad \bar{\Psi}_z(\tau)=v(\tau)-\mathrm{i}z(\tau), \quad (4.20)$$

$$\Psi_\varphi(\tau)=\beta(\tau)+\mathrm{i}w\varphi(\tau), \qquad \bar{\Psi}_\varphi(\tau)=\beta(\tau)-\mathrm{i}w\varphi(\tau), \quad (4.21)$$

are defined. The phase space coordinates $z(\tau)$, $v(\tau)$, $\varphi(\tau)$ and $\beta(\tau)$ can be expressed by the above complex functions and introduced to the equations of motion. This allows one to convert the problem (4.15)–(4.19) to its counterpart complex form

$$-\frac{1}{2}\mathrm{i}\left(\dot{\Psi}_z-\dot{\bar{\Psi}}_z\right)=\frac{1}{2}\left(\bar{\Psi}_z+\Psi_z\right), \quad (4.22)$$

$$-\frac{\mathrm{i}\left(\dot{\Psi}_\varphi-\dot{\bar{\Psi}}_\varphi\right)}{2w}=\frac{1}{2}\left(\bar{\Psi}_\varphi+\Psi_\varphi\right), \quad (4.23)$$

$$\begin{aligned}&\frac{1}{2}c_1\left(\bar{\Psi}_z+\Psi_z\right)-\frac{3}{4}\alpha z_e\left(\Psi_z-\bar{\Psi}_z\right)^2-\frac{3}{2}\mathrm{i}\alpha z_e^2\left(\Psi_z-\bar{\Psi}_z\right)-\\&\frac{1}{8}\left(\Psi_\varphi-\bar{\Psi}_\varphi\right)^2+\frac{1}{8}\mathrm{i}\alpha\left(\Psi_z-\bar{\Psi}_z\right)^3-\frac{1}{4}\left(\bar{\Psi}_\varphi+\Psi_\varphi\right)^2\left(1-\frac{1}{2}\mathrm{i}\left(\Psi_z-\bar{\Psi}_z\right)\right)-\\&\frac{1}{2}\mathrm{i}\left(\Psi_z-\bar{\Psi}_z+\frac{1}{2}\left(\dot{\Psi}_z+\dot{\bar{\Psi}}_z\right)\right)=f_1\cos\left(p_1\tau\right),\end{aligned} \quad (4.24)$$

$$\begin{aligned}&\frac{1}{2}\left(\bar{\Psi}_\varphi+\Psi_\varphi\right)\left(\bar{\Psi}_z+c_2+\Psi_z\right)+w^2\left(-\frac{\mathrm{i}(\Psi_\varphi-\bar{\Psi}_\varphi)^3}{48w^3}-\frac{\mathrm{i}\left(\Psi_\varphi-\bar{\Psi}_\varphi\right)}{2w}\right)+\\&\frac{1}{2}\left(1-\frac{1}{2}\mathrm{i}\left(\Psi_z-\bar{\Psi}_z\right)\right)\left(\dot{\Psi}_\varphi+\dot{\bar{\Psi}}_\varphi\right)=f_2\cos\left(p_2\tau\right),\end{aligned} \quad (4.25)$$

$$\begin{aligned}\Psi_z(0)&=v_0+\mathrm{i}z_0, \qquad \Psi_\varphi(0)=\beta_0+\mathrm{i}w\varphi_0,\\ \bar{\Psi}_z(0)&=v_0-\mathrm{i}z_0, \qquad \bar{\Psi}_\varphi(0)=\beta_0-\mathrm{i}w\varphi_0.\end{aligned} \quad (4.26)$$

Equations (4.22)–(4.25) can be written mutually as conjugated sets of two equations. For this purpose, time derivatives of the complex conjugated functions $\dot{\bar{\Psi}}_z$ and $\dot{\bar{\Psi}}_\varphi$ are determined from equations (4.22)–(4.23):

$$\dot{\bar{\Psi}}_z=\dot{\Psi}_z-\mathrm{i}\left(\bar{\Psi}_z+\Psi_z\right), \quad (4.27)$$

$$\dot{\bar{\Psi}}_\varphi=\dot{\Psi}_\varphi-\mathrm{i}\left(w\bar{\Psi}_\varphi+w\Psi_\varphi\right), \quad (4.28)$$

and introduced into equations (4.24)–(4.25). In this way, we obtain the set of two equations

$$\begin{gathered}\frac{1}{2}c_1\left(\bar{\Psi}_z+\Psi_z\right)-\frac{3}{4}\alpha z_e\left(\Psi_z-\bar{\Psi}_z\right)^2-\frac{3}{2}\mathrm{i}\alpha z_e^2\left(\Psi_z-\bar{\Psi}_z\right)-\\ \frac{1}{8}\left(\Psi_\varphi-\bar{\Psi}_\varphi\right)^2+\frac{1}{8}\mathrm{i}\alpha\left(\Psi_z-\bar{\Psi}_z\right)^3+\frac{1}{2}\left(2\dot{\Psi}_z+\mathrm{i}\left(-\bar{\Psi}_z-\Psi_z\right)\right)-\\ \frac{1}{4}\left(\bar{\Psi}_\varphi+\Psi_\varphi\right)^2\left(1-\frac{1}{2}\mathrm{i}\left(\Psi_z-\bar{\Psi}_z\right)\right)-\frac{1}{2}\mathrm{i}\left(\Psi_z-\bar{\Psi}_z\right)=f_1\cos\left(p_1\tau\right),\end{gathered}\tag{4.29}$$

$$\begin{gathered}\frac{1}{2}\left(\bar{\Psi}_\varphi+\Psi_\varphi\right)\left(\bar{\Psi}_z+c_2+\Psi_z\right)+w^2\left(-\frac{\mathrm{i}\left(\Psi_\varphi-\bar{\Psi}_\varphi\right)^3}{48w^3}-\frac{\mathrm{i}\left(\Psi_\varphi-\bar{\Psi}_\varphi\right)}{2w}\right)+\\ \frac{1}{2}\left(1-\frac{1}{2}\mathrm{i}\left(\Psi_z-\bar{\Psi}_z\right)\right)\left(2\dot{\Psi}_\varphi+\mathrm{i}\left(-w\bar{\Psi}_\varphi-w\Psi_\varphi\right)\right)=f_2\cos\left(p_2\tau\right).\end{gathered}\tag{4.30}$$

Determining derivatives $\dot{\Psi}_z$ and $\dot{\Psi}_\varphi$ and introducing them to equations (4.24)–(4.25) allows one to get conjugated equations of the system (4.29)–(4.30). They are not explicitly written here for greater clarity.

Afterward, the following exponential forms of the unknown functions are introduced

$$\begin{aligned}\Psi_z(\tau)&=\psi_z(\tau)e^{\mathrm{i}\tau}, &\bar{\Psi}_z(\tau)&=\bar{\psi}_z(\tau)e^{-\mathrm{i}\tau},\\ \Psi_\varphi(\tau)&=w\psi_\varphi(\tau)e^{\mathrm{i}w\tau}, &\bar{\Psi}_\varphi(\tau)&=w\bar{\psi}_\varphi(\tau)e^{-\mathrm{i}w\tau}.\end{aligned}\tag{4.31}$$

Next, taking into account (4.31) in equations (4.29)–(4.30) yields the following equations

$$\begin{gathered}\frac{1}{2}c_1e^{-2\mathrm{i}\tau}\left(\bar{\psi}_z+e^{2\mathrm{i}\tau}\psi_z\right)-\frac{3}{4}\alpha e^{-3\mathrm{i}\tau}z_e\left(\bar{\psi}_z-e^{2\mathrm{i}\tau}\psi_z\right)^2-\\ \frac{3}{2}\mathrm{i}\alpha z_e^2\left(\psi_z-e^{-2\mathrm{i}\tau}\bar{\psi}_z\right)+\dot{\psi}_z-\frac{3}{8}w^2e^{-\mathrm{i}(\tau+2\tau w)}\left(\bar{\psi}_\varphi^2+\psi_\varphi^2e^{4\mathrm{i}\tau w}\right)+\\ \frac{1}{8}\mathrm{i}w^2e^{-2\mathrm{i}\tau(w+1)}\left(-\bar{\psi}_z+e^{2\mathrm{i}\tau}\psi_z\right)\left(\psi_{\varphi s}^2+\psi_\varphi^2e^{4\mathrm{i}\tau w}\right)\\ +\frac{1}{4}\mathrm{i}e^{-2\mathrm{i}\tau}w^2\psi_\varphi\psi_{\varphi s}\left(\mathrm{i}e^{\mathrm{i}\tau}+e^{2\mathrm{i}\tau}\psi_z-\psi_{zs}\right)+\\ \frac{1}{8}\mathrm{i}\alpha e^{-4\mathrm{i}\tau}\left(-\psi_{zs}+e^{2\mathrm{i}\tau}\psi_z\right)^3=f_1e^{-\mathrm{i}\tau}\cos\left(p_1\tau\right),\end{gathered}\tag{4.32}$$

$$\begin{gathered}\frac{1}{2}c_2w\left(\psi_\varphi+\psi_{\varphi s}e^{-2\mathrm{i}\tau w}\right)-\frac{1}{48}\mathrm{i}w^2e^{-4\mathrm{i}\tau w}\left(-\psi_{\varphi s}+\psi_\varphi e^{2\mathrm{i}\tau w}\right)^3+\dot{\psi}_\varphi+\\ \frac{1}{4}e^{\mathrm{i}\tau}w(w+2)\psi_\varphi\psi_z-\frac{1}{4}(w-2)w\psi_{\varphi s}e^{\mathrm{i}\tau(1-2w)}\psi_z-\\ \frac{1}{2}\mathrm{i}e^{-\mathrm{i}\tau}w\dot{\psi}_\varphi\left(-\psi_{zs}+e^{2\mathrm{i}\tau}\psi_z\right)-\frac{1}{4}e^{-\mathrm{i}\tau}(w-2)w\psi_\varphi\psi_{zs}+\\ \frac{1}{4}w(w+2)\psi_{\varphi s}e^{-\mathrm{i}(\tau+2\tau w)}\psi_{zs}=f_2e^{-\mathrm{i}\tau w}\cos\left(p_2\tau\right),\end{gathered}\tag{4.33}$$

with the initial conditions

$$\psi_z(0) = v_0 + \mathrm{i}z_0, \qquad \psi_\varphi(0) = \frac{\beta_0}{w} + \mathrm{i}\varphi_0,$$
$$\bar{\psi}_z(0) = v_0 - \mathrm{i}z_0, \qquad \bar{\psi}_\varphi(0) = \frac{\beta_0}{w} - \mathrm{i}\varphi_0. \tag{4.34}$$

The functions $\bar{\psi}_z$ and $\bar{\psi}_\varphi$ should satisfy also the conjugated equations to equations (4.32)–(4.33).

However, it should be noted that the generalized coordinates have the form

$$z = \frac{\psi_z(\tau)e^{\mathrm{i}\tau} - \bar{\psi}_z(\tau)e^{-\mathrm{i}\tau}}{2\mathrm{i}}, \qquad \varphi = \frac{\psi_\varphi(\tau)e^{\mathrm{i}w\tau} - \bar{\psi}_\varphi(\tau)e^{-\mathrm{i}w\tau}}{2w\mathrm{i}}, \tag{4.35}$$

according to expressions (4.20), (4.21) and (4.31).

4.4 ASYMPTOTIC SOLUTION

Though the asymptotic analysis will be performed for equations (4.32)–(4.33), the parallel calculations can be performed for the complex conjugated system. The initial value problem (4.32)–(4.34) can be efficiently solved using the multiple scale method. According to the MSM, the time scales should be introduced instead of the actual time τ. In what follows, we adopt three time scales

$$\tau_0 = \tau, \qquad \tau_1 = \varepsilon\tau, \qquad \tau_2 = \varepsilon^2\tau, \tag{4.36}$$

where $0 < \varepsilon \ll 1$ is the dimensionless small (perturbation) parameter.

The solutions are searched in the form of the following power series

$$\psi_z(\tau;\varepsilon) = \sum_{k=1}^{3} \varepsilon^k \xi_{zk}(\tau_0, \tau_1, \tau_2) + O\left(\varepsilon^4\right), \tag{4.37}$$

$$\psi_\varphi(\tau;\varepsilon) = \sum_{k=1}^{3} \varepsilon^k \xi_{\varphi k}(\tau_0, \tau_1, \tau_2) + O\left(\varepsilon^4\right). \tag{4.38}$$

Due to the dependence of unknown functions on three time scales, the derivative operator takes the form

$$\frac{d}{d\tau} = \sum_{k=0}^{2} \varepsilon^k \frac{\partial}{\partial \tau_k} = \frac{\partial}{\partial \tau_0} + \varepsilon\frac{\partial}{\partial \tau_1} + \varepsilon^2\frac{\partial}{\partial \tau_2}. \tag{4.39}$$

We assume that the amplitudes of external forces acting on the pendulum are small. Moreover, slight damping and small static elongation of the spring are assumed. Therefore, we assume that the appropriate parameters are of the order of some powers of the small parameter, i.e. we take

$$c_1 = \varepsilon^2\tilde{c}_1, \quad c_2 = \varepsilon^2\tilde{c}_2, \quad f_1 = \varepsilon^3\tilde{f}_1, \quad f_2 = \varepsilon^3\tilde{f}_2, \quad z_e = \varepsilon^2\tilde{z}_e, \tag{4.40}$$

where parameters $\tilde{c}_1, \tilde{c}_2, \tilde{f}_1, \tilde{f}_2, \tilde{z}_e$ are of the order of 1. It should be noted that we do not assume the smallness of the nonlinearity parameter α.

Let us examine the case of two primary resonances appearing simultaneously, i.e. $p_1 \approx 1$, $p_2 \approx w$. To deal with this case, we introduce detuning parameters σ_1 and σ_2 which play a role of a measure of distance from strict resonance:

$$p_1 = 1 + \sigma_1, \qquad p_2 = w + \sigma_2. \tag{4.41}$$

To study motion near the resonance, we assume that the detuning parameters are also small, so we take $\sigma_1 = \varepsilon^2 \tilde{\sigma}_1$ and $\sigma_2 = \varepsilon^2 \tilde{\sigma}_2$.

Introducing (4.40), (4.41) into (4.32)–(4.33), expressing unknown functions by their power expansions (4.37)–(4.38) and replacing the ordinary derivatives by the differential operator (4.39) we obtain two equations with the small parameter ε of various powers. Since these equations should be satisfied for any value of the small parameter, after splitting them with respect to the powers of ε we get.

(i) Equations of order "1" (coefficient at ε^1)

$$\frac{\partial \xi_{z1}}{\partial \tau_0} = 0, \tag{4.42}$$

$$\frac{\partial \xi_{\varphi 1}}{\partial \tau_0} = 0. \tag{4.43}$$

(ii) Equations of order "2" (coefficient at ε^2)

$$\begin{gathered}\frac{\partial \xi_{z2}}{\partial \tau_0} + \frac{\partial \xi_{z1}}{\partial \tau_1} - \frac{1}{4} e^{-\mathrm{i}\tau_0} w^2 \xi_{\varphi 1} \bar{\xi}_{\varphi 1} - \\ \frac{3}{8} e^{-\mathrm{i}\tau_0 - 2\mathrm{i}w\tau_0} w^2 \left(e^{4\mathrm{i}w\tau_0} \xi_{\varphi 1}^2 + \bar{\xi}_{\varphi 1}^2\right) = 0,\end{gathered} \tag{4.44}$$

$$\begin{gathered}\frac{\partial \xi_{\varphi 2}}{\partial \tau_0} + \frac{\partial \xi_{\varphi 1}}{\partial \tau_1} - \frac{1}{2} \mathrm{i} e^{-\mathrm{i}\tau_0} \left(-\bar{\xi}_{z1} + e^{2\mathrm{i}\tau_0} \xi_{z1}\right) \frac{\partial \xi_{\varphi 1}}{\partial \tau_0} + \\ \frac{1}{4} e^{\mathrm{i}\tau_0} (w+2) \xi_{\varphi 1} \xi_{z1} - \frac{1}{4} (w-2) \bar{\xi}_{\varphi 1} e^{\mathrm{i}\tau_0(1-2w)} \xi_{z1} - \\ \frac{1}{4} e^{-\mathrm{i}\tau_0} (w-2) \xi_{\varphi 1} \bar{\xi}_{z1} + \frac{1}{4} (w+2) \bar{\xi}_{\varphi 1} e^{-\mathrm{i}(\tau_0 + 2\tau_0 w)} \bar{\xi}_{z1} = 0.\end{gathered} \tag{4.45}$$

(iii) Equations of order "3" (coefficient at ε^3)

$$\begin{gathered}\frac{\partial \xi_{z3}}{\partial \tau_0} + \frac{\partial \xi_{z2}}{\partial \tau_1} + \frac{\partial \xi_{z1}}{\partial \tau_2} + \frac{1}{2} \tilde{c}_1 \left(\xi_{z1} + e^{-2\mathrm{i}\tau_0} \bar{\xi}_{z1}\right) - \frac{3}{2} \mathrm{i} \tilde{\alpha} \tilde{z}_e^2 \left(\xi_{z1} - e^{-2\mathrm{i}\tau_0} \bar{\xi}_{z1}\right) - \\ \frac{3}{4} \tilde{\alpha} e^{-3\mathrm{i}\tau_0} \tilde{z}_e \left(\bar{\xi}_{z1} - e^{2\mathrm{i}\tau_0} \xi_{z1}\right)^2 + \frac{1}{8} \mathrm{i} \tilde{\alpha} e^{-4\mathrm{i}\tau_0} \left(-\bar{\xi}_{z1} + e^{2\mathrm{i}\tau_0} \xi_{z1}\right)^3 - \\ \frac{1}{2} \tilde{f}_1 e^{\mathrm{i}\tilde{\sigma}_1 \tau_2} - \frac{1}{2} \tilde{f}_1 e^{-2\mathrm{i}\tau_0 - \mathrm{i}\tilde{\sigma}_1 \tau_2} - \frac{1}{4} e^{-\mathrm{i}\tau_0} w^2 \left(\xi_{\varphi 1} \bar{\xi}_{\varphi 2} + \xi_{\varphi 2} \bar{\xi}_{\varphi 1}\right) -\end{gathered}$$

$$\frac{3}{4}w^2e^{-i(\tau_0+2\tau_0 w)}\left(\bar{\xi}_{\varphi1}\bar{\xi}_{\varphi2}+\xi_{\varphi1}\xi_{\varphi2}e^{4i\tau_0 w}\right)+\frac{1}{4}iw^2|\xi_{\varphi1}|^2\left(\xi_{z1}-e^{-2i\tau_0}\bar{\xi}_{z1}\right)+$$
$$\frac{1}{8}iw^2e^{-2i\tau_0(w+1)}\left(-\bar{\xi}_{z1}+e^{2i\tau_0}\xi_{z1}\right)\left(\bar{\xi}^2_{\varphi1}+\xi^2_{\varphi1}e^{4i\tau_0 w}\right)=0, \tag{4.46}$$

$$\frac{\partial\xi_{\varphi1}}{\partial\tau_2}+\frac{\partial\xi_{\varphi2}}{\partial\tau_1}+\frac{\partial\xi_{\varphi3}}{\partial\tau_0}-\frac{1}{2}ie^{-i\tau_0}\left(\left(e^{2i\tau_0}\xi_{z1}-\bar{\xi}_{z1}\right)\left(\frac{\partial\xi_{\varphi1}}{\partial\tau_1}+\frac{\partial\xi_{\varphi2}}{\partial\tau_0}\right)+\right.$$
$$\left.\left(e^{2i\tau_0}\xi_{z2}-\bar{\xi}_{z2}\right)\frac{\partial\xi_{\varphi1}}{\partial\tau_0}\right)+\frac{\tilde{c}_2}{2}\left(\xi_{\varphi1}+\bar{\xi}_{\varphi1}e^{-2i\tau_0 w}\right)-\frac{1}{2w}\tilde{f}_2e^{-2i\tau_0 w-i\tilde{\sigma}_2\tau_2}-$$
$$\frac{1}{2w}\tilde{f}_2e^{i\tilde{\sigma}_2\tau_2}-\frac{1}{48}iwe^{-4i\tau_0 w}\left(\xi_{\varphi1}e^{2i\tau_0 w}-\bar{\xi}_{\varphi1}\right)^3+$$
$$\frac{1}{4}e^{i\tau_0}(w+2)\left(\xi_{\varphi2}\xi_{z1}+\xi_{\varphi1}\xi_{z2}\right)-\frac{1}{4}(w-2)e^{i\tau_0(1-2w)}\left(\bar{\xi}_{\varphi2}\xi_{z1}+\bar{\xi}_{\varphi1}\xi_{z2}\right)- \tag{4.47}$$
$$\frac{1}{4}e^{-i\tau_0}(w-2)\left(\xi_{\varphi2}\bar{\xi}_{z1}+\xi_{\varphi1}\bar{\xi}_{z2}\right)+$$
$$\frac{1}{4}(w+2)e^{-i(\tau_0+2\tau_0 w)}\left(\bar{\xi}_{\varphi2}\bar{\xi}_{z1}+\bar{\xi}_{\varphi1}\bar{\xi}_{z2}\right)=0.$$

We have omitted the higher-order terms, i.e. coefficients of a higher order than ε^3.

The equations of the first approximation (4.42)–(4.43) imply that functions ξ_{z1} and $\xi_{\varphi1}$ do not depend on τ_0, so $\xi_{z1}=\xi_{z1}(\tau_1,\tau_2)$ and $\xi_{\varphi1}=\xi_{\varphi1}(\tau_1,\tau_2)$.

If there is no internal resonance $1-2w=0$, the secular terms will be removed from the equations of the second approximation when the following conditions are satisfied

$$\frac{\partial\xi_{z1}(\tau_1,\tau_2)}{\partial\tau_1}=0, \tag{4.48}$$

$$\frac{\partial\xi_{\varphi1}(\tau_1,\tau_2)}{\partial\tau_1}=0. \tag{4.49}$$

Based on the form of the secular terms (4.48) and (4.49) of the second approximation equations, we conclude that functions ξ_{z1} and ξ_{φ_1} depend only on the slowest time scale τ_2, so $\xi_{z1}=\xi_{z1}(\tau_2)$ and $\xi_{\varphi1}=\xi_{\varphi1}(\tau_2)$.

The particular integral of equations (4.44)–(4.45) reads

$$\xi_{z2}=\frac{3ie^{-i(1-2w)\tau_0}w^2\xi^2_{\varphi1}}{8(1-2w)}+\frac{1}{4}ie^{-i\tau_0}w^2\xi_{\varphi1}\bar{\xi}_{\varphi1}+\frac{3ie^{-i(1+2w)\tau_0}w^2\bar{\xi}^2_{\varphi1}}{8(1+2w)}, \tag{4.50}$$

$$\xi_{\varphi2}=\frac{1}{4}ie^{i\tau_0}(2+w)\xi_{z1}\xi_{\varphi1}+\frac{1}{4}ie^{-i\tau_0}(w-2)\bar{\xi}_{z1}\xi_{\varphi1}+$$
$$\frac{ie^{i(1-2w)\tau_0}(-2+w)\xi_{z1}\bar{\xi}_{\varphi1}}{-4+8w}-\frac{ie^{-i(1+2w)\tau_0}(2+w)\bar{\xi}_{z1}\bar{\xi}_{\varphi1}}{4(1+2w)}. \tag{4.51}$$

The solutions (4.50)–(4.51) are then introduced into equations of the third-order (4.46)–(4.47). This substitution awakes the terms leading to the infinite growth of the solution and these terms should be eliminated from the equations. Therefore, assuming that the system oscillates far from the internal resonance of the form $1-2w=0$, the requirement that the solutions should be limited yields the solvability conditions

$$\frac{\partial \xi_{z1}}{\partial \tau_2}+\frac{1}{2}\tilde{c}_1\xi_{z1}-\frac{3}{2}\mathrm{i}\tilde{\alpha}\tilde{z}_e^2\xi_{z1}-\frac{1}{2}\tilde{f}_1 e^{\mathrm{i}\tilde{\sigma}_1\tau_2}+ \frac{3\mathrm{i}\left(w^2-1\right)w^2\left|\xi_{\varphi 1}\right|^2\xi_{z1}}{4-16w^2}-\frac{3}{8}\mathrm{i}\tilde{\alpha}\xi_{z1}\left|\xi_{\varphi 1}\right|^2=0, \tag{4.52}$$

$$w\frac{\partial \xi_{\varphi 1}}{\partial \tau_2}+\frac{1}{2}\tilde{c}_2 w\xi_{\varphi 1}-\frac{1}{2}\tilde{f}_2 e^{\mathrm{i}\tilde{\sigma}_2\tau_2}+\frac{\mathrm{i}\left(8w^4-7w^2-1\right)w^2\xi_{\varphi 1}\left|\xi_{\varphi 1}\right|^2}{64w^2-16}- \frac{1}{4}\mathrm{i}w^2\xi_{\varphi 1}\left|\xi_{z1}\right|^2+\frac{\mathrm{i}\left(w^2+2\right)w^2\xi_{\varphi 1}\left|\xi_{z1}\right|^2}{16w^2-4}=0. \tag{4.53}$$

After considering the solutions (4.50) and (4.51), and the solvability conditions (4.52) and (4.53), the equations of the third approximation (4.46)–(4.47) have limited (physical) solutions of the following form

$$\begin{aligned}\xi_{z3}=&\frac{1}{4}\mathrm{i}e^{-\mathrm{i}(\tau_2\tilde{\sigma}_1+2\tau_0)}\tilde{f}_1-\frac{3}{4}\mathrm{i}e^{\mathrm{i}\tau_0}\tilde{\alpha}\tilde{z}_e\xi_{z1}^2-\frac{1}{16}e^{2\mathrm{i}\tau_0}\tilde{\alpha}\xi_{z1}^3-\\ &\frac{1}{4}\mathrm{i}e^{-2\mathrm{i}\tau_0}\tilde{c}_1\bar{\xi}_{z1}+\frac{3}{4}e^{-2\mathrm{i}\tau_0}\tilde{\alpha}\tilde{z}_e^2\bar{\xi}_{z1}-\frac{3}{2}\mathrm{i}e^{-\mathrm{i}\tau_0}\tilde{\alpha}\tilde{z}_e\left|\xi_{z1}\right|^2+\frac{1}{4}\mathrm{i}e^{-3\mathrm{i}\tau_0}\tilde{\alpha}\tilde{z}_e\bar{\xi}_{z1}^2+\\ &\frac{3}{16}e^{-2\mathrm{i}\tau_0}\tilde{\alpha}\xi_{z1}\bar{\xi}_{z1}^2-\frac{1}{32}e^{-4\mathrm{i}\tau_0}\tilde{\alpha}\bar{\xi}_{z1}^3+\frac{3e^{2\mathrm{i}w\tau_0}w(1+w)^2\xi_{z1}\xi_{\varphi 1}^2}{16+32w}+\\ &\frac{3e^{2\mathrm{i}(-1+w)\tau_0}\left(-1+w\right)w^2\bar{\xi}_{z1}\xi_{\varphi 1}^2}{-16+32w}+\frac{3e^{-2\mathrm{i}\tau_0}w^2\left(-1+w^2\right)\bar{\xi}_{z1}\left|\xi_{\varphi 1}\right|^2}{8\left(-1+4w^2\right)}+\\ &\frac{3e^{-2\mathrm{i}w\tau_0}\left(-1+w\right)^2w\xi_{z1}\bar{\xi}_{\varphi 1}^2}{16\left(-1+2w\right)}+\frac{3e^{-2\mathrm{i}(1+w)\tau_0}w^2\left(1+w\right)\bar{\xi}_{z1}\bar{\xi}_{\varphi 1}^2}{16\left(1+2w\right)},\end{aligned} \tag{4.54}$$

$$\begin{aligned}\xi_{\varphi 3}=&\frac{\mathrm{i}\tilde{c}_2\bar{\xi}_{\varphi 1}e^{-2\mathrm{i}\tau_0 w}}{4w}+\frac{\mathrm{i}\tilde{f}_2e^{-\mathrm{i}(\tilde{\sigma}_2\tau_2+2\tau_0 w)}}{4w^2}+\\ &\frac{\left(8w^4-7w^2-1\right)\left|\xi_{\varphi 1}\right|^2\bar{\xi}_{\varphi 1}e^{-2\mathrm{i}\tau_0 w}}{32\left(4w^2-1\right)}+\\ &\frac{\left(49w^2-1\right)\xi_{\varphi 1}^3e^{2\mathrm{i}\tau_0 w}}{96\left(4w^2-1\right)}+\frac{\left(49w^2-1\right)\bar{\xi}_{\varphi 1}^3e^{-4\mathrm{i}\tau_0 w}}{192\left(4w^2-1\right)}-\\ &\frac{e^{2\mathrm{i}\tau_0}\left(w^3+6w^2+11w+6\right)\xi_{\varphi 1}\xi_{z1}^2}{32w+16}+\frac{\left(w^2-5w+6\right)\bar{\xi}_{\varphi 1}e^{-2\mathrm{i}\tau_0(w-1)}\xi_{z1}^2}{32w-16}-\\ &-\frac{3\left(w^2-1\right)\bar{\xi}_{\varphi 1}e^{-2\mathrm{i}\tau_0 w}\left|\xi_{z1}\right|^2}{8\left(4w^2-1\right)}-\frac{e^{-2\mathrm{i}\tau_0}\left(w^3-6w^2+11w-6\right)\xi_{\varphi 1}\bar{\xi}_{z1}^2}{16\left(2w-1\right)}-\\ &\frac{\left(w^2+5w+6\right)\bar{\xi}_{\varphi 1}e^{-2\mathrm{i}\tau_0(w+1)}\bar{\xi}_{z1}^2}{16\left(2w+1\right)}.\end{aligned} \tag{4.55}$$

Let us express the complex functions appearing in the above solutions in polar form

$$\begin{aligned} \xi_{z1}(\tau_2) &= \tilde{a}_1(\tau_2)e^{i\delta_1(\tau_2)}, \quad \bar{\xi}_{z1}(\tau_2) = \tilde{a}_1(\tau_2)e^{-i\delta_1(\tau_2)}, \\ \xi_{\varphi 1}(\tau_2) &= \tilde{a}_2(\tau_2)e^{i\delta_2(\tau_2)}, \quad \bar{\xi}_{\varphi 1}(\tau_2) = \tilde{a}_2(\tau_2)e^{-i\delta_2(\tau_2)}, \end{aligned} \tag{4.56}$$

where $\tilde{a}_1(\tau_2) = \varepsilon a_1(\tau_2)$ and $\tilde{a}_2(\tau_2) = \varepsilon a_2(\tau_2)$, and functions $a_1(\tau_2)$, $a_2(\tau_2)$, $\delta_1(\tau_2)$ and $\delta_2(\tau_2)$ are real-valued. Introducing expressions (4.56) into solvability conditions (4.52)–(4.53) we get

$$\begin{aligned} &-\frac{1}{8}3i\tilde{\alpha}\tilde{a}_1^3e^{i\delta_1} + \frac{1}{2}\tilde{a}_1\tilde{c}_1e^{i\delta_1} + ia_1e^{i\delta_1}\frac{d\delta_1}{d\tau_2} - \frac{3i\tilde{a}_2^2\tilde{a}_1e^{i\delta_1}w^2\left(w^2-1\right)}{16w^2-4} - \\ &\frac{3}{2}i\tilde{\alpha}\tilde{a}_1e^{i\delta_1}\tilde{z}_e^2 + e^{i\delta_1}\frac{d\tilde{a}_1}{d\tau_2} - \frac{1}{2}\tilde{f}_1e^{i\tilde{\sigma}_1\tau_2} = 0, \end{aligned} \tag{4.57}$$

$$\begin{aligned} &\frac{1}{2}\tilde{a}_2\tilde{c}_2e^{i\delta_2}w + \frac{i\tilde{a}_2^3e^{i\delta_2}w^2\left(8w^4-7w^2-1\right)}{64w^2-16} + \frac{3i\tilde{a}_1^2\tilde{a}_2e^{i\delta_2}w^2\left(w^2-1\right)}{4-16w^2} + \\ &i\tilde{a}_2e^{i\delta_2}w\frac{d\delta_2}{d\tau_2} + e^{i\delta_2}w\frac{d\tilde{a}_2}{d\tau_2} - \frac{1}{2}\tilde{f}_2e^{i\tilde{\sigma}_2\tau_2} = 0. \end{aligned} \tag{4.58}$$

Equations (4.57)–(4.58) describe the behavior of functions $\tilde{a}_1(\tau_2)$, $\tilde{a}_2(\tau_2)$, $\delta_1(\tau_2)$ and $\delta_2(\tau_2)$ in the slowest time τ_2. It is convenient to write the equations for amplitudes and phases in time τ. For this purpose, we return to the original notations according to (4.40), and separate the real and imaginary parts of equations (4.57)–(4.58). Next, the separated equations are multiplied by the small parameters ε raised to the appropriate powers and then the differential operator (4.39) is employed. The latter procedure yields the following first-order ODEs:

$$\frac{da_1(\tau)}{d\tau} = -\frac{1}{2}c_1a_1(\tau) + \frac{f_1}{2}\cos\left(\sigma_1\tau - \delta_1(\tau)\right), \tag{4.59}$$

$$\begin{aligned} \frac{d\delta_1(\tau)}{d\tau} &= \sigma_1 - \frac{3}{8}\alpha a_1^2(\tau) - \frac{3}{2}\alpha z_e^2 + \frac{3a_2^2(\tau)w^2\left(w^2-1\right)}{4-16w^2} - \\ &\frac{1}{2a_1(\tau)}f_1\sin\left(\sigma_1\tau - \delta_1(\tau)\right), \end{aligned} \tag{4.60}$$

$$\frac{da_2(\tau)}{d\tau} = \frac{f_2\cos\left(\sigma_2\tau - \delta_2(\tau)\right)}{2w} - \frac{a_2(\tau)c_2}{2}, \tag{4.61}$$

$$\begin{aligned} \frac{d\delta_2(\tau)}{d\tau} &= \sigma_2 + \frac{a_2^2(\tau)w\left(8w^4-7w^2-1\right)}{64w^2-16} + \frac{3a_1^2(\tau)w\left(w^2-1\right)}{4-16w^2} - \\ &\frac{f_2\sin\left(\sigma_2\tau - \delta_2(\tau)\right)}{2a_2(\tau)w}. \end{aligned} \tag{4.62}$$

The set of equations (4.59)–(4.62) describe modulations of amplitudes and phases in time. These functions can be determined numerically for the specified initial conditions

$$a_1(0) = a_{10}, \qquad a_2(0) = a_{20}, \qquad \delta_1(0) = \delta_{10}, \qquad \delta_2(0) = \delta_{20}. \tag{4.63}$$

where the known values a_{10}, a_{20}, δ_{10}, δ_{20} must comply with the conditions (4.14).

We can now construct an approximate analytical form of the solution to the original system of equations of motion (4.12)–(4.13). For this purpose, the solutions of the second and third approximations (4.50), (4.51) and (4.54), (4.55) should be expressed by functions $a_1(\tau)$, $a_2(\tau)$, $\delta_1(\tau)$, $\delta_2(\tau)$ using formulas (4.56). Then, we substitute the solutions for ξ_{z1}, ξ_{z2}, ξ_{z3} and $\xi_{\varphi 1}$, $\xi_{\varphi 2}$, $\xi_{\varphi 3}$ into the power expansions (4.37)–(4.38). Finally, using expressions (4.35), we obtain

$$\begin{aligned}
&z(\tau) = a_1 \sin(\delta_1 + \tau) - \frac{1}{32}\alpha a_1^3 (6\sin(\delta_1+\tau) + \sin(3(\delta_1+\tau))) - \\
&\frac{1}{4} a_1 c_1 \cos(\delta_1+\tau) - \frac{3}{4}\alpha a_1 z_e^2 \sin(\delta_1+\tau) - \\
&\frac{1}{2}\alpha a_1^2 z_e (\cos(2(\delta_1+\tau)) + 3) - \frac{3a_2^2 a_1 w^2 (w^2-1)\sin(\delta_1+\tau)}{32w^2-8} - \\
&\frac{3a_2^2 w^2 \cos(2(\delta_2+\tau w))}{16w^2-4} + \frac{1}{4}a_2^2 w^2 + \\
&\frac{3a_2^2 a_1 w (w+1)\sin(\delta_1 + 2\delta_2 + \tau + 2\tau w)}{32w+16} - \\
&\frac{3a_2^2 a_1 (w-1) w \sin(\delta_1 - 2\delta_2 + \tau - 2\tau w)}{32w-16} + \frac{1}{4} f_1 \cos((\sigma_1+1)\tau),
\end{aligned} \tag{4.64}$$

$$\begin{aligned}
&\varphi(\tau) = -\frac{a_2 c_2 \cos(\delta_2+\tau w)}{4w} + \frac{a_2^3 (8w^4 - 7w^2 - 1)\sin(\delta_2+\tau w)}{32(4w^2-1)} + \\
&\frac{a_2^3 (49w^2-1)\sin(3(\delta_2+\tau w))}{192(4w^2-1)} + \\
&\frac{a_2(-3a_1^2(w^2-1) + 32w^2 - 8)\sin(\delta_2+\tau w)}{32w^2-8} - \\
&\frac{a_1^2 a_2 w (w^2 - 5w + 6)\sin(-2\delta_1 + \delta_2 + \tau(w-2))}{32w-16} - \\
&\frac{a_1^2 a_2 w (w^2 + 5w + 3)\sin(2\delta_1 + \delta_2 + \tau(w+2))}{32w+16} - \\
&\frac{3a_1^2 a_2 w \sin(2\delta_1 + \delta_2 + 2\tau + \tau w)}{16(2w+1)} + \\
&\frac{a_1 a_2 (w-2) w \cos(\delta_1 - \delta_2 + \tau + \tau(-w))}{4w-2} + \\
&\frac{a_1 a_2 w (w+2)\cos(\delta_1 + \delta_2 + \tau + \tau w)}{4w+2} + \frac{f_2 \cos(\tau(\sigma_2+w))}{4w^2},
\end{aligned} \tag{4.65}$$

where the functions $a_1(\tau)$, $a_2(\tau)$, $\delta_1(\tau)$, $\delta_2(\tau)$ satisfy modulation equations (4.59)–(4.62), and can be obtained numerically.

Let us take the following fixed parameters for: $f_2 = 0.0001$, $f_1 = 0.001$, $c_1 = 0.0005$, $c_2 = 0.002$, $w = 0.13$, $\sigma_1 = 0.001$, $\sigma_2 = -0.001$, $\alpha = 0.2$, and the initial conditions for the modulation $a_{10} = 0.1$, $a_{20} = 0.1$, $\delta_{10} = 0$, $\delta_{20} = 0$, which correspond to the conditions

$$z(0) = 0.000348, \quad \dot{z}(0) = 0.100036, \quad \varphi(0) = 0.003836, \quad \dot{\varphi}(0) = 0.012864.$$

The long time histories of the generalized coordinates obtained numerically based on the used initial value problem (4.12)–(4.14) are presented in Figure 4.2. Fast-changing vibrations are marked in gray, while the dark thick line depicts long-time amplitude modulation obtained numerically from equations (4.59)–(4.62). As can be seen from the diagram, this line perfectly reflects the amplitude of fast-changing oscillations. Long-lasting large changes in the vibration amplitude indicate an intense energy exchange between the system and the external excitation under the resonance conditions. The numerical results are obtained using the procedure *NDSolve* of the *Mathematica* software.

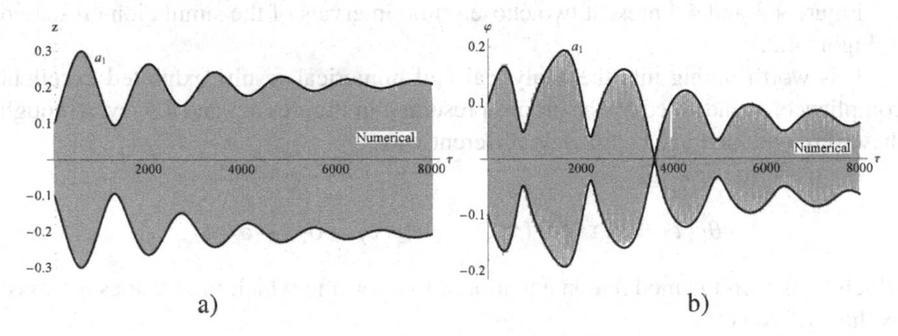

FIGURE 4.2 Time history of the longitudinal a) $z(\tau)$ and swing b) $\varphi(\tau)$ vibration for long-time simulation.

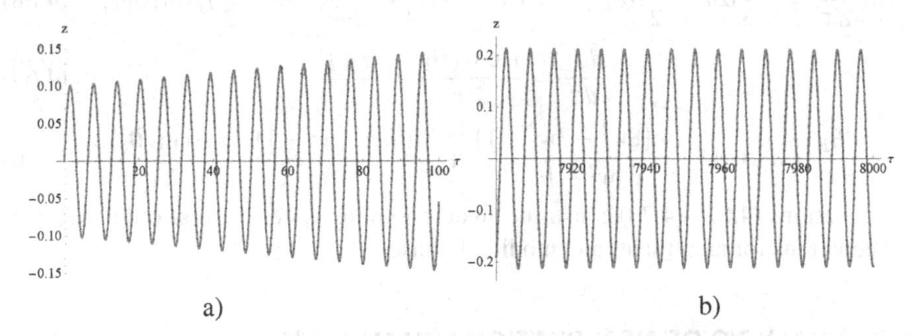

FIGURE 4.3 Time histories of the longitudinal vibration z for transient (a) and long-time (b) simulation obtained analytically (MSM), and numerically (dotted line).

The comparison between asymptotic solutions (4.64)–(4.65) and numerical results is presented in Figures 4.3 and 4.4. In Figure 4.3 the transient stage of motion and steady-state are reported separately for the longitudinal vibration, whereas similar time histories for swing vibrations are shown in Figure 4.4.

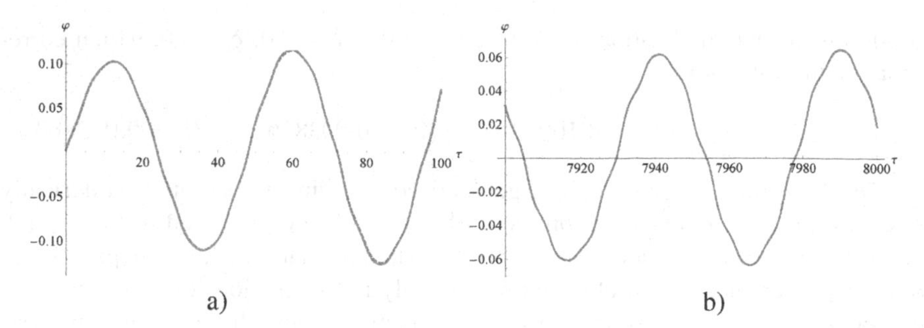

FIGURE 4.4 Time history of the swing vibration φ for transient (a) and long-time (b) simulation obtained analytically (MSM) and numerically (dotted line).

Figure 4.3 and 4.4 present two chosen time intervals of the simulation presented in Figure 4.2.

It is worth noting that the analytical and numerical results exhibited excellent compliance, as indicated by the graphs presented in Figures 4.3 and 4.4, even though they were obtained in a completely different way.

Let us define new functions

$$\theta_1(\tau) = \sigma_1 \tau - \delta_1(\tau), \qquad \theta_2(\tau) = \sigma_2 \tau - \delta_2(\tau), \tag{4.66}$$

which transform the modulation equation into a form in which time τ does not exist explicitly. We get

$$\frac{da_1}{d\tau} = -\frac{1}{2}c_1 a_1 + \frac{f_1}{2}\cos(\theta_1), \tag{4.67}$$

$$a_1\frac{d\theta_1}{d\tau} = -\frac{3}{8}\alpha a_1^3 - \frac{3}{2}\alpha a_1 z_e^2 + a_1\sigma_1 + \frac{3a_2^2 a_1 w^2(w^2-1)}{4-16w^2} - \frac{1}{2}f_1\sin(\theta_1), \tag{4.68}$$

$$\frac{da_2}{d\tau} = \frac{f_2\cos(\theta_2)}{2w} - \frac{a_2 c_2}{2}, \tag{4.69}$$

$$a_2\frac{d\theta_2}{d\tau} = a_2\sigma_2 + \frac{a_2^3 w(8w^4-7w^2-1)}{64w^2-16} + \frac{3a_1^2 a_2 w(w^2-1)}{4-16w^2} - \frac{f_2\sin(\theta_2)}{2w}. \tag{4.70}$$

Equations (4.67)–(4.70) constitute an autonomous system of first-order differential equations for amplitudes and modified phases.

4.5 ANALYSIS OF NON-STATIONARY MOTION

In the vicinity of resonance, the system is very sensitive to even small changes in parameter values. Let us examine the sensitivity of the system to changes with regard to the parameter α responsible for the nonlinear feature of the spring. It turns out that there is a certain critical value α_{cr} for this parameter. Exceeding this value leads to a sharp qualitative change in the behavior of the system. For $\alpha < \alpha_{cr}$, the oscillations can be considered as quasi-linear, while, for $\alpha > \alpha_{cr}$, they become strongly

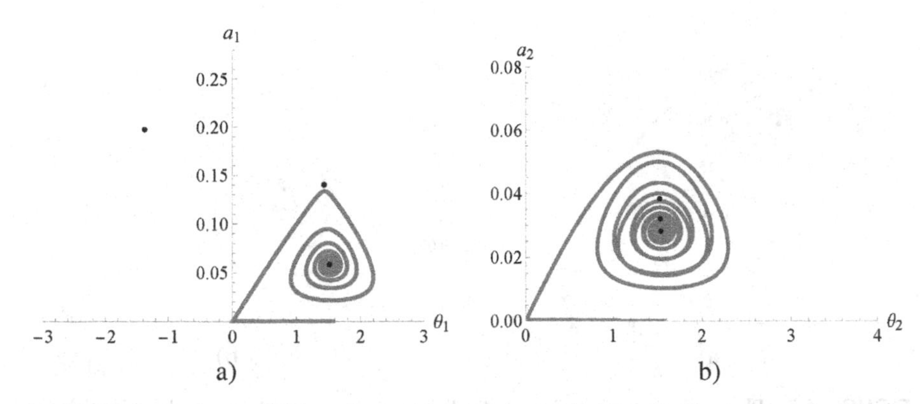

FIGURE 4.5 Phase plane trajectories of the longitudinal (a) and swing (b) vibration for $\alpha = 0.83 < \alpha_{cr}$ (just before the transition); black color refers to fixed points.

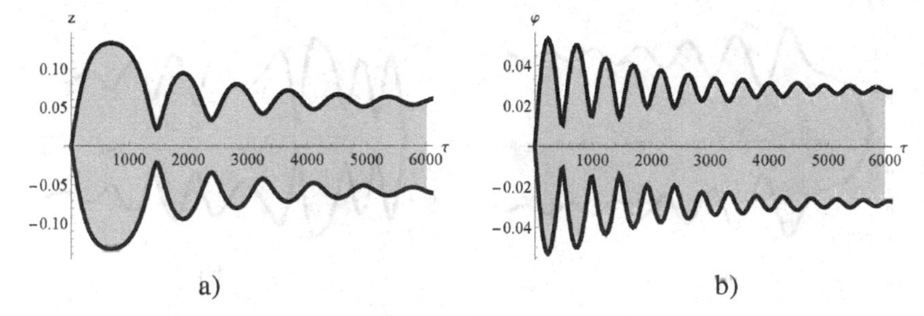

FIGURE 4.6 Amplitude modulation (thick lines) and time histories (gray line) of the longitudinal (a) and swing (b) vibration obtained for $\alpha = 0.83 < \alpha_{cr}$ (just before the transition).

nonlinear. With a further increase of the nonlinear parameter α, another qualitative change in dynamics occurs for $\alpha \approx \alpha_{cr2}$. It consists in a change of the number of fixed points of the system (4.67)–(4.70). In this case, a slight change in amplitude modulation is also observed. However, the spectacular qualitative change of the system dynamics has been detected only for $\alpha = \alpha_{cr}$. The critical value α_{cr} depends on all parameters of the system, but cannot be determined analytically due to the complicated nonlinear form of the modulation equations. It can be assessed only through numerical simulation. Results obtained in the above study are presented in Figures 4.5–4.10. The following fixed parameters are taken: $f_1 = 0.001$, $f_2 = 0.0001$, $w = 0.13$, $c_1 = 0.001$, $c_2 = 0.001$, $\sigma_1 = 0.01$, $\sigma_2 = 0.014$. For the assumed parameters, the value of $\alpha_{cr} \approx 0.835$. The behavior of the pendulum in the quasi-linear case for $\alpha < \alpha_{cr}$ is presented in Figures 4.5 and 4.6. The projection of the phase trajectory onto the phase planes $\{a_1, \theta_1\}$ and $\{a_2, \theta_2\}$ are shown in Figure 4.5. The fixed points of the modulation equations (4.67)–(4.70) are depicted in these graphs by the black points. One of these points at which trajectories are heading is stable.

Figure 4.6 shows the time histories of the generalized coordinates $z(\tau)$ and $\varphi(\tau)$. The bold line indicates slow variation in amplitudes, obtained by integrating the

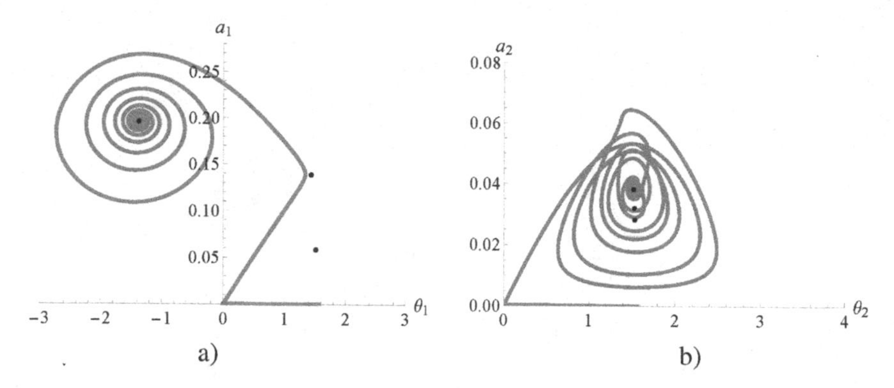

FIGURE 4.7 Phase plane trajectories of the longitudinal (a) and swing (b) vibration for $\alpha = 0.85 > \alpha_{cr}$ (just after the transition); black color refers to fixed points.

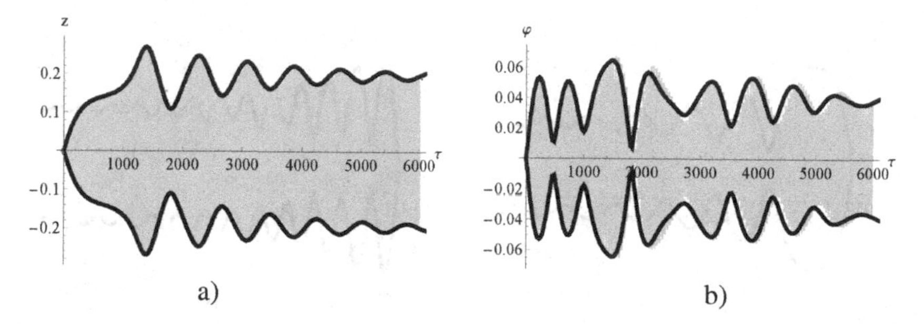

FIGURE 4.8 Amplitude modulation (thick lines) and time histories (gray line) of the longitudinal (a) and swing (b) vibration obtained for $\alpha = 0.85 > \alpha_{cr}$ (just after the transition).

modulation equations (4.67)–(4.70), while the gray ones are, indeed, the graphs of the fast-changing oscillations which are obtained by the numerical integration of the problem (4.12)–(4.14). The graphs show a high compliance of numerical and asymptotic/analytical results. In the examined case, the motion tends to steady oscillations with amplitudes $a_1 \approx 0.058$ and $a_2 \approx 0.028$ that correspond to the fixed points shown in Figure 4.5.

Figures 4.7 and 4.8 correspond to the situation when $\alpha = 0.85$ is slightly greater than the critical value α_{cr}, i.e. just after a spectacular qualitative change in the vibrations of the system. As a result of the qualitative transition, the vibration becomes strongly nonlinear. In this case, the vibrations stabilize at the values $a_1 \approx 0.196$ and $a_2 \approx 0.038$ which are much higher than for $\alpha < \alpha_{cr}$.

Another qualitative change in the dynamics of the pendulum is not as spectacular as the previous one. It is related to the disappearance of two unstable fixed points, and the phase trajectory tends to only one fixed point being stable. The amplitudes stabilize on the values $a_1 \approx 0.159$ and $a_2 \approx 0.0336$ and the amplitudes modulation takes a sawtooth shape which would be more visible for lower damping values.

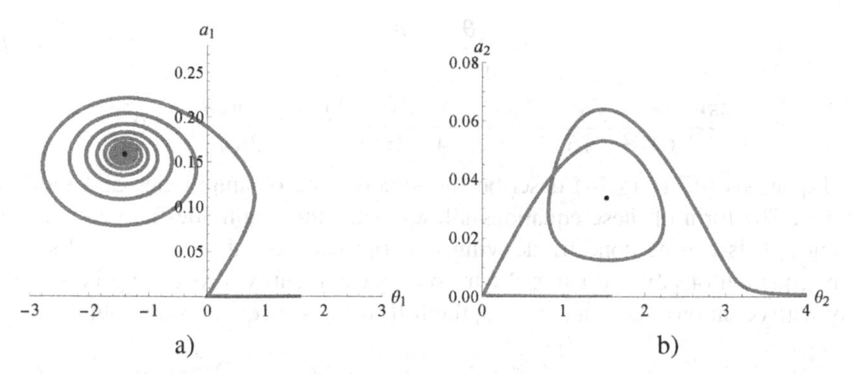

FIGURE 4.9 Phase plane trajectories of the longitudinal (a) and swing (b) vibration for $\alpha = 1.33 > \alpha_{cr2}$ (just after the second qualitative transition); black color refers to a fixed point.

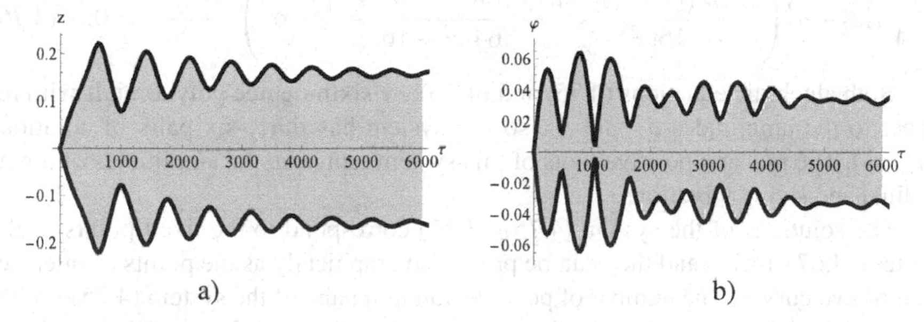

FIGURE 4.10 Amplitude modulation (thick lines) and time histories (gray line) of the longitudinal (a) and swing (b) vibration obtained for $\alpha = 1.33 > \alpha_{cr2}$ (just after the second qualitative transition).

A very good adjustment of the amplitude modulation curves to the vibration time courses reported in Figures 4.5 and 4.10 indicates the correctness of the results obtained by the multiple scale method and proves its high efficiency in both qualitative and quantitative analyzes of nonlinear problems.

4.6 STATIONARY MOTION

The autonomous form of the modulation equations (4.67)–(4.70) allows one to analyze the steady-state motion in a simple way. The fixed points of this set of differential equations of the first-order can be found by zeroing time derivatives. That leads to the following system of algebraic equations

$$-\frac{1}{2}c_1a_1 + \frac{f_1}{2}\cos(\theta_1) = 0, \tag{4.71}$$

$$-\frac{3}{8}\alpha a_1^2 - \frac{3}{2}\alpha z_e^2 + \sigma_1 + \frac{3a_2^2w^2(w^2-1)}{4-16w^2} - \frac{f_1}{2a_1}\sin(\theta_1) = 0, \tag{4.72}$$

$$\frac{f_2 \cos(\theta_2)}{2w} - \frac{a_2 c_2}{2} = 0, \tag{4.73}$$

$$\sigma_2 + \frac{a_2^2 w\left(8w^4 - 7w^2 - 1\right)}{64w^2 - 16} + \frac{3a_1^2 w\left(w^2 - 1\right)}{4 - 16w^2} - \frac{f_2 \sin(\theta_2)}{2wa_2} = 0, \tag{4.74}$$

Equations (4.71)–(4.74) describe the steady-state of amplitudes and modified phases. The form of these equations allows obtaining amplitude-frequency dependencies. This can be done by deriving $\cos(\theta_1)$ from equation (4.71) and $\sin(\theta_1)$ from equation (4.72), and using the trigonometric identity. Proceeding in a similar way with equations (4.73) and (4.74), the following two equations are obtained

$$\frac{1}{4}a_1^2 c_1^2 + \left(-\frac{3}{8}\alpha a_1^3 - \frac{3}{2}\alpha a_1 z_e^2 + a_1\sigma_1 + \frac{3a_2^2 a_1 w^2\left(w^2 - 1\right)}{4 - 16w^2}\right)^2 - \frac{f_1^2}{4} = 0, \tag{4.75}$$

$$\frac{1}{4}a_2^2 c_2^2 + a_2^2\left(\frac{3a_1^2 w\left(w^2 - 1\right)}{4 - 16w^2} + \frac{a_2^2 w\left(8w^4 - 7w^2 - 1\right)}{64w^2 - 16} + \sigma_2\right)^2 - \frac{f_2^2}{4w^2} = 0. \tag{4.76}$$

Both algebraic equations (4.75) and (4.76) are sixth-degree polynomials with respect to the amplitudes a_1 and a_2, so the system has thirty-six pairs of solutions (a_1, a_2). The real and positive roots of this system represent the amplitudes of longitudinal and swing vibrations.

The solutions of the system (4.75)–(4.76) correspond to the fixed points of the system (4.67)–(4.70) and they can be presented graphically as the points of intersection of two curves. The number of positive solution pairs of the system (4.75)–(4.76) may vary and depends on the physical parameters of the pendulum and the proximity of the resonance. The results for several values of σ_1 and σ_2 are shown in Figure 4.11. The analysis is performed for the following fixed parameters $w = 0.25, f_1 = 0.0001$, $f_2 = 0.0002$, $c_1 = 0.0012$, $c_2 = 0.0015$, $\alpha = 0.3$.

The way of presenting the solutions of the system (4.75)–(4.76) shown in Figure 4.11 allows for detection of the value and number of steady-state amplitudes. However, a separate analysis is necessary to investigate their stability. Figure 4.11 clearly shows that the number of real roots of the system (4.75)–(4.76) may vary from 1 to 7 depending on the parameters adopted.

An important result of the vibration analysis under resonance conditions is the resonance response curve, which shows the dependence of the vibration amplitude on the excitation frequency. The frequencies of the external loading are represented by the detuning parameters σ_1 and σ_2. The shape of the resonance response curves is very sensitive to even small changes in the system parameters, i.e. damping coefficients c_1, c_2, amplitudes of the external forces f_1, f_2, nonlinearity parameter α and characteristic frequency w. Some examples of the resonance response curves obtained solving equations (4.75)–(4.76) are presented in Figs. 4.12 and 4.13 for the same data as employed in getting drawing of Figure 4.11.

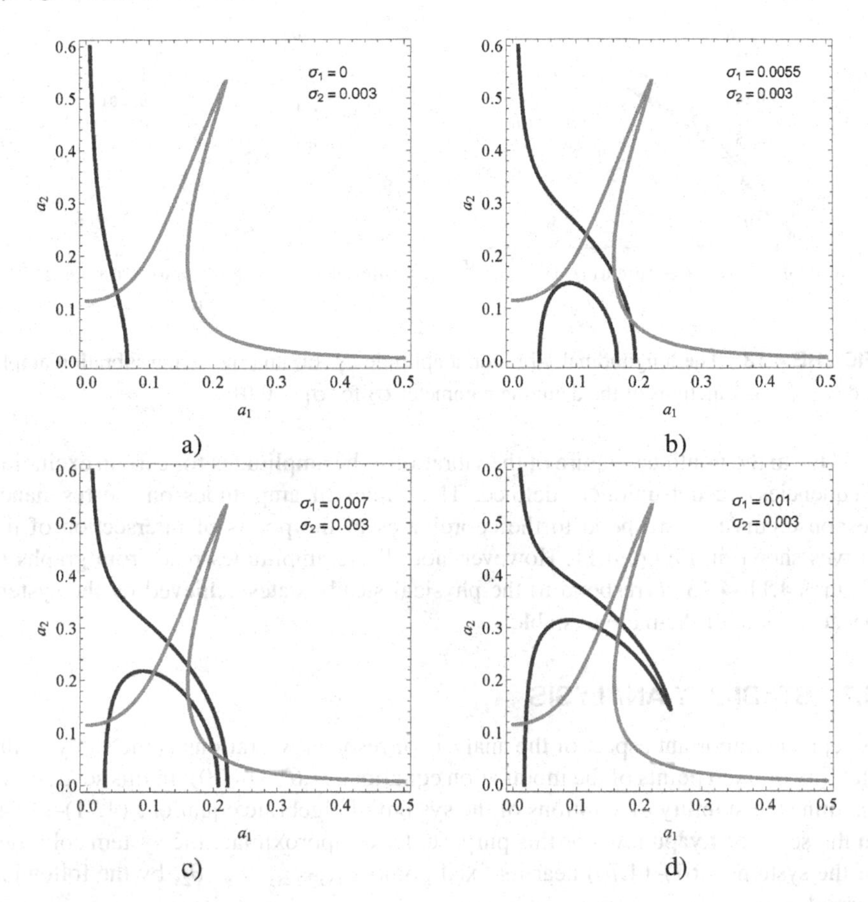

FIGURE 4.11 Graphical presentation of the solutions of the system (4.75) - (4.76) for various values of σ_1 and σ_2 where the intersection points of the curves identify the fixed points of the system (4.67)–(4.70): a) $\sigma_1 = 0,\ \sigma_2 = 0.003$; b) $\sigma_1 = 0.0055,\ \sigma_2 = 0.003$; c) $\sigma_1 = 0.007,\ \sigma_2 = 0.003$; d) $\sigma_1 = 0.01,\ \sigma_2 = 0.003$.

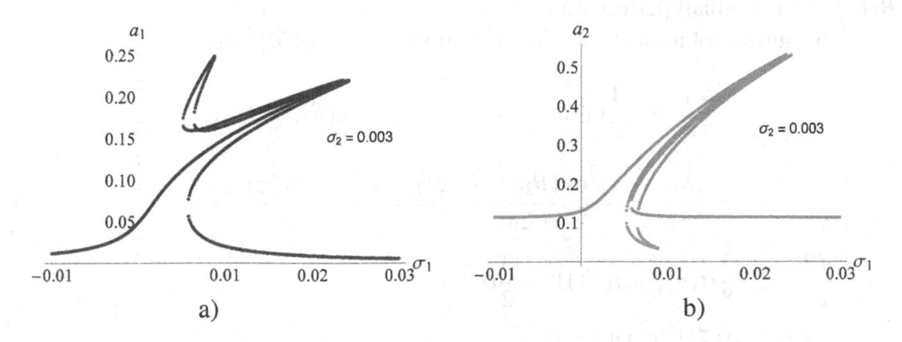

FIGURE 4.12 The longitudinal vibration amplitude a_1 (a) and the swing vibration amplitude a_2 (b) as functions of the detuning parameter σ_1 for $\sigma_2 = 0.003$.

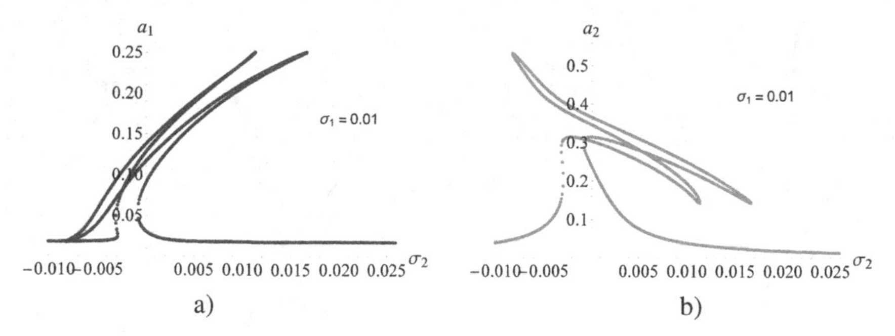

FIGURE 4.13 The longitudinal vibration amplitude a_1 (a) and the swing vibration amplitude a_2 (b) as functions of the detuning parameter σ_2 for $\sigma_1 = 0.01$.

Due to the nonlinear nature of the vibrations, the amplitudes for certain excitation frequencies are not uniquely defined. The values of amplitudes on the resonance response curves correspond to the coordinates of the points of intersection of the curves shown in Figure 4.11. However, not all the amplitudes read from graphs in Figures 4.11–4.13 correspond to the physical steady-states achieved by the system because some of them are unstable.

4.7 STABILITY ANALYSIS

A separate important aspect of the analysis of resonant vibrations is the study of the stability of fixed points of the modulation equations (4.67)–(4.70). In this section, we examine the stability of solutions of the system of algebraic equations (4.71)–(4.74) in the sense of Lyapunov. For this purpose, let us approximate the system solutions of the system (4.67)–(4.70) near its fixed points a_{1s}, a_{2s}, θ_{1s}, θ_{2s} by the following formulas

$$\begin{aligned} a_1(\tau) &= a_{1s} + \tilde{a}_1(\tau), \qquad & a_2(\tau) &= a_{2s} + \tilde{a}_2(\tau), \\ \theta_1(\tau) &= \theta_{1s} + \tilde{\theta}_1(\tau), \qquad & \theta_2(\tau) &= \theta_{2s} + \tilde{\theta}_2(\tau), \end{aligned} \tag{4.77}$$

where a_{1s}, a_{2s}, θ_{1s}, θ_{2s} satisfy equations (4.71)–(4.74), and $\tilde{a}_1(\tau)$, $\tilde{a}_2(\tau)$, $\tilde{\theta}_1(\tau)$ and $\tilde{\theta}_2(\tau)$ are their small perturbations.

Substituting solutions (4.77) into equations (4.67)–(4.70) yields

$$\frac{d\tilde{a}_1}{d\tau} = -\frac{1}{2}c_1\left(a_{1s} + \tilde{a}_1(\tau)\right) + \frac{f_1}{2}\cos\left(\theta_{1s} + \tilde{\theta}_1(\tau)\right), \tag{4.78}$$

$$\frac{d\tilde{a}_2}{d\tau} = \frac{f_2\cos\left(\theta_{2s} + \tilde{\theta}_2(\tau)\right)}{2w} - \frac{\left(a_{2s} + \tilde{a}_2(\tau)\right)c_2}{2}, \tag{4.79}$$

$$\begin{aligned} \frac{d\tilde{\theta}_1}{d\tau} &= -\frac{3}{8}\alpha(a_{1s} + \tilde{a}_1(\tau))^2 - \frac{3}{2}\alpha z_e^2 + \sigma_1 + \\ &\frac{3(a_{2s} + \tilde{a}_2(\tau))^2 w^2\left(w^2 - 1\right)}{4 - 16w^2} - \frac{1}{2\left(a_{1s} + \tilde{a}_1(\tau)\right)} f_1 \sin\left(\theta_{1s} + \tilde{\theta}_1(\tau)\right), \end{aligned} \tag{4.80}$$

$$\frac{d\tilde{\theta}_2}{d\tau} = \sigma_2 + \frac{(a_{2s}+\tilde{a}_2(\tau))^2 w\left(8w^4-7w^2-1\right)}{64w^2-16} + \frac{3(a_{1s}+\tilde{a}_1(\tau))^2 w\left(w^2-1\right)}{4-16w^2} - \frac{f_2\sin\left(\theta_{2s}+\tilde{\theta}_2(\tau)\right)}{2w\left(a_{2s}+\tilde{a}_2(\tau)\right)}. \tag{4.81}$$

The right-hand side of equations (4.78)–(4.81) is then expanded into power series and linearized with respect to disturbances $\tilde{a}_1(\tau)$, $\tilde{a}_2(\tau)$, $\tilde{\theta}_1(\tau)$, and $\tilde{\theta}_2(\tau)$ which leads to the system

$$\frac{d\tilde{a}_1}{d\tau} = -\frac{a_{1s}c_1}{2} - \frac{1}{2}c_1\tilde{a}_1(\tau) - \frac{1}{2}f_1\sin(\theta_{1s})\tilde{\theta}_1(\tau) + \frac{1}{2}f_1\cos(\theta_{1s}), \tag{4.82}$$

$$\frac{d\tilde{a}_2}{d\tau} = -\frac{a_{2s}c_2}{2} - \frac{1}{2}c_2\tilde{a}_2(\tau) - \frac{f_2\sin(\theta_{2s})\tilde{\theta}_1(\tau)}{2w} + \frac{f_2\cos(\theta_{2s})}{2w}, \tag{4.83}$$

$$\frac{d\tilde{\theta}_1}{d\tau} = -\frac{1}{8}3\alpha a_{1s}^2 + \frac{f_1\tilde{a}_1(\tau)\sin(\theta_{1s})}{2a_{1s}^2} - \frac{3}{4}\alpha a_{1s}\tilde{a}_1(\tau) - \frac{f_1\cos(\theta_{1s})\tilde{\theta}_1(\tau)}{2a_{1s}} - \frac{f_1\sin(\theta_{1s})}{2a_{1s}} + \frac{3a_{2s}^2 w^2\left(w^2-1\right)}{4-16w^2} + \frac{3a_{2s}w^2\left(w^2-1\right)\tilde{a}_2(\tau)}{2-8w^2} - \frac{3\alpha z_e^2}{2} + \sigma_1, \tag{4.84}$$

$$\frac{d\tilde{\theta}_2}{d\tau} = \frac{3a_{1s}^2 w\left(w^2-1\right)}{4-16w^2} + \frac{3a_{1s}w\left(w^2-1\right)\tilde{a}_1(\tau)}{2-8w^2} + \frac{f_2\tilde{a}_2(\tau)\sin(\theta_{2s})}{2a_{2s}^2 w} + \frac{a_{2s}^2 w\left(8w^4-7w^2-1\right)}{64w^2-16} + \frac{a_{2s}w\left(8w^4-7w^2-1\right)\tilde{a}_2(\tau)}{32w^2-8} - \frac{f_2\cos(\theta_{2s})\tilde{\theta}_2(\tau)}{2a_{2s}w} - \frac{f_2\sin(\theta_{2s})}{2a_{2s}w} + \sigma_2. \tag{4.85}$$

Taking into account that a_{1s}, a_{2s}, θ_{1s}, θ_{2s} satisfy, equations (4.71)–(4.74) yield the final linearized perturbation equations

$$\frac{d\tilde{a}_1}{d\tau} = -\frac{1}{2}c_1\tilde{a}_1(\tau) - \frac{1}{2}f_1\sin(\theta_{1s})\tilde{\theta}_1(\tau), \tag{4.86}$$

$$\frac{d\tilde{a}_2}{d\tau} = -\frac{1}{2}c_2\tilde{a}_2(\tau) - \frac{f_2\sin(\theta_{2s})\tilde{\theta}_2(\tau)}{2w}, \tag{4.87}$$

$$\frac{d\tilde{\theta}_1}{d\tau} = \frac{f_1\tilde{a}_1(\tau)\sin(\theta_{1s})}{2a_{1s}^2} - \frac{3}{4}\alpha a_{1s}\tilde{a}_1(\tau) - \frac{f_1\cos(\theta_{1s})\tilde{\theta}_1(\tau)}{2a_{1s}} + \frac{3a_{2s}\left(w^2-1\right)w^2\tilde{a}_2(\tau)}{2-8w^2}, \tag{4.88}$$

$$\frac{d\tilde{\theta}_2}{d\tau} = \frac{3a_{1s}w\left(w^2-1\right)\tilde{a}_1(\tau)}{2-8w^2} + \frac{f_2\tilde{a}_2(\tau)\sin(\theta_{2s})}{2a_{2s}^2 w} + \frac{a_{2s}w\left(8w^4-7w^2-1\right)\tilde{a}_2(\tau)}{32w^2-8} - \frac{f_2\cos(\theta_{2s})\tilde{\theta}_2(\tau)}{2a_{2s}w}. \tag{4.89}$$

The stability of the fixed points of the modulation equations (4.67)–(4.70) is tested using the standard procedure for the system of first-order homogeneous equations. According to this approach, the solutions are stable if all the roots λ_i of the characteristic equation

$$|\mathbf{A} - \lambda\mathbf{I}| = 0, \tag{4.90}$$

have negative values, where $\mathbf{A}$ stands for the matrix of the form

$$\mathbf{A} = \begin{bmatrix} -\frac{c_1}{2} & 0 & -\frac{f_1 \sin\theta_{1s}}{2} & 0 \\ 0 & -\frac{c_2}{2} & 0 & -\frac{f_2 \sin\theta_{2s}}{2w} \\ \frac{f_1 \sin\theta_{1s}}{2a_{1s}^2} - \frac{3\alpha a_{1s}}{4} & \frac{3a_{2s}w^2(w^2-1)}{2-8w^2} & -\frac{f_1 \cos\theta_{1s}}{2a_{1s}} & 0 \\ \frac{3a_{1s}w(w^2-1)}{2-8w^2} & \frac{f_2 \sin\theta_{2s}}{2a_{2s}^2 w} + \frac{a_{2s}w(8w^4-7w^2-1)}{32w^2-8} & 0 & -\frac{f_2 \cos\theta_{2s}}{2a_{2s}w} \end{bmatrix} \tag{4.91}$$

Figures 4.14–4.17 show the resonance response curves obtained for the amplitudes and phases from equations (4.71)–(4.74). The stable branches of these curves are colored in black. The data for construction of them are the same as earlier, i.e. $w = 0.25, f_1 = 0.0001, f_2 = 0.0002, c_1 = 0.0012, c_2 = 0.0015, \alpha = 0.3$.

The response curves indicate that in the analyzed case of resonance, for the assumed value of parameters, up to three stable fixed points of the system (4.67)–(4.70) can exist. Moreover, there are some ranges of the excitation frequency for which there are no stable branches of the resonant response at all. For example, in Figure 4.14 there are no stable solutions for $0.0013 < \sigma_1 < 0.005$. Confirmation of this conclusion is the time history and amplitude modulation presented in Figures 4.18, 4.19 obtained for the same data as previously.

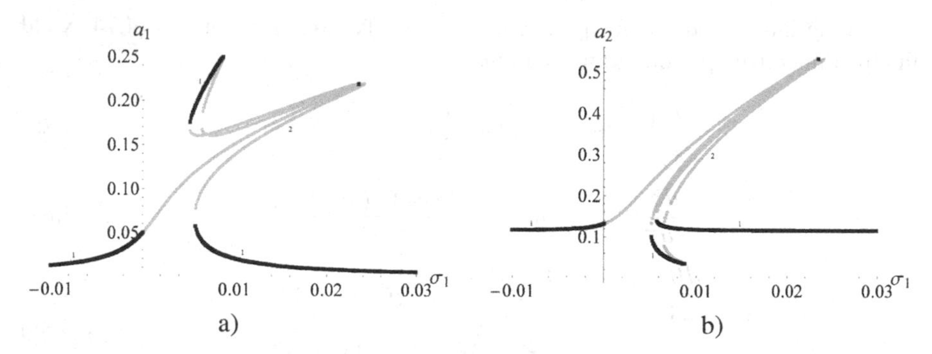

FIGURE 4.14 The oscillation amplitudes a_1 (a) and a_2 (b) for $\sigma_2 = 0.003$ vs. σ_1 obtained from Eqs (4.71)–(4.72); 1 – stable and 2 – unstable.

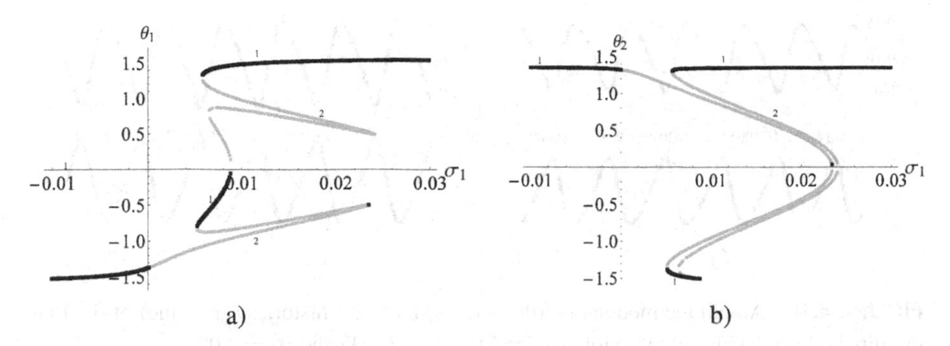

FIGURE 4.15 The modified phases θ_1 (a) and θ_2 (b) for $\sigma_2 = 0.003$ vs. σ_1 obtained from Eqs (4.71)–(4.72); 1 – stable and 2 – unstable branch.

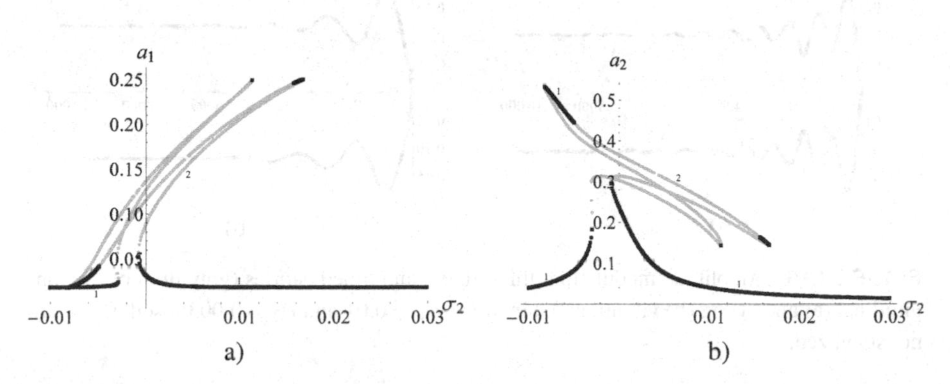

FIGURE 4.16 The oscillation amplitudes a_1 (a) and a_2 (b) for $\sigma_1 = 0.01$ vs. σ_2 obtained from Eqs (4.71)–(4.72); 1 – stable and 2 – unstable branch.

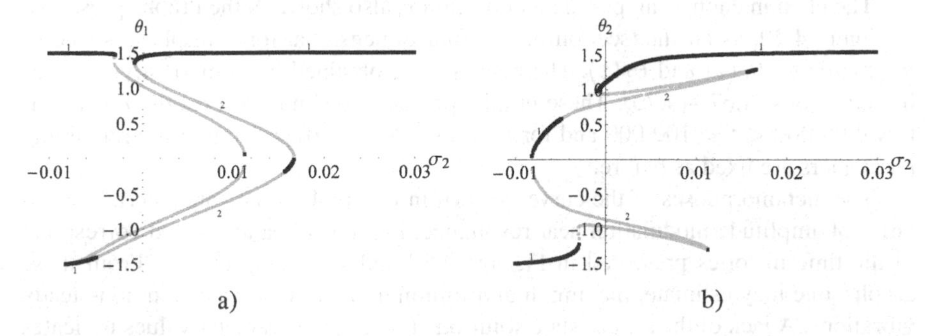

FIGURE 4.17 The modified phases θ_1 (a) and θ_2 (b) for $\sigma_1 = 0.01$ vs. σ_2 obtained from Eqs (4.46)–(4.47); 1 – stable and 2 – unstable branch.

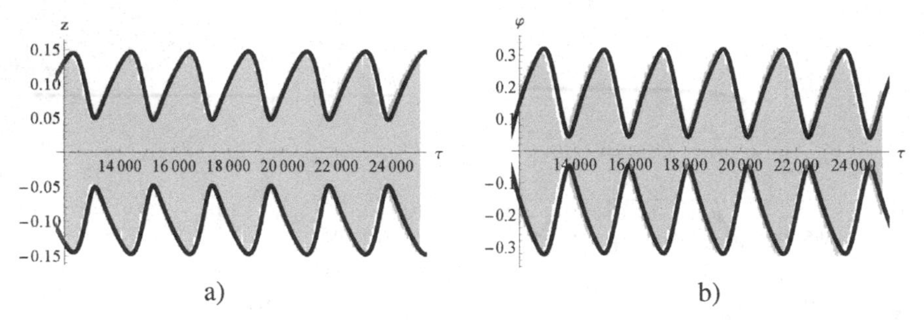

FIGURE 4.18 Amplitude modulation (thick lines) and time histories (gray line) of the longitudinal (a) and swing (b) vibration obtained for $\sigma_1 = 0.003$ and $\sigma_2 = 0.003$.

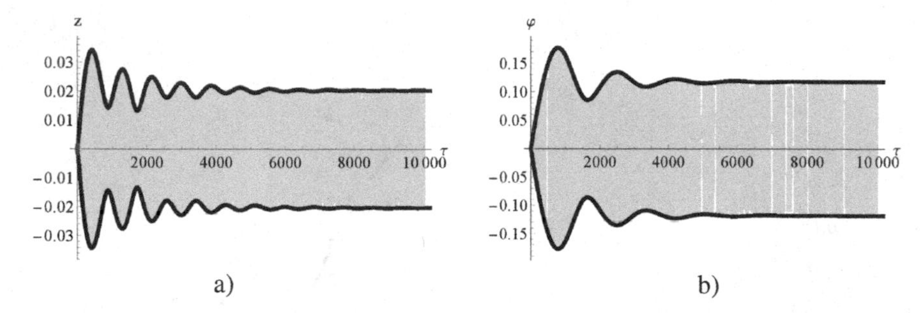

FIGURE 4.19 Amplitude modulation (thick lines) and time histories (gray line) of the longitudinal (a) and swing (b) vibration obtained for $\sigma_1 = 0.01$ and $\sigma_2 = 0.003$; oscillations are not stabilized.

Figure 4.18 presents the time interval of the oscillations far from the initial moment. It is possible to observe the phenomenon of amplitude modulation stabilization. On the other hand, in Figure 4.19 we can observe fast stabilization of the amplitude though exhibiting the fast-changing oscillations.

The phenomenon of amplitude modulation is also shown in the graphs presented in Figure 4.20, as an intersection of the four-dimensional torus in phase space of $a_1(\tau), a_2(\tau), \theta_1(\tau)$ and $\theta_2(\tau)$. The results were obtained by numerically integrating equations (4.67)–(4.70). These graphs present the simulation results for the interval $90\,000 < \tau < 100\,000$ and for several values of σ_1, assuming the remaining parameters are fixed as before.

The metamorphoses of the curves shown in Figure 4.20 indicate a certain regularity of amplitude modulation near resonance. Figures 4.20a and 4.20d correspond to the time histories presented in Figures 4.18 and 4.19, respectively. From these graphs, one may estimate, maximum and minimum amplitude values in non-steady vibrations. A lack of the steady-state solutions for certain parameter values indicates a rich and intensive energy exchange between the system and the external loading.

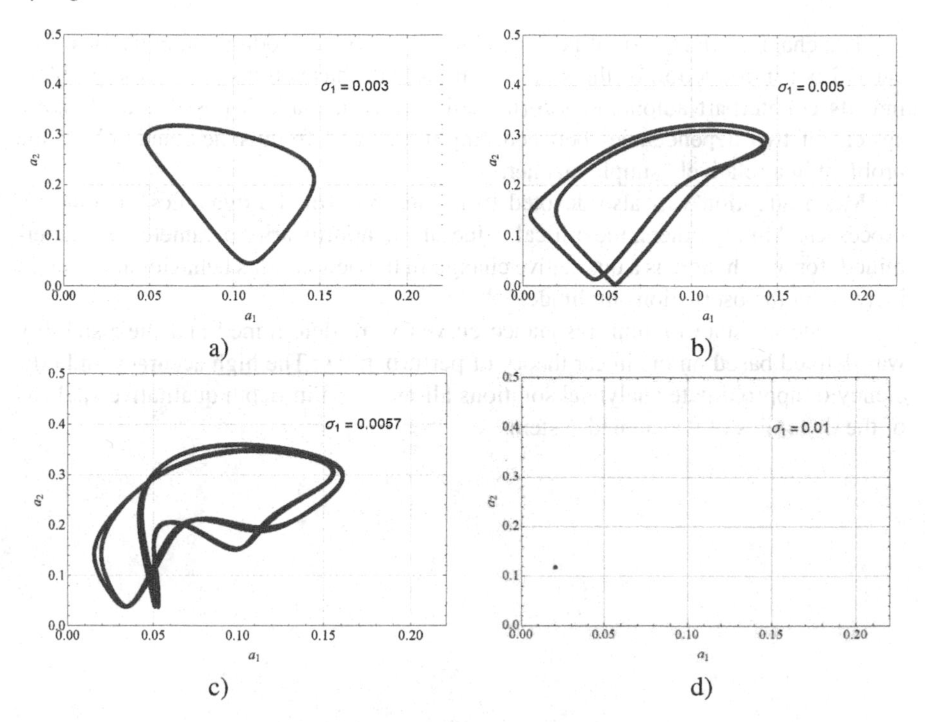

FIGURE 4.20 Projections of amplitudes modulation for $\sigma_2 = 0.003$ and various σ_1: a) $\sigma_1 = 0.003$; b) $\sigma_1 = 0.005$; c) $\sigma_1 = 0.0057$; d) $\sigma_1 = 0.01$.

4.8 CLOSING REMARKS

We have considered a spring pendulum whose longitudinal elasticity is of the cubic type. The pendulum has been excited harmonically and viscous damping has been taken into account. For such a physical model, the equations of motion were derived using the Lagrange formalism. The mathematical model was then transformed into a dimensionless form.

The next key step in the analysis was the introduction of complex functions which allowed us to write the mathematical model in the form of two first-order differential equations. The further analysis concerned the occurrence of two main resonances simultaneously. The equations of motion were then solved using the MSM in the time domain till the third-order of approximation. Modulation equations, which are an integral part of the applied method, were determined. The initial value problem for amplitudes and phases could not be solved analytically due to mathematical difficulties, and therefore a numerical approach was used for this purpose.

The comparison of the asymptotic solution of the problem with the numerical result showed a high agreement of the used approaches, which proves the correctness of the analysis. It should be emphasized that the approach used does not require the assumption that the parameter responsible for the nonlinearity of the spring is small.

The chapter offers a lot of new ideas while studying modulation equations. The carried out introduction of the so-called modified phases transformed the problem into its counterpart autonomous form. This, in turn, made it possible to derive a system of two dependencies between amplitudes and frequencies, and solved the problem in a relatively simple manner.

Much attention was also devoted to the analysis of the dynamics of transient processes. Among others, the critical value of the nonlinearity parameter was determined, for which there is a qualitative change in the pendulum's behavior and a sharp increase in the oscillation amplitude.

For steady-state motion, resonance curves were determined and their stability was defined based on the linear theory of perturbations. The high accuracy and efficiency of approximate analytical solutions allow for an in-depth qualitative analysis of the dynamics of the studied systems.

5 Physical Spring Pendulum

This chapter deals with analysis of dynamic responses of a harmonically excited 3 DoF planar physical pendulum. The problem is solved analytically with the solutions obtained up to the third-order by employing the MSM combined with AM, and all of them are validated numerically showing excellent agreements. The used method, in contrast to pure numerical approaches, allows one to detect parameters of the system responsible for occurrence of various resonances, whose prediction and understanding play a crucial role in engineering applications.

Time histories and backbone resonance curves are reported and stability of the periodic orbits are estimated. Three simultaneously resonance conditions are considered and the problem is satisfactorily solved showing the powerful use of the MSM/AM combined technique. The energy transfer from one mode of vibrations to another is illustrated and discussed, among others.

5.1 INTRODUCTION

In the previous chapter we have investigated elastic planar pendulum motion, i.e. the problem has been confined to two dimensions. In what follows, we consider either mechanical objects coupled with planar elastic pendulum governed by three second-order ODEs or the dynamics of 3D (spherical) pendula.

Miles [142, 143] belong to the first who studied stability of forced oscillations of a spherical pendulum and its resonance motion.

Cayton [50] analyzed 3D solutions though he met some numerical difficulties in numerical solutions while using the polar co-ordinates.

Krasnopolskaya and Shvets [108] reported chaotic oscillations of a spherical pendulum with emphasis to interaction with energy source.

Aston [16] analyzed bifurcations breaking the reflectonal symmetry of the horizontally forced spherical pendulums which yielded the nonplanar oscillations.

Lynch [132] considered the 3D motion with small amplitude of the elastic pendulum, and derived the linear normal modes. For unmodulated motion, elliptic-parabolic solutions were found. He detected the resonance phenomenon where energy was transferred periodically between predominantly vertical and predominantly horizontal oscillations, and he employed the multiple time-scale analysis. The approximate solutions were validated by numerical integrations.

Holm and Lynch [84] investigated 2:1:1 resonance arising at cubic order in approximate Lagrangian of the 3D elastic pendulum. The modulation equations (three-wave equations) were derived and the stepwise precession of the azimuthal angle was employed to follow characteristic features of pendulum dynamics.

Leung and Kuang [122] derived and studied equations of motion for a lightly damped spherical pendulum with harmonic excitation at its suspension point. The periodic orbits of the pendulum without base excitations were analyzed via the Jacobian

DOI: 10.1201/9781003270706-5

elliptic integral to detect homoclinic orbits, and then the homoclinic intersections of stable and unstable manifolds were performed. The Melnikov-Holmes-Marsden integral yielded the physical parameters leading to chaotic solutions. The predicted chaotic motion was validated by numerical simulations.

Pokorny [171] investigated stability and bifurcation problem for vertical oscillation of a 3D elastic pendulum.

Lee et al. [118] studied dynamics of a 3D elastic string pendulum attached to a rigid body, creating nontrivial coupling between wave propagation in the string and rigid body dynamics. The governing equations consisted of coupled oridinary and partial differential equations. The used computational methods were based on a finite element approximation and the use of variational techniques in a discrete-time setting. The problem was analyzed numerically including examples with a constant length string, string deployment, and string retrieval.

Perig et al. [169] derived the analytical solution of the natural frequencies problem of a spherical pendulum with a uniformly rotating suspension center. The numerical amplitude-frequency characteristics of the relative payload motion were presented and the analytical formula and numerical estimation for cable tension force were proposed.

Amer [6] considered dynamic behavior of a rigid body suspended on an elastic spring as a 3D pendulum model with emphasis put to its relative periodic motions. Computer codes were employed to obtain the graphical representations of the attached numerical solutions, and the stability problem was addressed.

Ismail [88] studied vertically planed pendulum motion governed by three second-order nonlinear ODEs. It was assumed that the suspended body moved in a rotating vertical plane uniformly with an arbitrary angular velocity. Periodic solutions were detected and their accuracy and stability were estimated numerically.

Freundlich and Sado [70] analyzed nonlinear dynamics of a 3D system with a spherical pendulum and damper of the fractional type in the vicinity of the internal and external resonances. In particular, the impact of a fractional order derivative on the system with a spherical pendulum was investigated. Time histories, bifurcation diagrams, Poincaré maps and Lyapunov exponents were reported for various orders of a fractional derivative. In addition, chaotic motion was revealed for some system parameters.

Litak et al. [125] studied oscillations of spherical pendulum with horizontal Lissajous excitation. The results were illustrated by the multi-colored maps of the largest Lyapunov exponent. It was pointed out that the used model can simulate overhead-traveling waves, with negligible small deflections and rail unevenness.

It should be noted that a similar system has been preliminary studied in reference (Awrejcewicz et al. [28]).

5.2 MATHEMATICAL MODEL

By the term physical planar spring pendulum, we mean a rigid body connected to an unmovable point O via an elastic-damping link. The plane motion is assumed, so the pendulum has three degrees of freedom. The total spring elongation $X(t)$ and two angles $\Phi(t)$ and $\Gamma(t)$ that are measured from the vertical lines, as is shown in Figure 5.1, are taken as the generalized coordinates.

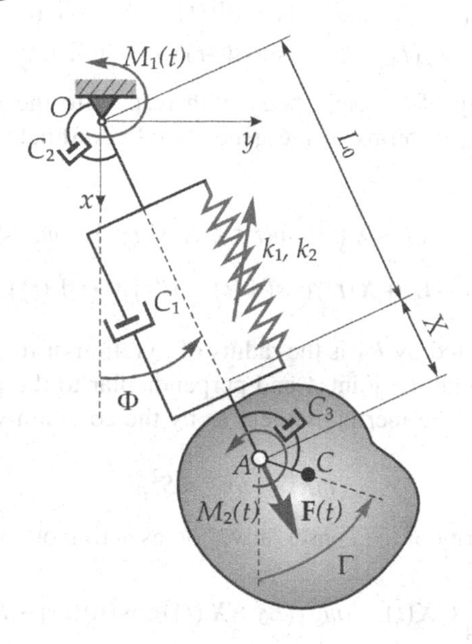

FIGURE 5.1 Physical planar spring pendulum.

In the physical spring/elastic pendulum, the spring is assumed to be massless and nonlinear with the nonlinearity of the cubic type, and k_1 and k_2 are the constant elastic coefficients. Both the coefficients are assumed to be positive. The spring and the viscous damper are arranged in parallel. The damping coefficient of the damper is equal to C_1. Moreover, two dampers of a purely viscous nature attenuate the swing vibration related to the angle Φ and the relative rotation between the body and the linker. The viscous coefficients of these dampers are denoted by C_2 and C_3. The rigid body of mass m is connected to the link by a pin joint A. The distance between point A and the mass center C of the body is denoted by S and called further the eccentricity. The body moment of inertia with respect to the axis that passes through the mass center C and perpendicular to the plane of motion is equal to I_C. In the direction of the main axis of the suspension system acts the known force **F**, the magnitude of which changes harmonically as $F(t) = F_0 \cos(\Omega_1 t)$. Besides, the system is loaded by two torques whose magnitudes change harmonically, i.e. $M_1(t) = M_{01}\cos(\Omega_2 t)$ and $M_2(t) = M_{02}\cos(\Omega_3 t)$. In the non-stretched state, the spring is of length L_0. The total spring elongation involves also the static elongation X_e which satisfies the following equilibrium condition

$$k_1 X_e + k_2 X_e^3 = mg, \tag{5.1}$$

where g is the gravity of Earth. The condition concerns the stable equilibrium position which the pendulum reaches when $\Phi_e = 0, \Gamma_e = 0$ and $X = X_e$.

The coordinates of the point denoted further as C in the reference frame OXY are as follows

$$\begin{aligned} X_C &= (L_0 + X(t))\cos(\Phi(t)) + S\cos(\Gamma(t)), \\ Y_C &= (L_0 + X(t))\sin(\Phi(t)) + S\sin(\Gamma(t)). \end{aligned} \tag{5.2}$$

The kinetic energy of the rigid body with respect to the immovable reference frame OXY expressed in terms of the generalized coordinates takes the following form

$$\begin{aligned} T = &\frac{m}{2}\left(\dot{X}(t)^2 + (L_0 + X(t))^2\dot{\Phi}(t)^2 + R_A^2\dot{\Gamma}(t)^2\right) + mS(\sin(\Phi(t) - \\ &\Gamma(t))\dot{X}(t)\dot{\Gamma}(t) + (L_0 + X(t))\cos(\Phi(t) - \Gamma(t))\dot{\Phi}(t)\dot{\Gamma}(t)). \end{aligned} \tag{5.3}$$

The quantity denoted by R_A is the radius of gyration of the body with respect to the axis passing through the joint A and perpendicular to the plane of motion. The radius R_A is related to the inertia moment I_C by the commonly known parallel axis theorem

$$mR_A^2 = I_C + mS^2. \tag{5.4}$$

The potential energy of the conservative forces acting on the rigid body is

$$V = \frac{1}{2}k_1X(t)^2 + \frac{1}{4}k_2X(t)^4 - mg((L_0 + X(t))\cos(\Phi(t)) + S\cos(\Gamma(t))). \tag{5.5}$$

All other forces and torques which act on the pendulum are introduced into consideration as the generalized forces corresponding to the generalized coordinates, so we have

$$\begin{aligned} Q_X &= F_0\cos(\Omega_1 t) - C_1\dot{X}(t), \\ Q_\Phi &= M_{01}\cos(\Omega_2 t) - C_2\dot{\Phi}(t) - C_3\left(\dot{\Phi}(t) - \dot{\Gamma}(t)\right), \\ Q_\Gamma &= M_{02}\cos(\Omega_3 t) - C_3\left(\dot{\Gamma}(t) - \dot{\Phi}(t)\right). \end{aligned} \tag{5.6}$$

Applying the Lagrange equations of the second kind yields the pendulum motion equations

$$\begin{aligned} &m\ddot{X}(t) + C_1\dot{X}(t) + k_1X(t) + k_2X(t)^3 - mg\cos\Phi(t) - m(L_0 + X(t))\dot{\Phi}(t)^2 + \\ &mS\left(\sin(\Phi(t) - \Gamma(t))\ddot{\Gamma}(t) - \cos(\Phi(t) - \Gamma(t))\dot{\Gamma}(t)^2\right) = F_0\cos(\Omega_1 t), \end{aligned} \tag{5.7}$$

$$\begin{aligned} &m(L_0 + X(t))^2\ddot{\Phi}(t) + C_2\dot{\Phi}(t) + C_3(\dot{\Phi}(t) - \dot{\Gamma}(t)) + \\ &mS(L_0 + X(t))(\cos(\Phi(t) - \Gamma(t))\ddot{\Gamma}(t) + \sin(\Phi(t) - \Gamma(t))\dot{\Gamma}(t)^2) + \\ &m(L_0 + X(t))(g\sin\Phi(t) + 2\dot{X}(t)\dot{\Phi}(t)) = M_{01}\cos(\Omega_2 t), \end{aligned} \tag{5.8}$$

$$\begin{aligned} &mR_A^2\ddot{\Gamma}(t) + C_3\left(\dot{\Gamma}(t) - \dot{\Phi}(t)\right) + mS\left(g\sin\Gamma(t) + 2\cos(\Phi(t) - \right. \\ &\left.\Gamma(t))\dot{X}(t)\dot{\Phi}(t)\right) + mS(L_0 + X(t))\left(\cos(\Phi(t) - \Gamma(t))\ddot{\Phi}(t) - \right. \\ &\left.\sin(\Phi(t) - \Gamma(t))\dot{\Phi}(t)^2\right) + mS\sin(\Phi(t) - \Gamma(t))\ddot{X}(t) = M_{02}\cos(\Omega_3 t). \end{aligned} \tag{5.9}$$

We aim to focus on the pendulum motion around the stable equilibrium position. Considering the condition given by (5.1), we get the following equations

$$\begin{aligned}
&m\ddot{X}_1(t)+C_1\dot{X}_1(t)+k_1X_1(t)+3k_2X_e^2X_1(t)+3k_2X_eX_1(t)^2+\\
&k_2X_1(t)^3+mg(1-\cos\Phi(t))-m(L+X_1(t))\dot{\Phi}(t)^2+\\
&mS\left(\sin(\Phi(t)-\Gamma(t))\ddot{\Gamma}(t)-\cos(\Phi(t)-\Gamma(t))\dot{\Gamma}(t)^2\right)=F_0\cos(\Omega_1 t),
\end{aligned} \tag{5.10}$$

$$\begin{aligned}
&m(L+X_1(t))^2\ddot{\Phi}(t)+C_2\dot{\Phi}(t)+C_3\left(\dot{\Phi}(t)-\dot{\Gamma}(t)\right)+\\
&mS(L+X_1(t))\left(\cos(\Phi(t)-\Gamma(t))\ddot{\Gamma}(t)+\sin(\Phi(t)-\Gamma(t))\dot{\Gamma}(t)^2\right)+\\
&m(L+X_1(t))\left(g\sin\Phi(t)+2\dot{X}_1(t)\dot{\Phi}(t)\right)=M_{01}\cos(\Omega_2 t),
\end{aligned} \tag{5.11}$$

$$\begin{aligned}
&mR_A^2\ddot{\Gamma}(t)+C_3\left(\dot{\Gamma}(t)-\dot{\Phi}(t)\right)+\ mS\left(g\sin\Gamma(t)+2\cos(\Phi(t)-\right.\\
&\left.\Gamma(t))\dot{X}_1(t)\dot{\Phi}(t)\right)+mS(L+X_1(t))\left(\cos(\Phi(t)-\Gamma(t))\ddot{\Phi}(t)-\right.\\
&\left.\sin(\Phi(t)-\Gamma(t))\dot{\Phi}(t)^2\right)+mS\sin(\Phi(t)-\Gamma(t))\ddot{X}_1(t)=M_{02}\cos(\Omega_3 t),
\end{aligned} \tag{5.12}$$

where $X_1(t)=X(t)-X_e$, $L=L_0+X_e$.

Differential equations (5.10)–(5.12) are supplemented with the following initial conditions

$$\begin{aligned}
&X_1(0)=U_{01},\quad \dot{X}_1(0)=U_{02},\quad \Phi(0)=U_{03},\\
&\dot{\Phi}(0)=U_{04},\quad \Gamma(0)=U_{05},\quad \dot{\Gamma}(0)=U_{06},
\end{aligned} \tag{5.13}$$

where $U_{01}, U_{02}, U_{03}, U_{04}, U_{05}, U_{06}$ are known quantities.

Let ω_1 denotes the frequency of the simple harmonic oscillator of mass m and stiffness k_1 , i.e. $\omega_1=\sqrt{\frac{k_1}{m}}$.

The length of the spring at the stable equilibrium position, i.e. L , and the characteristic time ω_1^{-1} have been chosen as the reference quantities. The dimensionless time

$$\tau=t\,\omega_1 \tag{5.14}$$

has been defined. The coordinate

$$\xi=\frac{X_1}{L} \tag{5.15}$$

understood as a function of τ is introduced. The functions $\varphi(\tau)$ and $\gamma(\tau)$ correspond the original angles $\Phi(t)$ and $\Gamma(t)$. Let $\omega_2^2=\frac{g}{L}$ and $\omega_3^2=\frac{g\,S}{R_A^2}$ denote two characteristic frequencies. Introducing the following dimensionless parameters

$$\begin{aligned}
&\xi_e=\frac{X_e}{L},\quad r_A=\frac{R_A}{L},\quad s=\frac{S}{L},\quad \alpha=\frac{k_2L^2}{m\,\omega_1^2},\quad w_2=\frac{\omega_2}{\omega_1},\quad w_3=\frac{\omega_3}{\omega_1},\\
&c_1=\frac{C_1}{m\omega_1},\quad c_2=\frac{C_2}{m\,L^2\,\omega_1},\quad c_3=\frac{C_3}{m\,L^2\,\omega_1},\quad f_1=\frac{F_0}{m\,L\,\omega_1^2},\\
&f_2=\frac{M_{01}}{m\,L^2\,\omega_1^2},\quad f_3=\frac{M_{02}}{m\,L^2\,\omega_1^2},\quad p_1=\frac{\Omega_1}{\omega_1},\quad p_2=\frac{\Omega_2}{\omega_1},\quad p_3=\frac{\Omega_3}{\omega_1}.
\end{aligned} \tag{5.16}$$

leads to the dimensionless form of the motion equations

$$\begin{aligned}&\ddot{\xi}(\tau)+c_1\dot{\xi}(\tau)+\xi(\tau)+\alpha\left(3\xi_e^2\xi(\tau)+3\xi_e\xi(\tau)^2+\xi(\tau)^3\right)-\\&(1+\xi(\tau))\dot{\varphi}(\tau)^2+w_2^2\,(1-\cos\varphi(\tau))+s\,(\sin(\varphi(\tau)-\gamma(\tau))\,\ddot{\gamma}(t)-\\&\cos(\varphi(\tau)-\gamma(\tau))\,\dot{\gamma}(\tau)^2\Big)=f_1\cos(p_1\tau),\end{aligned}\tag{5.17}$$

$$\begin{aligned}&(1+\xi(\tau))^2\ddot{\varphi}(\tau)+c_2\dot{\varphi}(\tau)+c_3(\dot{\varphi}(\tau)-\dot{\gamma}(t))+\\&(1+\xi(\tau))\left(w_2^2\sin\varphi(\tau)+2\dot{\xi}(\tau)\dot{\varphi}(\tau)\right)+s\,(1+\xi(\tau))\,(\cos(\varphi(\tau)-\\&\gamma(\tau))\,\ddot{\gamma}(\tau)+\sin(\varphi(\tau)-\gamma(\tau))\,\dot{\gamma}(\tau)^2\Big)=f_2\cos(p_2\tau)\end{aligned}\tag{5.18}$$

$$\begin{aligned}&\ddot{\gamma}(t)+\frac{c_3w_3^2}{sw_2^2}(\dot{\gamma}(t)-\dot{\varphi}(\tau))+\frac{w_3^2}{w_2^2}\cos(\varphi(\tau)-\gamma(\tau))\left(2\dot{\xi}(\tau)\dot{\varphi}(\tau)+\right.\\&(1+\xi(\tau))\,\ddot{\varphi}(\tau))+w_3^2\sin\gamma(\tau)+\frac{w_3^2}{w_2^2}\sin(\varphi(\tau)-\gamma(\tau))\left(\ddot{\xi}(\tau)-\right.\\&\left.(1+\xi(\tau))\,\dot{\varphi}(\tau)^2\right)=\frac{w_3^2}{sw_2^2}f_3\cos(p_3\tau).\end{aligned}\tag{5.19}$$

The dimensionless static elongation ξ_e of the spring satisfies the equation

$$\alpha\xi_e^3+\xi_e-w_2^2=0.\tag{5.20}$$

The initial conditions for equations (5.17)–(5.18) have the form

$$\begin{aligned}&\xi(0)=u_{01},\quad \dot{\xi}(0)=u_{02},\quad \varphi(0)=u_{03},\\&\dot{\varphi}(0)=u_{04},\quad \gamma(0)=u_{05},\quad \dot{\gamma}(0)=u_{06},\end{aligned}\tag{5.21}$$

where $u_{01}, u_{02}, u_{03}, u_{04}, u_{05}, u_{06}$ can be obtained directly from $U_{01}, U_{02}, U_{03}, U_{04}, U_{05}, U_{06}$ using formulae (5.14)–(5.15).

5.3 SOLUTION METHOD

The system's motion in a small neighbourhood of the stable equilibrium position is assumed. Therefore, the approximation of the trigonometric functions whose arguments are the generalized coordinates using the first three terms of the Taylor series expansion is reasonable. The following approximations hold

$$\begin{aligned}&\sin\varphi\approx\varphi-\frac{\varphi^3}{6},\quad cos\varphi\approx1-\frac{\varphi^2}{2},\quad \sin\gamma\approx\gamma-\frac{\gamma^3}{6},\quad \cos\gamma\approx1-\frac{\gamma^2}{2},\\&\sin(\varphi-\gamma)\approx\varphi-\gamma-\frac{(\varphi-\gamma)^3}{6},\quad \cos(\varphi-\gamma)\approx1-\frac{(\varphi-\gamma)^2}{2}.\end{aligned}\tag{5.22}$$

Substituting formulae (5.22) into equations (5.17)–(5.18) yields the approximate equations of motion

$$\ddot{\xi}+c_1\dot{\xi}+\xi+\alpha\left(3\,\xi_e^2\ \xi+3\,\xi_e\xi^2+\xi^3\right)-(1+\xi)\,\dot{\varphi}^2+w_2^2\,\frac{\varphi^2}{2}+$$
$$s\left(\left(\varphi-\gamma-\frac{(\varphi-\gamma)^3}{6}\right)\ddot{\gamma}-\left(1-\frac{(\varphi-\gamma)^2}{2}\right)\dot{\gamma}^2\right)=f_1\cos\left(p_1\tau\right), \tag{5.23}$$

$$(1+\xi)^2\ddot{\varphi}+c_2\dot{\varphi}+c_3\left(\dot{\varphi}-\dot{\gamma}\right)+(1+\xi)\left(w_2^2\left(\varphi-\frac{\varphi^3}{6}\right)+2\dot{\xi}\dot{\varphi}\right)+$$
$$s(1+\xi)\left(\left(1-\frac{(\varphi-\gamma)^2}{2}\right)\ddot{\gamma}+\left(\varphi-\gamma-\frac{(\varphi-\gamma)^3}{6}\right)\dot{\gamma}^2\right)=f_2\cos\left(p_2\right), \tag{5.24}$$

$$\ddot{\gamma}+\frac{c_3\,w_3^2}{s\,w_2^2}\left(\dot{\gamma}-\dot{\varphi}\right)+w_3^2\left(\gamma-\frac{\gamma^3}{6}\right)+\frac{w_3^2}{w_2^2}\left(1-\frac{(\varphi-\gamma)^2}{2}\right)\left(2\dot{\xi}\dot{\varphi}+(1+\xi)\,\ddot{\varphi}\right)+$$
$$\frac{w_3^2}{w_2^2}\left(\varphi-\gamma-\frac{(\varphi-\gamma)^3}{6}\right)\left(\ddot{\xi}-(1+\xi)\,\dot{\varphi}^2\right)=\frac{w_3^2}{s\,w_2^2}f_3\cos\left(p_3\tau\right). \tag{5.25}$$

The arguments of the unknown functions ξ, φ and γ have been omitted in (5.23)–(5.25).

The approximate analytical solution to the initial value problem (5.23)–(5.25) with the initial conditions (5.21) is obtained using the multiple scales method (MSM). Similarly, as in chapter 2, we adopt three variables: τ_0, τ_1, τ_2 that are related to the dimensionless time τ in the following way:

$$\tau_i=\varepsilon^i\tau,\quad i=0,1,2. \tag{5.26}$$

The small parameter ε which is non-dimension should satisfy the inequalities $0<\varepsilon\ll 1$.

According to the MSM, the ordinary derivatives with respect to the time τ are replaced by the partial derivatives

$$\frac{d}{d\tau}=\sum_{k=0}^{2}\varepsilon^k\frac{\partial}{\partial\tau_k}=\frac{\partial}{\partial\tau_0}+\varepsilon\frac{\partial}{\partial\tau_1}+\varepsilon^2\frac{\partial}{\partial\tau_2}, \tag{5.27}$$

$$\frac{d^2}{d\tau^2}=\frac{\partial^2}{\partial\tau_0^2}+2\varepsilon\frac{\partial^2}{\partial\tau_0\partial\tau_1}+\varepsilon^2\left(\frac{\partial^2}{\partial\tau_1^2}+2\frac{\partial^2}{\partial\tau_0\partial\tau_2}\right)+2\varepsilon^3\frac{\partial^2}{\partial\tau_1\partial\tau_2}+O\left(\varepsilon^4\right). \tag{5.28}$$

We assume the following asymptotic expansions of the functions $\xi\left(\tau\right)$, $\varphi\left(\tau\right)$ and $\gamma(\tau)$:

$$\xi\left(\tau;\varepsilon\right)=\sum_{k=1}^{3}\varepsilon^k x_k\left(\tau_0,\tau_1,\tau_2\right)+O\left(\varepsilon^4\right), \tag{5.29}$$

$$\varphi(\tau;\varepsilon)=\sum_{k=1}^{3}\varepsilon^k\,\phi_k(\tau_0,\tau_1,\tau_2)+O\left(\varepsilon^4\right), \tag{5.30}$$

$$\gamma(\tau;\varepsilon)=\sum_{k=1}^{3}\varepsilon^k\,\chi_k(\tau_0,\tau_1,\tau_2)+O\left(\varepsilon^4\right), \tag{5.31}$$

where functions $x_k(\tau_0,\tau_1,\tau_2)$, $\phi_k(\tau_0,\tau_1,\tau_2)$, $\chi_k(\tau_0,\tau_1,\tau_2)$, $k=1,\ \ldots,\ 3$ are to be found.

A few parameters characterizing the system are assumed to be small, which can be expressed using the small parameter as follows

$$\begin{aligned}&\alpha=\varepsilon^2\hat{\alpha},\quad s=\varepsilon\hat{s},\quad c_1=\varepsilon^2\hat{c}_1,\quad c_2=\varepsilon^2\hat{c}_2,\quad c_3=\varepsilon^2\hat{c}_3,\\&f_1=\varepsilon^3\hat{f}_1,\quad f_2=\varepsilon^3\hat{f}_2,\quad f_3=\varepsilon^4\hat{f}_3,\end{aligned} \tag{5.32}$$

where the coefficients $\hat{\alpha}$, $\hat{s}$, $\hat{c}_1$, $\hat{c}_2$, $\hat{c}_3$, $\hat{f}_1$, $\hat{f}_2$, $\hat{f}_3$ are understood as $O(1)$ when $\varepsilon\to 0$.

The above assumptions ensure that all nonlinear terms in the approximate equations of motion have the character of small perturbations. The pendulum is then a weakly nonlinear system.

Similarly, as in the case of the spring pendulum considered in chapter 2, we begin with the case of the non-resonant motion for which there is no need to formulate any additional relationships. The study allows one to detect the conditions at which the resonance may occur.

5.4 NON-RESONANT VIBRATION

Substituting (5.27)–(5.31) into equations (5.23)–(5.24) yields the equations in which the small parameter appears in a few different powers. All terms of the order $O\left(\varepsilon^4\right)$ are neglected in the equations following the assumed order of the approximation yielded by the asymptotic expansions. Each of the three equations should be satisfied for any value of ε. After ordering the terms in the equations according to the powers ε, ε^2 and ε^3, this requirement is realized by equating to zero all coefficients standing at these powers. In that manner, one can obtain the set of nine differential equations with unknown functions: $x_k(\tau_0,\ \tau_1,\ \tau_2)$, $\phi_k(\tau_0,\ \tau_1,\ \tau_2)$, and $\chi_k(\tau_0,\ \tau_1,\ \tau_2)$ for $k=1,\ \ldots,\ 3$. We organize the set of equations into three groups. To the first group belong the homogeneous equations that contain the terms standing at ε:

$$\frac{\partial^2 x_1}{\partial\tau_0^2}+x_1=0, \tag{5.33}$$

$$\frac{\partial^2\phi_1}{\partial\tau_0^2}+w_2^2\phi_1=0, \tag{5.34}$$

$$\frac{\partial^2\chi_1}{\partial\tau_0^2}+\frac{w_3^2}{w_2^2}\frac{\partial^2\phi_1}{\partial\tau_0^2}+w_3^2\chi_1=0. \tag{5.35}$$

The terms standing at ε^2 create the equations of the second-order approximation

$$\frac{\partial^2 x_2}{\partial \tau_0^2} + x_2 = -2\,\frac{\partial x_1^2}{\partial \tau_0 \partial \tau_1} + \left(\frac{\partial \phi_1}{\partial \tau_0}\right)^2 - \frac{w_2^2}{2}\phi_1^2, \tag{5.36}$$

$$\frac{\partial^2 \phi_2}{\partial \tau_0^2} + w_2^2 \phi_2 = -2\frac{\partial^2 \phi_1}{\partial \tau_0 \partial \tau_1} - 2x_1 \frac{\partial^2 \phi_1}{\partial \tau_0^2} - 2\frac{\partial x_1}{\partial \tau_0}\frac{\partial \phi_1}{\partial \tau_0} - \hat{s}\,\frac{\partial^2 \chi_1}{\partial \tau_0^2} - w_2^2 x_1 \phi_1, \tag{5.37}$$

$$\begin{aligned} &\frac{\partial^2 \chi_2}{\partial \tau_0^2} + \frac{w_3^2}{w_2^2}\,\frac{\partial^2 \phi_2}{\partial \tau_0^2} + w_3^2 \chi_2 = -2\frac{\partial^2 \chi_1}{\partial \tau_0 \partial \tau_1} - \frac{\hat{c}_3\, w_3^2}{\hat{s}\, w_2^2}\left(\frac{\partial\,\chi_1}{\partial \tau_0} - \frac{\partial\,\phi_1}{\partial \tau_0}\right) + \\ &\frac{w_3^2}{w_2^2}\left(\frac{\partial^2 x_1}{\partial \tau_0^2}(\chi_1 - \phi_1) - x_1\frac{\partial^2 \phi_1}{\partial \tau_0^2} - 2\,\frac{\partial^2 \phi_1}{\partial \tau_0 \partial \tau_1} - 2\,\frac{\partial\, x_1}{\partial \tau_0}\,\frac{\partial\, \phi_1}{\partial \tau_0}\right). \end{aligned} \tag{5.38}$$

The coefficients which are accompanied by ε^3 form the following equations of the third-order approximation

$$\begin{aligned} &\frac{\partial^2 x_3}{\partial \tau_0^2} + x_3 = \hat{f}_1 \cos(p_1 \tau_0) - \hat{c}_1 \frac{\partial x_1}{\partial \tau_0} - 3\hat{\alpha}\,\xi_e^2\, x_1 + \hat{s}\left(\frac{\partial \chi_1}{\partial \tau_0}\right)^2 + \\ &\hat{s}\frac{\partial^2 \chi_1}{\partial \tau_0^2}(\chi_1 - \phi_1) - \frac{\partial^2 x_1}{\partial \tau_1^2} - 2\left(\frac{\partial^2 x_1}{\partial \tau_0 \partial \tau_2} + \frac{\partial^2 x_2}{\partial \tau_0 \partial \tau_1}\right) + x_1\left(\frac{\partial \phi_1}{\partial \tau_0}\right)^2 + \\ &2\frac{\partial \phi_1}{\partial \tau_0}\left(\frac{\partial \phi_1}{\partial \tau_1} + \frac{\partial \phi_2}{\partial \tau_0}\right) - w_2^2 \phi_1 \phi_2, \end{aligned} \tag{5.39}$$

$$\begin{aligned} &\frac{\partial^2 \phi_3}{\partial \tau_0^2} + w_2^2 \phi_3 = \hat{f}_2 \cos(p_2 \tau_0) - \hat{c}_2\,\frac{\partial \phi_1}{\partial \tau_0} - \hat{c}_3\left(\frac{\partial \phi_1}{\partial \tau_0} - \frac{\partial \chi_1}{\partial \tau_0}\right) - \\ &\hat{s}\left(x_1\frac{\partial^2 \chi_1}{\partial \tau_0^2} + 2\frac{\partial^2 \chi_1}{\partial \tau_0 \partial \tau_1} + \frac{\partial^2 \chi_2}{\partial \tau_0^2}\right) - \frac{\partial^2 \phi_1}{\partial \tau_1^2} - \frac{\partial^2 \phi_1}{\partial \tau_0^2}(x_1^2 + 2x_2) - 2\frac{\partial^2 \phi_1}{\partial \tau_0 \partial \tau_2} - \\ &4\,x_1\,\frac{\partial^2 \phi_1}{\partial \tau_0 \partial \tau_1} - 2\frac{\partial^2 \phi_2}{\partial \tau_0 \partial \tau_1} - 2x_1\frac{\partial^2 \phi_2}{\partial \tau_0^2} - 2\,\frac{\partial \phi_1}{\partial \tau_0}\left(\frac{\partial x_1}{\partial \tau_1} + x_1\frac{\partial x_1}{\partial \tau_0} + \frac{\partial x_2}{\partial \tau_0}\right) - \\ &2\frac{\partial x_1}{\partial \tau_0}\left(\frac{\partial \phi_1}{\partial \tau_1} + \frac{\partial \phi_2}{\partial \tau_0}\right) + \frac{1}{6}w_2^2 \phi_1^3 - w_2^2\,(x_2\phi_1 + x_1\phi_2), \end{aligned} \tag{5.40}$$

$$\begin{aligned} &\frac{\partial^2 \chi_3}{\partial \tau_0^2} + \frac{w_3^2}{w_2^2}\,\frac{\partial^2 \phi_3}{\partial \tau_0^2} + w_3^2 \chi_3 = \frac{w_3^2}{\hat{s}\, w_2^2}\hat{f}_3 \cos(p_3 \tau_0) - \frac{\hat{c}_3\, w_3^2}{\hat{s}\, w_2^2}\left(\frac{\partial \chi_1}{\partial \tau_1} - \frac{\partial \phi_1}{\partial \tau_1} + \right. \\ &\left.\frac{\partial \chi_2}{\partial \tau_0} - \frac{\partial \phi_2}{\partial \tau_0}\right) + \frac{w_3^2}{w_2^2}\left((\chi_2 - \phi_2)\frac{\partial^2 x_1}{\partial \tau_0^2} + (\chi_1 - \phi_1)\frac{\partial^2 x_2}{\partial \tau_0^2}\right) + \\ &\frac{w_3^2}{w_2^2}\left(\left(\frac{1}{2}(\phi_1 - \chi_1)^2 - x_2\right)\frac{\partial^2 \phi_1}{\partial \tau_0^2} - x_1\frac{\partial^2 \phi_2}{\partial \tau_0^2}\right) + 2\frac{w_3^2}{w_2^2}\left((\chi_1 - \phi_1)\frac{\partial^2 x_1}{\partial \tau_0 \partial \tau_1} - \right. \end{aligned}$$

$$x_1 \frac{\partial^2 \phi_1}{\partial \tau_0 \partial \tau_1} - \frac{\partial^2 \phi_1}{\partial \tau_0 \partial \tau_2} - \frac{\partial^2 \phi_2}{\partial \tau_0 \partial \tau_1}\Bigg) - 2\left(\frac{\partial^2 \chi_1}{\partial \tau_0 \partial \tau_2} + \frac{\partial^2 \chi_2}{\partial \tau_0 \partial \tau_1}\right) - \frac{\partial^2 \chi_1}{\partial \tau_1^2} -$$
$$\frac{w_3^2}{w_2^2}\frac{\partial^2 \phi_1}{\partial \tau_1^2} - \frac{\partial^2 \chi_1}{\partial \tau_1^2} - 2\frac{w_3^2}{w_2^2}\left(\frac{\partial x_1}{\partial \tau_0}\left(\frac{\partial \phi_1}{\partial \tau_1} + \frac{\partial \phi_2}{\partial \tau_0}\right) + \frac{\partial \phi_1}{\partial \tau_0}\left(\frac{\partial x_2}{\partial \tau_0} + \frac{\partial x_1}{\partial \tau_1}\right)\right) - \tag{5.41}$$
$$\frac{w_3^2}{w_2^2}(\chi_1 - \phi_1)\left(\frac{\partial \phi_1}{\partial \tau_0}\right)^2 + \frac{1}{6}w_3^2 \chi_1^3 .$$

The system of equations (5.33)–(5.41) is solved recursively. The equations of motion of the pendulum indicate the existence of inertial couplings in the system. One can perceive the inertial coupling between the generalized coordinates ξ and γ in equation (5.17) and between φ and γ in equation (5.18). These couplings are conditioned by the parameter s that represents the eccentricity of point A. The coupling occurring in (5.19) involves all the coordinates. The assumptions about the weak character of the nonlinearities make it in such a way that there remains only one inertial coupling in the approximate model obtained as a result of applying the method of multiple scales. It is a coupling between the coordinates γ and φ that we observe in equations (5.35), (5.38) and (5.41). Occurrence of the derivatives of the functions ϕ_i , $i = 1,2,3$, in these equations requires modifying the recursive procedure for solving the system of equations of subsequent approximations. The procedure becomes more complex than in the case discussed in chapter 2.

We start with the solving of equations (5.33)–(5.34) that are mutually independent. The general solutions of the equations are expressed by the unknown functions $B_j(\tau_1, \tau_2)$, $(j = 1,2)$, and their complex conjugates $\bar{B}_j(\tau_1, \tau_2)$

$$x_1(\tau_0, \tau_1, \tau_2) = B_1(\tau_1, \tau_2)e^{i\tau_0} + \bar{B}_1(\tau_1, \tau_2)e^{-i\tau_0}, \tag{5.42}$$

$$\phi_1(\tau_0, \tau_1, \tau_2) = B_2(\tau_1, \tau_2)e^{iw_2\tau_0} + \bar{B}_2(\tau_1, \tau_2)e^{-iw_2\tau_0}, \tag{5.43}$$

where i denotes the imaginary unit.

Then, the solution (5.42) is substituted into the last of the equations of the first-order approximation. This leads to the nonhomogeneous equation

$$\frac{\partial^2 \chi_1}{\partial \tau_0^2} + w_3^2 \chi_1 = w_3^2\left(B_2(\tau_1, \tau_2)e^{iw_2\tau_0} + \bar{B}_2(\tau_1, \tau_2)e^{-iw_2\tau_0}\right). \tag{5.44}$$

The general solution of equation (5.44) have the form

$$\chi_1(\tau_0, \tau_1, \tau_2) = B_3(\tau_1, \tau_2)e^{iw_3\tau_0} + \bar{B}_3(\tau_1, \tau_2)e^{-iw_3\tau_0} +$$
$$\frac{w_3^2}{w_3^2 - w_2^2}\left(B_2(\tau_1, \tau_2)e^{iw_2\tau_0} + \bar{B}_2(\tau_1, \tau_2)e^{-iw_2\tau_0}\right), \tag{5.45}$$

where $B_3(\tau_1, \tau_2)$ and $\bar{B}_3(\tau_1, \tau_2)$ are unknown.

Then, solutions (5.42)–(5.43) and (5.45) are inserted into equations (5.36)–(5.37). This substitution makes that the secular terms appear among the inhomogeneous terms on the left sides of the equations because of the identical form of the differential operators in the equations of the subsequent orders of approximation. Rejecting the secular terms is necessary to obtain the limited solutions. In the case of equations (5.36)–(5.37) one can formulate the following solvability conditions

$$\frac{\partial B_1}{\partial \tau_1} = 0, \tag{5.46}$$

$$\frac{\partial \bar{B}_1}{\partial \tau_1} = 0, \tag{5.47}$$

$$2i\frac{\partial B_2}{\partial \tau_1} + \frac{\hat{s} w_2 w_3^2}{w_2^2 - w_3^2} B_2 = 0, \tag{5.48}$$

$$2i\frac{\partial \bar{B}_2}{\partial \tau_1} - \frac{\hat{s} w_2 w_3^2}{w_2^2 - w_3^2} \bar{B}_2 = 0. \tag{5.49}$$

Taking into account conditions (5.46)–(5.49) one can eliminate the secular terms from equations (5.36)–(5.37). The following equations are obtained then

$$\frac{\partial^2 x_2}{\partial \tau_0^2} + x_2 = w_2^2 \left(B_2 \bar{B}_2 - \frac{3}{2} \left(e^{2iw_2\tau_0} B_2^2 + e^{-2iw_2\tau_0} \bar{B}_2^2 \right) \right), \tag{5.50}$$

$$\begin{gathered} \frac{\partial^2 \phi_2}{\partial \tau_0^2} + w_2^2 \phi_2 = w_2 (w_2 + 2) \left(e^{i(w_2+1)\tau_0} B_1 B_2 + e^{-i(w_2+1)\tau_0} \bar{B}_1 \bar{B}_2 \right) + \\ w_2 (w_2 - 2) \left(e^{i(w_2-1)\tau_0} \bar{B}_1 B_2 + e^{-i(w_2-1)\tau_0} B_1 \bar{B}_2 \right) + \\ \hat{s} w_3^2 \left(B_3 e^{iw_3\tau_0} + \bar{B}_3 e^{-iw_3\tau_0} \right). \end{gathered} \tag{5.51}$$

The particular solutions to equations (5.50)–(5.51) are

$$x_2(\tau_0, \tau_1, \tau_2) = w_2^2 B_2 \bar{B}_2 + \frac{3 w_2^2 \left(e^{2iw_2\tau_0} B_2^2 + e^{-2iw_2\tau_0} \bar{B}_2^2 \right)}{2 \left(4w_2^2 - 1 \right)}, \tag{5.52}$$

$$\begin{gathered} \phi_2(\tau_0, \tau_1, \tau_2) = -\frac{w_2 (w_2 + 2) \left(e^{i(w_2+1)\tau_0} B_1 B_2 + e^{-i(w_2+1)\tau_0} \bar{B}_1 \bar{B}_2 \right)}{2w_2 + 1} + \\ \frac{w_2 (w_2 - 2) \left(e^{i(w_2-1)\tau_0} \bar{B}_1 B_2 + e^{-i(w_2-1)\tau_0} B_1 \bar{B}_2 \right)}{2w_2 - 1} + \\ \frac{\hat{s} w_3^2}{w_2^2 - w_3^2} \left(e^{iw_3\tau_0} B_3 + e^{-iw_3\tau_0} \bar{B}_3 \right). \end{gathered} \tag{5.53}$$

After substituting solutions (5.42), (5.43), (5.45) and (5.53) into equation (5.38), the secular terms should be detected and then eliminated. The solvability conditions corresponding to equation (5.38) are as follows

$$2i \frac{\partial B_3}{\partial \tau_1} + \frac{w_3^2}{w_2^2} \left(i\frac{\hat{c}_3}{\hat{s}} - \frac{\hat{s} w_3^3}{w_2^2 - w_3^2} \right) B_3 = 0, \tag{5.54}$$

$$2\mathrm{i}\,\frac{\partial \bar{B}_3}{\partial \tau_1}+\frac{w_3^2}{w_2^2}\left(\mathrm{i}\frac{\hat{c}_3}{\hat{s}}+\frac{\hat{s}w_3^3}{w_2^2-w_3^2}\right)\bar{B}_3=0. \tag{5.55}$$

After eliminating the secular terms from equations (5.36)–(5.38), we obtain

$$\begin{aligned}\frac{\partial^2 \chi_2}{\partial \tau_0^2}+w_3^2\chi_2 = {} & \frac{D_1 w_3^2\left(e^{\mathrm{i}(w_2+1)\tau_0}\,B_1B_2+e^{-\mathrm{i}(w_2+1)\tau_0}\,\bar{B}_1\bar{B}_2\right)}{(2w_2+1)\left(w_3^2-w_2^2\right)}+\\ & \frac{D_2 w_3^2\left(e^{\mathrm{i}(w_2-1)\tau_0}\bar{B}_1B_2+e^{-\mathrm{i}(w_2-1)\tau_0}\mathrm{B}_1\bar{B}_2\right)}{(2w_2-1)\left(w_3^2-w_2^2\right)}-\\ & \frac{w_3^2\left(\,w_3^4\hat{s}^2+\mathrm{i}w_2\left(w_3^2-w_2^2\right)\hat{c}_3\right)}{\left(w_3^2-w_2^2\right)^2\,\hat{s}}e^{\mathrm{i}w_2\tau_0}B_2-\\ & \frac{w_3^2\left(w_3^4\hat{s}^2-\mathrm{i}w_2\left(w_3^2-w_2^2\right)\hat{c}_3\right)}{\left(w_3^2-w_2^2\right)^2\hat{s}}e^{-\mathrm{i}w_2\tau_0}\bar{B}_2-\\ & \frac{w_3^2}{w_2^2}\left(e^{\mathrm{i}(w_3+1)\tau_0}B_1B_3+e^{-\mathrm{i}(w_3+1)\tau_0}\bar{B}_1\bar{B}_3+e^{\mathrm{i}(w_3-1)\tau_0}\bar{B}_1B_3+e^{-\mathrm{i}(w_3-1)\tau_0}B_1\bar{B}_3\right),\end{aligned} \tag{5.56}$$

where

$$D_1 = w_2^4+2\,w_2^3-w_2^2w_3^2-2w_2\left(w_3^2+1\right)-1,$$
$$D_2 = -w_2^4+2\,w_2^3+w_2^2w_3^2-2w_2\left(w_3^2+1\right)+1.$$

The particular solution to equation (5.56) has the form

$$\begin{aligned}\chi_2\left(\tau_0,\tau_1,\tau_2\right) = {} & \frac{D_1 w_3^2\left(e^{\mathrm{i}(w_2+1)\tau_0}B_1B_2+e^{-\mathrm{i}(w_2+1)\tau_0}\bar{B}_1\bar{B}_2\right)}{(2\,w_2+1)\left(w_3^2-w_2^2\right)\left(w_3^2-(w_2+1)^2\right)}+\\ & \frac{D_2 w_3^2\left(e^{\mathrm{i}(w_2-1)\tau_0}\bar{B}_1B_2+e^{-\mathrm{i}(w_2-1)\tau_0}\mathrm{B}_1\bar{B}_2\right)}{(2w_2-1)\left(w_3^2-w_2^2\right)\left(w_3^2-(w_2-1)^2\right)}-\\ & \frac{w_3^2\left(w_3^4\hat{s}^2+\mathrm{i}\,w_2\left(w_3^2-w_2^2\right)\hat{c}_3\right)}{\left(w_3^2-w_2^2\right)^3\hat{s}}e^{\mathrm{i}\,w_2\tau_0}B_2-\\ & \frac{w_3^2\left(w_3^4\hat{s}^2-\mathrm{i}\,w_2\left(w_3^2-w_2^2\right)\hat{c}_3\right)}{\left(w_3^2-w_2^2\right)^3\hat{s}}e^{-\mathrm{i}\,w_2\tau_0}\bar{B}_2+\\ & \frac{w_3^2\left(e^{\mathrm{i}(w_3+1)\tau_0}B_1B_3+e^{-\mathrm{i}(w_3+1)\tau_0}\bar{B}_1\bar{B}_3\right)}{w_2^2\left(2w_3+1\right)}-\frac{w_3^2\left(e^{\mathrm{i}(w_3-1)\tau_0}\bar{B}_1B_3+e^{-\mathrm{i}(w_3-1)\tau_0}B_1\bar{B}_3\right)}{w_2^2\left(2w_3-1\right)}.\end{aligned} \tag{5.57}$$

Solving the equations of the third-order approximation starts with equations (5.39)–(5.40). Inserting the previously obtained solutions (i.e. (5.42)–(5.43), (5.45),

(5.52)–(5.53) and (5.57)) into equations (5.39)–(5.40) generates again the secular terms. In accordance with the expectation that the solutions are limited, the secular terms should be eliminated. Consequently, we get the sequent solvability conditions

$$B_1\left(\frac{6w_2^2\left(w_2^2-1\right)B_2\bar{B}_2}{4w_2^2-1}+3\widehat{\alpha}\xi_e^2+\mathrm{i}\hat{c}_1\right)+2\mathrm{i}\frac{\partial B_1}{\partial\tau_2}=0, \tag{5.58}$$

$$\bar{B}_1\left(-\frac{6w_2^2\left(w_2^2-1\right)B_2\bar{B}_2}{4w_2^2-1}-3\widehat{\alpha}\xi_e^2+\mathrm{i}\hat{c}_1\right)+2\mathrm{i}\frac{\partial \bar{B}_1}{\partial\tau_2}=0, \tag{5.59}$$

$$\begin{aligned}&-2\mathrm{i}\,\frac{\partial B_2}{\partial\tau_2}-\frac{6w_2\left(w_2^2-1\right)B_1\bar{B}_1B_2}{4w_2^2-1}+\frac{w_2\left(8w_2^4-7w_2^2-1\right)B_2^2\bar{B}_2}{8w_2^2-2}-\\&\frac{4\mathrm{i}\left(w_3^2-w_2^2\right)^3\hat{c}_2+4\mathrm{i}w_2^4\left(w_3^2-w_2^2\right)\hat{c}_3+w_2w_3^4\left(7w_3^2-3w_2^2\right)\hat{s}^2}{4\left(w_3^2-w_2^2\right)^3}B_2=0,\end{aligned} \tag{5.60}$$

$$\begin{aligned}&2\mathrm{i}\,\frac{\partial \bar{B}_2}{\partial\tau_2}-\frac{6w_2\left(w_2^2-1\right)B_1\bar{B}_1\bar{B}_2}{4w_2^2-1}+\frac{w_2\left(8w_2^4-7w_2^2-1\right)B_2\bar{B}_2^2}{8w_2^2-2}+\\&\frac{4\mathrm{i}\left(w_3^2-w_2^2\right)^3\hat{c}_2+4\mathrm{i}w_2^4\left(w_3^2-w_2^2\right)\hat{c}_3-w_2w_3^4\left(7w_3^2-3w_2^2\right)\hat{s}^2}{4\left(w_3^2-w_2^2\right)^3}\bar{B}_2=0.\end{aligned} \tag{5.61}$$

After eliminating the secular terms we obtain two equations to be solved

$$\begin{aligned}\frac{\partial^2x_3}{\partial\tau_0^2}+x_3=\;&\hat{f}_1\cos\left(p_1\tau_0\right)-2w_3^2\hat{s}\left(e^{2\mathrm{i}w_3\tau_0}B_3^2+e^{-2\mathrm{i}w_3\tau_0}\bar{B}_3^2\right)+\\&\frac{3w_2^2(w_2+1)^2}{2w_2+1}\left(e^{\mathrm{i}(2w_2+1)\tau_0}B_1B_2^2+e^{-\mathrm{i}(2w_2+1)\tau_0}\bar{B}_1\bar{B}_2^2\right)-\\&\frac{3w_2^2(w_2-1)^2}{2w_2-1}\left(e^{\mathrm{i}(2w_2-1)\tau_0}\bar{B}_1B_2^2+e^{-\mathrm{i}(2w_2-1)\tau_0}B_1\bar{B}_2^2\right)-\\&2\hat{s}\frac{w_2^4w_3^2\left(3w_3^2+w_2^2-1\right)}{\left(w_2^2-w_3^2\right)^2\left(4w_2^2-1\right)}\left(e^{2\mathrm{i}w_2\tau_0}B_2^2+e^{-2\mathrm{i}w_2\tau_0}\bar{B}_2^2\right)+\\&\hat{s}\frac{w_2^2w_3^2}{w_2^2-w_3^2}\left(e^{\mathrm{i}(w_2+w_3)\tau_0}B_2B_3+e^{-\mathrm{i}\,(w_2+w_3)\tau_0}\bar{B}_2\bar{B}_3+\right.\\&\left.e^{\mathrm{i}(w_3-w_2)\tau_0}\bar{B}_2B_3+e^{-\mathrm{i}(w_3-w_2)\tau_0}B_2\bar{B}_3\right),\end{aligned} \tag{5.62}$$

$$
\begin{aligned}
&\frac{\partial^2 \phi_3}{\partial \tau_0^2}+w_2^2\phi_3=\hat{f}_2\cos(p_2\tau_0)+\frac{w_2^2\left(49w_2^2-1\right)}{6\left(4w_2^2-1\right)}\left(e^{3\mathrm{i}\,w_2\tau_0}B_2^3+e^{-3\mathrm{i}\,w_2\tau_0}\bar{B}_2^3\right)-\\
&\quad\frac{w_2\left(w_2^3+6w_2^2+11w_2+6\right)}{2w_2+1}\left(e^{\mathrm{i}(w_2+2)\tau_0}B_1^2B_2+e^{-\mathrm{i}(w_2+2)\tau_0}\bar{B}_1^2\bar{B}_2\right)+\\
&\quad\frac{w_2\left(w_2^3-6w_2^2+11w_2-6\right)}{2w_2-1}\left(e^{\mathrm{i}(w_2-2)\tau_0}\bar{B}_1^2B_2+e^{-\mathrm{i}(w_2-2)\tau_0}B_1^2\bar{B}_2\right)+\\
&\hat{s}\frac{w_3^2(w_2+1)^2\left(w_2^3+w_2^2-w_2\left(3w_3^2+1\right)-1\right)}{(2\,w_2+1)\left(w_2^2-w_3^2\right)\left((w_2+1)^2-w_3^2\right)}\left(e^{\mathrm{i}(w_2+1)\tau_0}B_1B_2+e^{-\mathrm{i}(w_2+1)\tau_0}\bar{B}_1\bar{B}_2\right)+\\
&\hat{s}\frac{w_3^2(w_2-1)^2\left(w_2^3-w_2^2-w_2\left(3w_3^2+1\right)+1\right)}{(2\,w_2-1)\left(w_2^2-w_3^2\right)\left((w_2-1)^2-w_3^2\right)}\left(e^{\mathrm{i}(w_2-1)\tau_0}\bar{B}_1B_2+e^{-\mathrm{i}(w_2-1)\tau_0}B_1\bar{B}_2\right)-\\
&\quad w_3\frac{\left(w_3^5\hat{s}^2-\mathrm{i}w_2^2\left(w_2^2-w_3^2\right)\hat{c}_3\right)e^{\mathrm{i}w_3\tau_0}B_3+\left(w_3^5\hat{s}^2+\mathrm{i}w_2^2\left(w_2^2-w_3^2\right)\hat{c}_3\right)e^{-\mathrm{i}w_3\tau_0}\bar{B}_3}{\left(w_3^2-w_2^2\right)^2}+\\
&\hat{s}\frac{w_3^2\left(w_3^2(w_3+1)^2-w_2^2\left(2w_3^3+6w_3^2+4w_3+1\right)\right)}{w_2^2\left(2w_3+1\right)\left(w_3^2-w_2^2\right)}\left(e^{\mathrm{i}(w_3+1)\tau_0}B_1B_3+e^{-\mathrm{i}(w_3+1)\tau_0}\bar{B}_1\bar{B}_3\right)-\\
&\hat{s}\frac{w_3^2\left(w_3^2(w_3-1)^2+w_2^2\left(2w_3^3-6w_3^2+4w_3-1\right)\right)}{w_2^2\left(2w_3-1\right)\left(w_3^2-w_2^2\right)}\left(e^{\mathrm{i}(w_3-1)\tau_0}\bar{B}_1B_3+e^{-\mathrm{i}(w_3-1)\tau_0}B_1\bar{B}_3\right).
\end{aligned}
\tag{5.63}
$$

Solving equations (5.62)–(5.63), we get

$$
\begin{aligned}
x_3(\tau_0,\tau_1,\tau_2)=&\frac{\hat{f}_1\cos(p_1\tau_0)}{1-p_1^2}+\frac{2\,w_3^2\,\hat{s}}{4w_3^2-1}\left(e^{2\mathrm{i}\,w_3\tau_0}B_3^2+e^{-2\mathrm{i}\,w_3\tau_0}\bar{B}_3^2\right)-\\
&\frac{3w_2(w_2+1)}{4(2w_2+1)}\left(e^{\mathrm{i}(2w_2+1)\tau_0}B_1B_2^2+e^{-\mathrm{i}(2w_2+1)\tau_0}\bar{B}_1\bar{B}_2^2\right)+\\
&\frac{3w_2(w_2-1)}{4(2w_2-1)}\left(e^{\mathrm{i}(2w_2-1)\tau_0}\bar{B}_1B_2^2+e^{-\mathrm{i}(2w_2-1)\tau_0}B_1\bar{B}_2^2\right)+\\
&\hat{s}\frac{2w_2^4w_3^2\left(3w_3^2+w_2^2-1\right)}{\left(w_3^2-w_2^2\right)^2\left(4w_2^2-1\right)^2}\left(e^{2\mathrm{i}w_2\tau_0}B_2^2+e^{-2\mathrm{i}w_2\tau_0}\bar{B}_2^2\right)-\\
&\hat{s}\frac{w_2^2w_3^2}{w_2^2-w_3^2}\left(\frac{e^{\mathrm{i}(w_2+w_3)\tau_0}B_2B_3+e^{-\mathrm{i}(w_2+w_3)\tau_0}\bar{B}_2\bar{B}_3}{(w_2+w_3)^2-1}+\right.\\
&\left.\frac{e^{\mathrm{i}(w_3-w_2)\tau_0}\bar{B}_2B_3+e^{-\mathrm{i}(w_3-w_2)\tau_0}B_2\bar{B}_3}{\left((w_3-w_2)^2-1\right)}\right),
\end{aligned}
\tag{5.64}
$$

$$\phi_3(\tau_0,\ \tau_1,\ \tau_2) = \frac{\hat{f}_2 \cos(p_2\tau_0)}{w_2^2 - p_2^2} - \frac{49w_2^2 - 1}{48\,(4w_2^2 - 1)}\left(e^{3\mathrm{i}\,w_2\tau_0} B_2^3 + e^{-3\mathrm{i}\,w_2\tau_0}\bar{B}_2^3\right) +$$

$$\frac{w_2\,(w_2^2 + 5w_2 + 6)}{4\,(2w_2 + 1)}\left(e^{\mathrm{i}(w_2+2)\tau_0} B_1^2 B_2 + e^{-\mathrm{i}(w_2+2)\tau_0}\bar{B}_1^2\bar{B}_2\right) +$$

$$\frac{w_2\,(w_2^2 - 5w_2 + 6)}{4\,(2w_2 - 1)}\left(e^{\mathrm{i}(w_2-2)\tau_0}\bar{B}_1^2 B_2 + e^{-\mathrm{i}(w_2-2)\tau_0} B_1^2\bar{B}_2\right) - \tag{5.65}$$

$$\hat{s}\frac{w_3^2(w_2+1)^2\,(w_2^3 + w_2^2 - w_2\,(3\,w_3^2 + 1) - 1)}{(2\,w_2+1)^2\,(w_2^2 - w_3^2)\left((w_2+1)^2 - w_3^2\right)}\left(e^{\mathrm{i}(w_2+1)\tau_0} B_1 B_2 + e^{-\mathrm{i}(w_2+1)\tau_0}\bar{B}_1\bar{B}_2\right) +$$

$$\hat{s}\frac{w_3^2(w_2-1)^2\,(w_2^3 - w_2^2 - w_2\,(3\,w_3^2 + 1) + 1)}{(2\,w_2-1)^2\,(w_2^2 - w_3^2)\left((w_2-1)^2 - w_3^2\right)}\left(e^{\mathrm{i}(w_2-1)\tau_0}\bar{B}_1 B_2 + e^{-\mathrm{i}(w_2-1)\tau_0} B_1\bar{B}_2\right) +$$

$$\hat{s}\frac{w_3^2\left(w_3^2(w_3+1)^2 - w_2^2\,(2w_3^3 + 6w_3^2 + 4w_3 + 1)\right)}{w_2^2\,(2w_3+1)\,(w_3^2 - w_2^2)\left(w_2^2 - (w_3+1)^2\right)}\left(e^{\mathrm{i}(w_3+1)\tau_0} B_1 B_3 + e^{-\mathrm{i}(w_3+1)\tau_0}\bar{B}_1\bar{B}_3\right) -$$

$$\hat{s}\frac{w_3^2\left(w_3^2(w_3-1)^2 + w_2^2\,(2w_3^3 - 6w_3^2 + 4w_3 - 1)\right)}{w_2^2\,(2w_3-1)\,(w_3^2 - w_2^2)\left(w_2^2 - (w_3-1)^2\right)}\left(e^{\mathrm{i}(w_3-1)\tau_0}\bar{B}_1 B_3 + e^{-\mathrm{i}(w_3-1)\tau_0} B_1\bar{B}_3\right) -$$

$$w_3\frac{\left(w_3^5\hat{s}^2 - \mathrm{i}w_2^2\,(w_2^2 - w_3^2)\,\hat{c}_3\right) e^{\mathrm{i}w_3\tau_0} B_3 + \left(w_3^5\hat{s}^2 + \mathrm{i}w_2^2\,(w_2^2 - w_3^2)\,\hat{c}_3\right) e^{-\mathrm{i}w_3\tau_0}\bar{B}_3}{\left(w_2^2 - w_3^2\right)^3}.$$

Then, solutions (5.42)–(5.43), (5.45), (5.52)–(5.53), (5.57) and (5.64)–(5.65) are inserted into equation (5.41). Elimination of the secular terms leads to the consecutive solvability conditions

$$-2\mathrm{i}\,\frac{\partial B_3}{\partial\tau_2} + \frac{2w_3^3 B_1\bar{B}_1 B_3}{w_2^4\,(4w_3^2 - 1)} + \frac{w_3^3\,(2w_2^2 - w_3^2)\,B_2\bar{B}_2 B_3}{\left(w_2^2 - w_3^2\right)^2} + \frac{w_3}{2}B_3^2\bar{B}_3 + \tag{5.66}$$

$$w_3^3\frac{\left(w_2^2 - w_3^2\right)^3\hat{c}_3^2 - 4\mathrm{i}\,w_3\,(w_2^2 - w_3^2)\,(w_3^4 - 2w_2^2)\,\hat{c}_3\hat{s}^2 - w_3^6\,(7w_2^2 - 3\,w_3^2)\,\hat{s}^4}{4w_2^4\left(w_2^2 - w_3^2\right)^3\hat{s}^2}B_3 = 0,$$

$$2\mathrm{i}\,\frac{\partial\bar{B}_3}{\partial\tau_2} + \frac{2w_3^3 B_1\bar{B}_1\bar{B}_3}{w_2^4\,(4w_3^2 - 1)} + \frac{w_3^3\,(2w_2^2 - w_3^2)\,B_2\bar{B}_2\bar{B}_3}{\left(w_2^2 - w_3^2\right)^2} + \frac{w_3}{2}B_3\bar{B}_3^2 + \tag{5.67}$$

$$w_3^4\frac{\left(w_2^2 - w_3^2\right)^3\hat{c}_3^2 + 4\mathrm{i}\,w_3\,(w_2^2 - w_3^2)\,(w_3^4 - 2w_2^2)\,\hat{c}_3\hat{s}^2 - w_3^6\,(7w_2^2 - 3\,w_3^2)\,\hat{s}^4}{4w_2^4\left(w_2^2 - w_3^2\right)^3\hat{s}^2}\bar{B}_3 = 0.$$

Taking into account (5.66), the following form of equation (5.41) that does not contain any secular terms are obtained

$$\frac{\partial^2 \chi_3}{\partial \tau_0^2} + w_3^2 \chi_3 = \frac{w_3^2}{\hat{s} w_2^2} \hat{f}_3 \cos(p_3 \tau_0) + \frac{p_2^2 w_3^2 \hat{f}_2}{w_2^2 (w_2^2 - p_2^2)} \cos(p_2 \tau_0) +$$
$$\frac{w_3^2}{6}\left(e^{3i\, w_3 \tau_0} B_3^3 + e^{-3i\, w_3 \tau_0} \bar{B}_3^3\right) - \frac{w_3^2 D_3 \left(e^{3i w_2 \tau_0} B_2^3 + e^{-3i w_2 \tau_0} \bar{B}_2^3\right)}{48 (w_2^2 - w_3^2)^3 (4 w_2^2 - 1)} +$$
$$\frac{D_4 w_2^4 w_3^2 \left(e^{i w_2 \tau_0} B_2^2 \bar{B}_2 + e^{-i w_2 \tau_0} B_2 \bar{B}_2^2\right)}{2 (4 w_2^2 - 1)(w_2^2 - w_3^2)^3} - \tag{5.68}$$
$$\frac{w_2^2 w_3^2 (e^{i(w_2 + 2 w_3)\tau_0} B_2 B_3^2 + e^{-i(w_2 + 2w_3)\tau_0} \bar{B}_2 \bar{B}_3^2)}{2 (w_2^2 - w_3^2)} +$$
$$\frac{w_2^2 w_3^2 (e^{i(w_2 - 2 w_3)\tau_0} B_2 \bar{B}_3^2 + e^{-i(w_2 - 2w_3)\tau_0} \bar{B}_2 B_3^2)}{2 (w_2^2 - w_3^2)} +$$
$$\frac{w_3^2 D_5 \left(e^{i(2w_2 + w_3)\tau_0} B_2^2 B_3 + e^{-i(2w_2 + w_3)\tau_0} \bar{B}_2^2 \bar{B}_3\right)}{2 (4 w_2^2 - 1)(w_2^2 - w_3^2)^2} -$$
$$\frac{w_3^2 D_5 (e^{i(2w_2 - w_3)\tau_0} B_2^2 \bar{B}_3 + e^{-i(2w_2 - w_3)\tau_0} \bar{B}_2^2 B_3)}{2 (4 w_2^2 - 1)(w_2^2 - w_3^2)^2} -$$
$$\frac{w_3^4 \left(e^{i(w_3+2)\tau_0} B_1^2 B_3 + e^{-i(w_3+2)\tau_0} \bar{B}_1^2 \bar{B}_3\right)}{w_2^4 (2 w_3 + 1)} + \frac{w_3^4 \left(e^{i(w_3-2)\tau_0} \bar{B}_1^2 B_3 + e^{-i(w_3-2)\tau_0} B_1^2 \bar{B}_3\right)}{w_2^4 (2 w_3 - 1)} +$$
$$\frac{w_3^2 D_6 \left(e^{i(w_2+2)\tau_0} B_1^2 B_2 + e^{-i(w_2+2)\tau_0} \bar{B}_1^2 \bar{B}_2\right)}{4 w_2^2 (2 w_2 + 1)(w_2^2 - w_3^2)\left((w_2+1)^2 - w_3^2\right)} +$$
$$\frac{D_7 w_3^2 \left(e^{i(w_2-2)\tau_0} \bar{B}_1^2 B_2 + e^{-i(w_2-2)\tau_0} B_1^2 \bar{B}_2\right)}{4 w_2^2 (2 w_2 - 1)(w_2^2 - w_3^2)\left((w_2-1)^2 - w_3^2\right)} -$$
$$\frac{w_3^2 \left(e^{i w_2 \tau_0} B_2 + e^{-i w_2 \tau_0} \bar{B}_2\right)}{(w_2^2 - w_3^2)} \left(w_2^2 B_3 \bar{B}_3 + \right.$$
$$\left. \frac{2 D_8 B_1 \bar{B}_1}{w_2^2 (4 w_2^2 - 1)\left((w_2 + w_3)^2 - 1\right)\left((w_2 - w_3)^2 - 1\right)} \right) +$$
$$\frac{\hat{s} D_9 \left(e^{i(w_3+1)\tau_0} B_1 B_3 + e^{-i(w_3+1)\tau_0} \bar{B}_1 \bar{B}_3\right)}{w_2^4 (2 w_3 + 1)(w_2^2 - w_3^2)\left((w_3+1)^2 - w_2^2\right)} -$$

$$\frac{\hat{s}D_{10}\left(e^{\mathrm{i}(w_3-1)\tau_0}\bar{B}_1B_3+e^{-\mathrm{i}(w_3-1)\tau_0}B_1\bar{B}_3\right)}{w_2^4(2w_3-1)\left(w_2^2-w_3^2\right)\left(w_2^2-(w_3-1)^2\right)}+$$

$$\frac{w_3^4\hat{s}D_{11}\left(B_1B_2e^{\mathrm{i}(w_2+1)\tau_0}+\bar{B}_1\bar{B}_2e^{-\mathrm{i}(w_2+1)\tau_0}\right)}{w_2^2(2w_2+1)^2\left(w_2^2-w_3^2\right)^3\left((w_2+1)^2-w_3^2\right)}+$$

$$\frac{\mathrm{i}\,\hat{c}_3w_3^2D_{12}\left(B_1B_2e^{\mathrm{i}(w_2+1)\tau_0}-\bar{B}_1\bar{B}_2e^{-\mathrm{i}(w_2+1)\tau_0}\right)}{\hat{s}w_2^2(2w_2+1)\left(w_2^2-w_3^2\right)^2\left((w_2+1)^2-w_3^2\right)}+$$

$$\frac{w_3^4\hat{s}D_{13}\left(\bar{B}_1B_2e^{\mathrm{i}(w_2-1)\tau_0}+B_1\bar{B}_2e^{-\mathrm{i}(w_2-1)\tau_0}\right)}{w_2^2(2w_2-1)^2\left(w_2^2-w_3^2\right)^3\left((w_2-1)^2-w_3^2\right)}+$$

$$\frac{\mathrm{i}\,\hat{c}_3w_3^2D_{14}\left(\bar{B}_1B_2e^{\mathrm{i}(w_2-1)\tau_0}-B_1\bar{B}_2e^{-\mathrm{i}(w_2-1)\tau_0}\right)}{\hat{s}w_2^2(2w_2-1)\left(w_2^2-w_3^2\right)^2\left((w_2-1)^2-w_3^2\right)}+$$

$$\frac{w_3^4B_2e^{\mathrm{i}w_2\tau_0}}{w_2^2-w_3^2}\left(\frac{2w_3^6\hat{s}^2}{\left(w_2^2-w_3^2\right)^3}-\frac{\hat{c}_3^2}{\hat{s}^2\left(w_2^2-w_3^2\right)}-\mathrm{i}\,\frac{\left(3w_2^4+w_2^2w_3^2+3w_3^4\right)\hat{c}_3}{2w_2\left(w_2^2-w_3^2\right)^2}-\mathrm{i}\frac{\hat{c}_2}{w_2}\right)+$$

$$\frac{w_3^4\bar{B}_2e^{-\mathrm{i}w_2\tau_0}}{w_2^2-w_3^2}\left(\frac{2w_3^6\hat{s}^2}{\left(w_2^2-w_3^2\right)^3}-\frac{\hat{c}_3^2}{\hat{s}^2\left(w_2^2-w_3^2\right)}+\mathrm{i}\frac{\left(3w_2^4+w_2^2w_3^2+3w_3^4\right)\hat{c}_3}{2w_2\left(w_2^2-w_3^2\right)^2}+\mathrm{i}\frac{\hat{c}_2}{w_2}\right),$$

where

$$D_3=81w_2^8-3w_2^6\left(49w_3^2+27\right)+147w_2^4w_3^2\left(w_3^2+1\right)-w_2^2w_3^4\left(49w_3^2+75\right)+w_3^6,$$

$$D_4=8w_2^6-w_2^4\left(16w_3^2+3\right)+w_2^2\left(8w_3^4-2w_3^2+1\right)+w_3^4,$$

$$D_5=4w_2^6-4w_2^4+6w_2^2w_3^2-3w_3^4,$$

$$D_6=w_2^9+7w_2^8-w_2^7\left(2w_3^2-17\right)-w_2^6\left(12w_3^2-13\right)+w_2^5\left(w_3^4-23w_3^2-10\right)+$$
$$w_2^4\left(5w_3^4-13w_3^2-20\right)+2w_2^3\left(3w_3^4+5w_3^2-4\right)+20w_2^2w_3^2+16w_2w_3^2+4w_3^2,$$

$$D_7=w_2^9-7w_2^8-w_2^7\left(2w_3^2-17\right)+w_2^6\left(12w_3^2-13\right)+w_2^5\left(w_3^4-23w_3^2-10\right)-$$
$$w_2^4\left(5w_3^4-13w_3^2-20\right)+2w_2^3\left(3w_3^4+5w_3^2-4\right)-20w_2^2w_3^2+16w_2w_3^2-4w_3^2,$$

$$D_8=3w_2^6\left(w_2^2-1\right)^2+w_3^2\left(1-6w_2^2+9w_2^4-10w_2^6-6w_2^8\right)+$$

$$w_3^4\left(3w_2^6+4w_2^4-3w_2^2-1\right),$$

$$D_9=w_3^4\left(w_3+1\right)\left(w_3^2(w_3+1)^2-w_2^2w_3^2-w_2^4\left(2w_3^2+3w_3+1\right)\right),$$

$$D_{10}=w_3^4\left(w_3-1\right)\left(w_3^2(w_3-1)^2-w_2^2w_3^2-w_2^4\left(2w_3^2-3w_3+1\right)\right),$$

$$D_{11}=-w_2^{10}\left(w_2+5\right)+w_2^9\left(7w_3^2-9\right)+5w_2^8\left(5w_3^2-1\right)-$$
$$w_2^7\left(11w_3^4-35w_3^2-5\right)-w_2^6\left(35w_3^4-15w_3^2-9\right)+w_2^5\left(5w_3^6-43w_3^4-16w_3^2+5\right)+$$
$$+w_2^4\left(15w_3^6-23w_3^4-23w_3^2+1\right)+w_2^3w_3^2\left(17w_3^4-3w_3^2-11\right)+$$
$$w_2^2w_3^2\left(13w_3^4+w_3^2-2\right)+w_3^6\left(6w_2+1\right),$$

$$D_{12}=w_2w_3^2\left(2w_2^6+10w_2^5+18w_2^4+18w_2^3+12w_2^2+5w_2+1\right)-$$
$$w_2^5\left(w_2^4+5w_2^3+9w_2^2+7w_2+2\right)-w_3^4\left(w_2^5+5w_2^4+9w_2^3+11w_2^2+6w_2+1\right),$$

$$D_{13}=w_2^{10}\left(w_2-5\right)-w_2^9\left(7w_3^2-9\right)+5w_2^8\left(5w_3^2-1\right)+w_2^7\left(11w_3^4-35w_3^2-5\right)-$$
$$w_2^6\left(35w_3^4-15w_3^2-9\right)-w_2^5\left(5w_3^6-43w_3^4-16w_3^2+5\right)+$$
$$w_2^4\left(15w_3^6-23w_3^4-23w_3^2+1\right)-w_2^3w_3^2\left(17w_3^4-3w_3^2-11\right)+$$
$$w_2^2w_3^2\left(13w_3^4+w_3^2-2\right)-w_3^6\left(6w_2-1\right),$$

$$D_{14}=w_2w_3^2\left(-2w_2^6+10w_2^5-18w_2^4+18w_2^3-12w_2^2+5w_2-1\right)+$$
$$w_2^5\left(w_2^4-5w_2^3+9w_2^2-7w_2+2\right)+w_3^4\left(w_2^5-5w_2^4+9w_2^3-11w_2^2+6w_2-1\right).$$

The particular solution to equation (5.68) is as follows

$$\chi_3\left(\tau_0,\tau_1,\tau_2\right)=\frac{w_3^2}{\hat{s}w_2^2}\hat{f}_3\frac{\cos\left(p_3\tau_0\right)}{w_3^2-p_3^2}+\frac{p_2^2w_3^2\hat{f}_2\cos\left(p_2\tau_0\right)}{w_2^2\left(w_2^2-p_2^2\right)\left(w_3^2-p_3^2\right)}-$$
$$\frac{e^{3\mathrm{i}w_3\tau_0}B_3^3+e^{-3\mathrm{i}w_3\tau_0}\bar{B}_3^3}{48}-\frac{w_3^2D_3\left(e^{3\mathrm{i}w_2\tau_0}B_2^3+e^{-3\mathrm{i}w_2\tau_0}\bar{B}_2^3\right)}{48\left(w_2^2-w_3^2\right)^3\left(4w_2^2-1\right)\left(w_3^2-9w_2^2\right)}-$$
$$\frac{D_4w_2^4w_3^2\left(e^{\mathrm{i}w_2\tau_0}\,B_2^2\bar{B}_2+e^{-\mathrm{i}w_2\tau_0}B_2\bar{B}_2^2\right)}{2\left(4\,w_2^2-1\right)\left(w_2^2-w_3^2\right)^4}+\tag{5.69}$$

$$\frac{w_2^2 w_3^2}{2}\left(\frac{e^{\mathrm{i}(w_2+2\,w_3)\tau_0}B_2B_3^2+e^{-\mathrm{i}(w_2+2w_3)\tau_0}\bar{B}_2\bar{B}_3^2}{(w_2+3w_3)(w_2+w_3)\left(w_2^2-w_3^2\right)}+\right.$$

$$\left.\frac{e^{\mathrm{i}(w_2+2w_3)\tau_0}B_2B_3^2+e^{-\mathrm{i}(w_2+2w_3)\tau_0}\bar{B}_2\bar{B}_3^2}{(w_2-3w_3)(w_2-w_3)\left(w_2^2-w_3^2\right)}\right)-$$

$$\frac{w_3^2 D_5}{8w_2}\left(\frac{e^{\mathrm{i}(2w_2+w_3)\tau_0}B_2^2B_3+e^{-\mathrm{i}(2w_2+w_3)\tau_0}\bar{B}_2^2\bar{B}_3}{(w_2+w_3)\left(4w_2^2-1\right)\left(w_2^2-w_3^2\right)^2}+\right.$$

$$\left.\frac{e^{\mathrm{i}(2w_2-w_3)\tau_0}B_2^2\bar{B}_3+e^{-\mathrm{i}(2w_2-w_3)\tau_0}\bar{B}_2^2B_3}{\left(4w_2^2-1\right)\left(w_2^2-w_3^2\right)^2(w_2-w_3)}\right)+$$

$$\frac{w_3^4\left(e^{\mathrm{i}(w_3+2)\tau_0}B_1^2B_3+e^{-\mathrm{i}(w_3+2)\tau_0}\bar{B}_1^2\bar{B}_3\right)}{4w_2^4(2w_3+1)(w_3+1)}+\frac{w_3^4\left(e^{\mathrm{i}(w_3-2)\tau_0}\bar{B}_1^2B_3+e^{-\mathrm{i}(w_3-2)\tau_0}B_1^2\bar{B}_3\right)}{4w_2^4(2w_3-1)(w_3-1)}+$$

$$\frac{w_3^2D_6\left(e^{\mathrm{i}(w_2+2)\tau_0}B_1^2B_2+e^{-\mathrm{i}(w_2+2)\tau_0}\bar{B}_1^2\bar{B}_2\right)}{4w_2^2(2w_2+1)\left(w_2^2-w_3^2\right)\left((w_2+1)^2-w_3^2\right)\left(w_3^2-(w_2+2)^2\right)}+$$

$$\frac{D_7w_3^2\left(e^{\mathrm{i}(w_2-2)\tau_0}\,\bar{B}_1^2B_2+e^{-\mathrm{i}(w_2-2)\tau_0}\,B_1^2\bar{B}_2\right)}{4w_2^2(2w_2-1)\left(w_2^2-w_3^2\right)\left((w_2-1)^2-w_3^2\right)\left(w_3^2-(w_2-2)^2\right)}+$$

$$\frac{w_3^2\left(e^{\mathrm{i}w_2\tau_0}\,B_2+e^{-\mathrm{i}w_2\tau_0}\bar{B}_2\right)}{\left(w_2^2-w_3^2\right)^2}\left(w_2^2B_3\bar{B}_3+\right.$$

$$\left.\frac{2D_8B_1\bar{B}_1}{w_2^2\left(4w_2^2-1\right)\left((w_2+w_3)^2-1\right)\left((w_2-w_3)^2-1\right)}\right)-$$

$$\frac{\hat{s}D_9\left(e^{\mathrm{i}(w_3+1)\tau_0}B_1B_3+e^{-\mathrm{i}(w_3+1)\tau_0}\bar{B}_1\bar{B}_3\right)}{w_2^4(2w_3+1)^2\left(w_2^2-w_3^2\right)\left((w_3+1)^2-w_2^2\right)}-$$

$$\frac{\hat{s}D_{10}\left(e^{\mathrm{i}(w_3-1)\tau_0}\bar{B}_1B_3+e^{-\mathrm{i}(w_3-1)\tau_0}B_1\bar{B}_3\right)}{w_2^4(2w_3-1)^2\left(w_2^2-w_3^2\right)\left(w_2^2-(w_3-1)^2\right)}-$$

$$\frac{w_3^4\hat{s}D_{11}\left(B_1B_2e^{\mathrm{i}(w_2+1)\tau_0}+\bar{B}_1\bar{B}_2e^{-\mathrm{i}(w_2+1)\tau_0}\right)}{w_2^2(2w_2+1)^2\left(w_2^2-w_3^2\right)^3\left((w_2+1)^2-w_3^2\right)^2}-$$

$$\frac{\mathrm{i}\,\hat{c}_3w_3^2D_{12}\left(B_1B_2e^{\mathrm{i}(w_2+1)\tau_0}-\bar{B}_1\bar{B}_2e^{-\mathrm{i}(w_2+1)\tau_0}\right)}{\hat{s}w_2^2(2w_2+1)\left(w_2^2-w_3^2\right)^2\left((w_2+1)^2-w_3^2\right)^2}-$$

$$\frac{w_3^4\hat{s}D_{13}\left(\bar{B}_1B_2e^{\mathrm{i}(w_2-1)\tau_0}+B_1\bar{B}_2e^{-\mathrm{i}(w_2-1)\tau_0}\right)}{w_2^2(2w_2-1)^2\left(w_2^2-w_3^2\right)^3\left((w_2-1)^2-w_3^2\right)^2}-$$

$$\frac{\mathrm{i}\,\hat{c}_3w_3^2D_{14}\left(\bar{B}_1B_2e^{\mathrm{i}(w_2-1)\tau_0}-B_1\bar{B}_2e^{-\mathrm{i}(w_2-1)\tau_0}\right)}{\hat{s}w_2^2\left(2w_2-1\right)\left(w_2^2-w_3^2\right)^2\left((w_2-1)^2-w_3^2\right)^2}-$$

$$\frac{w_3^4B_2e^{\mathrm{i}w_2\tau_0}}{\left(w_2^2-w_3^2\right)^2}\left(\frac{2w_3^6\hat{s}^2}{\left(w_2^2-w_3^2\right)^3}-\frac{\hat{c}_3^2}{\hat{s}^2\left(w_2^2-w_3^2\right)}-\right.$$

$$\left.\mathrm{i}\frac{\left(3w_2^4+w_2^2w_3^2+3w_3^4\right)\hat{c}_3}{2w_2\left(w_2^2-w_3^2\right)^2}-\mathrm{i}\frac{\hat{c}_2}{w_2}\right)-$$

$$\frac{w_3^4\bar{B}_2e^{-\mathrm{i}w_2\tau_0}}{\left(w_2^2-w_3^2\right)^2}\left(\frac{2w_3^6\hat{s}^2}{\left(w_2^2-w_3^2\right)^3}-\frac{\hat{c}_3^2}{\hat{s}^2\left(w_2^2-w_3^2\right)}+\right.$$

$$\left.\mathrm{i}\,\frac{\left(3w_2^4+w_2^2w_3^2+3w_3^4\right)\hat{c}_3}{2w_2\left(w_2^2-w_3^2\right)^2}+\mathrm{i}\frac{\hat{c}_2}{w_2}\right).$$

The unknown complex-valued functions $B_1(\tau_1,\tau_2)$, $B_2(\tau_1,\tau_2)$, $B_3(\tau_1,\tau_2)$ and their complex conjugates $\bar{B}_1(\tau_1,\tau_2)$, $\bar{B}_2(\tau_1,\tau_2)$, $\bar{B}_3(\tau_1,\tau_2)$ that occur in the solutions (5.42)–(5.43), (5.45), (5.52)–(5.53), (5.57), (5.64)–(5.65) and (5.69) are restricted by the solvability conditions. We present all the complex-valued functions in the exponential form

$$B_1(\tau_1,\tau_2)=\frac{1}{2}b_1(\tau_1,\tau_2)e^{\mathrm{i}\psi_1(\tau_1,\tau_2)},\quad \bar{B}_1(\tau_1,\tau_2)=\frac{1}{2}b_1(\tau_1,\tau_2)e^{-\mathrm{i}\psi_1(\tau_1,\tau_2)},\tag{5.70}$$

$$B_2(\tau_1,\tau_2)=\frac{1}{2}b_2(\tau_1,\tau_2)e^{\mathrm{i}\psi_2(\tau_1,\tau_2)},\quad \bar{B}_2(\tau_1,\tau_2)=\frac{1}{2}b_2(\tau_1,\tau_2)e^{-\mathrm{i}\psi_2(\tau_1,\tau_2)},\tag{5.71}$$

$$B_3(\tau_1,\tau_2)=\frac{1}{2}b_3(\tau_1,\tau_2)e^{\mathrm{i}\psi_3(\tau_1,\tau_2)},\quad \bar{B}_3(\tau_1,\tau_2)=\frac{1}{2}b_3(\tau_1,\tau_2)e^{-\mathrm{i}\psi_3(\tau_1,\tau_2)},\tag{5.72}$$

where the functions $b_i(\tau_1,\tau_2)$, $\psi_i(\tau_1,\tau_2)$ are real-valued for $i=1,2,3$.

Inserting relationships (5.70)–(5.72) into the solvability conditions, i.e. into equations (5.46)–(5.49), (5.52)–(5.53), (5.58)–(5.61) and (5.66)–(5.67) yields the system of twelve partial differential equations with unknown functions $b_i(\tau_1,\tau_2)$, $\psi_i(\tau_1,\tau_2)$, $i=1,2,3$. It is necessary to solve the system with respect to the derivatives

$$\frac{\partial b_i}{\partial\tau_1},\quad \frac{\partial b_i}{\partial\tau_2},\quad \frac{\partial\psi_i}{\partial\tau_1},\quad \frac{\partial\psi_i}{\partial\tau_2},\qquad i=1,2,3.\tag{5.73}$$

The solution follows

$$\frac{\partial b_1}{\partial \tau_1}=0, \qquad \frac{\partial b_1}{\partial \tau_2}=-\frac{\hat{c}_1}{2}b_1,$$

$$\frac{\partial b_2}{\partial \tau_1}=0, \qquad \frac{\partial b_2}{\partial \tau_2}=-\frac{\left(w_3^2-w_2^2\right)^2\hat{c}_2+w_2^4\hat{c}_3}{2\left(w_3^2-w_2^2\right)^2}b_2,$$

$$\frac{\partial b_3}{\partial \tau_1}=-\frac{w_3^2\hat{c}_3}{2w_2^2\hat{s}}b_3, \qquad \frac{\partial b_3}{\partial \tau_2}=-\frac{w_3^4\left(w_3^4-2w_2^4\right)\hat{c}_3}{2w_2^4\left(w_3^2-w_2^2\right)^2}b_3,$$

$$\frac{\partial \psi_1}{\partial \tau_1}=0, \qquad \frac{\partial \psi_1}{\partial \tau_2}=-\frac{3\left(\left(w_2^4-w_2^2\right)b_2^2+2\xi_e^2\hat{\alpha}\left(4w_2^2-1\right)\right)}{4\left(4w_2^2-1\right)}, \tag{5.74}$$

$$\frac{\partial \psi_2}{\partial \tau_1}=-\frac{w_2w_3^2\hat{s}}{2\left(w_3^2-w_2^2\right)},$$

$$\frac{\partial \psi_2}{\partial \tau_2}=\frac{w_2}{16}\left(-\frac{\left(w_2^2-1\right)\left(-12b_1^2+\left(8w_2^2+1\right)b_2^2\right)}{4w_2^2-1}-\frac{2w_3^4\left(3w_2^2-7w_3^2\right)\hat{s}^2}{\left(w_3^2-w_2^2\right)^3}\right),$$

$$\frac{\partial \psi_3}{\partial \tau_1}=\frac{w_3^5\hat{s}}{2w_2^2\left(w_3^2-w_2^2\right)},$$

$$\frac{\partial \psi_3}{\partial \tau_2}=-\frac{w_3^3b_1^2}{4w_2^4\left(4w_3^2-1\right)}+\frac{w_3^3\left(w_3^2-2w_2^2\right)b_2^2}{8\left(w_3^2-w_2^2\right)^2}-$$
$$\frac{w_3b_3^2}{16}-\frac{w_3^3\hat{c}_3^2}{8w_2^4\hat{s}^2}-\frac{w_3^9\left(7w_2^2-3w_3^2\right)\hat{s}^2}{8w_2^4\left(w_3^2-w_2^2\right)^3}.$$

The functions $b_i(\tau_1,\tau_2)$, $\psi_i(\tau_1,\tau_2)$, $i=1,2,3$ depend only on the slow time variables. It means that considering definition (5.27) one can write formally that their derivatives with respect to the primary time variable τ take the following form

$$\frac{db_i}{d\tau}=\varepsilon\frac{\partial b_i}{\partial \tau_1}+\varepsilon^2\frac{\partial b_i}{\partial \tau_2}, \qquad i=1,2,3, \tag{5.75}$$

$$\frac{d\psi_i}{d\tau}=\varepsilon\frac{\partial \psi_i}{\partial \tau_1}+\varepsilon^2\frac{\partial \psi_i}{\partial \tau_2}, \qquad i=1,2,3. \tag{5.76}$$

The problem of determining the functions $b_i(\tau_1,\tau_2)$, $\psi_i(\tau_1,\tau_2)$ is originally formulated in the slow time variables. Identities (5.75)–(5.76) allow for converting this formulation into the description in which only the primary time τ occurs. Moreover, the system of twelve partial differential equations has been substituted by the system of six ordinary differential equations as a result of applying the identities. After substituting relationships (5.74) in equations (5.75)–(5.76) and taking into account the relations (5.32), we get

$$\frac{da_1}{d\tau}=-\frac{1}{2}c_1a_1(\tau), \tag{5.77}$$

$$\frac{da_2}{d\tau} = -\frac{c_3 w_2^4 + c_2\left(w_3^2 - w_2^2\right)^2}{2\left(w_3^2 - w_2^2\right)^2} a_2(\tau), \tag{5.78}$$

$$\frac{da_3}{d\tau} = -\frac{c_3 w_3^2\left(w_2^6 - 2(s+1) w_2^4 w_3^2 + w_2^2 w_3^4 + s w_3^6\right)}{2 s w_2^4\left(w_3^2 - w_2^2\right)^2} a_3(\tau), \tag{5.79}$$

$$\frac{d\psi_1}{d\tau} = \frac{3 w_2^2 (w_2^2 - 1)}{4\left(4 w_2^2 - 1\right)} a_2^2(\tau) + \frac{3 \xi_e^2 \alpha}{2}, \tag{5.80}$$

$$\begin{aligned}\frac{d\psi_2}{d\tau} = {} & \frac{w_2\left(w_2^2 - 1\right)\left(\left(1 + 8 w_2^2\right) a_2^2(\tau) - 12 a_1^2(\tau)\right)}{16\left(4\, w_2^2 - 1\right)} + \\ & \frac{s w_2 w_3^2\left(4 w_2^4 + (3s - 8) w_2^2 w_3^3 + (4 - 7s) w_3^4\right)}{8\left(w_2^2 - w_3^2\right)^3},\end{aligned} \tag{5.81}$$

$$\begin{aligned}\frac{d\psi_3}{d\tau} = {} & -\frac{c_3^2 w_3^3}{8 s^2 w_2^4} - \frac{s w_3^5\left(4 w_2^6 - 8 w_2^4 w_3^2 + (4 - 7s) w_2^2 w_3^4 + 3s\, w_3^6\right)}{8 w_2^4\left(w_2^2 - w_3^2\right)^3} + \\ & \frac{w_3^3 a_1^2(\tau)}{4 w_2^4\left(1 - 4 w_3^2\right)} + \frac{w_3^3\left(w_3^2 - 2 w_2^2\right) a_2^2(\tau)}{8\left(w_2^2 - w_3^2\right)^2} - \frac{w_3 a_3^2(\tau)}{16},\end{aligned} \tag{5.82}$$

where $a_i(\tau) = \varepsilon b_i(\tau)$, $i = 1,\ 2, 3$.

The functions $a_1(\tau)$, $a_2(\tau)$, $a_3(\tau)$ and $\psi_1(\tau)$, $\psi_2(\tau)$, $\psi_3(\tau)$ are the amplitudes and the phases, respectively, of the main components of the pendulum vibration that are described by the solutions of the first-order approximation (5.42)–(5.43) and (5.45). Equations (5.77)–(5.82) govern the evolution of these functions in the slow time variables. The modulation equations should be supplemented by the initial conditions

$$\begin{aligned} a_1(0) &= a_{10}, \qquad a_2(0) = a_{20}, \qquad a_3(0) = a_{30}, \\ \psi_1(0) &= \psi_{10}, \qquad \psi_2(0) = \psi_{20}, \qquad \psi_3(0) = \psi_{30}. \end{aligned} \tag{5.83}$$

where known quantities a_{10}, a_{20}, a_{30}, ψ_{10}, ψ_{20}, ψ_{30} are compatible with the primary initial values u_{01}, u_{02}, u_{03}, u_{04}, u_{05}, u_{06} .

The exact solution of the initial value problem (5.77)–(5.82) with (5.83) has the form

$$a_1(\tau) = a_{10} e^{-\frac{c_1 \tau}{2}}, \tag{5.84}$$

$$a_2(\tau) = a_{20} e^{-\frac{v_1}{2}\tau}, \tag{5.85}$$

$$a_3(\tau) = a_{30} e^{-\frac{v_2}{2}\tau}, \tag{5.86}$$

$$\psi_1(\tau) = \frac{3}{2}\alpha \xi_e^2 \tau + \frac{3 a_{20}^2 w_2^2\left(w_2^2 - 1\right)\left(1 - e^{-v_1 \tau}\right)}{4\left(4 w_2^2 - 1\right) v_1} + \psi_{10}, \tag{5.87}$$

$$\psi_2(\tau) = \frac{3a_{10}^2 w_2 \left(w_2^2-1\right)\left(1-e^{-c_1\tau}\right)}{4c_1\left(4w_2^2-1\right)} - \frac{a_{20}^2 w_2 \left(8w_2^4-7w_2^2-1\right)\left(1-e^{-v_1\tau}\right)}{16v_1\left(4w_2^2-1\right)} + \frac{sw_2w_3^2\tau}{2(w_2^2-w_3^2)} + \frac{s^2w_3^4\left(3w_2^3-7w_2w_3^2\right)\tau}{8(w_2^2-w_3^2)} + \psi_{20}, \tag{5.88}$$

$$\psi_3(\tau) = -\frac{a_{10}^2 w_3^3\left(1-e^{-c_1\tau}\right)}{4c_1\left(4w_3^2-1\right)} - \frac{a_{20}^2 w_3^3\left(2w_2^2-w_3^2\right)\left(1-e^{-v_1\tau}\right)}{8v_1\left(w_2^2-w_3^2\right)^2} - \frac{a_{30}^2 w_3\left(1-e^{-v_2\tau}\right)}{16v_2\left(4w_3^2-1\right)} + \frac{c_3^2w_3^3\left(w_3^2-w_2^2\right)^3 + s^4w_3^9\left(7w_2^2-3w_3^2\right) - 4s^3w_3^5w_2^2\left(w_2^2-w_3^2\right)^2}{8s^2w_2^4\left(w_2^2-w_3^2\right)^3}\tau + \psi_{30}, \tag{5.89}$$

where

$$v_1 = \frac{c_3w_2^4 + c_2\left(w_2^2-w_3^2\right)^2}{\left(w_2^2-w_3^2\right)^2},$$

$$v_2 = \frac{c_3w_3^2\left(w_2^6 - 2(s+1)w_2^4w_3^2 + w_2^2w_3^4 + sw_3^6\right)}{sw_2^4\left(w_2^2-w_3^2\right)^2},$$

Assembling the subsequent solutions in accordance with (5.29)–(5.31) and omitting all terms of order $O(4)$, one can obtain the approximate solution of the considered problem

$$\xi(\tau) = \frac{f_1\cos(p_1\tau)}{1-p_1^2} + a_1\cos(\tau+\psi_1) + \frac{sw_3^2a_3^2\cos(2w_3\tau+2\psi_3)}{4w_3^2-1} + \frac{w_2^2}{4}a_2^2 + 3w_2a_1a_2^2\left(\frac{(w_2-1)\cos((1-2w_2)\tau+\psi_1-2\psi_2)}{32w_2-16} - \frac{(w_2+1)\cos((1+2w_2)\tau+\psi_1+2\psi_2)}{32w_2+16}\right) - \frac{sw_2^2w_3^2(w_2-1)a_2a_3}{2\left(w_2^2-w_3^2\right)}\left(\frac{\cos((w_2-w_3)\tau+\psi_2-\psi_3)}{(w_2-w_3)^2-1} + \frac{\cos((w_2+w_3)\tau+\psi_2+\psi_3)}{(w_2+w_3)^2-1}\right) + \frac{w_2^2(12w_2^6-3w_3^4+w_2^4\left(4w_3^2(s-6)-3\right)}{4\left(w_2^2-1\right)^2\left(w_2^2-w_3^2\right)^2} + \tag{5.90}$$

$$\frac{2w_2^2w_3^2\left(3-2s+6\left(s+1\right)w_3^2\right)a_2^2\cos\left(2w_2\tau+2\psi_2\right)}{4\left(w_2^2-1\right)^2\left(w_2^2-w_3^2\right)^2},$$

$$\varphi\left(\tau\right)=\frac{f_2\cos\left(p_2\tau\right)}{w_2^2-p_2^2}+a_2\cos\left(w_2\tau+\psi_2\right)$$

$$\frac{sw_3^2\left(\left(s-1\right)w_3^4+2w_2^2w_3^2-w_2^4\right)a_3}{\left(w_3^2-w_2^2\right)^3}\cos\left(w_3\tau+\psi_3\right)-$$

$$\frac{c_3w_2^2w_3a_3}{\left(w_3^2-w_2^2\right)^2}\sin\left(w_3\tau+\psi_3\right)-\frac{\left(49w_2^2-1\right)a_2^3}{192\left(4w_2^2-1\right)}\cos\left(3w_2\tau+3\psi_2\right)+ \tag{5.91}$$

$$\frac{w_2\left(w_2+2\right)\left(w_2+3\right)a_1^2a_2}{32w_2+16}\cos\left(\left(w_2+2\right)\tau+2\psi_1+\psi_2\right)+$$

$$\frac{w_2\left(w_2-2\right)\left(w_2-3\right)a_1^2a_2}{32w_2-16}\cos\left(\left(2-w_2\right)\tau+2\psi_1-\psi_2\right)+$$

$$\frac{sw_3^2\left(w_2^2\left(2w_3\left(w_3+1\right)\left(w_3+2\right)+1\right)-w_3^2\left(w_3+1\right)^2\right)a_1a_3}{2w_2^2\left(2w_3+1\right)\left(w_2^2-w_3^2\right)\left(w_2^2-\left(w_3+1\right)^2\right)}\cos\left(\left(w_3+1\right)\tau+\psi_1+\psi_3\right)+$$

$$\frac{sw_3^2\left(w_2^2\left(2w_3\left(w_3-1\right)\left(w_3-2\right)-1\right)+w_3^2\left(w_3-1\right)^2\right)a_1a_3}{2w_2^2\left(2w_3-1\right)\left(w_2^2-w_3^2\right)\left(w_2^2-\left(w_3-1\right)^2\right)}\cos\left(\left(w_3-1\right)\tau-\psi_1+\psi_3\right)+$$

$$\frac{D_{13}a_1a_2}{2(2w_2+1)^2\left(w_2^2-w_3^2\right)\left(\left(w_2+1\right)^2-w_3^2\right)}\cos\left(\left(w_2+1\right)\tau+\psi_1+\psi_2\right)+$$

$$\frac{D_{14}a_1a_2}{2(2w_2-1)^2\left(w_2^2-w_3^2\right)\left(\left(w_2-1\right)^2-w_3^2\right)}\cos\left(\left(w_2-1\right)\tau+\psi_2-\psi_1\right),$$

$$\gamma(\tau)=\frac{w_3^2}{sw_2^2}\frac{f_3\cos\left(p_3\tau_0\right)}{w_3^2-p_3^2}+\frac{p_2^2w_3^2f_2\cos\left(p_2\tau_0\right)}{w_2^2\left(w_2^2-p_2^2\right)\left(w_3^2-p_3^2\right)}+$$

$$a_3\cos\left(w_3\tau+\psi_3\right)-\frac{a_3^3}{192}\cos\left(3w_3\tau+3\psi_3\right)-$$

$$\frac{w_3^2D_3a_2^3\cos\left(3\left(w_2\tau+\psi_2\right)\right)}{192\left(w_2^2-w_3^2\right)^3\left(4w_2^2-1\right)\left(w_3^2-9w_2^2\right)}-\frac{w_2^4w_3^2D_4a_2^3\cos\left(w_2\tau+\psi_2\right)}{8\left(4w_2^2-1\right)\left(w_2^2-w_3^2\right)^4}+$$

$$\frac{w_2^2w_3^2a_2a_3^2}{8\left(w_2^2-w_3^2\right)}\left(\frac{\cos\left(\left(w_2+2w_3\right)\tau+\psi_2+2\psi_3\right)}{\left(w_2+3w_3\right)\left(w_2+w_3\right)}+\right. \tag{5.92}$$

$$\left.\frac{\cos\left(\left(w_2-2w_3\right)\tau+\psi_2-2\psi_3\right)}{\left(w_2-3w_3\right)\left(w_2-w_3\right)}\right)-$$

$$\frac{w_3^2D_5a_2^2a_3}{32w_2\left(4w_2^2-1\right)}\left(\frac{\cos\left(\left(2w_2+w_3\right)\tau+2\psi_2+\psi_3\right)}{\left(w_2+w_3\right)\left(w_2^2-w_3^2\right)^2}+\right.$$

$$\left.\frac{\cos\left((2w_2-w_3)\,\tau+2\psi_2-\psi_3\right)}{\left(w_2^2-w_3^2\right)^2(w_2-w_3)}\right)+$$

$$\frac{w_3^4a_1^2\,a_3}{16w_2^4}\left(\frac{\cos\left((2+w_3)\,\tau+2\psi_1+\psi_3\right)}{(2w_3+1)\,(w_3+1)}+\right.$$

$$\left.\frac{\cos\left((w_3-2)\,\tau+\psi_3-2\psi_1\right)}{(2w_3-1)\,(w_3-1)}\right)+$$

$$\frac{w_3^2D_6a_1^2a_2\cos\left((w_2+2)\,\tau+\psi_2+2\psi_1\right)}{16w_2^2\left(w_2^2-w_3^2\right)(2w_2+1)\left((w_2+1)^2-w_3^2\right)\left(w_3^2-(w_2+2)^2\right)}+$$

$$\frac{w_3^2D_7a_1^2a_2\cos\left((w_2-2)\,\tau+\psi_2-2\psi_1\right)}{16w_2^2\left(w_2^2-w_3^2\right)(2w_2-1)\left((w_2-1)^2-w_3^2\right)\left(w_3^2-(w_2-2)^2\right)}+$$

$$\frac{w_3^2a_2\cos\left(w_2\tau+\psi_2\right)}{4\left(w_2^2-w_3^2\right)^2}\left(w_2^2a_3^2+\frac{2D_8a_1^2}{w_2^2\left(4w_2^2-1\right)\left((w_2+w_3)^2-1\right)\left((w_2-w_3)^2-1\right)}\right)+$$

$$\left(w_3^2-\frac{sD_9}{w_2^2\,(2w_3+1)\left(w_2^2-w_3^2\right)\left((w_3+1)^2-w_2^2\right)}\right)$$

$$\frac{a_1a_3\cos\left((w_3+1)\,\tau+\psi_3+\psi_1\right)}{2w_2^2\,(2w_3+1)}-$$

$$\left(w_3^2-\frac{sD_{10}}{w_2^2\,(2w_3-1)\left(w_2^2-w_3^2\right)\left((w_3-1)^2-w_2^2\right)}\right)$$

$$\frac{a_1a_3\cos\left((w_3-1)\,\tau+\psi_3-\psi_1\right)}{2w_2^2\,(2w_3-1)}+$$

$$\left(D_1-\frac{w_3^2sD_{11}}{w_2^2\,(2w_2+1)\left(w_2^2-w_3^2\right)^2\left((w_2+1)^2-w_3^2\right)}\right)$$

$$\frac{w_3^2\,a_1a_2\cos\left((w_2+1)\,\tau+\psi_2+\psi_1\right)}{2\,(2w_2+1)\left(w_2^2-w_3^2\right)\left((w_2+1)^2-w_3^2\right)}+$$

$$\left(D_2-\frac{w_3^2sD_{13}}{w_2^2\,(2w_2-1)\left(w_2^2-w_3^2\right)^2\left((w_2-1)^2-w_3^2\right)}\right)$$

$$\frac{w_3^2a_1a_2\cos\left((w_2-1)\,\tau+\psi_2-\psi_1\right)}{2\,(2w_2-1)\left(w_2^2-w_3^2\right)\left((w_2-1)^2-w_3^2\right)}+$$

$$\frac{c_3 w_3^2 D_{12} a_1 a_2 \sin\left(\left(w_2+1\right)\tau+\psi_2+\psi_1\right)}{2 s w_2^2\left(2w_2+1\right)\left(w_2^2-w_3^2\right)^2\left(\left(w_2+1\right)^2-w_3^2\right)^2}+$$

$$\frac{c_3 w_3^2 D_{14} a_1 a_2 \sin\left(\left(w_2-1\right)\tau+\psi_2-\psi_1\right)}{2 s w_2^2\left(2w_2-1\right)\left(w_2^2-w_3^2\right)^2\left(\left(w_2-1\right)^2-w_3^2\right)^2}+$$

$$\left(\frac{w_3^4 s}{\left(w_2^2-w_3^2\right)^2}-\frac{2w_3^8 s^2}{\left(w_2^2-w_3^2\right)^4}+\frac{w_3^2 c_3^2}{s^2\left(w_2^2-w_3^2\right)^2}-1\right)\frac{w_3^2 a_2 \cos\left(w_2\tau+\psi_2\right)}{w_2^2-w_3^2}+$$

$$\left(\frac{w_2 c_3}{s}-\frac{w_3^2\left(3w_2^4+w_2^2 w_3^2+3w_3^4\right)c_3}{2w_2\left(w_2^2-w_3^2\right)^2}-\frac{w_3^2 c_2}{w_2}\right)\frac{w_3^2 a_2 \sin\left(w_2\tau+\psi_2\right)}{\left(w_2^2-w_3^2\right)^2},$$

where a_1, a_2, a_3 and ψ_1, ψ_2, ψ_3 are determined by (5.84)–(5.89).

The approximate solution given by (5.90)–(5.92) fails when any of the denominators occurring in the formulae is equal to zero. We exclude the cases when $s=0$ or $w_2=0$ since they coutradict the introduced assumptions. The resonant vibration appears in the system when any of the denominators is close to zero due to a specific relationship between the frequencies. Observe that the approximate solution based on the approach with three time scales allows one to recognize the following resonant cases:

(i) primary external resonances, when $p_1 \approx 1$, $p_2 \approx w_2$, $p_3 \approx w_3$;
(ii) internal resonance, when $2w_2 \approx 1$, $w_2 \approx 1$, $w_2 \approx 1$, $2w_3 \approx 1$, $w_2 \approx w_3$, $3w_2 \approx w_3$, $w_2 \approx 3w_3$, $w_2 \approx 1-w_3$, $w_2-w_3 \approx 1$, $w_2+w_3 \approx 1$.

The resonant cases should be considered separately based on additionally formulated assumptions. Excluding the resonant cases, the approximate solution (5.90)–(5.92) together with (5.84)–(5.89) govern the vibrational motion of the spring physical pendulum in the non-resonant conditions. We assume the following example values of the parameters of the pendulum: $w_2=0.39$, $w_3=0.11$, $\alpha=0.01$, $s=0.013$, $c_1=0.001$, $c_2=0.001$, $c_3=0.001$, $f_1=0.0001$, $f_2=0.0001$, $f_3=0.00001$, $p_1=4.15$, $p_2=0.03$, $p_3=0.17$ to present and test the solution.

The initial values for the modulation problem (5.84)–(5.87) are as follows: $a_{10}=0.01$, $a_{20}=0.10$, $a_{30}=0.08$, $\psi_{10}=0$, $\psi_{20}=0$, $\psi_{30}=0$.

The following initial values for the motion equation (5.17)–(5.19): $u_{01}=0.0074664$, $u_{02}=5.04325\cdot10^{-7}$, $u_{03}=0.10200$, $u_{04}=-1.18927\cdot10^{-4}$, $u_{05}=0.067700$, $u_{06}=4.91297\cdot10^{-4}$ remain in compliance with the initial values of amplitudes and phases.

The time histories of the generalized coordinates $\xi(\tau)$, $\varphi(\tau)$ and $\gamma(\tau)$ are presented in Figures 5.2–5.7. In each of this group of graphs, two curves are depicted, i.e. the approximate analytical solution obtained using MSM (solid black line), and the numerical solution obtained using the standard *NDSolve* of the *Mathematica* software. Figures 5.2–5.4 exhibit the initial stage of the movement in which the transient vibration is realised, whereas Figures 5.5–5.7 present the stationary pendulum oscillations. High compliance between the numerical and the approximate analytical solution is achieved.

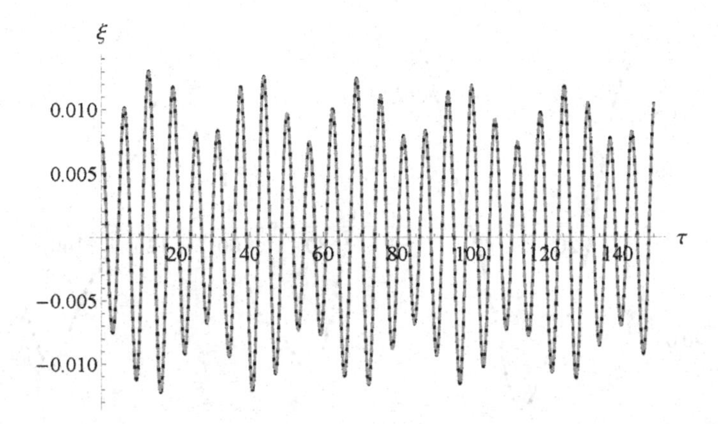

FIGURE 5.2 The transient longitudinal oscillations $\xi(\tau)$ estimated based on numerical (dotted line) and analytical method (MSM).

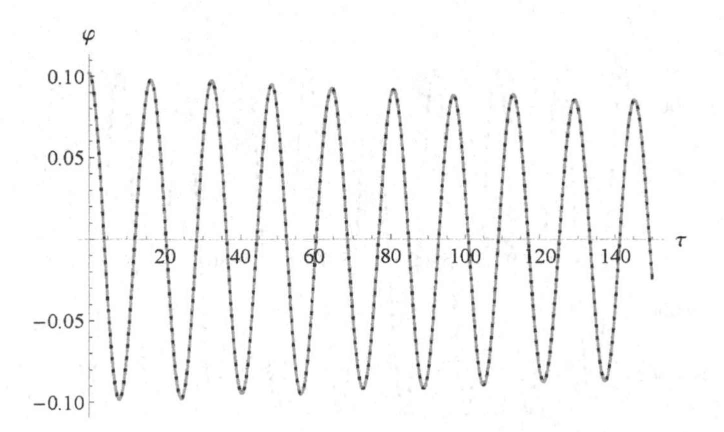

FIGURE 5.3 The transient oscillations for the coordinate $\varphi(\tau)$ estimated based on numerical (dotted line) and analytical method (MSM).

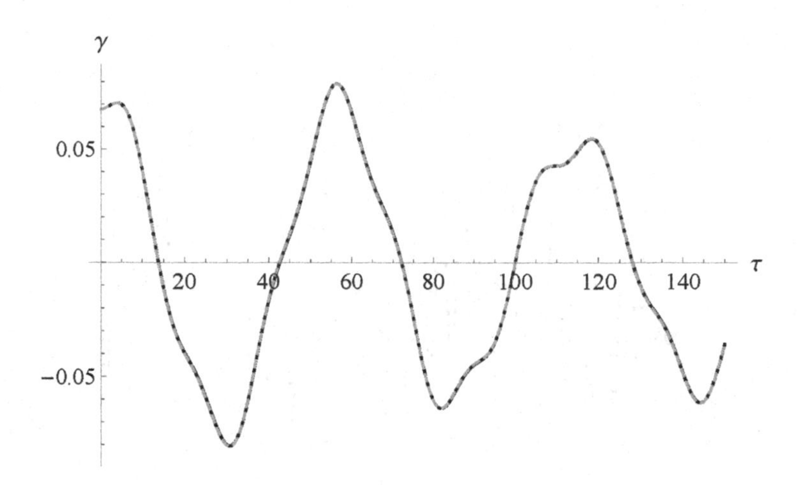

FIGURE 5.4 The transient oscillations for the coordinate $\gamma(\tau)$ estimated based on numerical (dotted line) and analytical method (MSM).

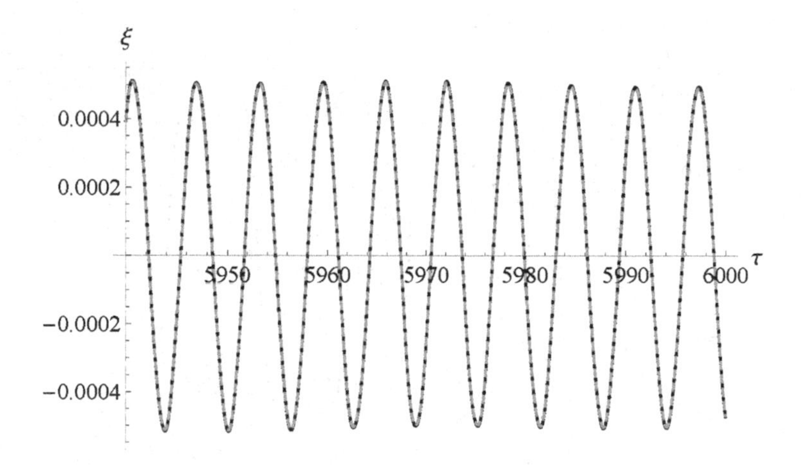

FIGURE 5.5 The stationary longitudinal oscillations $\xi(\tau)$ estimated based on numerical (dotted line) and analytical method (MSM).

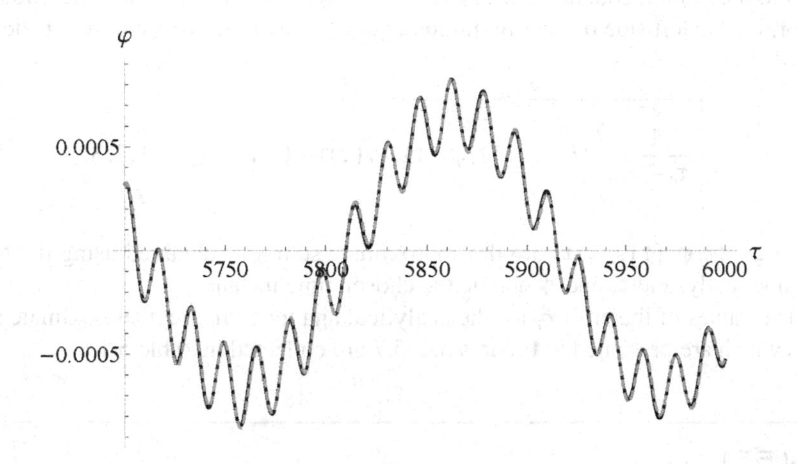

FIGURE 5.6 The stationary oscillations for the coordinate $\varphi(\tau)$ estimated based on numerical (dotted line) and analytical method (MSM).

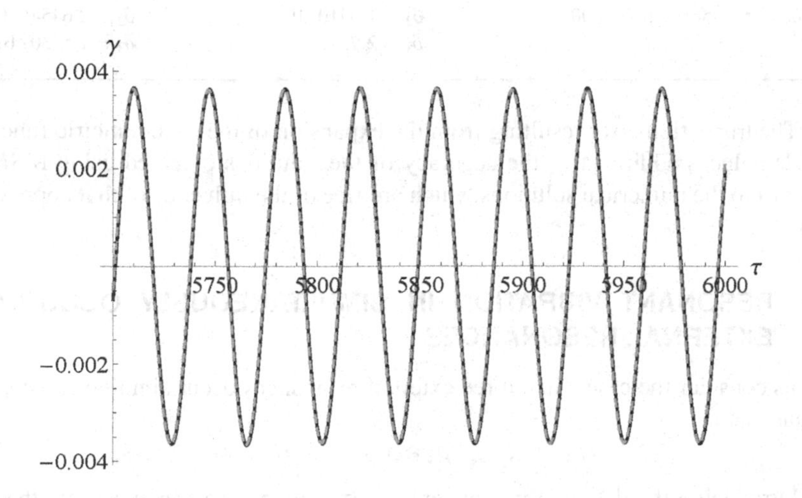

FIGURE 5.7 The stationary oscillations for the coordinate $\gamma(\tau)$ estimated based on numerical (dotted line) and analytical method (MSM).

The error satisfying the motion equations (5.17)–(5.19) is assumed to be a measure of accuracy of the approximate solution. Let us assign the numbers from $i = 1$ to $i = 3$ to the motion equations (5.17)–(5.19). If $H_i(\ldots)$ stands for the differential operator, i.e. the left side of the i-th motion equation, then the error measure is defined as follows

$$\delta_i = \sqrt{\frac{1}{\tau_e - \tau_s}\int_{\tau_s}^{\tau_e} \left(H_i\left(\xi_a(\tau), \varphi_a(\tau), \gamma_a(\tau)\right) - 0\right)^2 d\tau}, \quad i = 1,2,3, \tag{5.93}$$

where $\xi_a(\tau), \varphi_a(\tau), \gamma_a(\tau)$ are the approximate solutions obtained using the MSM or numerically, and τ_s and τ_e denote the chosen time instants.

The values of the error δ_i for the analytical and the numerical approximate solutions which are presented in Figures 5.2–5.7 are collected in Table 5.1.

TABLE 5.1
The values of the error measuring satisfaction to the governing equations

	MSM solution	numerical solution
transient stage (Figures 5.2 – 5.4) here: $\tau_0 = 0, \tau_e = 45$	$\delta_1 = 1.2489 \cdot 10^{-5}$ $\delta_2 = 5.0846 \cdot 10^{-6}$ $\delta_3 = 3.2672 \cdot 10^{-6}$	$\delta_1 = 6.9576 \cdot 10^{-5}$ $\delta_2 = 1.3214 \cdot 10^{-4}$ $\delta_3 = 6.4894 \cdot 10^{-6}$
transient stage (Figures 5.5 – 5.7) here: $\tau_0 = 5880,\ \tau_e = 6000$	$\delta_1 = 2.8977 \cdot 10^{-8}$ $\delta_2 = 1.0410 \cdot 10^{-6}$ $\delta_3 = 2.7398 \cdot 10^{-6}$	$\delta_1 = 4.6486 \cdot 10^{-7}$ $\delta_2 = 2.6454 \cdot 10^{-11}$ $\delta_3 = 1.5730 \cdot 10^{-11}$

The truncation error resulting from the expansion of the trigonometric functions (5.22) reduces additionally, the accuracy of the solutions obtained using MSM, in contrast to the numerical solutions which are free of the influence of that approximation.

5.5 RESONANT VIBRATION IN SIMULTANEOUSLY OCCURRING EXTERNAL RESONANCES

Let us consider the case when three external resonances occur simultaneously. This means that

$$p_1 \approx 1, \qquad p_2 \approx w_2, \qquad p_3 \approx w_3. \tag{5.94}$$

Introducing the detuning parameters σ_1, σ_2 and σ_3, we can substitute the relations (5.96) by the following equalities

$$p_1 = 1 + \sigma_1, \qquad p_2 = w_2 + \sigma_2, \qquad p_2 = w_3 + \sigma_2. \tag{5.95}$$

The detuning parameters are assumed to be small numbers of the order $O\left(\varepsilon^2\right)$

$$\sigma_1 = \varepsilon^2 \hat{\sigma}_1, \qquad \sigma_2 = \varepsilon^2 \hat{\sigma}_2, \qquad \sigma_3 = \varepsilon^2 \hat{\sigma}_3. \tag{5.96}$$

The coefficients $\hat{\sigma}_1$, $\hat{\sigma}_2$, $\hat{\sigma}_3$ are understood as $O(1)$ when $\varepsilon^2 \to 0$. We substitute relationships (5.95)–(5.96) into equations (5.23)–(5.25) which are the approximate form of the original motion equations (5.17)–(5.19). Using the multiple scales method, we approximate the sought solution by the asymptotic expansions (5.29)–(5.31). The first and the second time derivatives are replaced by the partial differential operators according to (5.27)–(5.28). Next, we take into account the assumptions (5.32) which concern the smallness of some parameters. All these operations result in occurence of the small parameter ε in the obtained equations. After ordering the terms of the equations with respect to the powers of the small parameter, we derive a set of equations from the requirement that each equation should be satisfied for any value of ε. Omitting the terms accompanied by ε in powers higher than three, one can obtain the system of nine partial differential equations with regard to unknown functions: $x_k\left(\tau_0, \tau_1, \tau_2\right)$, $\phi_k\left(\tau_0, \tau_1, \tau_2\right)$, and $\chi_k\left(\tau_0, \tau_1, \tau_2\right)$ for $k = 1, \ldots, 3$. As previously, we organize the set of the equations into three groups. The first group contains the homogeneous equations that include the terms standing at ε:

$$\frac{\partial^2 x_1}{\partial \tau_0^2} + x_1 = 0, \tag{5.97}$$

$$\frac{\partial^2 \phi_1}{\partial \tau_0^2} + w_2^2 \phi_1 = 0, \tag{5.98}$$

$$\frac{\partial^2 \chi_1}{\partial \tau_0^2} + \frac{w_3^2}{w_2^2} \frac{\partial^2 \phi_1}{\partial \tau_0^2} + w_3^2 \chi_1 = 0. \tag{5.99}$$

The terms accompanied by ε^2 create the equations of the second-order approximation

$$\frac{\partial^2 x_2}{\partial \tau_0^2} + x_2 = -2\, \frac{\partial x_1^2}{\partial \tau_0 \partial \tau_1} + \left(\frac{\partial \phi_1}{\partial \tau_0}\right)^2 - \frac{w_2^2}{2} \phi_1^2, \tag{5.100}$$

$$\frac{\partial^2 \phi_2}{\partial \tau_0^2} + w_2^2 \phi_2 = -2 \frac{\partial^2 \phi_1}{\partial \tau_0 \partial \tau_1} - 2 x_1 \frac{\partial^2 \phi_1}{\partial \tau_0^2} - 2 \frac{\partial x_1}{\partial \tau_0} \frac{\partial \phi_1}{\partial \tau_0} - \hat{s}\, \frac{\partial^2 \chi_1}{\partial \tau_0^2} - w_2^2 x_1 \phi_1, \tag{5.101}$$

$$\begin{aligned} \frac{\partial^2 \chi_2}{\partial \tau_0^2} + \frac{w_3^2}{w_2^2} \frac{\partial^2 \phi_2}{\partial \tau_0^2} + w_3^2 \chi_2 = -2 \frac{\partial^2 \chi_1}{\partial \tau_0 \partial \tau_1} - \frac{\hat{c}_3 w_3^2}{\hat{s} w_2^2} \left(\frac{\partial \chi_1}{\partial \tau_0} - \frac{\partial \phi_1}{\partial \tau_0}\right) + \\ \frac{w_3^2}{w_2^2} \left(\frac{\partial^2 x_1}{\partial \tau_0^2} (\chi_1 - \phi_1) - x_1 \frac{\partial^2 \phi_1}{\partial \tau_0^2} - 2 \frac{\partial^2 \phi_1}{\partial \tau_0 \partial \tau_1} - 2 \frac{\partial x_1}{\partial \tau_0} \frac{\partial \phi_1}{\partial \tau_0}\right). \end{aligned} \tag{5.102}$$

The terms which are accompanied by ε^3 form the following equations of the third-order approximation

$$\frac{\partial^2 x_3}{\partial \tau_0^2}+x_3=\hat{f}_1\cos\left(\left(1+\varepsilon^2\hat{\sigma}_1\right)\tau_0\right)-\hat{c}_1\frac{\partial x_1}{\partial \tau_0}-3\hat{\alpha}\xi_e^2 x_1+\hat{s}\left(\frac{\partial \chi_1}{\partial \tau_0}\right)^2+ \\ \hat{s}\frac{\partial^2 \chi_1}{\partial \tau_0^2}(\chi_1-\phi_1)-\frac{\partial^2 x_1}{\partial \tau_1^2}-2\left(\frac{\partial^2 x_1}{\partial \tau_0\partial \tau_2}+\frac{\partial^2 x_2}{\partial \tau_0\partial \tau_1}\right)+ \\ x_1\left(\frac{\partial \phi_1}{\partial \tau_0}\right)^2+2\frac{\partial \phi_1}{\partial \tau_0}\left(\frac{\partial \phi_1}{\partial \tau_1}+\frac{\partial \phi_2}{\partial \tau_0}\right)-w_2^2\phi_1\phi_2, \tag{5.103}$$

$$\frac{\partial^2 \phi_3}{\partial \tau_0^2}+w_2^2\phi_3=\hat{f}_2\cos\left(\left(w_2+\varepsilon^2\hat{\sigma}_2\right)\tau_0\right)-\hat{c}_2\frac{\partial \phi_1}{\partial \tau_0}-\hat{c}_3\left(\frac{\partial \phi_1}{\partial \tau_0}-\frac{\partial \chi_1}{\partial \tau_0}\right)- \\ \hat{s}\left(x_1\frac{\partial^2 \chi_1}{\partial \tau_0^2}+2\frac{\partial^2 \chi_1}{\partial \tau_0\partial \tau_1}+\frac{\partial^2 \chi_2}{\partial \tau_0^2}\right)-\frac{\partial^2 \phi_1}{\partial \tau_1^2}-\frac{\partial^2 \phi_1}{\partial \tau_0^2}(x_1^2+2x_2)- \\ 2\frac{\partial^2 \phi_1}{\partial \tau_0\partial \tau_2}-4x_1\frac{\partial^2 \phi_1}{\partial \tau_0\partial \tau_1}-2\frac{\partial^2 \phi_2}{\partial \tau_0\partial \tau_1}-2x_1\frac{\partial^2 \phi_2}{\partial \tau_0^2}- \\ 2\frac{\partial \phi_1}{\partial \tau_0}\left(\frac{\partial x_1}{\partial \tau_1}+x_1\frac{\partial x_1}{\partial \tau_0}+\frac{\partial x_2}{\partial \tau_0}\right)-2\frac{\partial x_1}{\partial \tau_0}\left(\frac{\partial \phi_1}{\partial \tau_1}+\frac{\partial \phi_2}{\partial \tau_0}\right)+ \\ \frac{1}{6}w_2^2\phi_1^3-w_2^2(x_2\phi_1+x_1\phi_2), \tag{5.104}$$

$$\frac{\partial^2 \chi_3}{\partial \tau_0^2}+\frac{w_3^2}{w_2^2}\frac{\partial^2 \phi_3}{\partial \tau_0^2}+w_3^2\chi_3=\frac{w_3^2}{\hat{s}\,w_2^2}\hat{f}_3\cos\left(\left(w_3+\varepsilon^2\hat{\sigma}_3\right)\tau_0\right)- \\ \frac{\hat{c}_3 w_3^2}{\hat{s}\,w_2^2}\left(\frac{\partial \chi_1}{\partial \tau_1}-\frac{\partial \phi_1}{\partial \tau_1}+\frac{\partial \chi_2}{\partial \tau_0}-\frac{\partial \phi_2}{\partial \tau_0}\right)+\frac{w_3^2}{w_2^2}\left((\chi_2-\phi_2)\frac{\partial^2 x_1}{\partial \tau_0^2}+(\chi_1-\phi_1)\frac{\partial^2 x_2}{\partial \tau_0^2}\right)+ \\ \frac{w_3^2}{w_2^2}\left(\left(\frac{1}{2}(\phi_1-\chi_1)^2-x_2\right)\frac{\partial^2 \phi_1}{\partial \tau_0^2}-x_1\frac{\partial^2 \phi_2}{\partial \tau_0^2}\right)+ \\ 2\frac{w_3^2}{w_2^2}\left((\chi_1-\phi_1)\frac{\partial^2 x_1}{\partial \tau_0\partial \tau_1}-x_1\frac{\partial^2 \phi_1}{\partial \tau_0\partial \tau_1}-\frac{\partial^2 \phi_1}{\partial \tau_0\partial \tau_2}-\frac{\partial^2 \phi_2}{\partial \tau_0\partial \tau_1}\right)- \\ 2\left(\frac{\partial^2 \chi_1}{\partial \tau_0\partial \tau_2}+\frac{\partial^2 \chi_2}{\partial \tau_0\partial \tau_1}\right)-\frac{\partial^2 \chi_1}{\partial \tau_1^2}-\frac{w_3^2}{w_2^2}\frac{\partial^2 \phi_1}{\partial \tau_1^2}-\frac{\partial^2 \chi_1}{\partial \tau_1^2}- \\ 2\frac{w_3^2}{w_2^2}\left(\frac{\partial x_1}{\partial \tau_0}\left(\frac{\partial \phi_1}{\partial \tau_1}+\frac{\partial \phi_2}{\partial \tau_0}\right)+\frac{\partial \phi_1}{\partial \tau_0}\left(\frac{\partial x_2}{\partial \tau_0}+\frac{\partial x_1}{\partial \tau_1}\right)\right)- \\ \frac{w_3^2}{w_2^2}(\chi_1-\phi_1)\left(\frac{\partial \phi_1}{\partial \tau_0}\right)^2+\frac{1}{6}w_3^2\chi_1^3. \tag{5.105}$$

The system (5.97)–(5.105) is solved recursively starting from equations of the first-order approximation (5.97) and (5.98) that are mutually independent. Their general solutions are

$$x_1(\tau_0,\tau_1,\tau_2) = B_1(\tau_1,\tau_2)e^{i\tau_0} + \bar{B}_1(\tau_1,\tau_2)e^{-i\tau_0}, \tag{5.106}$$

$$\phi_1(\tau_0,\tau_1,\tau_2) = B_2(\tau_1,\tau_2)e^{iw_2\tau_0} + \bar{B}_2(\tau_1,\tau_2)e^{-iw_2\tau_0}, \tag{5.107}$$

where the functions $B_1(\tau_1,\tau_2)$, $B_2(\tau_1,\tau_2)$ and their complex conjugates $\bar{B}_1(\tau_1,\tau_2)$, $\bar{B}_2(\tau_1,\tau_2)$ are unknown.

Then, the solution (5.107) is substituted into the last solution from the equations of the first-order approximation. This leads to the nonhomogeneous equation

$$\frac{\partial^2 \chi_1}{\partial \tau_0^2} + w_3^2 \chi_1 = w_3^2\left(B_2(\tau_1,\tau_2)e^{iw_2\tau_0} + \bar{B}_2(\tau_1,\tau_2)e^{-iw_2\tau_0}\right). \tag{5.108}$$

The general solution of equation (5.108) has the form

$$\begin{aligned}\chi_1(\tau_0,\tau_1,\tau_2) &= B_3(\tau_1,\tau_2)e^{iw_3\tau_0} + \bar{B}_3(\tau_1,\tau_2)e^{-iw_3\tau_0} + \\ &\frac{w_3^2}{w_3^2 - w_2^2}\left(B_2(\tau_1,\tau_2)e^{iw_2\tau_0} + \bar{B}_2(\tau_1,\tau_2)e^{-iw_2\tau_0}\right),\end{aligned} \tag{5.109}$$

where $B_3(\tau_1,\tau_2)$ and $\bar{B}_3(\tau_1,\tau_2)$ are unknown.

Inserting the general solutions (5.106)–(5.107) and (5.109) into (5.100)–(5.101) allows one to solve the equations of the second-order approximation. The secular terms appear then among the inhomogeneous terms on the left hand sides of the equations because of the identical form of the differential operators in the equations of the subsequent orders of approximation. Elimination of the secular terms by equalizing them to zero is necessary to obtain the limited solutions. The equations obtained in that way are called the solvability conditions because they impose some restrictions on the unknown functions. In the case of the functions $B_1(\tau_1,\tau_2)$ and $B_2(\tau_1,\tau_2)$, the following restrictions are formulated

$$\frac{\partial B_1}{\partial \tau_1} = 0, \tag{5.110}$$

$$\frac{\partial \bar{B}_1}{\partial \tau_1} = 0, \tag{5.111}$$

$$2i\frac{\partial B_2}{\partial \tau_1} + \frac{\hat{s}w_2 w_3^2}{w_2^2 - w_3^2}B_2 = 0, \tag{5.112}$$

$$2i\frac{\partial \bar{B}_2}{\partial \tau_1} - \frac{\hat{s}w_2 w_3^2}{w_2^2 - w_3^2}\bar{B}_2 = 0. \tag{5.113}$$

After elimination the secular terms, equations (5.100)–(5.101) have the form

$$\frac{\partial^2 x_2}{\partial \tau_0^2} + x_2 = w_2^2 \left(B_2 \bar{B}_2 - \frac{3}{2} \left(e^{2 i w_2 \tau_0} B_2^2 + e^{-2 i w_2 \tau_0} \bar{B}_2^2 \right) \right), \tag{5.114}$$

$$\begin{aligned} \frac{\partial^2 \phi_2}{\partial \tau_0^2} + w_2^2 \phi_2 = {} & w_2 (w_2 + 2) \left(e^{i(w_2+1)\tau_0} B_1 B_2 + e^{-i(w_2+1)\tau_0} \bar{B}_1 \bar{B}_2 \right) + \\ & w_2 (w_2 - 2) \left(e^{i(w_2-1)\tau_0} \bar{B}_1 B_2 + e^{-i(w_2-1)\tau_0} B_1 \bar{B}_2 \right) + \\ & \hat{s} w_3^2 \left(B_3 e^{i w_3 \tau_0} + \bar{B}_3 e^{-i w_3 \tau_0} \right). \end{aligned} \tag{5.115}$$

The particular solutions to equations (5.114)–(5.115) are as follows

$$x_2 (\tau_0, \tau_1, \tau_2) = w_2^2 B_2 \bar{B}_2 + \frac{3 w_2^2 \left(e^{2 i w_2 \tau_0} B_2^2 + e^{-2 i w_2 \tau_0} \bar{B}_2^2 \right)}{2 \left(4 w_2^2 - 1 \right)}, \tag{5.116}$$

$$\begin{aligned} \phi_2 (\tau_0, \tau_1, \tau_2) = {} & - \frac{w_2 (w_2 + 2) \left(e^{i(w_2+1)\tau_0} B_1 B_2 + e^{-i(w_2+1)\tau_0} \bar{B}_1 \bar{B}_2 \right)}{2 w_2 + 1} + \\ & \frac{w_2 (w_2 - 2) \left(e^{i(w_2-1)\tau_0} \bar{B}_1 B_2 + e^{-i(w_2-1)\tau_0} B_1 \bar{B}_2 \right)}{2 w_2 - 1} + \\ & \frac{\hat{s} w_3^2}{w_2^2 - w_3^2} \left(e^{i w_3 \tau_0} B_3 + e^{-i w_3 \tau_0} \bar{B}_3 \right). \end{aligned} \tag{5.117}$$

Inserting solutions (5.106)–(5.107), (5.109) and (5.117) into equation (5.102) generate the secular terms that should be detected and then eliminated. As a result of the procedure, we obtain the subsequent solvability conditions

$$2 i \frac{\partial B_3}{\partial \tau_1} + \frac{w_3^2}{w_2^2} \left(i \frac{\hat{c}_3}{\hat{s}} - \frac{\hat{s} w_3^3}{w_2^2 - w_3^2} \right) B_3 = 0, \tag{5.118}$$

$$2 i \frac{\partial \bar{B}_3}{\partial \tau_1} + \frac{w_3^2}{w_2^2} \left(i \frac{\hat{c}_3}{\hat{s}} + \frac{\hat{s} w_3^3}{w_2^2 - w_3^2} \right) \bar{B}_3 = 0. \tag{5.119}$$

Equation (5.99) with inserted solutions (5.106)–(5.107), (5.109) and (5.117) and with the eliminated secular terms takes the form

$$\begin{aligned} \frac{\partial^2 \chi_2}{\partial \tau_0^2} + w_3^2 \chi_2 = {} & \frac{D_1 w_3^2 \left(e^{i(w_2+1)\tau_0} B_1 B_2 + e^{-i(w_2+1)\tau_0} \bar{B}_1 \bar{B}_2 \right)}{(2 w_2 + 1) \left(w_3^2 - w_2^2 \right)} + \\ & \frac{D_2 w_3^2 \left(e^{i(w_2-1)\tau_0} \bar{B}_1 B_2 + e^{-i(w_2-1)\tau_0} \mathrm{B}_1 \bar{B}_2 \right)}{(2 w_2 - 1) \left(w_3^2 - w_2^2 \right)} - \end{aligned}$$

$$\frac{w_3^2\left(w_3^4\hat{s}^2+\mathrm{i}w_2\left(w_3^2-w_2^2\right)\hat{c}_3\right)}{\left(w_3^2-w_2^2\right)^2\hat{s}}e^{\mathrm{i}w_2\tau_0}B_2- \qquad (5.120)$$

$$\frac{w_3^2\left(w_3^4\hat{s}^2-\mathrm{i}w_2\left(w_3^2-w_2^2\right)\hat{c}_3\right)}{\left(w_3^2-w_2^2\right)^2\hat{s}}e^{-\mathrm{i}w_2\tau_0}\bar{B}_2-$$

$$\frac{w_3^2}{w_2^2}\left(e^{\mathrm{i}(w_3+1)\tau_0}B_1B_3+e^{-\mathrm{i}(w_3+1)\tau_0}\bar{B}_1\bar{B}_3+e^{\mathrm{i}(w_3-1)\tau_0}\bar{B}_1B_3+e^{-\mathrm{i}(w_3-1)\tau_0}B_1\bar{B}_3\right).$$

The coefficients D_1 and D_2 are depicted in the footnotes to equation (5.56). The particular solution to equation (5.120) has the form

$$\chi_2(\tau_0,\tau_1,\tau_2)=\frac{D_1w_3^2\left(e^{\mathrm{i}(w_2+1)\tau_0}B_1B_2+e^{-\mathrm{i}(w_2+1)\tau_0}\bar{B}_1\bar{B}_2\right)}{(2w_2+1)\left(w_3^2-w_2^2\right)\left(w_3^2-(w_2+1)^2\right)}+$$

$$\frac{D_2w_3^2\left(e^{\mathrm{i}(w_2-1)\tau_0}\bar{B}_1B_2+e^{-\mathrm{i}(w_2-1)\tau_0}B_1\bar{B}_2\right)}{(2w_2-1)\left(w_3^2-w_2^2\right)\left(w_3^2-(w_2-1)^2\right)}-$$

$$\frac{w_3^2\left(w_3^4\hat{s}^2+\mathrm{i}w_2\left(w_3^2-w_2^2\right)\hat{c}_3\right)}{\left(w_3^2-w_2^2\right)^3\hat{s}}e^{\mathrm{i}w_2\tau_0}B_2- \qquad (5.121)$$

$$\frac{w_3^2\left(w_3^4\hat{s}^2-\mathrm{i}w_2\left(w_3^2-w_2^2\right)\hat{c}_3\right)}{\left(w_3^2-w_2^2\right)^3\hat{s}}e^{-\mathrm{i}w_2\tau_0}\bar{B}_2+$$

$$\frac{w_3^2\left(e^{\mathrm{i}(w_3+1)\tau_0}B_1B_3+e^{-\mathrm{i}(w_3+1)\tau_0}\bar{B}_1\bar{B}_3\right)}{w_2^2(2w_3+1)}-$$

$$\frac{w_3^2\left(e^{\mathrm{i}(w_3-1)\tau_0}\bar{B}_1B_3+e^{-\mathrm{i}(w_3-1)\tau_0}B_1\bar{B}_3\right)}{w_2^2(2w_3-1)}$$

Inserting solutions (5.106)–(5.107), (5.109), (5.116)–(5.117) and (5.120) into equations (5.103)–(5.104) generates again the secular terms. The presence of the secular terms in the equations is in contradiction of the expectation that the oscillations are limited. Eliminating those terms, we get the sequent solvability conditions

$$2\mathrm{i}\frac{\partial B_1}{\partial\tau_2}+B_1\left(\frac{6w_2^2\left(w_2^2-1\right)B_2\bar{B}_2}{4w_2^2-1}+3\widehat{\alpha}\xi_e^2+\mathrm{i}\hat{c}_1\right)-\frac{1}{2}\hat{f}_1e^{\mathrm{i}\varepsilon^2\hat{\sigma}_1\tau_0}=0, \qquad (5.122)$$

$$2\mathrm{i}\frac{\partial\bar{B}_1}{\partial\tau_2}+\bar{B}_1\left(-\frac{6w_2^2\left(w_2^2-1\right)B_2\bar{B}_2}{4w_2^2-1}-3\widehat{\alpha}\xi_e^2+\mathrm{i}\hat{c}_1\right)+\frac{1}{2}\hat{f}_1e^{-\mathrm{i}\varepsilon^2\hat{\sigma}_1\tau_0}=0, \qquad (5.123)$$

$$-2\mathrm{i}w_2\frac{\partial B_2}{\partial\tau_2}-\frac{6w_2^2\left(w_2^2-1\right)B_1\bar{B}_1B_2}{4w_2^2-1}+\frac{w_2^2\left(8w_2^4-7w_2^2-1\right)B_2^2\bar{B}_2}{2\left(4w_2^2-1\right)}-$$

$$w_2\frac{4\mathrm{i}\left(w_2^2-w_3^2\right)^3\hat{c}_2+4\mathrm{i}\,w_2^4\left(w_2^2-w_3^2\right)\hat{c}_3-w_2w_3^4\left(7w_3^2-3w_2^2\right)\hat{s}^2}{4\left(w_2^2-w_3^2\right)^3}B_2+ \tag{5.124}$$
$$\frac{1}{2}\hat{f}_2e^{\mathrm{i}\varepsilon^2\hat{\sigma}_2\tau_0}=0,$$

$$2\mathrm{i}w_2\frac{\partial\bar{B}_2}{\partial\tau_2}-\frac{6w_2^2\left(w_2^2-1\right)B_1\bar{B}_1\bar{B}_2}{4w_2^2-1}+\frac{w_2^2\left(8w_2^4-7w_2^2-1\right)B_2\bar{B}_2^2}{2\left(4w_2^2-1\right)}-$$
$$w_2\frac{4\mathrm{i}\left(w_2^2-w_3^2\right)^3\hat{c}_2+4\mathrm{i}w_2^4\left(w_2^2-w_3^2\right)\hat{c}_3+w_2w_3^4\left(7w_3^2-3w_2^2\right)\hat{s}^2}{4\left(w_2^2-w_3^2\right)^3}\bar{B}_2+ \tag{5.125}$$
$$\frac{1}{2}\hat{f}_2e^{-\mathrm{i}\varepsilon^2\hat{\sigma}_2\tau_0}=0,$$

Equations (5.103)–(5.104) with the insertion of all previously obtained solutions and without the secular terms, take the following form

$$\frac{\partial^2x_3}{\partial\tau_0^2}+x_3=-2w_3^2\hat{s}\left(e^{2\mathrm{i}w_3\tau_0}B_3^2+e^{-2\mathrm{i}w_3\tau_0}\bar{B}_3^2\right)+$$
$$\frac{3w_2^2(w_2+1)^2}{2w_2+1}\left(e^{\mathrm{i}(2w_2+1)\tau_0}B_1B_2^2+e^{-\mathrm{i}(2w_2+1)\tau_0}\bar{B}_1\bar{B}_2^2\right)-$$
$$\frac{3w_2^2(w_2-1)^2}{2w_2-1}\left(e^{\mathrm{i}(2w_2-1)\tau_0}\bar{B}_1B_2^2+e^{-\mathrm{i}(2w_2-1)\tau_0}B_1\bar{B}_2^2\right)- \tag{5.126}$$
$$2\hat{s}\frac{w_2^4w_3^2\left(3w_3^2+w_2^2-1\right)}{\left(w_2^2-w_3^2\right)^2\left(4w_2^2-1\right)}\left(e^{2\mathrm{i}w_2\tau_0}B_2^2+e^{-2\mathrm{i}w_2\tau_0}\bar{B}_2^2\right)+$$
$$\hat{s}\frac{w_2^2w_3^2}{w_2^2-w_3^2}\left(e^{\mathrm{i}(w_2+w_3)\tau_0}B_2B_3+e^{-\mathrm{i}\,(w_2+w_3)\tau_0}\bar{B}_2\bar{B}_3+\right.$$
$$\left.e^{\mathrm{i}(w_3-w_2)\tau_0}\bar{B}_2B_3+e^{-\mathrm{i}(w_3-w_2)\tau_0}B_2\bar{B}_3\right),$$

$$\frac{\partial^2\phi_3}{\partial\tau_0^2}+w_2^2\phi_3=\frac{w_2^2\left(49w_2^2-1\right)}{6\left(4w_2^2-1\right)}\left(e^{3\mathrm{i}w_2\tau_0}B_2^3+e^{-3\mathrm{i}w_2\tau_0}\bar{B}_2^3\right)-$$
$$\frac{w_2\left(w_2^3+6w_2^2+11w_2+6\right)}{2w_2+1}\left(e^{\mathrm{i}(w_2+2)\tau_0}B_1^2B_2+e^{-\mathrm{i}(w_2+2)\tau_0}\bar{B}_1^2\bar{B}_2\right)+$$
$$\frac{w_2\left(w_2^3-6w_2^2+11w_2-6\right)}{2w_2-1}\left(e^{\mathrm{i}(w_2-2)\tau_0}\bar{B}_1^2B_2+e^{-\mathrm{i}(w_2-2)\tau_0}B_1^2\bar{B}_2\right)+ \tag{5.127}$$
$$\hat{s}\frac{w_3^2(w_2+1)^2\left(w_2^3+w_2^2-w_2\left(3w_3^2+1\right)-1\right)}{\left(2\,w_2+1\right)\left(w_2^2-w_3^2\right)\left(\left(w_2+1\right)^2-w_3^2\right)}$$

$$\left(e^{i(w_2+1)\tau_0}B_1B_2+e^{-i(w_2+1)\tau_0}\bar{B}_1\bar{B}_2\right)+$$

$$\hat{s}\frac{w_3^2(w_2-1)^2\left(w_2^3-w_2^2-w_2\left(3w_3^2+1\right)+1\right)}{(2\,w_2-1)\left(w_2^2-w_3^2\right)\left((w_2-1)^2-w_3^2\right)}$$

$$\left(e^{i(w_2-1)\tau_0}\bar{B}_1B_2+e^{-i(w_2-1)\tau_0}B_1\bar{B}_2\right)+$$

$$\hat{s}\frac{w_3^2\left(w_3^2(w_3+1)^2-w_2^2\left(2w_3^3+6w_3^2+4w_3+1\right)\right)}{w_2^2\left(2w_3+1\right)\left(w_3^2-w_2^2\right)}$$

$$\left(e^{i(w_3+1)\tau_0}B_1B_3+e^{-i(w_3+1)\tau_0}\bar{B}_1\bar{B}_3\right)-$$

$$\hat{s}\frac{w_3^2\left(w_3^2(w_3-1)^2+w_2^2\left(2w_3^3-6w_3^2+4w_3-1\right)\right)}{w_2^2\left(2w_3-1\right)\left(w_3^2-w_2^2\right)}$$

$$\left(e^{i(w_3-1)\tau_0}\bar{B}_1B_3+e^{-i(w_3-1)\tau_0}B_1\bar{B}_3\right)-$$

$$w_3\frac{\left(w_3^5\hat{s}^2-iw_2^2\left(w_2^2-w_3^2\right)\hat{c}_3\right)e^{iw_3\tau_0}B_3+\left(w_3^5\hat{s}^2+iw_2^2\left(w_2^2-w_3^2\right)\hat{c}_3\right)e^{-iw_3\tau_0}\bar{B}_3}{\left(w_3^2-w_2^2\right)^2}.$$

The particular solutions to equations (5.126)–(5.127) are as follows

$$x_3\left(\tau_0,\tau_1,\tau_2\right)=\frac{2w_3^2\hat{s}}{4w_3^2-1}\left(e^{2iw_3\tau_0}B_3^2+e^{-2i\,w_3\tau_0}\bar{B}_3^2\right)-$$

$$\frac{3w_2\left(w_2+1\right)}{4\left(2w_2+1\right)}\left(e^{i(2w_2+1)\tau_0}B_1B_2^2+e^{-i(2w_2+1)\tau_0}\bar{B}_1\bar{B}_2^2\right)+$$

$$\frac{3w_2\left(w_2-1\right)}{4\left(2w_2-1\right)}\left(e^{i(2w_2-1)\tau_0}\bar{B}_1B_2^2+e^{-i(2w_2-1)\tau_0}B_1\bar{B}_2^2\right)+ \tag{5.128}$$

$$\hat{s}\frac{2w_2^4w_3^2\left(3w_3^2+w_2^2-1\right)}{\left(w_3^2-w_2^2\right)^2\left(4w_2^2-1\right)^2}\left(e^{2iw_2\tau_0}B_2^2+e^{-2iw_2\tau_0}\bar{B}_2^2\right)-$$

$$\hat{s}\frac{w_2^2w_3^2}{w_2^2-w_3^2}\left(\frac{e^{i(w_2+w_3)\tau_0}B_2B_3+e^{-i(w_2+w_3)\tau_0}\bar{B}_2\bar{B}_3}{\left(w_2+w_3\right)^2-1}+\right.$$

$$\left.\frac{e^{i(w_3-w_2)\tau_0}\bar{B}_2B_3+e^{-i(w_3-w_2)\tau_0}B_2\bar{B}_3}{\left(\left(w_3-w_2\right)^2-1\right)}\right),$$

$$\phi_3\left(\tau_0,\tau_1,\tau_2\right)=-\frac{49w_2^2-1}{48\left(4w_2^2-1\right)}\left(e^{3iw_2\tau_0}B_2^3+e^{-3iw_2\tau_0}\bar{B}_2^3\right)+$$

$$\frac{w_2\left(w_2^2+5w_2+6\right)}{4\left(2w_2+1\right)}\left(e^{i(w_2+2)\tau_0}B_1^2B_2+e^{-i(w_2+2)\tau_0}\bar{B}_1^2\bar{B}_2\right)+$$

$$\frac{w_2\left(w_2^2-5w_2+6\right)}{4\left(2w_2-1\right)}\left(e^{\mathrm{i}(w_2-2)\tau_0}\bar{B}_1^2B_2+e^{-\mathrm{i}(w_2-2)\tau_0}B_1^2\bar{B}_2\right)- \tag{5.129}$$

$$\hat{s}\frac{w_3^2(w_2+1)^2\left(w_2^3+w_2^2-w_2\left(3w_3^2+1\right)-1\right)}{(2w_2+1)^2\left(w_2^2-w_3^2\right)\left((w_2+1)^2-w_3^2\right)}$$

$$\left(e^{\mathrm{i}(w_2+1)\tau_0}B_1B_2+e^{-\mathrm{i}(w_2+1)\tau_0}\bar{B}_1\bar{B}_2\right)+$$

$$\hat{s}\frac{w_3^2(w_2-1)^2\left(w_2^3-w_2^2-w_2\left(3w_3^2+1\right)+1\right)}{(2w_2-1)^2\left(w_2^2-w_3^2\right)\left((w_2-1)^2-w_3^2\right)}$$

$$\left(e^{\mathrm{i}(w_2-1)\tau_0}\bar{B}_1B_2+e^{-\mathrm{i}(w_2-1)\tau_0}B_1\bar{B}_2\right)+$$

$$\hat{s}\frac{w_3^2\left(w_3^2(w_3+1)^2-w_2^2\left(2w_3^3+6w_3^2+4w_3+1\right)\right)}{w_2^2\left(2w_3+1\right)\left(w_3^2-w_2^2\right)\left(w_2^2-(w_3+1)^2\right)}$$

$$\left(e^{\mathrm{i}(w_3+1)\tau_0}B_1B_3+e^{-\mathrm{i}(w_3+1)\tau_0}\bar{B}_1\bar{B}_3\right)-$$

$$\hat{s}\frac{w_3^2\left(w_3^2(w_3-1)^2+w_2^2\left(2w_3^3-6w_3^2+4w_3-1\right)\right)}{w_2^2\left(2w_3-1\right)\left(w_3^2-w_2^2\right)\left(w_2^2-(w_3-1)^2\right)}$$

$$\left(e^{\mathrm{i}(w_3-1)\tau_0}\bar{B}_1B_3+e^{-\mathrm{i}(w_3-1)\tau_0}B_1\bar{B}_3\right)-$$

$$w_3\frac{\left(w_3^5\hat{s}^2-\mathrm{i}w_2^2\left(w_2^2-w_3^2\right)\hat{c}_3\right)e^{\mathrm{i}w_3\tau_0}B_3+\left(w_3^5\hat{s}^2+\mathrm{i}w_2^2\left(w_2^2-w_3^2\right)\hat{c}_3\right)e^{-\mathrm{i}w_3\tau_0}\bar{B}_3}{\left(w_2^2-w_3^2\right)^3}.$$

Solutions (5.106)–(5.107), (5.109), (5.116)–(5.117), (5.121) and (5.129) should be inserted into equation (5.105). Elimination of the secular terms that appear in the result of this operation leads to the consecutive solvability conditions

$$-2\mathrm{i}w_3\frac{\partial B_3}{\partial\tau_2}+\frac{w_3^2\hat{f}_3}{2\hat{s}\,w_2^2}e^{\mathrm{i}\varepsilon^2\hat{\sigma}_3\tau_0}+\frac{2w_3^4B_1\bar{B}_1B_3}{w_2^4\left(4w_3^2-1\right)}+$$

$$\frac{w_3^4\left(2w_2^2-w_3^2\right)B_2\bar{B}_2B_3}{\left(w_2^2-w_3^2\right)^2}+\frac{w_3^2}{2}B_3^2\bar{B}_3+$$

$$w_3^4\frac{\left(w_2^2-w_3^2\right)^3\hat{c}_3^2-4\mathrm{i}w_3\left(w_2^2-w_3^2\right)\left(w_3^4-2w_2^4\right)\hat{c}_3\hat{s}^2}{4w_2^4\left(w_2^2-w_3^2\right)^3\hat{s}^2}- \tag{5.130}$$

$$\frac{w_3^6\left(7w_2^2-3\,w_3^2\right)\hat{s}^4}{4w_2^4\left(w_2^2-w_3^2\right)^3\hat{s}^2}B_3=0,$$

$$2\mathrm{i}w_3\frac{\partial \bar{B}_3}{\partial \tau_2}+\frac{w_3^2\hat{f}_3}{2\hat{s}w_2^2}e^{-\mathrm{i}\varepsilon^2\hat{\sigma}_3\tau_0}+\frac{2w_3^4B_1\bar{B}_1\bar{B}_3}{w_2^4\left(4w_3^2-1\right)}+$$
$$\frac{w_3^4\left(2w_2^2-w_3^2\right)B_2\bar{B}_2\bar{B}_3}{\left(w_2^2-w_3^2\right)^2}+\frac{w_3^2}{2}B_3\bar{B}_3^2+$$
$$w_3^4\frac{\left(w_2^2-w_3^2\right)^3\hat{c}_3^2+4\mathrm{i}w_3\left(w_2^2-w_3^2\right)\left(w_3^4-2w_2^4\right)\hat{c}_3\hat{s}^2}{4w_2^4\left(w_2^2-w_3^2\right)^3\hat{s}^2}- \tag{5.131}$$
$$\frac{w_3^6\left(7w_2^2-3\,w_3^2\right)\hat{s}^4}{4w_2^4\left(w_2^2-w_3^2\right)^3\hat{s}^2}\bar{B}_3=0.$$

After elimination of the secular terms, equation (5.105) in which have been already taken into account, all necessary solutions obtained as a result of using the recursively solving procedure, takes the following form

$$\frac{\partial^2\chi_3}{\partial\tau_0^2}+w_3^2\chi_3=\frac{w_3^4\hat{f}_2\cos\left(\left(w_2+\varepsilon^2\hat{\sigma}_2\right)\tau_0\right)}{w_2^2\left(w_2^2-w_3^2\right)}+$$
$$\frac{w_3^2}{6}\left(e^{3\mathrm{i}\,w_3\tau_0}B_3^3+e^{-3\mathrm{i}\,w_3\tau_0}\bar{B}_3^3\right)-\frac{w_3^2D_3\left(e^{3\mathrm{i}w_2\tau_0}B_2^3+e^{-3\mathrm{i}w_2\tau_0}\bar{B}_2^3\right)}{48\left(w_2^2-w_3^2\right)^3\left(4w_2^2-1\right)}+$$
$$\frac{D_4w_2^4w_3^2\left(e^{\mathrm{i}w_2\tau_0}B_2^2\bar{B}_2+e^{-\mathrm{i}w_2\tau_0}B_2\bar{B}_2^2\right)}{2\left(4w_2^2-1\right)\left(w_2^2-w_3^2\right)^3}- \tag{5.132}$$
$$\frac{w_2^2w_3^2\left(e^{\mathrm{i}(w_2+2\,w_3)\tau_0}B_2B_3^2+e^{-\mathrm{i}(w_2+2w_3)\tau_0}\bar{B}_2\bar{B}_3^2\right)}{2\left(w_2^2-w_3^2\right)}+$$
$$\frac{w_2^2w_3^2\left(e^{\mathrm{i}(w_2-2\,w_3)\tau_0}B_2\bar{B}_3^2+e^{-\mathrm{i}(w_2-2w_3)\tau_0}\bar{B}_2B_3^2\right)}{2\left(w_2^2-w_3^2\right)}+$$
$$\frac{w_3^2D_5\left(e^{\mathrm{i}(2w_2+w_3)\tau_0}B_2^2B_3+e^{-\mathrm{i}(2w_2+w_3)\tau_0}\bar{B}_2^2\bar{B}_3\right)}{2\left(4w_2^2-1\right)\left(w_2^2-w_3^2\right)^2}+$$
$$\frac{w_3^2D_5\left(e^{\mathrm{i}(2w_2-w_3)\tau_0}B_2^2\bar{B}_3+e^{-\mathrm{i}(2w_2-w_3)\tau_0}\bar{B}_2^2B_3\right)}{2\left(4w_2^2-1\right)\left(w_2^2-w_3^2\right)^2}-$$
$$\frac{w_3^4\left(e^{\mathrm{i}(w_3+2)\tau_0}B_1^2B_3+e^{-\mathrm{i}(w_3+2)\tau_0}\bar{B}_1^2\bar{B}_3\right)}{w_2^4\left(2w_3+1\right)}+$$
$$\frac{w_3^4\left(e^{\mathrm{i}(w_3-2)\tau_0}\bar{B}_1^2B_3+e^{-\mathrm{i}(w_3-2)\tau_0}B_1^2\bar{B}_3\right)}{w_2^4\left(2w_3-1\right)}+$$

$$\frac{w_3^2 D_6\left(e^{\mathrm{i}(w_2+2)\tau_0} B_1^2 B_2 + e^{-\mathrm{i}(w_2+2)\tau_0}\,\bar{B}_1^2\bar{B}_2\right)}{4w_2^2\,(2\,w_2+1)\left(w_2^2-w_3^2\right)\left((w_2+1)^2-w_3^2\right)}+$$

$$\frac{D_7 w_3^2\left(e^{\mathrm{i}(w_2-2)\tau_0}\,\bar{B}_1^2 B_2 + e^{-\mathrm{i}(w_2-2)\tau_0} B_1^2\bar{B}_2\right)}{4w_2^2\,(2w_2-1)\left(w_2^2-w_3^2\right)\left((w_2-1)^2-w_3^2\right)}-$$

$$\frac{w_3^2\left(e^{\mathrm{i}w_2\tau_0} B_2 + e^{-\mathrm{i}w_2\tau_0}\bar{B}_2\right)}{\left(w_2^2-w_3^2\right)}\left(w_2^2 B_3\bar{B}_3+\right.$$

$$\left.\frac{2D_8\,B_1\bar{B}_1}{w_2^2\left(4w_2^2-1\right)\left((w_2+w_3)^2-1\right)\left((w_2-w_3)^2-1\right)}\right)+$$

$$\frac{\hat{s}D_9\left(e^{\mathrm{i}(w_3+1)\tau_0} B_1 B_3 + e^{-\mathrm{i}(w_3+1)\tau_0}\bar{B}_1\bar{B}_3\right)}{w_2^4\,(2w_3+1)\left(w_2^2-w_3^2\right)\left((w_3+1)^2-w_2^2\right)}-$$

$$\frac{\hat{s}D_{10}\left(e^{\mathrm{i}(w_3-1)\tau_0}\bar{B}_1 B_3 + e^{-\mathrm{i}(w_3-1)\tau_0} B_1\bar{B}_3\right)}{w_2^4\,(2w_3-1)\left(w_2^2-w_3^2\right)\left(w_2^2-(w_3-1)^2\right)}+$$

$$\frac{w_3^4\hat{s}D_{11}\left(B_1 B_2 e^{\mathrm{i}(w_2+1)\tau_0} + \bar{B}_1\bar{B}_2 e^{-\mathrm{i}(w_2+1)\tau_0}\right)}{w_2^2(2w_2+1)^2\left(w_2^2-w_3^2\right)^3\left((w_2+1)^2-w_3^2\right)}+$$

$$\frac{\mathrm{i}\,\hat{c}_3 w_3^2 D_{12}\left(B_1 B_2 e^{\mathrm{i}(w_2+1)\tau_0} - \bar{B}_1\bar{B}_2 e^{-\mathrm{i}(w_2+1)\tau_0}\right)}{\hat{s}w_2^2\,(2w_2+1)\left(w_2^2-w_3^2\right)^2\left((w_2+1)^2-w_3^2\right)}+$$

$$\frac{w_3^4\hat{s}D_{13}\left(\bar{B}_1 B_2 e^{\mathrm{i}(w_2-1)\tau_0} + B_1\bar{B}_2 e^{-\mathrm{i}(w_2-1)\tau_0}\right)}{w_2^2(2w_2-1)^2\left(w_2^2-w_3^2\right)^3\left((w_2-1)^2-w_3^2\right)}+$$

$$\frac{\mathrm{i}\,\hat{c}_3 w_3^2 D_{14}\left(\bar{B}_1 B_2 e^{\mathrm{i}(w_2-1)\tau_0} - B_1\bar{B}_2 e^{-\mathrm{i}(w_2-1)\tau_0}\right)}{\hat{s}w_2^2\,(2w_2-1)\left(w_2^2-w_3^2\right)^2\left((w_2-1)^2-w_3^2\right)}+$$

$$\frac{w_3^4 B_2 e^{\mathrm{i}w_2\tau_0}}{w_2^2-w_3^2}\left(\frac{2w_3^6\hat{s}^2}{\left(w_2^2-w_3^2\right)^3}-\frac{\hat{c}_3^2}{\hat{s}^2\left(w_2^2-w_3^2\right)}-\right.$$

$$\left.\mathrm{i}\frac{\left(3w_2^4+w_2^2w_3^2+3w_3^4\right)\hat{c}_3}{2w_2\left(w_2^2-w_3^2\right)^2}-\mathrm{i}\frac{\hat{c}_2}{w_2}\right)+$$

$$\frac{w_3^4\bar{B}_2e^{-\mathrm{i}w_2\tau_0}}{w_2^2-w_3^2}\left(\frac{2w_3^6\hat{s}^2}{\left(w_2^2-w_3^2\right)^3}-\frac{\hat{c}_3^2}{\hat{s}^2\left(w_2^2-w_3^2\right)}+\right.$$
$$\left.\mathrm{i}\frac{\left(3w_2^4+w_2^2w_3^2+3w_3^4\right)\hat{c}_3}{2w_2\left(w_2^2-w_3^2\right)^2}+\mathrm{i}\frac{\hat{c}_2}{w_2}\right),$$

where the coefficients D_k, $k=3,...,14$, are the same as in the footnotes to equation (5.68).

The particular solution to equation (5.132) is as follows

$$\chi_3(\tau_0,\tau_1,\tau_2)=\frac{w_3^4\hat{f}_2\cos\left((w_2+\varepsilon^2\hat{\sigma}_2)\tau_0\right)}{w_2^2\left(w_2^2-w_3^2\right)\left(w_3^2-(w_2+\varepsilon^2\hat{\sigma}_2)^2\right)}-$$
$$\frac{e^{3\mathrm{i}\,w_3\tau_0}B_3^3+e^{-3\mathrm{i}\,w_3\tau_0}\bar{B}_3^3}{48}-\frac{w_3^2D_3\left(e^{3\mathrm{i}w_2\tau_0}B_2^3+e^{-3\mathrm{i}w_2\tau_0}\bar{B}_2^3\right)}{48\left(w_2^2-w_3^2\right)^3\left(4w_2^2-1\right)\left(w_3^2-9w_2^2\right)}-$$
$$\frac{D_4w_2^4w_3^2\left(e^{\mathrm{i}w_2\tau_0}B_2^2\bar{B}_2+e^{-\mathrm{i}w_2\tau_0}B_2\bar{B}_2^2\right)}{2\left(4w_2^2-1\right)\left(w_2^2-w_3^2\right)^4}+ \tag{5.133}$$
$$\frac{w_2^2w_3^2}{2}\left(\frac{e^{\mathrm{i}(w_2+2\,w_3)\tau_0}B_2B_3^2+e^{-\mathrm{i}(w_2+2\,w_3)\tau_0}\bar{B}_2\bar{B}_3^2}{(w_2+3w_3)(w_2+w_3)\left(w_2^2-w_3^2\right)}+\right.$$
$$\left.\frac{e^{\mathrm{i}(w_2+2w_3)\tau_0}B_2B_3^2+e^{-\mathrm{i}(w_2+2w_3)\tau_0}\bar{B}_2\bar{B}_3^2}{(w_2-3w_3)(w_2-w_3)\left(w_2^2-w_3^2\right)}\right)-$$
$$\frac{w_3^2D_5}{8w_2}\left(\frac{e^{\mathrm{i}(2w_2+w_3)\tau_0}B_2^2B_3+e^{-\mathrm{i}(2w_2+w_3)\tau_0}\bar{B}_2^2\bar{B}_3}{(w_2+w_3)\left(4w_2^2-1\right)\left(w_2^2-w_3^2\right)^2}+\right.$$
$$\left.\frac{e^{\mathrm{i}(2w_2-w_3)\tau_0}B_2^2\bar{B}_3+e^{-\mathrm{i}(2w_2-w_3)\tau_0}\bar{B}_2^2B_3}{\left(4w_2^2-1\right)\left(w_2^2-w_3^2\right)^2(w_2-w_3)}\right)+$$
$$\frac{w_3^4\left(e^{\mathrm{i}(w_3+2)\tau_0}B_1^2B_3+e^{-\mathrm{i}(w_3+2)\tau_0}\bar{B}_1^2\bar{B}_3\right)}{4w_2^4(2w_3+1)(w_3+1)}+$$
$$\frac{w_3^4\left(e^{\mathrm{i}(w_3-2)\tau_0}\bar{B}_1^2B_3+e^{-\mathrm{i}(w_3-2)\tau_0}B_1^2\bar{B}_3\right)}{4w_2^4(2w_3-1)(w_3-1)}+$$
$$\frac{w_3^2D_6\left(e^{\mathrm{i}(w_2+2)\tau_0}B_1^2B_2+e^{-\mathrm{i}(w_2+2)\tau_0}\,\bar{B}_1^2\bar{B}_2\right)}{4w_2^2(2w_2+1)\left(w_2^2-w_3^2\right)\left((w_2+1)^2-w_3^2\right)\left(w_3^2-(w_2+2)^2\right)}+$$
$$+\frac{D_7w_3^2\left(e^{\mathrm{i}(w_2-2)\tau_0}\bar{B}_1^2B_2+e^{-\mathrm{i}(w_2-2)\tau_0}B_1^2\bar{B}_2\right)}{4w_2^2(2w_2-1)\left(w_2^2-w_3^2\right)\left((w_2-1)^2-w_3^2\right)\left(w_3^2-(w_2-2)^2\right)}+$$

$$\frac{w_3^2\left(e^{\mathrm{i}w_2\tau_0}B_2+e^{-\mathrm{i}w_2\tau_0}\bar{B}_2\right)}{\left(w_2^2-w_3^2\right)^2}\left(w_2^2B_3\bar{B}_3+\right.$$

$$\left.\frac{2D_8\,B_1\bar{B}_1}{w_2^2\left(4\,w_2^2-1\right)\left(\left(w_2+w_3\right)^2-1\right)\left(\left(w_2-w_3\right)^2-1\right)}\right)-$$

$$\frac{\hat{s}D_9\left(e^{\mathrm{i}(w_3+1)\tau_0}B_1B_3+e^{-\mathrm{i}(w_3+1)\tau_0}\bar{B}_1\bar{B}_3\right)}{w_2^4(2w_3+1)^2\left(w_2^2-w_3^2\right)\left(\left(w_3+1\right)^2-w_2^2\right)}-$$

$$\frac{\hat{s}D_{10}\left(e^{\mathrm{i}(w_3-1)\tau_0}\bar{B}_1B_3+e^{-\mathrm{i}(w_3-1)\tau_0}B_1\bar{B}_3\right)}{w_2^4(2w_3-1)^2\left(w_2^2-w_3^2\right)\left(w_2^2-\left(w_3-1\right)^2\right)}-$$

$$\frac{w_3^4\hat{s}D_{11}\left(B_1B_2e^{\mathrm{i}(w_2+1)\tau_0}+\bar{B}_1\bar{B}_2e^{-\mathrm{i}(w_2+1)\tau_0}\right)}{w_2^2(2w_2+1)^2\left(w_2^2-w_3^2\right)^3\left(\left(w_2+1\right)^2-w_3^2\right)^2}-$$

$$\frac{\mathrm{i}\,\hat{c}_3w_3^2D_{12}\left(B_1B_2e^{\mathrm{i}(w_2+1)\tau_0}-\bar{B}_1\bar{B}_2e^{-\mathrm{i}(w_2+1)\tau_0}\right)}{\hat{s}w_2^2\left(2w_2+1\right)\left(w_2^2-w_3^2\right)^2\left(\left(w_2+1\right)^2-w_3^2\right)^2}-$$

$$\frac{w_3^4\hat{s}D_{13}\left(\bar{B}_1B_2e^{\mathrm{i}(w_2-1)\tau_0}+B_1\bar{B}_2e^{-\mathrm{i}(w_2-1)\tau_0}\right)}{w_2^2(2w_2-1)^2\left(w_2^2-w_3^2\right)^3\left(\left(w_2-1\right)^2-w_3^2\right)^2}-$$

$$\frac{\mathrm{i}\,\hat{c}_3w_3^2D_{14}\left(\bar{B}_1B_2e^{\mathrm{i}(w_2-1)\tau_0}-B_1\bar{B}_2e^{-\mathrm{i}(w_2-1)\tau_0}\right)}{\hat{s}w_2^2\left(2w_2-1\right)\left(w_2^2-w_3^2\right)^2\left(\left(w_2-1\right)^2-w_3^2\right)^2}-$$

$$-\frac{w_3^4B_2e^{\mathrm{i}w_2\tau_0}}{\left(w_2^2-w_3^2\right)^2}\left(\frac{2w_3^6\hat{s}^2}{\left(w_2^2-w_3^2\right)^3}-\frac{\hat{c}_3^2}{\hat{s}^2\left(w_2^2-w_3^2\right)}-\right.$$

$$\left.\mathrm{i}\frac{\left(3w_2^4+w_2^2w_3^2+3w_3^4\right)\hat{c}_3}{2w_2\left(w_2^2-w_3^2\right)^2}-\mathrm{i}\frac{\hat{c}_2}{w_2}\right)-$$

$$\frac{w_3^4\bar{B}_2e^{-\mathrm{i}w_2\tau_0}}{\left(w_2^2-w_3^2\right)^2}\left(\frac{2w_3^6\hat{s}^2}{\left(w_2^2-w_3^2\right)^3}-\frac{\hat{c}_3^2}{\hat{s}^2\left(w_2^2-w_3^2\right)}+\right.$$

$$\left.\mathrm{i}\frac{\left(3w_2^4+w_2^2w_3^2+3w_3^4\right)\hat{c}_3}{2w_2\left(w_2^2-w_3^2\right)^2}+\mathrm{i}\frac{\hat{c}_2}{w_2}\right).$$

In equations (5.106)–(5.107), (5.109), (5.116)–(5.117), (5.128)–(5.129) and (5.133), there are unknown complex-valued functions $B_1(\tau_1,\tau_2)$, $B_2(\tau_1,\tau_2)$,

$B_3(\tau_1,\tau_2)$ and their complex conjugates $\bar{B}_1(\tau_1,\tau_2)$, $\bar{B}_2(\tau_1,\tau_2)$, $\bar{B}_3(\tau_1,\tau_2)$. As before, we present the sought complex-valued functions in the exponential form

$$B_1(\tau_1,\tau_2) = \frac{1}{2}b_1(\tau_1,\tau_2)e^{i\psi_1(\tau_1,\tau_2)}, \quad \bar{B}_1(\tau_1,\tau_2) = \frac{1}{2}b_1(\tau_1,\tau_2)e^{-i\psi_1(\tau_1,\tau_2)}, \tag{5.134}$$

$$B_2(\tau_1,\tau_2) = \frac{1}{2}b_2(\tau_1,\tau_2)e^{i\psi_2(\tau_1,\tau_2)}, \quad \bar{B}_2(\tau_1,\tau_2) = \frac{1}{2}b_2(\tau_1,\tau_2)e^{-i\psi_2(\tau_1,\tau_2)}, \tag{5.135}$$

$$B_3(\tau_1,\tau_2) = \frac{1}{2}b_3(\tau_1,\tau_2)e^{i\psi_3(\tau_1,\tau_2)}, \quad \bar{B}_3(\tau_1,\tau_2) = \frac{1}{2}b_3(\tau_1,\tau_2)e^{-i\psi_3(\tau_1,\tau_2)}, \tag{5.136}$$

where the functions $b_i(\tau_1,\tau_2)$, $\psi_i(\tau_1,\tau_2)$ are real-valued for $i = 1,2,3$.

Next, we substitute relationships (5.134)–(5.136) into solvability conditions (5.110)–(5.113), (5.118)–(5.119), (5.122)–(5.125) and (5.130)–(5.131), and we get

$$\left(\frac{\partial b_1}{\partial \tau_1} + ib_1\frac{\partial \psi_1}{\partial \tau_1}\right)e^{i\psi_1(\tau_1,\tau_2)} = 0, \tag{5.137}$$

$$\left(\frac{\partial b_1}{\partial \tau_1} - ib_1\frac{\partial \psi_1}{\partial \tau_1}\right)e^{-i\psi_1(\tau_1,\tau_2)} = 0, \tag{5.138}$$

$$\left(2\,i\frac{\partial b_2}{\partial \tau_1} - 2\,b_2\frac{\partial \psi_2}{\partial \tau_1} + \frac{\hat{s}\,w_2 w_3^2}{w_2^2 - w_3^2}b_2\right)e^{i\psi_2(\tau_1,\tau_2)} = 0, \tag{5.139}$$

$$\left(2\,i\frac{\partial b_2}{\partial \tau_1} + 2b_2\frac{\partial \psi_2}{\partial \tau_1} - \frac{\hat{s}w_2 w_3^2}{w_2^2 - w_3^2}b_2\right)e^{-i\psi_2(\tau_1,\tau_2)} = 0, \tag{5.140}$$

$$\left(2i\,\frac{\partial b_3}{\partial \tau_1} - 2b_3\frac{\partial \psi_3}{\partial \tau_1} + \frac{w_3^2}{w_2^2}\left(i\frac{\hat{c}_3}{\hat{s}} - \frac{\hat{s}w_3^3}{w_2^2 - w_3^2}\right)b_3\right)e^{i\psi_3(\tau_1,\tau_2)} = 0, \tag{5.141}$$

$$\left(2i\,\frac{\partial b_3}{\partial \tau_1} + 2b_3\frac{\partial \psi_3}{\partial \tau_1} + \frac{w_3^2}{w_2^2}\left(i\frac{\hat{c}_3}{\hat{s}} + \frac{\hat{s}w_3^3}{w_2^2 - w_3^2}\right)b_3\right)e^{-i\psi_3(\tau_1,\tau_2)} = 0, \tag{5.142}$$

$$\left(2i\frac{\partial b_1}{\partial \tau_2} - 2b_1\frac{\partial \psi_1}{\partial \tau_2} + \left(\frac{3w_2^2\left(w_2^2-1\right)b_2^2}{2\left(4w_2^2-1\right)} + 3\hat{\alpha}\xi_e^2 + i\hat{c}_1\right)b_1\right)e^{i\psi_1(\tau_1,\tau_2)} - \hat{f}_1 e^{i\varepsilon^2\hat{\sigma}_1\tau_0} = 0, \tag{5.143}$$

$$\left(2\mathrm{i}\frac{\partial b_1}{\partial \tau_2}+2b_1\frac{\partial \psi_1}{\partial \tau_2}+\left(-\frac{3w_2^2\left(w_2^2-1\right)b_2^2}{2\left(4w_2^2-1\right)}-3\widehat{\alpha}\xi_{\mathrm{e}}^2+\mathrm{i}\hat{c}_1\right)b_1\right)e^{-\mathrm{i}\psi_1(\tau_1,\tau_2)}+$$

$$\hat{f}_1 e^{-\mathrm{i}\varepsilon^2\hat{\sigma}_1\tau_0}=0, \tag{5.144}$$

$$w_2\left(2\mathrm{i}\frac{\partial b_2}{\partial \tau_2}-2\,b_2\frac{\partial \psi_2}{\partial \tau_2}+\frac{3w_2^2\left(w_2^2-1\right)}{2\left(4w_2^2-1\right)}b_1^2b_2-\right.$$

$$\left.-\frac{w_2^2\left(8w_2^4-7w_2^2-1\right)b_2^3}{8\left(4w_2^2-1\right)}\right)e^{\mathrm{i}\psi_2(\tau_1,\tau_2)}+ \tag{5.145}$$

$$w_2\frac{4\mathrm{i}\,\left(w_2^2-w_3^2\right)^3\hat{c}_2+4\mathrm{i}\,w_2^4\left(w_2^2-w_3^2\right)\hat{c}_3-w_2w_3^4\left(7w_3^2-3w_2^2\right)\hat{s}^2}{4\,\left(w_2^2-w_3^2\right)^3}b_2e^{\mathrm{i}\psi_2(\tau_1,\tau_2)}-$$

$$\hat{f}_2e^{\mathrm{i}\varepsilon^2\hat{\sigma}_2\tau_0}=0,$$

$$w_2\left(2\mathrm{i}\frac{\partial b_2}{\partial \tau_2}+2\,b_2\frac{\partial \psi_2}{\partial \tau_2}-\frac{3w_2^2\left(w_2^2-1\right)}{2\left(4w_2^2-1\right)}b_1^2b_2+\right.$$

$$\left.\frac{w_2^2\left(8w_2^4-7w_2^2-1\right)b_2^3}{8\left(4w_2^2-1\right)}\right)e^{-\mathrm{i}\psi_2(\tau_1,\tau_2)}+ \tag{5.146}$$

$$w_2\frac{4\mathrm{i}\,\left(w_2^2-w_3^2\right)^3\hat{c}_2+4\mathrm{i}\,w_2^4\left(w_2^2-w_3^2\right)\hat{c}_3+w_2w_3^4\left(7w_3^2-3\,w_2^2\right)\hat{s}^2}{4\,\left(w_2^2-w_3^2\right)^3}b_2e^{-\mathrm{i}\psi_2(\tau_1,\tau_2)}+$$

$$\hat{f}_2e^{-\mathrm{i}\varepsilon^2\hat{\sigma}_2\tau_0}=0,$$

$$w_3\left(2\mathrm{i}\frac{\partial b_3}{\partial \tau_2}-2b_3\frac{\partial \psi_3}{\partial \tau_2}-\frac{w_3^4b_1^2b_3}{2w_2^4\left(4w_3^2-1\right)}-\right.$$

$$\left.\frac{w_3^4\left(2w_2^2-w_3^2\right)b_2^2b_3}{4\left(w_2^2-w_3^2\right)^2}-\frac{w_3^2}{8}b_3^3\right)e^{\mathrm{i}\psi_3(\tau_1,\tau_2)}- \tag{5.147}$$

$$w_3^4\frac{\left(w_2^2-w_3^2\right)^3\hat{c}_3^2+4\mathrm{i}\,w_3\left(w_2^2-w_3^2\right)\left(2w_2^4-w_3^4\right)\hat{c}_3\hat{s}^2-w_3^6\left(7w_2^2-3w_3^2\right)\hat{s}^4}{4w_2^4\left(w_2^2-w_3^2\right)^3\hat{s}^2}b_3e^{\mathrm{i}\psi_3(\tau_1,\tau_2)}-$$

$$\frac{w_3^2\hat{f}_3}{\hat{s}w_2^2}e^{\mathrm{i}\varepsilon^2\hat{\sigma}_3\tau_0}=0,$$

$$w_3\left(2\mathrm{i}\frac{\partial b_3}{\partial\tau_2}+2b_3\frac{\partial\psi_3}{\partial\tau_2}+\frac{w_3^4b_1^2b_3}{2w_2^4\left(4w_3^2-1\right)}+\right.$$

$$\left.\frac{w_3^4\left(2w_2^2-w_3^2\right)b_2^2b_3}{4\left(w_2^2-w_3^2\right)^2}+\frac{w_3^2}{8}b_3^3\right)e^{-\mathrm{i}\psi_3(\tau_1,\tau_2)}+ \tag{5.148}$$

$$w_3^4\frac{\left(w_2^2-w_3^2\right)^3\hat{c}_3^2-4\mathrm{i}\,w_3\left(w_2^2-w_3^2\right)\left(2w_2^4-w_3^4\right)\hat{c}_3\hat{s}^2-w_3^6\left(7w_2^2-3w_3^2\right)\hat{s}^4}{4w_2^4\left(w_2^2-w_3^2\right)^3\hat{s}^2}b_3e^{-\mathrm{i}\psi_3(\tau_1,\tau_2)}+$$

$$\frac{w_3^2\hat{f}_3}{\hat{s}\,w_2^2}e^{-\mathrm{i}\varepsilon^2\hat{\sigma}_3\tau_0}=0.$$

It is convenient to formulate the problem of determining the functions $b_i(\tau_1,\tau_2)$, $\psi_i(\tau_1,\tau_2)$, $i=1,2,3$, with respect to the original time variable τ because we have then to solve the ordinary differential equations instead of the partial ones. For this purpose, we solve the system of equations (5.137)–(5.148) with respect to the partial derivatives of the functions $b_i(\tau_1,\tau_2)$, $\psi_i(\tau_1,\tau_2)$, $i=1,2,3$. Referring to definition (5.27), one can write formally that the derivatives of the sought functions with respect to the time variable τ are of the following form

$$\frac{db_i}{d\tau}=\varepsilon\frac{\partial b_i}{\partial\tau_1}+\varepsilon^2\frac{\partial b_i}{\partial\tau_2},\quad i=1,2,3, \tag{5.149}$$

$$\frac{d\psi_i}{d\tau}=\varepsilon\frac{\partial\psi_i}{\partial\tau_1}+\varepsilon^2\frac{\partial\psi_i}{\partial\tau_2},\quad i=1,2,3. \tag{5.150}$$

After inserting solutions of system (5.137)–(5.147) into equations (5.149)–(5.150) and taking into account the relationships (5.32) and (5.96), six ordinary differential equations of the first-order are obtained

$$\frac{da_1}{d\tau}=-\frac{c_1}{2}a_1(\tau)+\frac{f_1}{2}\sin\left(\sigma_1\tau-\psi_1(\tau)\right), \tag{5.151}$$

$$\frac{da_2}{d\tau}=-\frac{c_3w_2^4+c_2\left(w_2^2-w_3^2\right)^2}{2\left(w_2^2-w_3^2\right)^2}a_2(\tau)+\frac{f_2}{2w_2}\sin\left(\sigma_2\tau-\psi_2(\tau)\right), \tag{5.152}$$

$$\frac{da_3}{d\tau}=-\frac{c_3w_3^2\left(w_2^6-2(s+1)w_2^4w_3^2+w_2^2w_3^4+sw_3^6\right)}{2sw_2^4\left(w_2^2-w_3^2\right)^2}a_3(\tau)$$

$$+\frac{w_3f_3\sin\left(\sigma_3\tau-\psi_3(\tau)\right)}{2sw_2^2}, \tag{5.153}$$

$$\frac{d\psi_1}{d\tau}=\frac{3w_2^2(w_2^2-1)}{4\left(4w_2^2-1\right)}a_2^2(\tau)+\frac{3\xi_e^2\alpha}{2}-\frac{f_1}{2a_1(\tau)}\cos\left(\sigma_1\tau-\psi_1(\tau)\right), \tag{5.154}$$

$$\frac{d\psi_2}{d\tau} = \frac{w_2\left(w_2^2-1\right)\left(12a_1^2(\tau)-\left(1+8w_2^2\right)a_2^2(\tau)\right)}{16\left(4w_2^2-1\right)} + \frac{sw_2w_3^2\left(4w_2^4+(3s-8)w_2^2w_3^3+(4-7s)w_3^4\right)}{8\left(w_2^2-w_3^2\right)^3} - \frac{f_2}{2w_2a_2(\tau)}\cos\left(\sigma_2\tau-\psi_2(\tau)\right), \tag{5.155}$$

$$\frac{d\psi_3}{d\tau} = -\frac{c_3^2w_3^3}{8s^2w_2^4} - \frac{sw_3^5\left(4w_2^6-8w_2^4w_3^2+(4-7s)w_2^2w_3^4+3sw_3^6\right)}{8w_2^4\left(w_2^2-w_3^2\right)^3} - \frac{w_3^3a_1^2(\tau)}{4w_2^4\left(4w_3^2-1\right)} + \frac{w_3^3\left(w_3^2-2w_2^2\right)a_2^2(\tau)}{8\left(w_2^2-w_3^2\right)^2} - \frac{w_3a_3^2(\tau)}{16} - \frac{w_3f_3\cos\left(\sigma_3\tau-\psi_3(\tau)\right)}{2sw_2^2a_3(\tau)}, \tag{5.156}$$

where $a_i(\tau) = \varepsilon b_i(\tau)$, $i = 1,2,3$.

The system consisting of equations (5.151)–(5.156) is supplemented by the initial conditions

$$a_1(0) = a_{10}, \quad a_2(0) = a_{20}, \quad a_3(0) = a_{30},$$
$$\psi_1(0) = \psi_{10}, \quad \psi_2(0) = \psi_{20}, \quad \psi_3(0) = \psi_{30}, \tag{5.157}$$

where known quantities $a_{10}, a_{20}, a_{30}, \psi_{10}, \psi_{20}, \psi_{30}$ are compatible with the following initial values: $u_{01}, u_{02}, u_{03}, u_{04}, u_{05}, u_{06}$.

The functions $a_1(\tau)$, $a_2(\tau)$, $a_3(\tau)$ and $\psi_1(\tau)$, $\psi_2(\tau)$, $\psi_3(\tau)$ are the amplitudes and the phases, respectively, of the main components of the pendulum vibration. The components are determined by the solutions of the first-order approximation, i.e. (5.106)–(5.107) and (5.109). The evolution of the amplitudes and the phases governed by equations (5.151)–(5.156) is related to the slow time variables. The main difference between the modulation equations concerning the resonant cases and the previously discussed modulation equations (5.77)–(5.82) lies in the occurrence of the unknown phases $\psi_1(\tau)$, $\psi_2(\tau)$, $\psi_3(\tau)$ in the expressions being the arguments of the trigonometric functions. That circumstance shows that system (5.151)–(5.156) with (5.157) cannot be solved exactly.

Substituting solutions (5.106)–(5.107), (5.109), (5.116)–(5.117), (5.121), (5.128)–(5.129) and (5.133) into the asymptotic expansions (5.29)–(5.30) yields the approximate solution describing the problem of the spring pendulum vibration at assumed external resonance. This solution consists of the following three functions

$$\xi(\tau) = a_1\cos(\tau+\psi_1) + \frac{sw_3^2a_3^2\cos(2w_3\tau+2\psi_3)}{4w_3^2-1} + \frac{w_2^2}{4}a_2^2 + 3w_2a_1a_2^2\left(\frac{(w_2-1)\cos((1-2w_2)\tau+\psi_1-2\psi_2)}{16(w_2-1)} - \right.$$

$$\frac{(w_2+1)\cos\left((1+2w_2)\tau+\psi_1+2\psi_2\right)}{16\,(w_2+1)}\Bigg)- \tag{5.158}$$

$$\frac{s w_2^2 w_3^2\,(w_2-1)\,a_2 a_3}{2\left(w_2^2-w_3^2\right)}\left(\frac{\cos\left((w_2-w_3)\,\tau+\psi_2-\psi_3\right)}{(w_2-w_3)^2-1}+\right.$$

$$\left.\frac{\cos\left((w_2+w_3)\,\tau+\psi_2+\psi_3\right)}{(w_2+w_3)^2-1}\right)+$$

$$\frac{w_2^2(12w_2^6-3w_3^4)+w_2^4\left(4w_3^2\,(s-6)-3\right)}{4\left(w_2^2-1\right)^2\left(w_2^2-w_3^2\right)^2}+$$

$$\frac{2w_2^2w_3^2\left(3-2s+6\,(s+1)\,w_3^2\right)a_2^2\cos\left(2w_2\tau+2\psi_2\right)}{4\left(w_2^2-1\right)^2\left(w_2^2-w_3^2\right)^2},$$

$$\varphi(\tau)=a_2\cos\left(w_2\tau+\psi_2\right)-\frac{s w_3^2\left((s-1)\,w_3^4+2w_2^2w_3^2-w_2^4\right)a_3}{\left(w_2^2-w_3^2\right)^3}\cos\left(w_3\tau+\psi_3\right)-$$

$$\frac{c_3 w_2^2 w_3 a_3}{\left(w_2^2-w_3^2\right)^2}\sin\left(w_3\tau+\psi_3\right)-\frac{\left(49w_2^2-1\right)a_2^3}{192\left(4w_2^2-1\right)}\cos\left(3w_2\tau+3\psi_2\right)+$$

$$\frac{w_2\,(w_2+2)\,(w_2+3)\,a_1^2a_2}{16\,(w_2+1)}\cos\left((w_2+2)\,\tau+2\psi_1+\psi_2\right)+ \tag{5.159}$$

$$\frac{w_2\,(w_2-2)\,(w_2-3)\,a_1^2a_2}{16\,(w_2-1)}\cos\left((2-w_2)\,\tau+2\psi_1-\psi_2\right)+$$

$$\frac{s w_3^2\left(w_2^2\,(2w_3\,(w_3+1)\,(w_3+2)+1)-w_3^2(w_3+1)^2\right)a_1a_3}{2w_2^2\,(2w_3+1)\left(w_2^2-w_3^2\right)\left(w_2^2-(w_3+1)^2\right)}\cos\left((w_3+1)\,\tau+\psi_1+\psi_3\right)+$$

$$\frac{s w_3^2\left(w_2^2\,(2w_3\,(w_3-1)\,(w_3-2)-1)+w_3^2(w_3-1)^2\right)a_1a_3}{2w_2^2\,(2w_3-1)\left(w_2^2-w_3^2\right)\left(w_2^2-(w_3-1)^2\right)}\cos\left((w_3-1)\,\tau-\psi_1+\psi_3\right)+$$

$$\frac{D_{13}a_1a_2}{2(2w_2+1)^2\left(w_2^2-w_3^2\right)\left((w_2+1)^2-w_3^2\right)}\cos\left((w_2+1)\,\tau+\psi_1+\psi_2\right)+$$

$$\frac{D_{14}a_1a_2}{2(2w_2-1)^2\left(w_2^2-w_3^2\right)\left((w_2-1)^2-w_3^2\right)}\cos\left((w_2-1)\,\tau+\psi_2-\psi_1\right),$$

$$\gamma(\tau)=\frac{p_2^2w_3^2f_2\cos\left(p_2\tau_0\right)}{w_2^2\left(w_2^2-p_2^2\right)\left(w_3^2-p_3^2\right)}+a_3\cos\left(w_3\tau+\psi_3\right)-\frac{a_3^3}{192}\cos\left(3w_3\tau+3\psi_3\right)+$$

$$
\begin{aligned}
&-\frac{w_3^2 D_3 a_2^3 \cos\left(3\left(w_2\tau+\psi_2\right)\right)}{192\left(w_2^2-w_3^2\right)^3\left(4w_2^2-1\right)\left(w_3^2-9w_2^2\right)}-\frac{w_2^4 w_3^2 D_4 a_2^3 \cos\left(w_2\tau+\psi_2\right)}{8\left(4\,w_2^2-1\right)\left(w_2^2-w_3^2\right)^4}+\\
&\frac{w_2^2 w_3^2 a_2 a_3^2}{8\left(w_2^2-w_3^2\right)}\left(\frac{\cos\left(\left(w_2+2w_3\right)\tau+\psi_2+2\psi_3\right)}{\left(w_2+3w_3\right)\left(w_2+w_3\right)}+\right.\\
&\left.\frac{\cos\left(\left(w_2-2w_3\right)\tau+\psi_2-2\psi_3\right)}{\left(w_2-3w_3\right)\left(w_2-w_3\right)}\right)- \qquad (5.160)\\
&\frac{w_3^2 D_5 a_2^2 a_3}{32w_2\left(4w_2^2-1\right)}\left(\frac{\cos\left(\left(2w_2+w_3\right)\tau+2\psi_2+\psi_3\right)}{\left(w_2+w_3\right)\left(w_2^2-w_3^2\right)^2}+\right.\\
&\left.\frac{\cos\left(\left(2w_2-w_3\right)\tau+2\psi_2-\psi_3\right)}{\left(w_2^2-w_3^2\right)^2\left(w_2-w_3\right)}\right)+\\
&\frac{w_3^4 a_1^2\, a_3}{16w_2^4}\left(\frac{\cos\left(\left(2+w_3\right)\tau+2\psi_1+\psi_3\right)}{\left(2w_3+1\right)\left(w_3+1\right)}+\frac{\cos\left(\left(w_3-2\right)\tau+\psi_3-2\psi_1\right)}{\left(2w_3-1\right)\left(w_3-1\right)}\right)+\\
&\frac{w_3^2 D_6 a_1^2 a_2\,\cos\left(\left(w_2+2\right)\tau+\psi_2+2\psi_1\right)}{16w_2^2\left(w_2^2-w_3^2\right)\left(2w_2+1\right)\left(\left(w_2+1\right)^2-w_3^2\right)\left(w_3^2-\left(w_2+2\right)^2\right)}+\\
&\frac{w_3^2 D_7 a_1^2 a_2 \cos\left(\left(w_2-2\right)\tau+\psi_2-2\psi_1\right)}{16w_2^2\left(w_2^2-w_3^2\right)\left(2w_2-1\right)\left(\left(w_2-1\right)^2-w_3^2\right)\left(w_3^2-\left(w_2-2\right)^2\right)}+\\
&\frac{w_3^2 a_2 \cos\left(w_2\tau+\psi_2\right)}{4\left(w_2^2-w_3^2\right)^2}\left(w_2^2 a_3^2+\right.\\
&\left.\frac{2D_8 a_1^2}{w_2^2\left(4w_2^2-1\right)\left(\left(w_2+w_3\right)^2-1\right)\left(\left(w_2-w_3\right)^2-1\right)}\right)+\\
&\left(w_3^2-\frac{sD_9}{w_2^2\left(2w_3+1\right)\left(w_2^2-w_3^2\right)\left(\left(w_3+1\right)^2-w_2^2\right)}\right)\\
&\frac{a_1 a_3 \cos\left(\left(w_3+1\right)\tau+\psi_3+\psi_1\right)}{2w_2^2\left(2w_3+1\right)}-\\
&\left(w_3^2-\frac{sD_{10}}{w_2^2\left(2w_3-1\right)\left(w_2^2-w_3^2\right)\left(\left(w_3-1\right)^2-w_2^2\right)}\right)\\
&\frac{a_1 a_3 \cos\left(\left(w_3-1\right)\tau+\psi_3-\psi_1\right)}{2w_2^2\left(2w_3-1\right)}+\\
&\left(D_1-\frac{w_3^2 s D_{11}}{w_2^2\left(2w_2+1\right)\left(w_2^2-w_3^2\right)^2\left(\left(w_2+1\right)^2-w_3^2\right)}\right)
\end{aligned}
$$

$$\frac{w_3^2 a_1 a_2 \cos\left((w_2+1)\tau+\psi_2+\psi_1\right)}{2(2w_2+1)\left(w_2^2-w_3^2\right)\left((w_2+1)^2-w_3^2\right)}+$$

$$\left(D_2-\frac{w_3^2 s D_{13}}{w_2^2(2w_2-1)\left(w_2^2-w_3^2\right)^2\left((w_2-1)^2-w_3^2\right)}\right)$$

$$\frac{w_3^2 a_1 a_2 \cos\left((w_2-1)\tau+\psi_2-\psi_1\right)}{2(2w_2-1)\left(w_2^2-w_3^2\right)\left((w_2-1)^2-w_3^2\right)}+$$

$$\frac{c_3 w_3^2 D_{12} a_1 a_2 \sin\left((w_2+1)\tau+\psi_2+\psi_1\right)}{2sw_2^2(2w_2+1)\left(w_2^2-w_3^2\right)^2\left((w_2+1)^2-w_3^2\right)^2}+$$

$$\frac{c_3 w_3^2 D_{14}\, a_1 a_2 \sin\left((w_2-1)\tau+\psi_2-\psi_1\right)}{2sw_2^2(2w_2-1)\left(w_2^2-w_3^2\right)^2\left((w_2-1)^2-w_3^2\right)^2}+$$

$$\left(\frac{w_3^4 s}{\left(w_2^2-w_3^2\right)^2}-\frac{2w_3^8 s^2}{\left(w_2^2-w_3^2\right)^4}+\frac{w_3^2 c_3^2}{s^2\left(w_2^2-w_3^2\right)^2}-1\right)\frac{w_3^2 a_2 \cos\left(w_2\tau+\psi_2\right)}{w_2^2-w_3^2}+$$

$$\left(\frac{w_2 c_3}{s}-\frac{w_3^2\left(3w_2^4+w_2^2 w_3^2+3w_3^4\right)c_3}{2w_2\left(w_2^2-w_3^2\right)^2}-\frac{w_3^2 c_2}{w_2}\right)\frac{w_3^2 a_2 \sin\left(w_2\tau+\psi_2\right)}{\left(w_2^2-w_3^2\right)^2},$$

where the coefficients D_k, $k=3,\ldots,14$, are the same as in the footnotes to equation (5.68).

The symbols $a_1, a_2, a_3, \psi_1, \psi_2, \psi_3$ with omitted arguments are used as the shorted denotation for the functions $a_1(\tau)$, $a_2(\tau)$, $a_3(\tau)$, $\psi_1(\tau)$, $\psi_2(\tau)$, $\psi_3(\tau)$. The functions are numerically obtained, and serve as solutions of the initial value problem and (5.151)–(5.157) which depict the slowly changing amplitudes and phases of the main vibration components, i.e. $\xi_1(\tau)=a_1\cos(\tau+\psi_1)$, $\varphi_1(\tau)=a_2\cos(w_2\tau+\psi_2)$ and $\gamma_1(\tau)=a_3\cos(w_3\tau+\psi_3)$.

To present and test the solution concerning the three simultaneously occurring external resonances, we assume the following values of the parameters of the physical pendulum: $w_2=0.275$, $w_3=0.105$, $\alpha=0.01$, $s=0.013$, $c_1=0.005$, $c_2=0.003$, $c_3=0.0012$, $f_1=0.0006$, $f_2=0.0006$, $f_3=0.00006$, $\sigma_1=-0.001$, $\sigma_2=-0.001$, $\sigma_3=-0.001$. The initial values for the modulation problem are as follows: $a_{10}=0.01$, $a_{20}=0.10$, $a_{30}=0.08$, $\psi_{10}=0$, $\psi_{20}=0$, $\psi_{30}=0$.

The following initial values for the motion equations (5.17)–(5.19)

$$u_{01}=0.0093769,\quad u_{02}=2.21141\cdot 10^{-5},\quad u_{03}=0.10052,$$

$$u_{04}=-2.46929\cdot 10^{-4},\quad u_{05}=0.062865,\quad u_{06}=1.2649\cdot 10^{-3}$$

remain in compliance with the initial values of the amplitudes and phases.

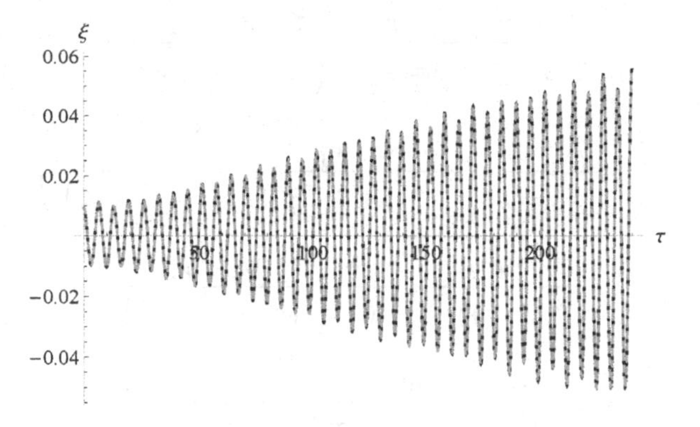

FIGURE 5.8 The transient oscillations for the coordinate $\xi(\tau)$ in the triple external resonance estimated based on the numerical (dotted line) and analytical (MSM) method.

The time histories of the generalized coordinates $\xi(\tau)$, $\varphi(\tau)$ and $\gamma(\tau)$ are presented in Figures 5.8–5.13. Figures 5.8–5.10 show the transient stage of the oscillations. Figures 5.11–5.13 present the pendulum oscillations at the end of the simulation. The solid black line presents the approximate analytical solution obtained using MSM, whereas the numerical solution obtained using the standard procedure *NDSolve* of the *Mathematica* software is drawn by the dotted line. One can observe high compliance between the numerical solution and the approximate analytical solution at the transient period of the pendulum motion.

The values of the error calculated according to formula (5.93) for the analytical and numerical approximate solutions presented in Figures 5.8–5.13 are gathered in Table 5.2.

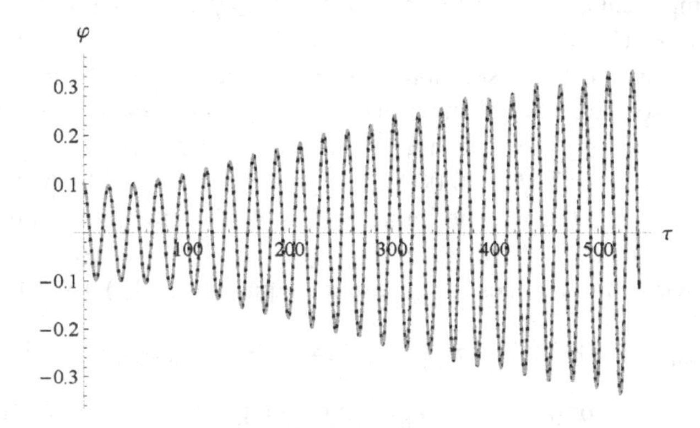

FIGURE 5.9 The transient oscillations for the coordinate $\varphi(\tau)$ in the triple external resonance estimated based on the numerical (dotted line) and analytical(MSM) method.

One can observe that after the transient stage, the oscillations of the generalized coordinates fix themselves with a different effect. The stationary and periodic oscillations are visible only in the case of the swing motion described by the coordinate $\varphi(\tau)$. The resonant oscillations related to others coordinates do not set in the simulation time and are less regular, but still quasi-periodic.

The confirmation of these conclusions are found in Figures 5.14–5.16. The figures denoted additionally by the letter "a" show the time histories of the main vibration components, i.e. $\xi_1(\tau) = a_1(\tau)\cos(\tau + \psi_1(\tau))$, $\varphi_1(\tau) = a_2(\tau)\cos(w_2\tau + \psi_2(\tau))$ and $\gamma_1(\tau) = a_3(\tau)\cos(w_3\tau + \psi_3(\tau))$, respectively. The figures denoted by the letter "b" present the whole approximate solutions obtained using the method of multiple scales. In each of the six graphs, the plots of the amplitudes $a_i(\tau), -a_i(\tau)$, where $i = 1,2,3$, are drawn. All the amplitudes and also the main vibration components become set in the simulation time. However, the whole asymptotic solutions which are much more complex do not fix themselves.

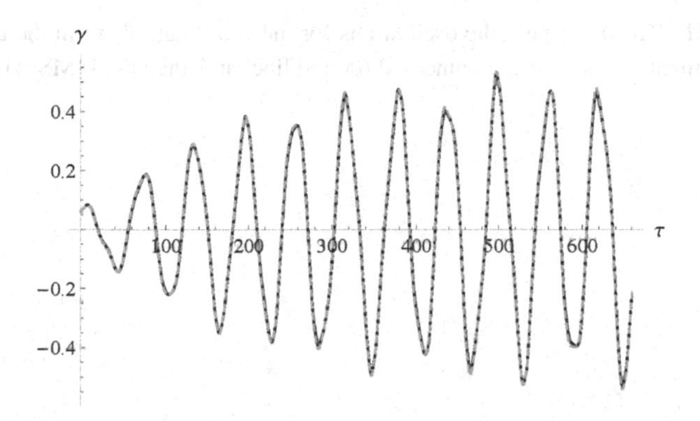

FIGURE 5.10 The transient oscillations for the coordinate $\gamma(\tau)$ in the triple external resonance estimated based on the numerical (dotted line) and analytical (MSM) method.

TABLE 5.2
The values of error of the fulfilment of the governing equations for the resonant vibration

	MSM solution	**numerical solution**
transient stage	$\delta_1 = 4.6255 \cdot 10^{-5}$	$\delta_1 = 9.2844 \cdot 10^{-5}$
$\tau_s = 0,\ \tau_e = 150$	$\delta_2 = 2.9869 \cdot 10^{-5}$	$\delta_2 = 7.1533 \cdot 10^{-5}$
	$\delta_3 = 2.6063 \cdot 10^{-5}$	$\delta_3 = 4.1154 \cdot 10^{-7}$
stationary stage	$\delta_1 = 7.4074 \cdot 10^{-5}$	$\delta_1 = 1.0783 \cdot 10^{-7}$
$\tau_s = 2850,\ \tau_e = 3000$	$\delta_2 = 5.3530 \cdot 10^{-5}$	$\delta_2 = 6.0496 \cdot 10^{-8}$
	$\delta_3 = 1.4676 \cdot 10^{-4}$	$\delta_3 = 1.3545 \cdot 10^{-8}$

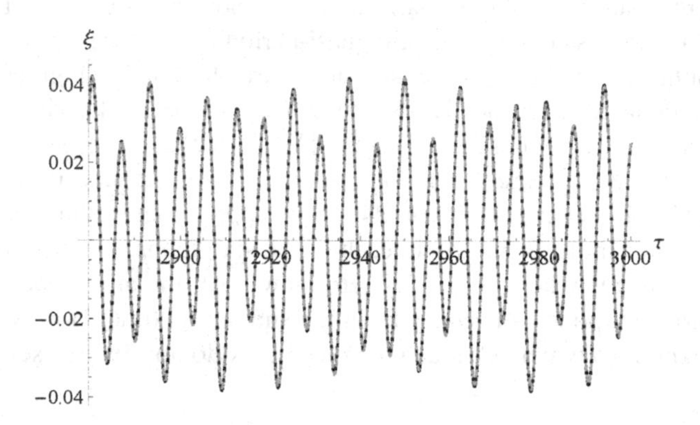

FIGURE 5.11 The quasi-periodic oscillations for the coordinate $\xi(\tau)$ in the triple external resonance estimated based on the numerical (dotted line) and analytical (MSM) method.

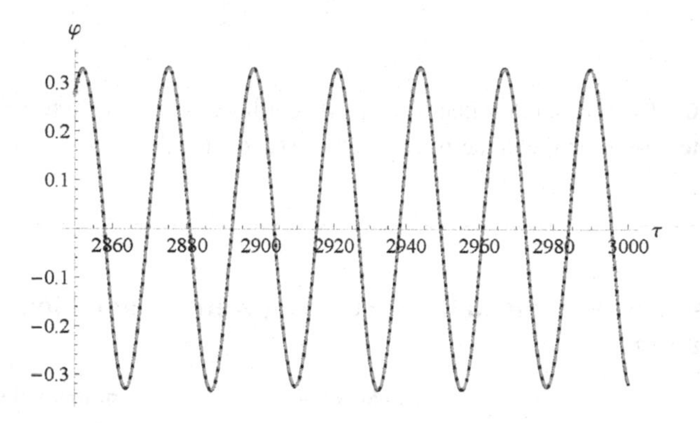

FIGURE 5.12 The stationary oscillations for the coordinate $\varphi(\tau)$ in the triple external resonance estimated based on the numerical (dotted line) and analytical (MSM) method.

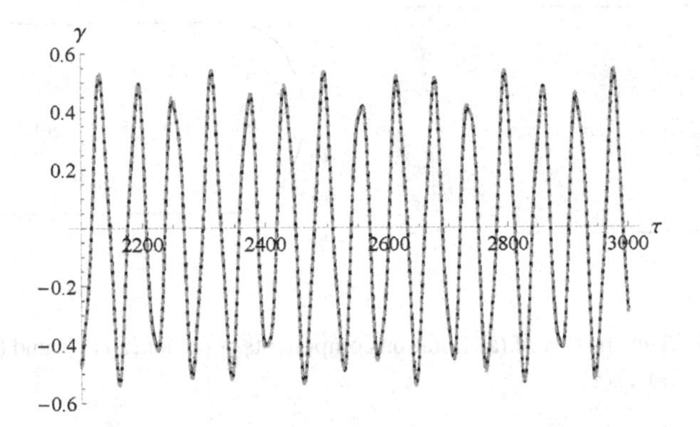

FIGURE 5.13 The quasi-periodic oscillations for the coordinate $\gamma(\tau)$ in the triple external resonance estimated based on the numerical (dotted line) and analytical (MSM) method.

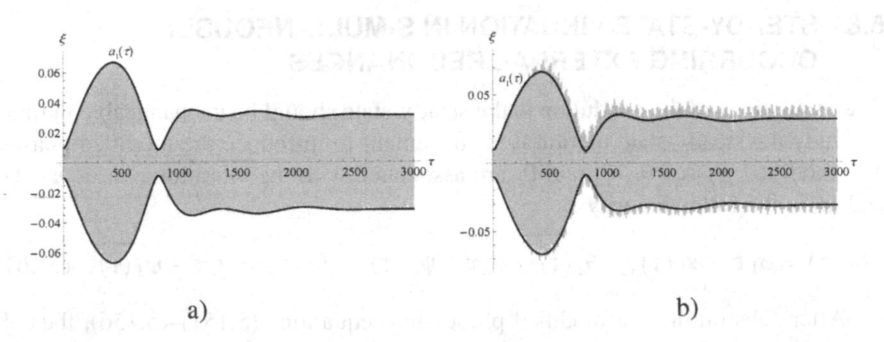

FIGURE 5.14 Time history of (a) vibration components $\xi_1(\tau)$ and $a_1(\tau)$, and (b) asymptotic solutions $\xi(\tau)$ and $a_1(\tau)$.

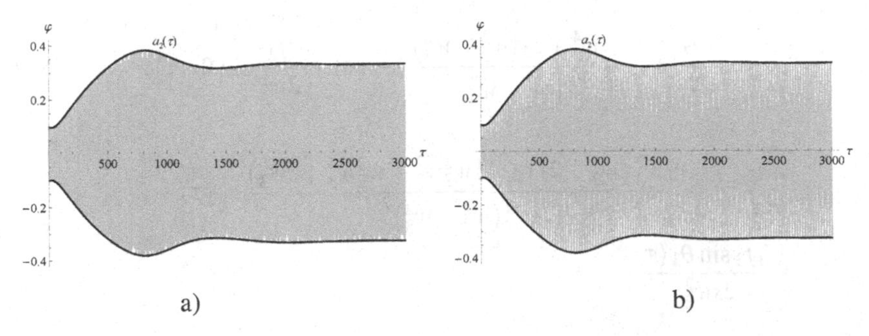

FIGURE 5.15 Time history of (a) vibration components $\varphi_1(\tau)$ and $a_2(\tau)$, and (b) asymptotic solutions $\varphi(\tau)$ and $a_2(\tau)$.

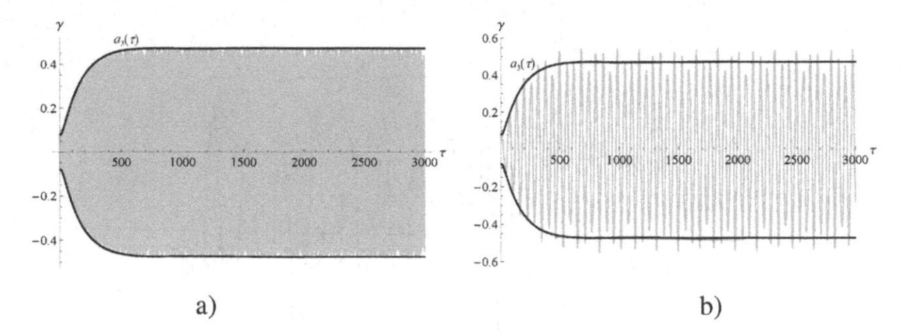

FIGURE 5.16 Time history of (a) vibration components $\gamma_1(\tau)$ and $a_3(\tau)$, and (b) asymptotic solutions $\gamma(\tau)$ and $a_3(\tau)$.

Despite the different course of vibration for the particular coordinates, the solution obtained using MSM shows a high agreement with the numerical solution of the motion equations in the original form.

5.6 STEADY-STATE VIBRATION IN SIMULTANEOUSLY OCCURRING EXTERNAL RESONANCES

The oscillations of the pendulum in the steady-state should be unchangeable in time. To study the steady-state motion it is convenient to introduce the modified phases. The modified phases θ_1, θ_2 and θ_3 are associated with the vibration phases ψ_1, ψ_2 and ψ_3 in the following way

$$\theta_1(\tau) = \sigma_1\tau - \psi_1(\tau), \quad \theta_2(\tau) = \sigma_2\tau - \psi_2(\tau), \quad \theta_3(\tau) = \sigma_3\tau - \psi_3(\tau). \tag{5.161}$$

After substituting the modified phases into equations (5.151)–(5.156), the following system of six differential equations is obtained

$$\frac{da_1}{d\tau} = -\frac{c_1}{2}a_1(\tau) + \frac{f_1}{2}\sin\theta_1(\tau), \tag{5.162}$$

$$\frac{da_2}{d\tau} = -\frac{c_3w_2^4 + c_2\left(w_2^2 - w_3^2\right)^2}{2\left(w_2^2 - w_3^2\right)^2}a_2(\tau) + \frac{f_2}{2w_2}\sin\theta_2(\tau), \tag{5.163}$$

$$\frac{da_3}{d\tau} = -\frac{c_3w_3^2\left(w_2^6 - 2(s+1)w_2^4w_3^2 + w_2^2w_3^4 + sw_3^6\right)}{2sw_2^4\left(w_2^2 - w_3^2\right)^2}a_3(\tau) + \frac{w_3f_3\sin\theta_3(\tau)}{2sw_2^2}, \tag{5.164}$$

$$\sigma_1 - \frac{d\theta_1}{d\tau} = \frac{3w_2^2(w_2^2-1)}{4\left(4w_2^2-1\right)}a_2^2(\tau) + \frac{3\xi_e^2\alpha}{2} - \frac{f_1}{2a_1(\tau)}\cos\theta_1(\tau), \tag{5.165}$$

$$\sigma_2 - \frac{d\theta_2}{d\tau} = w_2\left(w_2^2-1\right)\frac{12a_1^2(\tau)-\left(1+8w_2^2\right)a_2^2(\tau)}{16\left(4\,w_2^2-1\right)}+ \\ \frac{sw_2w_3^2\left(4w_2^4+(3s-8)\,w_2^2w_3^3+(4-7s)\,w_3^4\right)}{8\left(w_2^2-w_3^2\right)^3}-\frac{f_2\cos\theta_2(\tau)}{2w_2a_2(\tau)}, \tag{5.166}$$

$$\sigma_3 - \frac{d\theta_3}{d\tau} = -\frac{c_3^2w_3^3}{8s^2w_2^4}-\frac{sw_3^5\left(4w_2^6-8w_2^4w_3^2+(4-7s)\,w_2^2w_3^4+3s\,w_3^6\right)}{8w_2^4\left(w_2^2-w_3^2\right)^3}- \\ \frac{w_3^3a_1^2(\tau)}{4w_2^4\left(4w_3^2-1\right)}+\frac{w_3^3\left(w_3^2-2w_2^2\right)a_2^2(\tau)}{8\left(w_2^2-w_3^2\right)^2}-\frac{w_3a_3^2(\tau)}{16}-\frac{w_3f_3\cos\theta_3(\tau)}{2sw_2^2a_3(\tau)}. \tag{5.167}$$

Differential equations (5.162)–(5.167) present the autonomous dynamical system in the standard form. The approximate asymptotic solutions will have a fixed character when the variability of the amplitudes and the modified phases disappears. According to this expectation, we postulate that the time derivatives of the amplitudes and the modified phases become equal to zero. In this case, the amplitudes a_1, a_2, a_3 and the modified phases $\theta_1, \theta_2, \theta_3$ stand as constant quantities. The steady-state vibration is governed by the following set of six equations

$$-c_1a_1+f_1\sin\theta_1=0, \tag{5.168}$$

$$-\frac{c_3w_2^4+c_2\left(w_2^2-w_3^2\right)^2}{\left(w_2^2-w_3^2\right)^2}a_2+\frac{f_2}{w_2}\sin\theta_2=0, \tag{5.169}$$

$$-\frac{c_3w_3^2\left(w_2^6-2(s+1)\,w_2^4w_3^2+w_2^2w_3^4+sw_3^6\right)}{w_2^2\left(w_2^2-w_3^2\right)^2}a_3+w_3f_3\sin\theta_3=0, \tag{5.170}$$

$$\frac{3w_2^2(w_2^2-1)}{2\left(4w_2^2-1\right)}a_2^2+3\xi_e^2\alpha-\frac{f_1\cos\theta_1}{a_1}-2\sigma_1=0, \tag{5.171}$$

$$w_2\left(w_2^2-1\right)\frac{12a_1^2-\left(1+8w_2^2\right)a_2^2}{8\left(4\,w_2^2-1\right)}+ \\ \frac{sw_2w_3^2\left(4w_2^4+(3s-8)\,w_2^2w_3^3+(4-7s)\,w_3^4\right)}{4\left(w_2^2-w_3^2\right)^3}-\frac{f_2\cos\theta_2}{w_2a_2}-2\sigma_2=0, \tag{5.172}$$

$$-\frac{c_3^2w_3^3}{4s^2w_2^4}-\frac{sw_3^5\left(4w_2^6-8w_2^4w_3^2+(4-7s)\,w_2^2w_3^4+3s\,w_3^6\right)}{4w_2^4\left(w_2^2-w_3^2\right)^3}- \\ \frac{w_3^3a_1^2}{2w_2^4\left(4w_3^2-1\right)}+\frac{w_3^3\left(w_3^2-2w_2^2\right)a_2^2}{4\left(w_2^2-w_3^2\right)^2}-\frac{w_3a_3^2}{8}-\frac{w_3f_3\cos\theta_3}{sw_2^2a_3}-2\sigma_3=0. \tag{5.173}$$

The values of the variables a_1, a_2, a_3 and θ_1, θ_2, θ_3 that satisfy the system consisting of equations (5.168)–(5.173) are the fixed amplitudes and modified phases of the pendulum vibration in the case of three external resonances occurring simultaneously.

There are algebraic and also trigonometric nonlinearities in the system. The modified phases can be eliminated from equations (5.168)–(5.173). Indeed, using the commonly known trigonometric identities allows one to express the amplitude-frequency dependencies as a set of three algebraic equations of the following form

$$\left(\frac{c_1 a_1}{f_1}\right)^2 + \frac{a_1^2\left(3\left(w_2^4 - w_2^2\right)a_2^2 - 6\alpha\,\xi_e^2 + 24 w_2^2 \alpha \xi_e^2 + 4\sigma_1 - 16 w_2^2 \sigma_1\right)^2}{4 f_1^2 \left(4 w_2^2 - 1\right)^2} - 1 = 0, \tag{5.174}$$

$$\begin{aligned}
&\frac{3w_2^3\left(w_2^2-1\right)\left(s w_2 w_3^2\left(4w_2^4+(3s-8)\,w_2^2w_3^2-(7s-4)\,w_3^4\right)-8\left(w_2^2-w_3^2\right)^3\sigma_2\right)}{4f_2^2\left(4w_2^2-1\right)\left(w_2^2-w_3^2\right)^3}a_1^2a_2^2+\\
&\frac{9w_2^4a_1^4a_2^2}{4f_2^2\left(4w_2^2-1\right)^2}-\frac{3w_2^4\left(w_2^2-1\right)^2\left(8w_2^2+1\right)a_1^2a_2^4}{8f_2^2\left(4w_2^2-1\right)^2}+\\
&\frac{w_2^4\left(8w_2^4-7w_2^2-1\right)^2}{64f_2^2\left(4w_2^2-1\right)^2}a_2^6+\\
&\frac{w_2^3\left(8w_2^4-7w_2^2-1\right)\left(s w_2 w_3^2\left(-4w_2^4-(3s-8)\,w_2^2w_3^2+(7s-4)\,w_3^4\right)\right)}{16f_2^2\left(4w_2^2-1\right)\left(w_2^2-w_3^2\right)^3}a_2^4+\\
&\frac{w_2^3\left(8w_2^4-7w_2^2-1\right)\left(8\left(w_2^2-w_3^2\right)^3\sigma_2\right)}{16f_2^2\left(4w_2^2-1\right)\left(w_2^2-w_3^2\right)^3}a_2^4+\\
&\frac{w_2^2c_2^2a_2^2}{f_2^2}+\frac{w_2^{10}c_3^2a_2^2}{f_2^2\left(w_2^2-w_3^2\right)^4}+\frac{2w_2^6c_2c_3a_2^2}{f_2^2\left(w_2^2-w_3^2\right)^2}+\\
&\frac{4w_2^2\sigma_2^2a_2^2}{f_2^2}+\frac{w_2^4w_3^8\left(3w_2^2-7w_3^2\right)^2s^4a_2^2}{16f_2^2\left(w_2^2-w_3^2\right)^6}+\\
&\frac{w_2^3w_3^4\left(w_2^3-w_2w_3^2+\left(7w_3^2-3w_2^2\right)\sigma_2\right)s^2a_2^2}{f_2^2\left(w_2^2-w_3^2\right)^3}-\\
&\frac{w_2^4w_3^6\left(3w_2^2-7w_3^2\right)s^3a_2^2}{2f_2^2\left(w_2^2-w_3^2\right)^4}+\frac{4w_2^3w_3^2sa_2^2}{f_2^2\left(w_2^2-w_3^2\right)}-1=0,
\end{aligned} \tag{5.175}$$

$$\frac{s^2 w_2^4 a_3^2}{4 f_3^2}\left(\frac{w_3 a_1^2}{w_2^4\left(4w_3^2-1\right)}-\frac{(w_3^2-2w_2^2)w_3^3 a_2^2}{2\left(w_2^2-w_3^2\right)^2}+\frac{a_3^2}{4}+D_{15}+\frac{4}{w_3}\sigma_3\right)^2+$$
$$\frac{w_3^2 c_3^2\left(w_2^6-2\left(s+1\right)w_2^4 w_3^2+w_2^2 w_3^4+s w_3^6\right)^2 a_3^2}{f_3^2 w_2^4\left(w_2^2-w_3^2\right)^4}-1=0, \tag{5.176}$$

where

$$D_{15}=\frac{(3w_3^2-7w_2^2)w_3^8\, s^2}{2w_2^4\left(w_2^2-w_3^2\right)^3}+\frac{2w_3^4 s}{w_2^2\left(w_2^2-w_3^2\right)}+\frac{w_3 c_3^2}{2s^2 w_2^4}.$$

To draw the resonance response curves concerning the assumed case of the triple external resonance, we have to solve many times the system of equations (5.174)–(5.176) for various values of the detuning parameters. The system has been solved numerically using the standard procedure *NSolve* of the *Mathematica*.

Let us write equations (5.174)–(5.176) in the shortened form

$$G_1\left(a_1,a_2,a_3\right)=0, \tag{5.177}$$

$$G_2\left(a_1,a_2,a_3\right)=0, \tag{5.178}$$

$$G_3\left(a_1,a_2,a_3\right)=0, \tag{5.179}$$

where $G_i\left(a_1,a_2,a_3\right)$, $i=1,2,3$, denotes the polynomial that occurs on the left side of the appropriately chosen equation. Each of the polynomials is of degree six, hence the total number of the solutions of the system (5.177)–(5.179), understood as an ordered triple (a_1,a_2,a_3), equals two hundred and sixteen. Not all of them are represented by real numbers. We focus on the finding such ordered triples that consist only of the real values. Since the polynomials $G_i\left(a_1,a_2,a_3\right), i=1,2,3$, are even functions with respect to their arguments, the solution of the system (5.177)–(5.179) can be restricted only to the region where the values a_1,a_2,a_3 are positive.

To exemplify the relationships between the magnitudes of the amplitudes in the resonant steady-state and the frequencies of the acting harmonic forces, the following values of the parameters are assumed: $w_2=0.275$, $w_3=0.105$, $\alpha=0.01$, $s=0.013$, $c_1=0.005$, $c_2=0.003$, $c_3=0.0012$, $f_1=0.0006$, $f_2=0.0006$, $f_3=0.00006$.

The resonance response curves shown in Figures 5.17a–5.17c illustrate the dependence of the detuning parameter σ_1 on the steady-state amplitudes a_1, a_2 and a_3. The values of the other detuning parameters have been determined as small constant numbers. They are $\sigma_2=-0.002$, $\sigma_3=-0.001$. The main resonant effect concerns the coordinate ξ whose amplitude increases about six times. The resonant response is of the hard-type. The amplitudes of the other two coordinates are also slightly increased, and are also of the hard-type responses. The additional solutions of the system (5.174)–(5.176) appear in the range of the detuning parameter values from about 0.002 to 0.008. On each of the graphs, they form a closed curve that is separated from the main branch of the response curve.

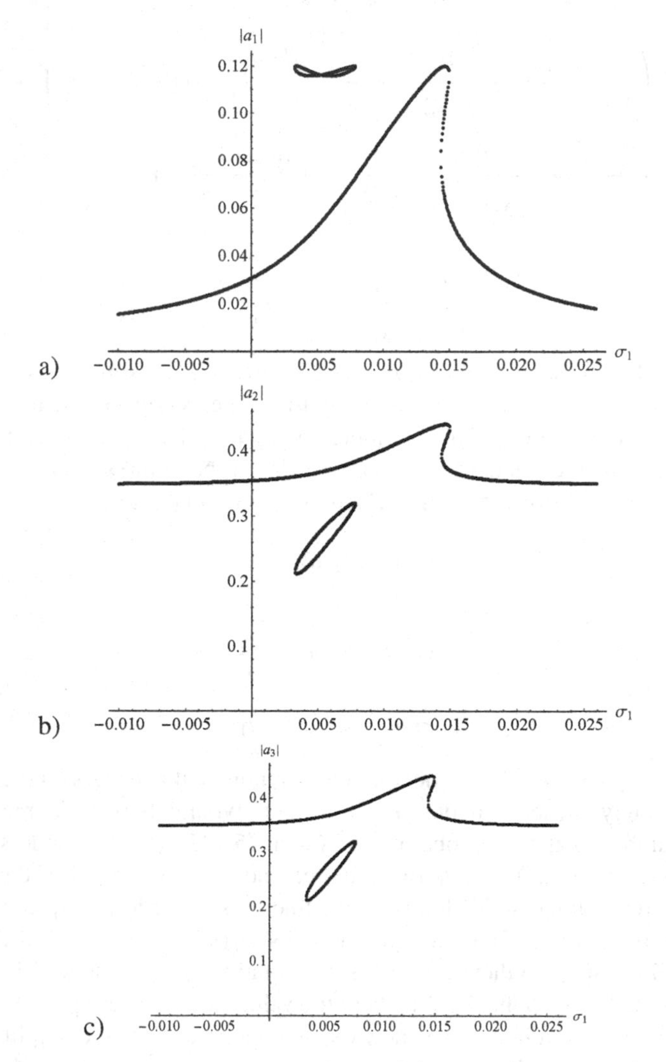

FIGURE 5.17 Dependencies $a_1(\sigma_1)$ (a), $a_2(\sigma_1)$ (b), $a_3(\sigma_1)$ (c) for the following fixed parameters: $\sigma_2 = -0.002$, $\sigma_3 = -0.001$.

The influence of the detuning parameter σ_2 on the steady-state resonant amplitudes is depicted in Figures 5.18a–5.18c. It has been assumed that $\sigma_1 = 0.001$ and $\sigma_3 = -0.001$. The main resonant effect concerns the angle φ, and is shown in Figure 5.18b. The resonance response curve is of the soft-type and the maximal increase of the amplitude a_2 is more than four times. The values of the other amplitudes decrease, while the amplitude a_3 is slightly changed.

The relationships between the steady-state resonant amplitudes and the detuning parameter σ_3 are shown in Figures 5.19a–5.19c. The graphs have been made assuming that $\sigma_1 = 0.001$ and $\sigma_2 = -0.002$. The main and the only effect concerns the

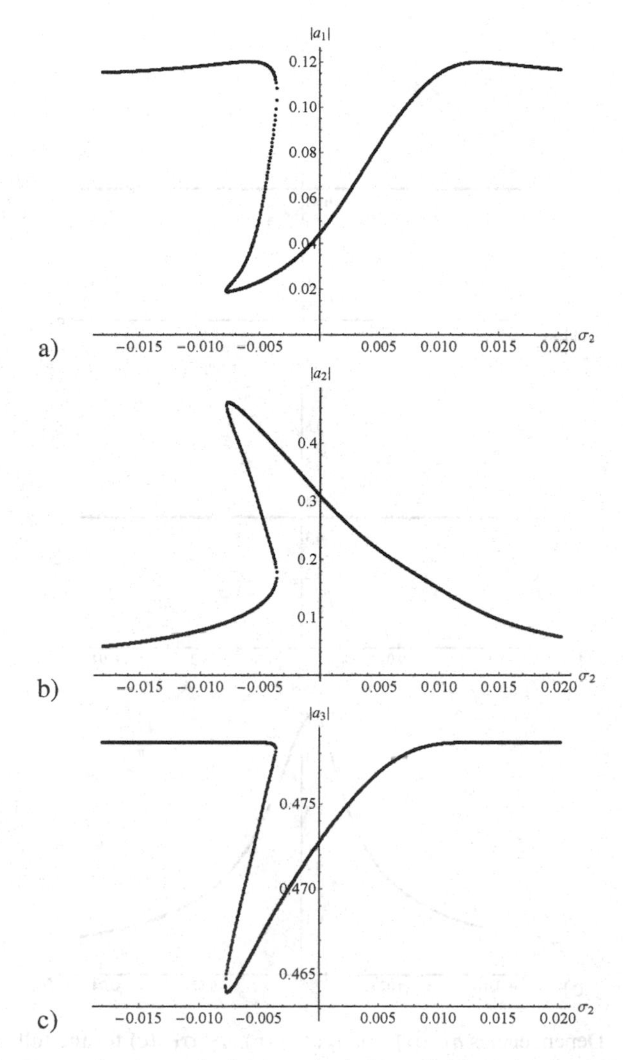

FIGURE 5.18 Dependencies $a_1(\sigma_2)$ (a), $a_2(\sigma_2)$ (b), $a_3(\sigma_2)$ (c) for the following fixed parameters: $\sigma_1 = 0.001$, $\sigma_3 = -0.001$.

coordinate γ. The width of the resonant frequencies interval is relatively large compared to the response curves previously discussed. The increase of the amplitude is significant, and the curve is unambiguous for any value of the detuning parameter. The influence of the detuning parameter σ_3 on the amplitudes a_1 and a_2 is not noticed. This observation is in accordance with the structure of the system of equations (5.174)–(5.176) that govern the steady-sate. There is neither the detuning parameter σ_3 nor the amplitude a_3 in equations (5.174)–(5.175). Therefore, the impact is excluded on the level of the modulation equations, which is the consequence of the adopted model assuming the weak couplings and the method of solution. To assess

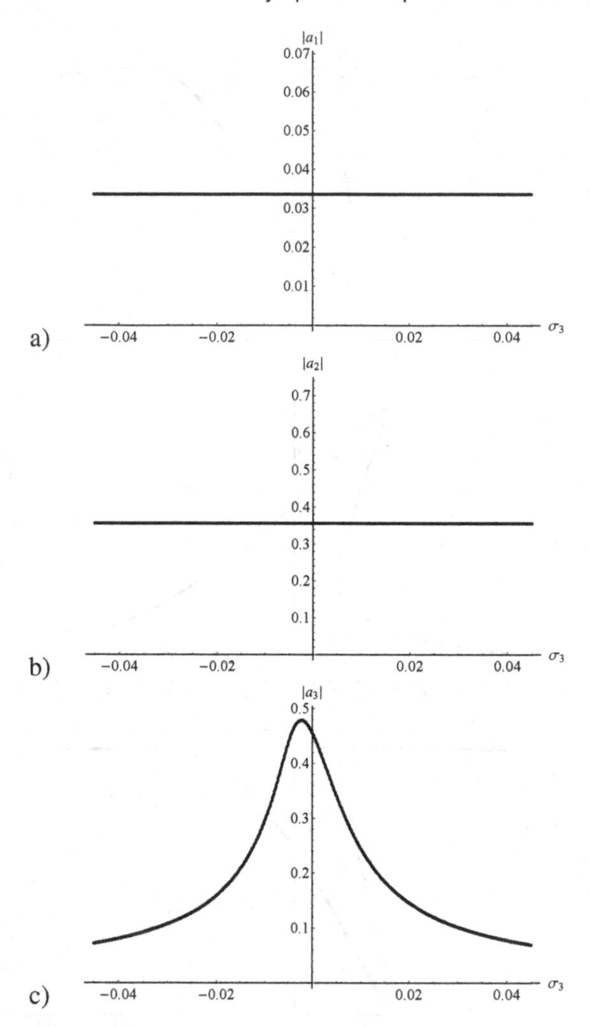

FIGURE 5.19 Dependencies $a_1(\sigma_3)$ (a), $a_2(\sigma_3)$ (b), $a_3(\sigma_3)$ (c) for the following fixed parameters: $\sigma_1 = 0.001,\ \sigma_2 = -0.002$.

how much these simplifications change the image of the resonant case discussed here, we investigated the system behavior through direct numerical simulations. Assuming the same values of the parameters as for the case illustrated in Figure 5.19b, the original equations of motion (5.17)–(5.19) are solved for some small values of the detuning parameter by employing the standard method *NDSolve* of the *Mathematica*. The results in the form of the time history of the coordinate φ in the steady-state are shown in Figure 5.20. There are no significant changes in the value of the amplitude a_2 depending on the detuning parameter σ_3. There is also no significant effect of the parameter σ_3 on the steady-state vibrations for the coordinate ξ.

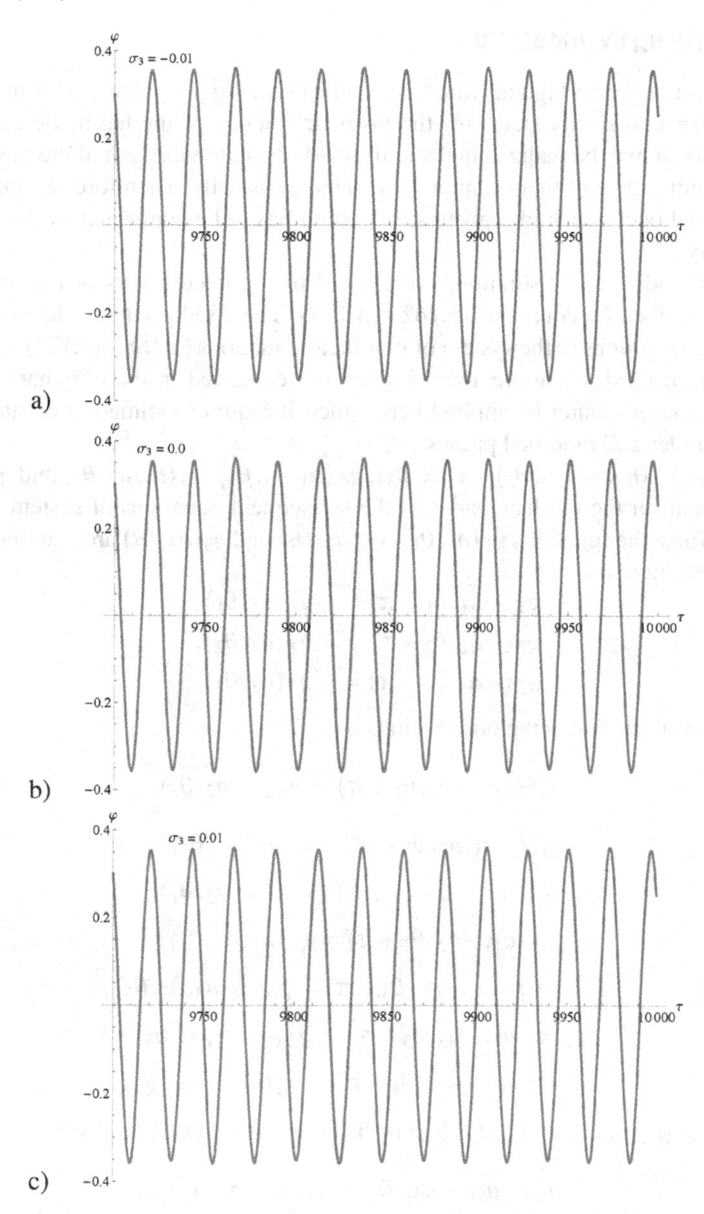

FIGURE 5.20 The steady state oscillations for the coordinate $\varphi(\tau)$: a) numerical solution for $\sigma_3 = -0.01$; b) numerical solution for $\sigma_3 = 0.0$; c) numerical solution for $\sigma_3 = 0.01$.

5.7 STABILITY ANALYSIS

The curves may be ambiguous functions of the detuning parameters. This means that there are several steady states for the given frequency of the harmonic excitation. Some of them may be realized in the real world resonant vibration of the mechanical system, while some of them cannot be realized physically. Therefore, the possibility of the actual occurrence of a given steady state has to be confirmed by the study of its stability.

The periodic steady solutions correspond to the fixed points of the dynamical system described by equations (5.162)–(5.167). The fixed points of the system can be found as solutions to the system of nonlinear equations (5.168)–(5.173). However, the approach used in the previous chapter and consisted in the elimination of the modified phases cannot be applied here, since it requires estimation of stability of the amplitudes and modified phases.

Let $g_1(a_1,\theta_1)$, $g_2(a_2,\theta_2)$, $g_3(a_3,\theta_3)$, $g_4(a_1,a_2,\theta_1)$, $g_5(a_1,a_2,\theta_2)$ and $g_6(a_1,a_2,a_3,\theta_3)$ stand for the left hand sides of the subsequent equations of system (5.168)–(5.173). Since the functions $g_1(a_1,\theta_1)$, $g_2(a_2,\theta_2)$ and $g_3(a_3,\theta_3)$ are odd, the following property holds

$$\begin{aligned} g_1(-a_1,\theta_1+\pi) &= -g_1(a_1,\theta_1),\\ g_2(-a_2,\theta_2+\pi) &= -g_2(a_2,\theta_2),\\ g_3(-a_3,\theta_3+\pi) &= -g_3(a_3,\theta_3). \end{aligned} \tag{5.180}$$

Moreover, the following observations hold

$$\begin{aligned} g_4(-a_1,-a_2,\theta_1+\pi) &= g_4(a_1,a_2,\theta_1),\\ g_4(-a_1,a_2,\theta_1+\pi) &= g_4(a_1,a_2,\theta_1),\\ g_5(-a_1,-a_2,\theta_2+\pi) &= g_5(a_1,a_2,\theta_1),\\ g_5(a_1,-a_2,\theta_2+\pi) &= g_5(a_1,a_2,\theta_1),\\ g_6(-a_1,-a_2,-a_3,\theta_3+\pi) &= g_6(a_1,a_2,a_3,\theta_3),\\ g_6(a_1,a_2,-a_3,\theta_3+\pi) &= g_6(a_1,a_2,a_3,\theta_3),\\ g_6(-a_1,a_2,-a_3,\theta_3+\pi) &= g_6(a_1,a_2,a_3,\theta_3). \end{aligned} \tag{5.181}$$

Thus, if $(a_{1s},a_{2s},a_{3s},\theta_{1s},\theta_{2s},\theta_{3s})$ is the solution of system (5.168)–(5.173), then

$$(-a_{1s},-a_{2s},-a_{3s},\theta_{1s}+\pi,\theta_{2s}+\pi,\theta_{3s}+\pi),$$

$$\begin{gathered} (-a_{1s},a_{2s},a_{3s},\theta_{1s}+\pi,\theta_{2s},\theta_{3s}),\quad (a_{1s},-a_{2s},a_{3s},\theta_{1s},\theta_{2s}+\pi,\theta_{3s}),\\ (a_{1s},a_{2s},-a_{3s},\theta_{1s},\theta_{2s},\theta_{3s}+\pi),\ (-a_{1s},-a_{2s},a_{3s},\theta_{1s}+\pi,\theta_{2s}+\pi,\theta_{3s}),\\ (-a_{1s},a_{2s},-a_{3s},\theta_{1s}+\pi,\theta_{2s},\theta_{3s}+\pi),\quad (a_{1s},-a_{2s},-a_{3s},\theta_{1s},\theta_{2s}+\pi,\theta_{3s}+\pi) \end{gathered}$$

are also its solutions. Changing the sign of the values of the functions in relations (5.180) is insignificant because we are dealing with the equations of the form $g_i(\ldots)=0,\quad i=1,2,3.$

Both the sine and the cosine of the modified phases occur in equations (5.168)–(5.173). Eliminating one of them using the Pythagorean identity is associated with the ambiguity of the sign. These difficulties are removed through the following relations

$$\sin(\theta_i) = \frac{2t_i}{1+t_i^2}, \quad \cos(\theta_i) = \frac{1-t_i^2}{1+t_i^2}, \quad i=1,2,3, \tag{5.182}$$

where

$$t_i = \tan\left(\frac{\theta_i}{2}\right), \quad i=1,2,3.$$

Employing the identities (5.182) in equations (5.168)–(5.173) yields

$$c_1 a_1 - \frac{2f_1 t_1}{1+t_1^2} = 0, \tag{5.183}$$

$$\frac{c_3 w_2^4 + c_2\left(w_2^2 - w_3^2\right)^2}{\left(w_2^2 - w_3^2\right)^2} a_2 - \frac{2f_2 t_2}{w_2\left(1+t_2^2\right)} = 0, \tag{5.184}$$

$$\frac{c_3 w_3^2\left(w_2^6 - 2(s+1)\, w_2^4 w_3^2 + w_2^2 w_3^4 + s w_3^6\right)}{w_2^2\left(w_2^2 - w_3^2\right)^2} a_3 - \frac{2w_3 f_3 t_3}{1+t_3^2} = 0, \tag{5.185}$$

$$\frac{3w_2^2(w_2^2-1)}{4w_2^2-1} a_2^2 - \frac{2f_1}{a_1}\frac{1-t_1^2}{1+t_1^2} = 4\sigma_1 - 6\xi_e^2\alpha, \tag{5.186}$$

$$\begin{aligned} w_2\left(w_2^2-1\right)\frac{12a_1^2 - \left(1+8w_2^2\right)a_2^2}{4\,w_2^2-1} - \frac{8f_2}{w_2 a_2}\frac{1-t_2^2}{1+t_2^2} = 16\sigma_2 - \\ \frac{2sw_2w_3^2\left(4w_2^4 + (3s-8)\,w_2^2w_3^3 + (4-7s)\,w_3^4\right)}{\left(w_2^2 - w_3^2\right)^3}, \end{aligned} \tag{5.187}$$

$$\begin{aligned} \frac{2w_3^3\left(w_3^2 - 2w_2^2\right)a_2^2}{\left(w_2^2-w_3^2\right)^2} - \frac{4w_3^3 a_1^2}{w_2^4\left(4w_3^2-1\right)} - w_3 a_3^2 - \frac{8w_3 f_3}{s w_2^2 a_3}\frac{1-t_3^2}{1+t_3^2} = \\ 16\sigma_3 + \frac{2c_3^2 w_3^3}{s^2 w_2^4} + \frac{2sw_3^5\left(4w_2^6 - 8w_2^4w_3^2 + (4-7s)\,w_2^2w_3^4 + 3s\,w_3^6\right)}{w_2^4\left(w_2^2-w_3^2\right)^3}. \end{aligned} \tag{5.188}$$

The system of six algebraic equations is solved numerically using the standard procedure *NSolve* of the *Mathematica*. The resonance response curves understood as the solutions of equations (5.183)–(5.188) are determined assuming the same values of the parameters as in the example considered in section 5.5, i.e. $w_2 = 0.275$, $w_3 = 0.105$, $\alpha = 0.01$, $s = 0.013$, $c_1 = 0.005$, $c_2 = 0.003$, $c_3 = 0.0012$, $f_1 = 0.0006$, $f_2 = 0.0006$, $f_3 = 0.00006$.

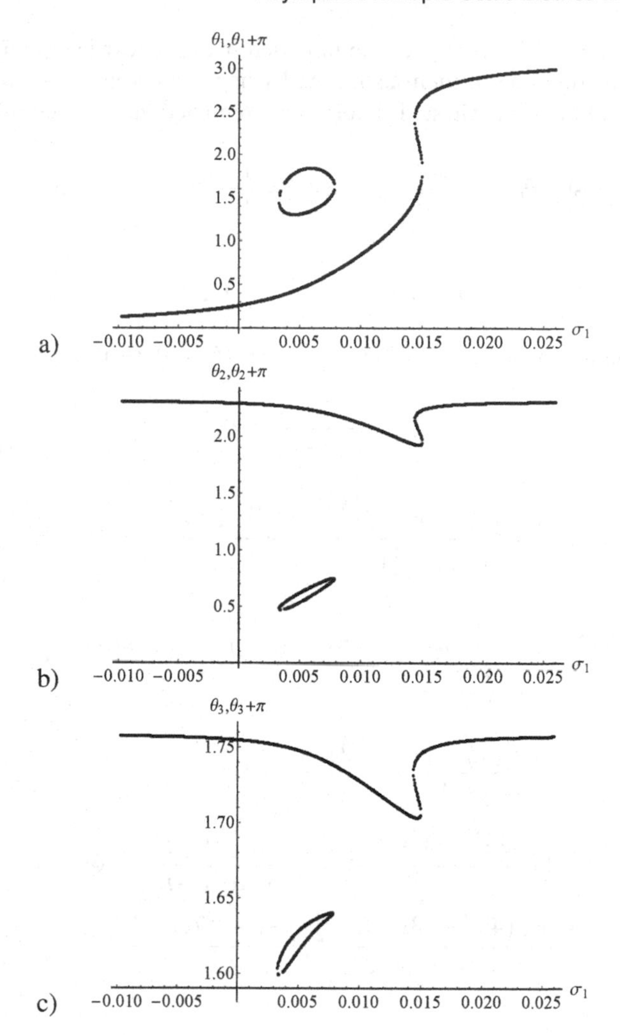

FIGURE 5.21 Dependencies $\theta_1(\sigma_1)$ (a), $\theta_2(\sigma_1)$ (b), $\theta_3(\sigma_1)$ (c) for the following fixed parameters: $\sigma_2 = -0.002$, $\sigma_3 = -0.001$.

The graphs presenting the dependence between the amplitudes and the detuning parameters are identical to the ones discussed in the mentioned example. Therefore, we present here only the resonance response curves exhibiting the impact of the external forces frequencies on the modified phases. The curves are shown in Figures 5.21–5.23. Making use of the symmetry properties written by relations (5.181), only the main branches of these curves are plotted, i.e. when $a_i > 0$ and $0 < \theta_i < \pi$, $i = 1,2,3$.

The resonance response curves shown in Figures 5.21a–5.21c depict the influence of the detuning parameter σ_1 on the steady-state modified phases θ_1, θ_2 and θ_3.

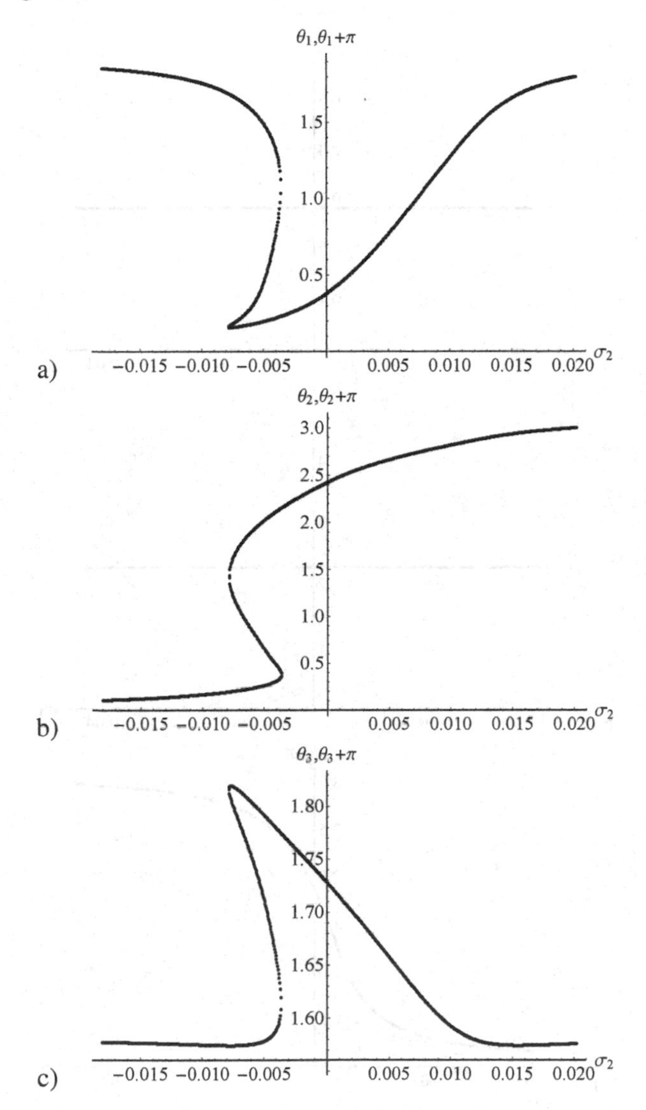

FIGURE 5.22 Dependencies $\theta_1(\sigma_2)$ (a), $\theta_2(\sigma_2)$ (b), $\theta_3(\sigma_2)$ (c) for the following fixed parameters: $\sigma_1 = 0.001$, $\sigma_3 = -0.001$.

The values of the other detuning parameters are constant, and $\sigma_2 = -0.002$, $\sigma_3 = -0.001$. The main effect concerns the modified phase θ_1 which in the resonant frequencies region varies about by π. The other two modified phases slightly decrease in that region. In the range of the detuning parameter values from about 0.002 to 0.008, the closed curves appear, which are separated from the main branches of the resonance response curves.

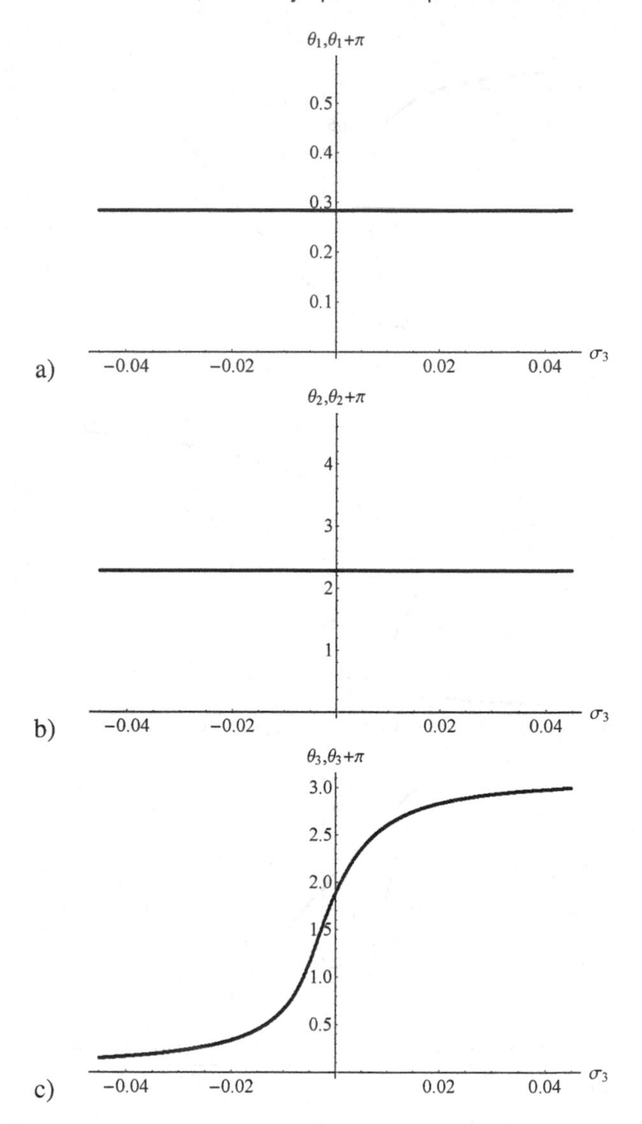

FIGURE 5.23 Dependencies $\theta_1(\sigma_3)$ (a), $\theta_2(\sigma_3)$ (b), $\theta_3(\sigma_3)$ (c) for the following fixed parameters: $\sigma_1 = 0.001$, $\sigma_2 = -0.002$.

The influence of the detuning parameter σ_2 on the steady-state modified phases θ_1, θ_2 and θ_3 is shown in Figures 5.22a–5.22c. For $\sigma_1 = 0.001$, $\sigma_3 = -0.001$. The main effect is related to the modified phase θ_2 which in the interval of the resonant frequencies, varies about by π. Quite a deep decrease is noticed in the interval of the resonant frequencies in the case of the modified phase θ_1. On the other hand, the increase in the values of the modified phase θ_3 is relatively small.

The graphs presenting the relationships between the detuning parameter σ_3 and the modified phases of the steady-state have been constructed assuming that $\sigma_1 = 0.001$ and $\sigma_2 = -0.002$. Only the modified phase θ_3 is affected by the changes in the frequencies of the torque M_2. The increase of about π is observed, and the response curve is unambiguous in the whole range of the detuning parameter variability. Similarly as in the case shown in Figures 5.19a–5.19b, the influence of the detuning parameter σ_3 on the phases θ_1 and θ_2 is not noticed.

Let $(a_{1s}, a_{2s}, a_{3s}, \theta_{1s}, \theta_{2s}, \theta_{3s})$ denotes any fixed point of the dynamical system equations (5.162)–(5.167). To examine the stability in the sense of Lyapunov of this point, we focus on the non-stationary solutions of the system (5.162)–(5.167) that are close to the steady-state solution $(a_{1s}, a_{2s}, a_{3s}, \theta_{1s}, \theta_{2s}, \theta_{3s})$. Introducing the functions $\tilde{a}_1(\tau), \tilde{a}_2(\tau), \tilde{a}_3(\tau), \tilde{\theta}_1(\tau), \tilde{\theta}_2(\tau), \tilde{\theta}_3(\tau)$ that can be treated as small perturbations, one can write the perturbed solutions in the form

$$\begin{aligned} a_1(\tau) &= a_{1s} + \tilde{a}_1(\tau), \quad a_2(\tau) = a_{2s} + \tilde{a}_2(\tau), \quad a_3(\tau) = a_{3s} + \tilde{a}_3(\tau), \\ \theta_1(\tau) &= \theta_{1s} + \tilde{\theta}_1(\tau), \quad \theta_2(\tau) = \theta_{2s} + \tilde{\theta}_2(\tau), \quad \theta_3(\tau) = \theta_{3s} + \tilde{\theta}_3(\tau), \end{aligned} \tag{5.189}$$

assuming that the introduced perturbations are small.

Substituting (5.189) into equations (5.162)–(5.167) yields

$$\frac{d\tilde{a}_1}{d\tau} = -\frac{1}{2}c_1\left(a_{1s} + \tilde{a}_1(\tau)\right) + \frac{f_1}{2}\sin\left(\theta_{1s} + \tilde{\theta}_1(\tau)\right), \tag{5.190}$$

$$\frac{d\tilde{a}_2}{d\tau} = -\frac{c_3 w_2^4 + c_2\left(w_2^2 - w_3^2\right)^2}{2\left(w_2^2 - w_3^2\right)^2}\left(a_{2s} + \tilde{a}_2(\tau)\right) + \frac{f_2}{2w_2}\sin\left(\theta_{2s} + \tilde{\theta}_2(\tau)\right), \tag{5.191}$$

$$\begin{aligned} \frac{d\tilde{a}_3}{d\tau} = &-\frac{c_3 w_3^2\left(w_2^6 - 2(s+1)w_2^4 w_3^2 + w_2^2 w_3^4 + s w_3^6\right)}{2 s w_2^4\left(w_2^2 - w_3^2\right)^2}\left(a_{3s} + \tilde{a}_3(\tau)\right) + \\ &\frac{w_3 f_3 \sin\left(\theta_{3s} + \tilde{\theta}_3(\tau)\right)}{2 s w_2^2}, \end{aligned} \tag{5.192}$$

$$\frac{d\tilde{\theta}_1}{d\tau} = \sigma_1 - \frac{3w_2^2(w_2^2 - 1)}{4\left(4w_2^2 - 1\right)}\left(a_{2s} + \tilde{a}_2(\tau)\right)^2 - \frac{3\xi_e^2\alpha}{2} + \frac{f_1\cos\left(\theta_{1s} + \tilde{\theta}_1(\tau)\right)}{2\left(a_{1s} + \tilde{a}_1(\tau)\right)}, \tag{5.193}$$

$$\begin{aligned} \frac{d\tilde{\theta}_2}{d\tau} = &\ \sigma_2 - w_2\left(w_2^2 - 1\right)\frac{12\left(a_{1s} + \tilde{a}_1(\tau)\right)^2 - \left(1 + 8w_2^2\right)\left(a_{2s} + \tilde{a}_2(\tau)\right)^2}{16\left(4w_2^2 - 1\right)} - \\ &\frac{s w_2 w_3^2\left(4w_2^4 + (3s-8)w_2^2 w_3^3 + (4-7s)w_3^4\right)}{8\left(w_2^2 - w_3^2\right)^3} + \frac{f_2\cos\left(\theta_{2s} + \tilde{\theta}_2(\tau)\right)}{2w_2\left(a_{2s} + \tilde{a}_2(\tau)\right)}, \end{aligned} \tag{5.194}$$

$$
\begin{aligned}
\frac{d\tilde{\theta}_3}{d\tau} &= \sigma_3 + \frac{c_3^2 w_3^3}{8s^2 w_2^4} + \frac{s w_3^5 \left(4w_2^6 - 8w_2^4 w_3^2 + (4-7s)\, w_2^2 w_3^4 + 3s\, w_3^6\right)}{8w_2^4 \left(w_2^2 - w_3^2\right)^3} + \\
&\frac{w_3^3 (a_{1s} + \tilde{a}_1(\tau))^2}{4w_2^4 \left(4w_3^2 - 1\right)} - \frac{w_3^3 \left(w_3^2 - 2w_2^2\right) (a_{2s} + \tilde{a}_2(\tau))^2}{8\left(w_2^2 - w_3^2\right)^2} + \\
&\frac{w_3 (a_{3s} + \tilde{a}_3(\tau))^2}{16} + \frac{w_3 f_3 \cos(\theta_{3s} + \tilde{\theta}_3(\tau))}{2s w_2^2 (a_{3s} + \tilde{a}_3(\tau))} .
\end{aligned} \tag{5.195}
$$

The functions on the right hand sides of equations (5.190)–(5.195) are expanded in the Taylor series around the fixed point $(a_{1s}, a_{2s}, a_{3s}, \theta_{1s}, \theta_{2s}, \theta_{3s})$. Keeping only the terms of Taylor's series to order $n = 1$, the following linearized form of equations (5.190)–(5.195) is obtained

$$
\frac{d\tilde{a}_1}{d\tau} = -\frac{c_1}{2}(a_{1s} + \tilde{a}_1(\tau)) + \frac{f_1}{2}\left(\sin\theta_{1s} + \cos\theta_{1s}\tilde{\theta}_1(\tau)\right), \tag{5.196}
$$

$$
\begin{aligned}
\frac{d\tilde{a}_2}{d\tau} &= -\frac{c_3 w_2^4 + c_2 \left(w_2^2 - w_3^2\right)^2}{2\left(w_2^2 - w_3^2\right)^2}(a_{2s} + \tilde{a}_2(\tau)) + \\
&\frac{f_2}{2w_2}(\sin\theta_{2s} + \cos\theta_{2s}\tilde{\theta}_2(\tau)),
\end{aligned} \tag{5.197}
$$

$$
\begin{aligned}
\frac{d\tilde{a}_3}{d\tau} &= -\frac{c_3 w_3^2 \left(w_2^6 - 2(s+1)\, w_2^4 w_3^2 + w_2^2 w_3^4 + s w_3^6\right)}{2s w_2^4 \left(w_2^2 - w_3^2\right)^2}(a_{3s} + \tilde{a}_3(\tau)) + \\
&\frac{w_3 f_3}{2s w_2^2}\left(\sin\theta_{3s} + \cos\theta_{3s}\tilde{\theta}_3(\tau)\right),
\end{aligned} \tag{5.198}
$$

$$
\begin{aligned}
\frac{d\tilde{\theta}_1}{d\tau} &= \sigma_1 - \frac{3\xi_e^2 \alpha}{2} - \frac{3w_2^2 (w_2^2 - 1)}{4\left(4w_2^2 - 1\right)} a_{2s} (a_{2s} + 2\tilde{a}_2(\tau)) + \\
&\frac{f_1 \cos\theta_{1s}}{2a_{1s}^2}(a_{1s} - \tilde{a}_1(\tau)) - \frac{f_1 \sin\theta_{1s}}{2a_{1s}}\tilde{\theta}_1(\tau),
\end{aligned} \tag{5.199}
$$

$$
\begin{aligned}
\frac{d\tilde{\theta}_2}{d\tau} &= \sigma_2 - w_2\left(w_2^2 - 1\right)\frac{12\left(a_{1s}^2 + 2a_{1s}\tilde{a}_1(\tau)\right) - \left(1 + 8w_2^2\right)\left(a_{2s}^2 + 2a_{2s}\tilde{a}_2(\tau)\right)}{16\left(4\, w_2^2 - 1\right)} - \\
&\frac{s w_2 w_3^2 \left(4w_2^4 + (3s-8)\, w_2^2 w_3^3 + (4-7s)\, w_3^4\right)}{8\left(w_2^2 - w_3^2\right)^3} + \\
&\frac{f_2 \cos\theta_{2s}}{2w_2 a_{2s}^2}(a_{2s} - \tilde{a}_2(\tau)) - \frac{f_2 \sin\theta_{2s}}{2w_2 a_{2s}}\tilde{\theta}_2(\tau),
\end{aligned} \tag{5.200}
$$

$$\frac{d\tilde{\theta}_3}{d\tau} = \sigma_3 + \frac{c_3^2 w_3^3}{8s^2 w_2^4} + \frac{s w_3^5 \left(4w_2^6 - 8w_2^4 w_3^2 + (4-7s) w_2^2 w_3^4 + 3s\, w_3^6\right)}{8w_2^4 \left(w_2^2 - w_3^2\right)^3} +$$
$$\frac{w_3^3 \left(a_{1s}^2 + 2a_{1s}\tilde{a}_1(\tau)\right)}{4w_2^4 \left(4w_3^2 - 1\right)} - \frac{w_3^3 \left(w_3^2 - 2w_2^2\right) \left(a_{2s}^2 + 2a_{2s}\tilde{a}_2(\tau)\right)}{8\left(w_2^2 - w_3^2\right)^2} + \tag{5.201}$$
$$\frac{w_3 \left(a_{3s}^2 + 2a_{3s}\tilde{a}_3(\tau)\right)}{16} + \frac{w_3 f_3 \cos\theta_{3s}}{2s w_2^2 a_{3s}^2} \left(a_{3s} - \tilde{a}_3(\tau)\right) - \frac{w_3 f_3 \sin\theta_{3s}}{2s w_2^2 a_{3s}} \tilde{\theta}_3(\tau).$$

The fixed point $(a_{1s}, a_{2s}, a_{3s}, \theta_{1s}, \theta_{2s}, \theta_{3s})$ satisfies the steady-state equations (5.168)–(5.173), and hence equations (5.196)–(5.201) can be simplified to the following form

$$\frac{d\tilde{a}_1}{d\tau} = -\frac{c_1}{2}\tilde{a}_1(\tau) + \frac{f_1}{2}\cos\theta_{1s}\tilde{\theta}_1(\tau), \tag{5.202}$$

$$\frac{d\tilde{a}_2}{d\tau} = -\frac{c_3 w_2^4 + c_2\left(w_2^2 - w_3^2\right)^2}{2\left(w_2^2 - w_3^2\right)^2}\tilde{a}_2(\tau) + \frac{f_2}{2w_2}\cos\theta_{2s}\tilde{\theta}_2(\tau), \tag{5.203}$$

$$\frac{d\tilde{a}_3}{d\tau} = -\frac{c_3 w_3^2 \left(w_2^6 - 2(s+1) w_2^4 w_3^2 + w_2^2 w_3^4 + s w_3^6\right)}{2s w_2^4 \left(w_2^2 - w_3^2\right)^2}\tilde{a}_3(\tau) + \tag{5.204}$$
$$\frac{w_3 f_3}{2s w_2^2}\cos\theta_{3s}\tilde{\theta}_3(\tau),$$

$$\frac{d\tilde{\theta}_1}{d\tau} = -\frac{f_1\cos\theta_{1s}}{2a_{1s}^2}\tilde{a}_1(\tau) - \frac{3w_2^2(w_2^2-1)}{2\left(4w_2^2-1\right)}a_{2s}\tilde{a}_2(\tau) - \frac{f_1\sin\theta_{1s}}{2a_{1s}}\tilde{\theta}_1(\tau), \tag{5.205}$$

$$\frac{d\tilde{\theta}_2}{d\tau} = -w_2\left(w_2^2 - 1\right)\frac{12a_{1s}\tilde{a}_1(\tau) - \left(1 + 8w_2^2\right)a_{2s}\tilde{a}_2(\tau)}{8\left(4w_2^2 - 1\right)} - \tag{5.206}$$
$$\frac{f_2\cos\theta_{2s}}{2w_2 a_{2s}^2}\tilde{a}_2(\tau) - \frac{f_2\sin\theta_{2s}}{2w_2 a_{2s}}\tilde{\theta}_2(\tau),$$

$$\frac{d\tilde{\theta}_3}{d\tau} = \frac{w_3^3 a_{1s}\tilde{a}_1(\tau)}{2w_2^4\left(4w_3^2 - 1\right)} - \frac{w_3^3\left(w_3^2 - 2w_2^2\right)a_{2s}\tilde{a}_2(\tau)}{4\left(w_2^2 - w_3^2\right)^2} + \tag{5.207}$$
$$\frac{w_3 a_{3s}\tilde{a}_3(\tau)}{8} - \frac{w_3 f_3\cos\theta_{3s}}{2s w_2^2 a_{3s}^2}\tilde{a}_3(\tau) - \frac{w_3 f_3\sin\theta_{3s}}{2s w_2^2 a_{3s}}\tilde{\theta}_3(\tau).$$

Provided the assumed small perturbations in the dynamical system (5.202)–(5.207) are linear with respect to the functions $\tilde{a}_1(\tau), \tilde{a}_2(\tau), \tilde{a}_3(\tau)\, \tilde{\theta}_1(\tau), \tilde{\theta}_2(\tau),$

$\tilde{\theta}_3(\tau)$. Moreover, equations (5.202)–(5.205) are homogeneous and have the constant coefficients. The general solution of the system can be presented as the linear combination of the exponential functions

$$\tilde{a}_i(\tau) = \sum_{j=1}^{6} D_j^{(i)} e^{\lambda_j \tau}, \qquad i = 1,2,3,$$
$$\tilde{\theta}_i(\tau) = \sum_{j=1}^{6} D_j^{(i+3)} e^{\lambda_j \tau}, \qquad i = 1,2,3, \tag{5.208}$$

where $\lambda_j,\ j = 1,..,6$ are the roots of the following characteristic equation

$$|\mathbf{A} - \lambda\,\mathbf{I}| = 0. \tag{5.209}$$

The symbol $\mathbf{I}$ denotes the identity matrix, and $\mathbf{A}$ is the characteristic matrix with regard to the differential equations (5.202)–(5.207). The square matrix $\mathbf{A}$ has the following structure

$$\mathbf{A} = \begin{bmatrix} A_{11} & 0 & 0 & A_{14} & 0 & 0 \\ 0 & A_{22} & 0 & 0 & A_{25} & 0 \\ 0 & 0 & A_{33} & 0 & 0 & 0 \\ A_{41} & A_{42} & 0 & A_{44} & 0 & 0 \\ A_{51} & A_{52} & 0 & 0 & A_{55} & 0 \\ A_{61} & A_{62} & A_{63} & 0 & 0 & A_{66} \end{bmatrix} \tag{5.210}$$

where

$$A_{11} = -\frac{c_1}{2}, \quad A_{14} = \frac{f_1 \cos\theta_{1s}}{2},$$

$$A_{22} = -\frac{c_3 w_2^4 + c_2\left(w_2^2 - w_3^2\right)^2}{2\left(w_2^2 - w_3^2\right)^2}, \quad A_{25} = \frac{f_2 \cos\theta_{2s}}{2w_2},$$

$$A_{33} = -\frac{c_3 w_3^2\left(w_2^6 - 2(s+1)\,w_2^4 w_3^2 + w_2^2 w_3^4 + s w_3^6\right)}{2s w_2^4\left(w_2^2 - w_3^2\right)^2},$$

$$A_{41} = -\frac{f_1 \cos\theta_{1s}}{2a_{1s}^2}, \quad A_{42} = -\frac{3w_2^2(w_2^2-1)}{2\left(4w_2^2-1\right)}, \quad A_{44} = -\frac{f_1 \sin\theta_{1s}}{2a_{1s}},$$

$$A_{51} = -\frac{3w_2\left(w_2^2-1\right)a_{1s}}{2\left(4w_2^2-1\right)}, \quad A_{52} = \frac{w_2\left(w_2^2-1\right)\left(1+8w_2^2\right)a_{2s}}{8\left(4\,w_2^2-1\right)} - \frac{f_2\cos\theta_{2s}}{2w_2 a_{2s}^2},$$

$$A_{51} = -\frac{f_2 \sin\theta_{2s}}{2w_2 a_{2s}}, \quad A_{61} = \frac{w_3^3 a_{1s}}{2w_2^4\left(4w_3^2-1\right)},$$

$$A_{62} = -\frac{w_3^3\left(w_3^2 - 2w_2^2\right)a_{2s}}{4\left(w_2^2 - w_3^2\right)^2}, \quad A_{63} = \frac{w_3 a_{3s}}{8}, \quad A_{66} = \frac{w_3 f_3 \sin\theta_{3s}}{2s w_2^2 a_{3s}}.$$

If the real parts of all eigenvalues of the matrix $\mathbf{A}$ are negative, then the fixed point $(a_{1s}, a_{2s}, a_{3s}, \theta_{1s}, \theta_{2s}, \theta_{3s})$ understood as the steady-state solution in the considered resonance is asymptotically stable in the sense of Lyapunov. This statement is justified only based on the introduced linear approximation.

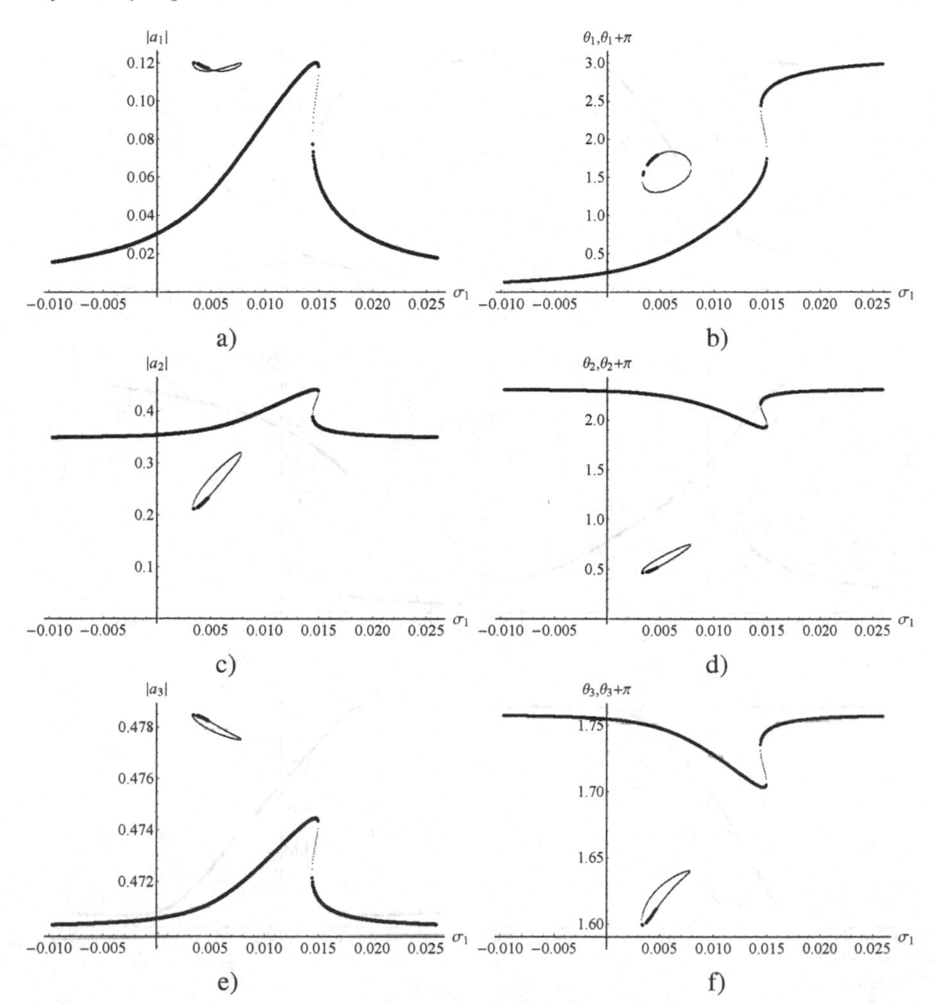

FIGURE 5.24 The resonance response curves (black) and their stable branches (gray) versus the detuning parameter σ_1: a) $a_1(\sigma_1)$, b) $\theta_1(\sigma_1)$, c) $a_2(\sigma_1)$, d) $\theta_2(\sigma_1)$, e) $a_3(\sigma_1)$, f) $\theta_3(\sigma_1)$ for fixed parameters $\sigma_2 = -0.002$, $\sigma_3 = -0.001$. Stable points are drawn with larger dots.

The stability of the resonance response curves obtained for the following fixed values of the system parameters: $w_2 = 0.275$, $w_3 = 0.105$, $\alpha = 0.01$, $s = 0.013$, $c_1 = 0.005$, $c_2 = 0.003$, $c_3 = 0.0012$, $f_1 = 0.0006$, $f_2 = 0.0006$, $f_3 = 0.00006$ have been examined. The resonance response curves related to these data are shown in Figures 5.17–5.19 and Figures 5.21–5.23.

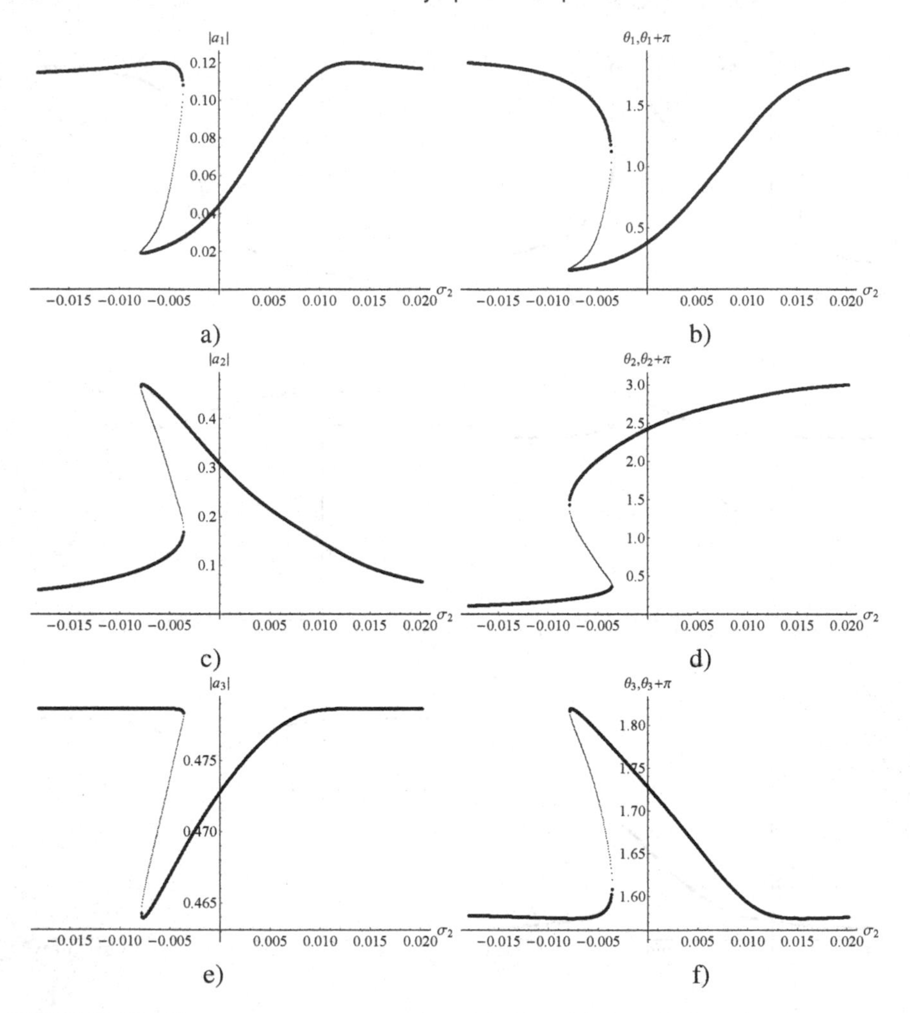

FIGURE 5.25 The resonance response curves (black) and their stable branches (gray) versus the detuning parameter σ_2: a) $a_1(\sigma_2)$, b) $\theta_1(\sigma_2)$, c) $a_2(\sigma_2)$, d) $\theta_2(\sigma_2)$, e) $a_3(\sigma_2)$, f) $\theta_3(\sigma_2)$ for fixed parameters $\sigma_1 = 0.001$, $\sigma_3 = -0.001$. Stable points are drawn with larger dots.

The results of the stability analysis are presented in Figures 5.24–5.26. By incrementing with a constant step the value of one detuning parameter with the set values of the other two detuning parameters, the system of equations (5.168)–(5.173) was solved. The real roots are analysed for stability. Fixed points of the dynamical system (5.162)–(5.167) that are asymptotically stable in the sense of Lyapunov are drawn with points of greater size. The unstable points are represented by finer points. As previously, taking into account symmetry properties of the resonance response curves, their main branches are only plotted, i.e. for $a_i > 0$ and $0 < \theta_i < \pi$, $i = 1, 2, 3$. The eigenvalues of matrix $\mathbf{A}$ are determined using the standard procedure *Eigenvalues* of the *Mathematica* software.

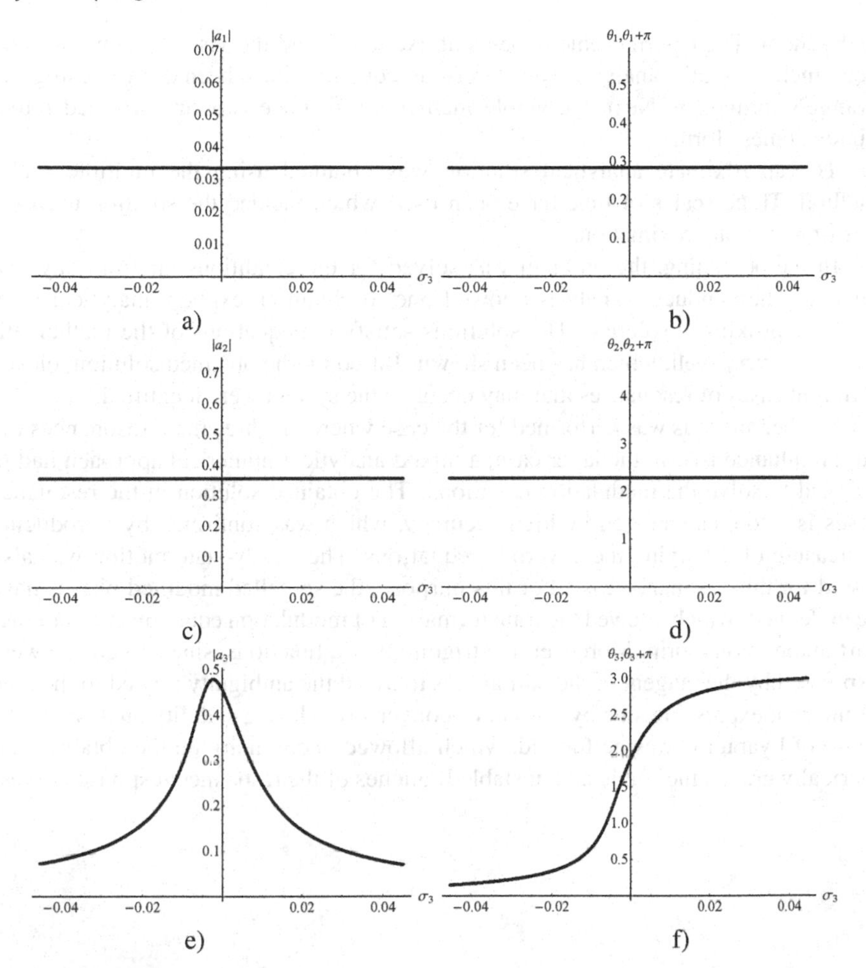

FIGURE 5.26 The resonance response curves (black) and their stable branches (gray) versus the detuning parameter σ_3: a) $a_1(\sigma_3)$, b) $\theta_1(\sigma_3)$, c) $a_2(\sigma_3)$, d) $\theta_2(\sigma_3)$, e) $a_3(\sigma_3)$, f) $\theta_3(\sigma_3)$ for fixed parameters $\sigma_1 = 0.001$, $\sigma_2 = -0.002$. Stable points are drawn with larger dots.

The stability study excludes the fixed points which are impossible to be achieved after going through the transient motion. The analysis confirms that every point of each of the resonance response curves which are shown in Figure 5.26 is stable. In the case of the curves which present the dependence on the detuning parameters σ_1 and σ_2, there are unstable branches in the regions of the ambiguity.

5.8 CLOSING REMARKS

The dynamics of the physical pendulum suspended on the nonlinear spring have been examined. The motion is assumed to be planar, so the system has three degrees of freedom. The external time-dependent excitation and the viscous damping have been included in the analysis. The governing equations have been derived using Lagrange's formalism. Next, the whole mathematical model was transformed into a dimensionless form.

The approximate analytical solution was obtained using the multiple scales method. Three scales of time have been used which yielded the solution until the third-order of approximation.

In the beginning, the problem was solved for the conditions far from any resonance. The conducted analysis allowed one to obtain an explicit analytical form of the approximate solution. The solutions satisfy the equations of the mathematical model very well, which has been shown. Based on the obtained solution, eleven different cases of resonances that may occur in the system were identified.

Further analysis was performed for the case where the three main resonances occur simultaneously. In the latter case, a mixed analytical-numerical approach had to be used to solve the modulation equations. The obtained solution in the resonance cases is also characterized by high accuracy, which was confirmed by introducing a measure of satisfying the governing equations. The steady-state motion was also tested for this resonance case. For this purpose, the so-called modified phases have been defined, which allowed the transformation of modulation equations to a counter part autonomous form. Moreover, the trigonometric functions, sine and cosine were expressed by the tangent of the half angles to avoid the ambiguity related to the sign of the root expressing sine by cosine, or conversely. Then, a stability analysis in the sense of Lyapunov was performed, which allowed one to mark on the obtained numerically graphs the stable and unstable branches of the resonance response curves.

6 Nonlinear Torsional Micromechanical Gyroscope

This chapter deals with the analytical (MSM) study of the micromechanical gyroscope constructed for measuring one component of angular velocity. The sensing plate is supported by the Cardan suspension, whereas the gimbal and the plate with sensors are linked via torsional joints. The Coriolis effect yields vibration of the sensing plate. The motion equations are governed by coupled two nonlinear second-order ODEs derived from the Lagrange equations of the second kind. The main and internal resonances occurring simultaneously are investigated using the MSM which allows one to derive and study the amplitude-phase differential equations. In addition, the non-resonance vibrations are analyzed. The useful engineering oriented conclusions are provided. The obtained approximate analytical solutions are validated numerically and the errors of their approximation with regard to the numerical results are estimated.

6.1 INTRODUCTION

Gyroscopes are present in a broad range of engineering systems such as air vehicles, automobiles, and satellites to track their orientation and control their path. Besides the directional gyroscopes, there are varieties of gyroscopes (e.g., mechanical, optical and vibrating) that are being used to measure the angular velocity. The critical part of the conventional mechanical gyroscope is a wheel spinning at a high speed. Therefore, conventional gyroscopes, although accurate, are bulky and very expensive and they are applicable mainly in the navigation systems of large vehicles, such as ships, airplanes, space crafts, etc.

Micromechanical gyroscopes and angular rate sensors allow for signification miniaturization in contrast to solid-state gyroscopes, laser ring and fiber optic gyroscopes.

Progress in micromachining technology embraces the development of the miniaturized gyroscopes with improved performance and low power consumption that allows the integration with electronic circuits. Their manufacturing cost is also significantly lower ([91, 235]). Such type of gyroscopes belongs to broad class of microelectromechanical systems (MEMS). Practically, any device fabricated using photo-lithography-based techniques with micrometer scale features that utilizes both electrical and mechanical functions could be considered as MEMS.

The operating principle of vibrating gyroscopes is based on the transfer of the mechanical energy among two vibrations modes via the Coriolis effect which occurs

DOI: 10.1201/9781003270706-6

in the presence of a combination of rotational motions about two orthogonal axes. The drive mode is mainly generated employing the electrostatic actuation mechanism.

However, it is widely recognized that miniaturization achieved via fabrication technologies requires detailed studies from a point of view of nonlinear dynamical systems in order to understand and control some unexpected behavior of microcomponents, micromachines and MEMS/NEMS, and in particular of micromechanical gyroscopes. Reliable modeling of the micromechanical vibratory gyroscopes allows for improvement in sensitive elements and circuit design, and hence, it has an important impact on achieving high performances of the mentioned micromechanical systems. In other words, the micro- and nanotechnologies require support of theoretical approaches based on theory of vibrations and nonlinear phenomena. In what follows, we briefly describe state of the art approaches in the recent achievement in modeling and analysis of some chosen MEMS/NEMS and micro-/nanogyroscopes.

Turner et al. [214] pointed out importance of parametric resonances in a micromechanical system.

Lifshitz and Cross [124] investigated a response of the microring gyroscope under combined external forcing and parametric excitation to achieve required parametric amplification.

Nayfeh and Younis [156] investigated dynamics of MEMS resonators under superharmonic/subharmonic excitations.

Gallacher et al. [73] proposed a control scheme for a MEMS electrostatic resonant gyroscope subjected to both harmonic forcing and parametric excitation.

Braghin et al. [44] considered a MEMS translational gyroscope with the nonlinear hardening characteristic of the supporting beams. Its nonlinear vibrations were investigated both experimentally and numerically based on a simple lumped parameter model. The semi-analytical integration model allowed for determination of stable and unstable branches of the system dynamic response.

Rhoads et al. [181] reviewed works devoted to the nonlinear dynamics of resonant micro- and nano- electromechanical systems (MEMS/NEMS). The paper addresses important nonlinear effects exhibited by micro/nanoresonators including direct and parametric resonances, parametric amplification, impact, self-excited oscillations as well as localization and synchronization phenomena.

Kacem et al. [93] improved the performance of NEMS sensors based on employment of theoretical approaches of modeling nonlinear dynamics of nanomechanical beam resonators.

Lestev and Tikhonov [121] analyzed nonlinear dynamical behavior of micromechanical gyroscopes using the method of averaging. They pointed out that even though the parameters of the microstructural components are chosen in a way to provide a linear response, it cannot be achieved due to the fabrication errors. They investigated stable steady-state modes of vibratory micromechanical gyroscopes, and they presented the corresponding resonance curves.

Nonlinear dynamics and chaos of electrostatically actuated MEMS resonators under two-frequency external and parametric excitations were analyzed by Zhang et al. [242]. In particular, they illustrated effects of nonlinear square damping on the frequency response. Resonance frequencies and nonlinear dynamic characteristics were also reported. However, their investigation concerned relatively simple model consisting of a mass–spring–damper system.

Martynenko et al. [137] studied nonlinear phenomena of a vibrating micromechanical gyroscope with a ring resonator flexibly supported. The Krylov–Bogolubov averaging method was employed to predict fabrication errors, unstable branches of resonance curves, and quenching phenomenon.

Yoon et al. [240] modeled vibratory ring gyroscopes by four vibration models (two flexural and two translation). The developed model consisted of the ring structure, the support-string structure, and the electrodes. It was shown that the developed model becomes vibration sensitive in the presence of both non-proportional damping and the sense electrodes capacitive nonlinearity.

Matheny et al. [138] studied nonlinear mode-coupling in nanomechanical systems. They demonstrated measurement protocol and design rules for getting accurate in situ characterization of nonlinear properties of NEMS resonators. In particular, the employment of the Euler–Bernoulli beam model was validated through the carried out laboratory measurements.

Ovchinnikova et al. [163] developed a model of micromechanical gyroscope using inertia properties of standing elastic waves providing maximum vibration amplitude with minimum control. The employed schemes of stabilization of the excited amplitude reduced nonlinear transformation characteristics. The method allowed for computation of the envelope of the fundamental mode of vibration of the governing two second-order ODEs, yielded by the Bubnov–Galerkin approach.

Yoon et al. [241] studied a micromechanical vibrating ring gyroscope under high shocks based on mathematical analysis supported by the finite element method. They suggested employment of the developed vibrating ring gyroscope in navigation systems when both performance and high shock resistance are crucial in getting proper measurements.

Lestev [120] investigated combination resonances of sensitive elements of micromechanical gyros under translational and angular motions of the platform. The governing nonlinear ODEs were derived and the obtained results were validated experimentally.

Nitzan et al. [157] considered parametric amplification of a micromechanical resonating disk gyroscope taking into account the self-induced parametric excitation and Coriolis forces. The parametric self-induced amplification was yielded by nonlinear stiffness coupling between degenerate orthogonal vibration modes in a high-quality-factor micromechanical resonator.

Defoort et al. [60] analyzed occurrence of synchronization between two degenerate resonance modes of a microdisk resonator gyroscope. The carried out consideration were based on two second-order ODEs including a geometric nonlinearity of a cubic type. They demonstrated how mutual synchronization between modes was robust over temperature variation.

An impact of a cubic nonlinearity on the operation of a rate-integrating gyroscope was studied by Nitzan et al. [158]. It was shown how below the bifurcation threshold of cubic nonlinearity, a splitting of angle-depending frequency between two resonant gyroscope modes occurred which impacted angle-dependent bias, quadrature error and controller efficency. The method of compensating for angle-dependent frequency error was proposed and was experimentally validated.

A useful overview of gyroscopic technology including mechanical and optical principles, at macro- and microscale was given by Passaro et al. [165].

Král and Straka [109] studied the problem of nonlinear parameter estimation of a MEMS gyroscope. Novel method approximating the angular rate and using the first-order Gauss-Markov process was utilized. Properties and validity of the proposed estimation method were reported.

Liang et al. [123] investigated linear/nonlinear dynamic responses of a vibratory ring gyroscope based on two coupled second-order ODEs. Nonlinear responses were compared with linear ones, and gyroscopy sensitivity was estimated.

Hosseini-Pishrobat and Keighobadi [85] developed a nonlinear disturbance-based controller for robust output regulation of a triaxial MEMS vibratory gyroscope. It allowed stabilization of a zero-error invariant manifold in the tracking error space. A nonlinear extended state observer was designed to estimate the total disturbance, and its convergence was studied in a Lyapunov-Lurie framework. The stability of the overall closed loop system was guaranteed.

Hu and Gallacher [86] investigated the influence of nonlinearity on the procession angle dependent bias error of the electrostatic MEMS Coriolis vibratory gyroscopes. It was shown that high-order angular-modulated drift error was implied by nonlinear damping while the stiffness nonlinearity was responsible for the 4th-harmonics frequency component. The proposed nonlinearity correction to the capacitive displacement detection reduced significantly, the higher-order drift error.

Nabholz et al. [149] constructed a model of two nonlinearly coupled mechanical modes and used it to clarify measured drive mode instabilities in MEMS gyroscopes. It was illustrated how the 3:1 internal resonance implied energy transfer within the conservative system.

In this chapter, we conduct an analysis of the dynamics of a MEMS gyroscope. This microdevice is a torsional resonator. Resonance is the desirable state of work of this sensor, so the elastic properties should be appropriately matched. In designing the resonator, only linear elasticity was taken into account. There arises the question of what is the significance of the nonlinear properties of resilient resonator elements. Therefore, we propose the mathematical model describing motion of the MEMS gyroscope taking into account the nonlinear effects generated by the elastic properties of the suspension elements (Starosta et al. [202]). The main objective of this study is to obtain and to examine the resonant responses of the considered system.

The preliminary results of this chapter are reported in (Awrejcewicz et al. [32]).

6.2 OPERATION PRINCIPLE OF MICROMECHANICAL GYROSCOPE

The torsional micromechanical gyroscope belonging to the broad class of microelectromechanical systems (MEMS) is employed to measure one component of the angular velocity of the object on which it is mounted. The demonstrative drawing of the torsional micromechanical gyroscope is presented in Figure 6.1. The anchor (1) serves as the point for fastening the micro device on the body whose angular velocity is measured. The Cardan suspension concept is applied to connect the sensor plate (3) that is the main part of the system with the anchors. Exactly, the version of the suspension with two mutually orthogonal axes is applied. However, in contrast to the standard Cardan suspension, the torsional members are used instead of the pivot axes. Two torsional joints aligned along with a common direction, link the intermediate frame (2), called also the gimbal, with the anchors. In turn, two other torsional joints, having also a common line, connect the frame (2) and the sensor plate (3).

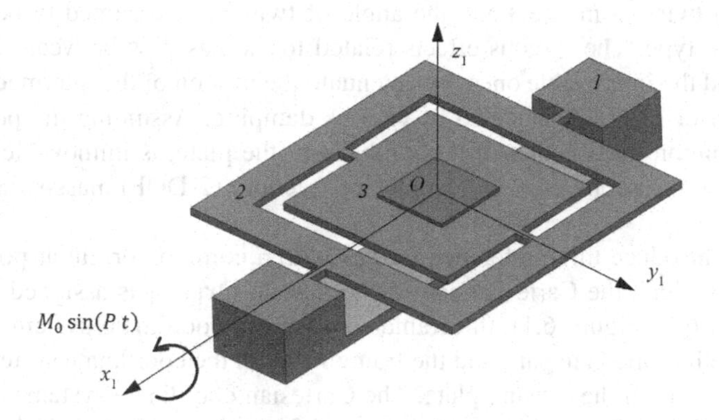

FIGURE 6.1 Scheme of the torsional micromechanical gyroscope; 1 – anchor, 2 – gimbal, 3 – sense plate.

The gimbal and the sensing plate can rotate independently around the axes designated by both pairs of the torsional joints. The rotation of the gimbal around the axis determined by the connectors between the anchors and the gimbal is excited by the external torque, the magnitude of which changes harmonically, i.e. $M(t) = M_0 \sin(Pt)$. Wherefore, the axis is called the drive axis. Due to this forced rotation, a coupling effect appears in the system. When the anchors are subjected to the rotation about the z-axis caused by the substrate motion, then the sinusoidal Coriolis torque with the frequency P, equal to the frequency of the drive-mode excitation, is induced in the direction perpendicular to the drive axis. Through the torsional members, this torque is transferred to the sensing plate. This is the reason for which the axis determined by the joints between the gimbal and the central plate is named the sense axis. The Coriolis torque excites the proof-mass, placed on the sensor plate, to oscillate around the sense axis. The response caused by the Coriolis effect, which is proportional to the angular velocity being measured, is registered by the detection electrodes. To attain the maximum response of the proof-mass, the micromechanical gyroscope is designed with such an intention that the simultaneous resonance in vibration around both axes is its normal work mode. When designing the system, the inertial properties of its parts and the elastic features of the torsional joints should be properly determined to guarantee the coincidence of both resonances. It seems to be advisable to consider the influence of the nonlinear elastic properties of the torsional links on the micromechanical system operation. Important factors disturbing the optimal work state are material parameter uncertainties, inevitable manufacturing errors, and the influence of external conditions. It all affects the accuracy and sensitivity of the micro gyroscope.

6.3 MATHEMATICAL MODEL

The anchors, the intermediate frame, and the sensing plate are assumed to be rigid bodies. The joints between the rigid parts are considered as weightless and elastic members, where only "pure" torsion is taken into account. The relationship between the internal twisting moment and the angle of twisting is assumed to be nonlinear of the cubic type. The viscous effects related to the gas flow between the rotating surfaces and the immovable ones that attenuate the motion of the intermediate frame and the sensor plate are modeled as viscous damping. Assuming the point O, being the common mass center of the gimbal and the plate, is immovable we model the torsional gyroscope as a two degrees-of-freedom (2-DOF) mass-spring-damper system.

Let us introduce three reference frames with a common origin at point O. The frame F_1 to which the Cartesian coordinate system $Ox_1y_1z_1$ is assigned is fixed on the anchors (see Figure 6.1), the frame F_2 with the coordinate system $Ox_2y_2z_2$ is fixed on the intermediate part, and the frame F_3 with the coordinate system $Ox_3y_3z_3$ is, in turn, fixed on the sensing plate. The Cartesian coordinate systems assigned to the frames F_2 and F_3 are depicted in Figure 6.2. At the stable equilibrium position shown in Figure 6.1, all these coordinate systems overlap themselves.

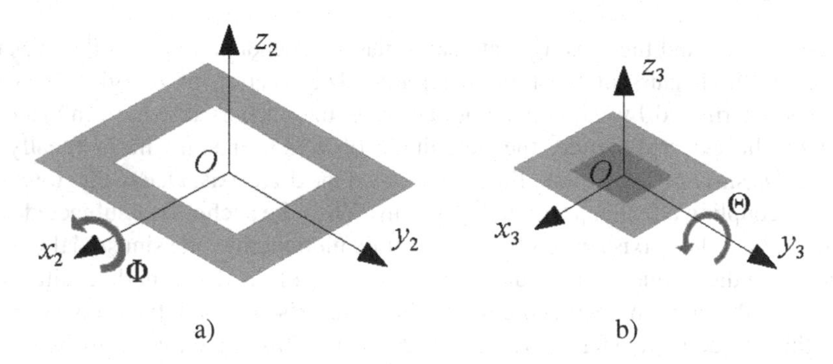

FIGURE 6.2 The frame (a) and the sensing plate (b) with their coordinate systems.

None of the introduced frames can be treated as an inertial reference frame when the substrate rotates. The frame F_1 rotates together with the substrate around the axis z_0. The frame F_2 can oscillate about the drive axis $x_2 = x_1$ which is also moving. The frame F_3 can oscillate around the movable axis $y_3 = y_2$.

The meaning of the rotation angles Φ and Θ is clarified in Figure 6.3. In Figure 6.3a, the frame F_2 is presented in the position rotated by the angle Φ. The position of the frame F_3 which is a result of the composition of two rotations is depicted in Figure 6.3b. The first of them is the rotation by the angle Φ, the second one is the rotation by

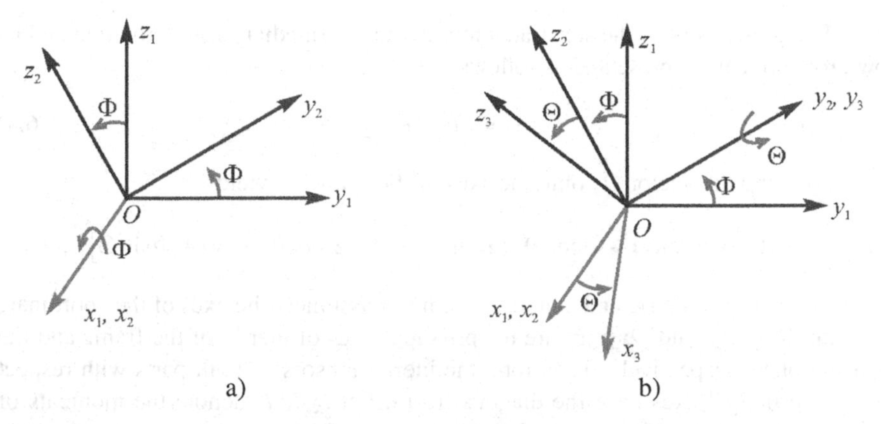

FIGURE 6.3 The angles of rotation Φ and Θ: (a) the rotation of the frame F_2 by Φ about x_1-axis; (b) the composition of two rotations of the frame F_3 by Φ about x_1-axis and by Θ around y_2 axis.

the angle Θ. The location of the frame F_2 relative to the frame F_1 is unambiguously determined by the angle Φ. In turn, the position of the frame F_3 within the reference frame F_1 is unambiguously established by two angles Φ and Θ. Both angles are regarded as positive when the rotations related to them are counter-clockwise. Concluding, the considered MEMS gyroscope has two degrees of freedom in its motion relative to the body whose angular velocity is measured. The angles $\Phi(t)$ and $\Theta(t)$ are taken as the generalized coordinates.

Let us assume the substrate and the anchors rotate only around the fixed pivot axis z_0 with the absolute angular velocity Ω_1. Within the coordinate system $Ox_1y_1z_1$, the vector Ω_1 can be written in terms of its components as

$$\Omega_1 = [0, 0, \Omega_z]^T, \tag{6.1}$$

where Ω_z is to be measured and the superscript T denotes the vector transposition.

The angular velocity of the own intermediate frame rotation about the drive axis expressed in terms of the components within the coordinate system $Ox_2y_2z_2$ is equal $\Omega_{2/1} = [\dot{\Phi}, 0, 0]^T$. The absolute angular velocity of the intermediate gimbal, Ω_2, is equal to the sum of the vectors Ω_1 and $\Omega_{2/1}$

$$\Omega_2 = \Omega_1 + \Omega_{2/1}, \tag{6.2}$$

When projecting the vector Ω_2 onto the axes of the frame F_2, we obtain

$$\Omega_2 = \left[\dot{\Phi}, \Omega_z \sin\Phi(t), \Omega_z \cos\Phi(t)\right]^T. \tag{6.3}$$

In turn, the angular velocity of the sensing plate in its relative motion about the sensing axis written in terms of the components within the coordinate system $Ox_3y_3z_3$ is equal $\Omega_{3/2} = \left[0, \dot{\Theta}, 0\right]^T$. The absolute angular velocity Ω_3 of the sensing plate being the result of the substrate motion, the intermediary frame rotation and its own rotation can be presented as follows

$$\Omega_3 = \Omega_2 + \Omega_{3/2}. \tag{6.4}$$

Projecting the vector Ω_3 onto the axes of the frame F_3 yields

$$\Omega_3 = \left[-\Omega_z \cos\Phi \sin\Theta + \dot{\Phi}\cos\Theta,\ \Omega_z \sin\Phi + \dot{\Theta},\ \Omega_z \cos\Phi \cos\Theta + \dot{\Phi}\sin\Theta\right]^T. \tag{6.5}$$

Due to the geometric and material symmetry assumed, the axes of the coordinate systems $Ox_2y_2z_2$ and $Ox_3y_3z_3$ are the principal axes of inertia of the frame and the sensing plate, respectively. Therefore, the inertia tensors of both parts with respect to their principal axes have the diagonal form. Let I_x, I_y, I_z denote the moments of inertia of the intermediate frame related to the axes of the coordinate system $Ox_2y_2z_2$ whereas J_x, J_y, J_z stand for the principal moments of inertia of the sensing plate with respect to the axes of the coordinate system $Ox_3y_3z_3$. The inertia tensors of the frame and the sensor plate can be written as follows

$$\hat{\mathbf{I}} = \begin{bmatrix} I_x & 0 & 0 \\ 0 & I_y & 0 \\ 0 & 0 & I_z \end{bmatrix}, \quad \hat{\mathbf{J}} = \begin{bmatrix} J_x & 0 & 0 \\ 0 & J_y & 0 \\ 0 & 0 & J_z \end{bmatrix}. \tag{6.6}$$

According to the Koenig theorem, the kinetic energy of the MEMS gyroscope, whose mass center O remains constantly at rest, with respect to an inertial reference frame can be expressed as a sum of two symmetric quadratic forms

$$T = \frac{1}{2}\left(\Omega_2^T \cdot \hat{\mathbf{I}} \cdot \Omega_2 + \Omega_3^T \cdot \hat{\mathbf{J}} \cdot \Omega_3\right), \tag{6.7}$$

where symbol $(\cdot)$ denotes the inner product.

Substitution of equations (6.3) and (6.5)–(6.6) into equation (6.7) yields the kinetic energy expressed in terms of the generalized coordinates and their derivatives

$$\begin{aligned} T = {} & \frac{I_x}{2}\dot{\Phi}^2 + \left(I_y \sin^2\Phi + I_z \cos^2\Phi\right)\frac{\Omega_z^2}{2} + \frac{J_x}{2}\left(\Omega_z \cos\Phi \sin\Theta - \dot{\Phi}\cos\Theta\right)^2 + \\ & \frac{J_y}{2}\left(\Omega_z \sin\Phi + \dot{\Theta}\right)^2 + \frac{J_z}{2}\left(\Omega_z \cos\Phi \cos\Theta + \dot{\Phi}\sin\Theta\right)^2. \end{aligned} \tag{6.8}$$

Assuming the nonlinear relationships between the internal twisting moment and the angle of twisting one can write the potential energy of the elastic forces as follows

$$V = \frac{k_{11}}{2}\Phi(t)^2 - \frac{k_{12}}{4}\Phi(t)^4 + \frac{k_{21}}{2}\Theta(t)^2 - \frac{k_{22}}{4}\Theta(t)^4, \tag{6.9}$$

where k_{11}, k_{12} and k_{21}, k_{22} are the constant elastic coefficients of the torsional members, respectively for the anchors-frame and the frame-sensing plate connections. All the coefficients are assumed to be positive, so the nonlinear features of the joints are assumed to be of the soft type.

The external loading and the damping moments of the purely viscous nature are introduced into the consideration as the generalized forces

$$Q_\Phi = M_0 \sin(Pt) - C_1\dot{\Phi}(t), \quad Q_\Theta = -C_2\dot{\Theta}(t), \tag{6.10}$$

where C_1 and C_2 are the constant damping coefficients.

Employing the Lagrange formalism yields the following motion equations

$$\begin{aligned}&\frac{1}{2}(2I_x + J_x + J_z + (J_x - J_z)\cos(2\Theta))\ddot{\Phi} + C_1\dot{\Phi} + k_{11}\Phi - k_{12}\Phi^3 - \\ &(J_x - J_z)\sin(2\Theta)\dot{\Phi}\dot{\Theta} - (J_y + (J_x - J_z)\cos(2\Theta))\cos(\Phi)\Omega_z\dot{\Theta} + \\ &\frac{\Omega_z^2}{4}(2(I_z - I_y) + J_x - 2J_y + J_z - (J_x - J_z)\cos(2\Theta))\sin(2\Theta) = M_0\sin(Pt),\end{aligned} \tag{6.11}$$

$$\begin{aligned}&J_y\ddot{\Theta} + C_2\dot{\Theta} + k_{21}\Theta - k_{22}\Theta^3 + (J_x - J_z)\cos(\Theta)\sin(\Theta)\dot{\Phi}^2 + \\ &(J_y + (J_x - J_z)\cos(2\Theta))\cos(\Phi)\Omega_z\dot{\Phi} - \\ &(J_x - J_z)\cos(\Theta)\sin(\Theta)\cos^2(\Phi)\Omega_z^2 = 0.\end{aligned} \tag{6.12}$$

Differential equations (6.11)–(6.12) are supplemented with the following initial conditions

$$\Phi(0) = U_{01}, \qquad \dot{\Phi}(0) = U_{02}, \qquad \Theta(0) = U_{03}, \qquad \dot{\Theta}(0) = U_{04}, \tag{6.13}$$

where U_{01}, U_{02}, U_{03}, U_{04} are known quantities describing the initial kinematic state of the micro gyroscope.

6.4 APPROXIMATE MOTION EQUATIONS FOR SMALL VIBRATION

The micro gyroscope on which no external forces act reaches its stable equilibrium position when $\Phi = 0$ and $\Theta = 0$. We are aimed to study the gyroscope motion within a small neighborhood of the stable equilibrium position. Accordingly, we approximate the trigonometric functions using the first term of the Taylor series, i.e.

$$\sin\Phi \approx \Phi, \qquad \cos\Phi \approx 1, \qquad \sin\Theta \approx \Theta, \qquad \cos\Theta \approx 1, \tag{6.14}$$

This linear approximation may seem too coarse. But expecting the elements of the system to vibrate on the ranges of very small values of the angles Φ and Θ, it is yet justified. Making use of the assumptions (6.14), we get the following approximate form of motion equations (6.11)–(6.13)

$$\begin{aligned}&(I_x + J_x)\ddot{\Phi} + C_1\dot{\Phi} + (k_{11} + (I_z + J_z - I_y - J_y)\Omega_z^2)\Phi - k_{12}\Phi^3 - \\ &2(J_x - J_z)\Theta\dot{\Phi}\dot{\Theta} + (J_z - J_x - J_y)\Omega_z\dot{\Theta} = M_0\sin(Pt),\end{aligned} \tag{6.15}$$

$$\begin{aligned}&J_y\ddot{\Theta} + C_2\dot{\Theta} + \left(k_{21} - (J_x - J_z)\Omega_z^2\right)\Theta - k_{22}\Theta^3 + \\ &(J_x - J_z)\Theta(t)\dot{\Phi}^2 + (J_x + J_y - J_z)\Omega_z\dot{\Phi} = 0.\end{aligned} \tag{6.16}$$

The initial conditions in the unchanged form (6.13) complete the approximate equations of motion. Equations (6.15)–(6.16) are valid only for the case of the small vibration around the stable equilibrium position.

It is convenient to transform the initial value problem given by equations (6.15)–(6.16) and conditions (6.13) into the non-dimensional form. Let us assume that the quantity ω_1 understood as follows

$$\omega_1 = \sqrt{\frac{k_{11}}{I_x + J_x}} \tag{6.17}$$

is the reference frequency. Employing the frequency ω_1 we define the dimensionless time

$$\tau = \omega_1 t. \tag{6.18}$$

Let

$$\omega_2 = \sqrt{\frac{k_{21}}{J_y}} \tag{6.19}$$

denotes the characteristic frequency related to equation (6.16). Introducing the following dimensionless parameters

$$w = \frac{\omega_2}{\omega_1}, \quad p = \frac{P}{\omega_1}, \quad \omega_z = \frac{\Omega_z}{\omega_1}, \quad f_0 = \frac{M_0}{(I_x + J_x)\,\omega_1^2},$$

$$c_1 = \frac{C_1}{(I_x + J_x)\,\omega_1}, \quad c_2 = \frac{C_2}{J_y \omega_1}, \quad \alpha_1 = \frac{k_{12} L^2}{(I_x + J_x)\,\omega_1^2}, \quad \alpha_2 = \frac{k_{22} L^2}{J_y \omega_1^2}, \tag{6.20}$$

$$j_1 = \frac{I_z + J_z - I_y - J_y}{I_x + J_x}, \quad j_2 = \frac{J_z - J_x}{I_x + J_x}, \quad j_3 = \frac{J_y}{I_x + J_x}, \quad j_4 = \frac{J_z - J_x}{J_y}$$

leads to the following dimensionless form of the motion equations

$$\begin{aligned} &\ddot{\varphi} + \left(1 + j_1 \omega_z^2\right) \varphi - \alpha_1 \varphi^3 + \left(c_1 + 2 j_2 \vartheta \dot{\vartheta}\right) \dot{\varphi} + \\ &(j_2 - j_3)\, \omega_z \dot{\vartheta} - f_0 \sin(p\tau) = 0, \end{aligned} \tag{6.21}$$

$$\ddot{\vartheta} + \left(w^2 + j_4 \omega_z^2 - j_4 \dot{\varphi}^2\right) \vartheta - \alpha_2 \vartheta^3 + c_2 \dot{\vartheta} + (1 - j_4)\, \omega_z \dot{\varphi} = 0. \tag{6.22}$$

Within the non-dimensional representation, the functions $\varphi(\tau)$ and $\vartheta(\tau)$ replace the primary generalized coordinates $\Phi(t)$ and $\Theta(t)$, respectively and the dots over symbols denote the derivatives respecting the dimensionless time .

The initial conditions related to equations (6.21)–(6.22) take the following form

$$\varphi(0) = u_{01}, \quad \dot{\varphi}(0) = u_{02}, \quad \vartheta(0) = u_{03}, \quad \dot{\vartheta}(0) = u_{04}, \tag{6.23}$$

where u_{01}, u_{02}, u_{03}, u_{04} can be obtained directly from equations (6.13) making use of definitions (6.17)–(6.18).

6.5 ASYMPTOTIC SOLUTION FOR NON-RESONANT VIBRATION

We start our study by determining the approximate solution for the non-resonant motion. This case of the motion is unambiguously described by equations (6.21)–(6.22) with initial conditions (6.23) and does not require formulating any additional relationships. The approximate analytical solution to this initial value problem is obtained using the method of multiple scales in the time domain (MSM). In accordance with the MSM, the system evolution in time is described using several variables. We adopt three variables: $\tau_0,\ \tau_1,\ \tau_2$ that are related to the original dimensionless time τ in the following manner

$$\tau_i = \varepsilon^i \tau, \qquad i = 0, 1, 2. \tag{6.24}$$

The non-dimensional small parameter ε satisfies the condition $0 < \varepsilon \ll 1$.

As a result of assumption (6.24), all functions dependent on time τ automatically become the functions of the new time variables τ_0, τ_1, τ_2. Similarly, the derivatives with respect to the original time τ are replaced by the partial derivatives according to

$$\frac{d}{d\tau} = \sum_{k=0}^{2} \varepsilon^k \frac{\partial}{\partial \tau_k} = \frac{\partial}{\partial \tau_0} + \varepsilon \frac{\partial}{\partial \tau_1} + \varepsilon^2 \frac{\partial}{\partial \tau_2}, \tag{6.25}$$

$$\frac{d^2}{d\tau} = \frac{\partial^2}{\partial \tau_0^2} + 2\varepsilon \frac{\partial^2}{\partial \tau_0 \partial \tau_1} + \varepsilon^2 \left(\frac{\partial^2}{\partial \tau_1^2} + 2 \frac{\partial^2}{\partial \tau_0 \partial \tau_2} \right) + 2\varepsilon^3 \frac{\partial^2}{\partial \tau_1 \partial \tau_2} + O\left(\varepsilon^4\right). \tag{6.26}$$

The solution to the initial-value problem, given by equations (6.21), (6.22) and (6.23), is sought in the form of the asymptotic expansions of the functions $\varphi(\tau)$ and $\vartheta(\tau)$ as follows

$$\varphi(\tau;\varepsilon) = \sum_{k=1}^{3} \varepsilon^k \phi_k(\tau_0, \tau_1, \tau_2) + O\left(\varepsilon^4\right), \tag{6.27}$$

$$\vartheta(\tau;\varepsilon) = \sum_{k=1}^{3} \varepsilon^k \theta_k(\tau_0, \tau_1, \tau_2) + O\left(\varepsilon^4\right), \tag{6.28}$$

where functions $\phi_k(\tau_0,\ \tau_1,\ \tau_2),\ \theta_k(\tau_0,\ \tau_1,\ \tau_2),\ k = 1,\ \ldots,\ 3$ are unknown.

Some parameters characterizing the features of the micromechanical gyroscope and its loading are assumed to be small, which can be expressed in terms of the powers of the small parameter

$$c_1 = \varepsilon^2 \hat{c}_1, \quad c_2 = \varepsilon^2 \hat{c}_2, \quad f_0 = \varepsilon^3 \hat{f}_0, \quad \omega_z = \varepsilon \hat{\omega}_z, \tag{6.29}$$

where the quantities $\hat{c}_1,\ \hat{c}_2,\ \hat{c}_3,\ \hat{f}_0$ and $\hat{\omega}_z$ are recognized as $O(1)$ when $\varepsilon \to 0$.

The last of the assumptions (6.29) means, in practice, that the substrate angular velocity is much smaller than the frequency with which the movable gyroscope elements oscillate. Assumptions (6.29) and also the approximate form of the solutions cause that all nonlinear terms in equations (6.21)–(6.22) have the character of small perturbations. The mathematical model of the gyroscope motion belongs then to the class of weakly nonlinear systems.

Substitution of equations (6.25)–(6.28) into equations (6.21)–(6.22) yields the equations in which the small parameter occurs in a few different powers. According to the assumed order of the asymptotic expansions (6.27)–(6.28), all terms of the order $O\left(\varepsilon^4\right)$ and higher orders will be neglected. Equations obtained in this way have to be satisfied for any value of the small parameter. After rearranging the terms of both equations according to the powers ε, ε^2 and ε^3, this requirement is fulfilled by equating to zero all coefficients standing at these powers. In that manner, one can obtain the system of six differential equations with unknown functions: $\phi_k\left(\tau_0,\ \tau_1,\ \tau_2\right)$, and $\theta_k\left(\tau_0,\ \tau_1,\ \tau_2\right)$, where $k = 1,2,3$. Organizing the equations into three groups corresponding to the respective approximation orders, we solve the system recursively. The first group consists of the equations obtained in the result of equating to zero the terms standing at the first power of ε. The equations are homogeneous and have the following form

$$\frac{\partial^2\phi_1}{\partial\tau_0^2} + \phi_1 = 0, \tag{6.30}$$

$$\frac{\partial^2\theta_1}{\partial\tau_0^2} + w^2\theta_1 = 0. \tag{6.31}$$

Equations (6.30)–(6.31) are named the equations of the first-order approximation. Equations belonging to the second group, termed the equations of the second-order approximation, contain the terms standing at the ε^2. They are as follows

$$\frac{\partial^2\phi_2}{\partial\tau_0^2} + \phi_2 = -2\frac{\partial^2\phi_1}{\partial\tau_0\partial\tau_1} + \left(j_3 - j_2\right)\hat{\omega}_z\frac{\partial\theta_1}{\partial\tau_0}, \tag{6.32}$$

$$\frac{\partial^2\theta_2}{\partial\tau_0^2} + w^2\theta_2 = -2\frac{\partial^2\theta_1}{\partial\tau_0\partial\tau_1} + \left(j_4 - 1\right)\hat{\omega}_z\frac{\partial\phi_1}{\partial\tau_0}. \tag{6.33}$$

On the right sides of equations (6.32)–(6.33) occur solutions to the equations of the first-order approximation, which ensures the recursive manner of the solution procedure.

Similarly, by gathering the terms that are accompanied by ε^3 and then equating their sum to zero, we get the equations of the third-order approximation

$$\begin{gathered}\frac{\partial^2\phi_3}{\partial\tau_0^2} + \phi_3 = \hat{f}_0\sin\left(p\tau_0\right) - j_1\hat{\omega}_z^2\phi_1 + \alpha_1\phi_1^3 - \hat{c}_1\frac{\partial\phi_1}{\partial\tau_0} - \frac{\partial^2\phi_1}{\partial\tau_1^2} - \\ 2\left(\frac{\partial^2\phi_1}{\partial\tau_0\partial\tau_2} + \frac{\partial^2\phi_2}{\partial\tau_0\partial\tau_1}\right) + \hat{\omega}_z\left(j_3 - j_2\right)\left(\frac{\partial\theta_1}{\partial\tau_1} + \frac{\partial\theta_2}{\partial\tau_0}\right) - 2j_2\theta_1\frac{\partial\theta_1}{\partial\tau_0}\frac{\partial\phi_1}{\partial\tau_0},\end{gathered} \tag{6.34}$$

$$\begin{gathered}\frac{\partial^2\theta_3}{\partial\tau_0^2} + w^2\theta_3 = -\hat{c}_2\frac{\partial\theta_1}{\partial\tau_0} - j_4\hat{\omega}_z^2\theta_1 + \alpha_2\theta_1^3 - \frac{\partial^2\theta_1}{\partial\tau_1^2} - \\ 2\left(\frac{\partial^2\theta_1}{\partial\tau_0\partial\tau_2} + \frac{\partial^2\theta_2}{\partial\tau_0\partial\tau_1}\right) + \hat{\omega}_z\left(j_4 - 1\right)\left(\frac{\partial\phi_1}{\partial\tau_1} + \frac{\partial\phi_2}{\partial\tau_0}\right) + j_4\theta_1\left(\frac{\partial\phi_1}{\partial\tau_0}\right)^2.\end{gathered} \tag{6.35}$$

The linear differential operators of equations (6.30)–(6.35) are the same on each level of approximation. Therefore, starting with equations (6.32)–(6.33) after introducing the solutions of the lower levels of approximation onto the right sides, it is inevitable that among the solutions the secular terms will appear. However, in the case of the damped small vibration, both generalized coordinates must be bounded on an arbitrary time range, hence all the secular terms should be eliminated from each of equations (6.32)–(6.35) before we start solving them.

The recursive procedure of the solution begins with the equations of the first-order approximation. Their general solutions can be written as follows

$$\phi_1(\tau_0,\tau_1,\tau_2) = B_1(\tau_1,\tau_2)\,e^{\mathrm{i}\tau_0} + \bar{B}_1(\tau_1,\tau_2)\,e^{-\mathrm{i}\tau_0}, \tag{6.36}$$

$$\theta_1(\tau_0,\tau_1,\tau_2) = B_2(\tau_1,\tau_2)\,e^{\mathrm{i}w\tau_0} + \bar{B}_2(\tau_1,\tau_2)\,e^{-\mathrm{i}w\tau_0}, \tag{6.37}$$

where i denotes the imaginary unit and $\bar{B}_j(\tau_1,\ \tau_2)$ stands for the complex conjugate of the function $B_j(\tau_1,\ \tau_2)$, where $j = 1,2$.

Subsequently, solutions (6.36)–(6.37) are inserted into equations (6.32)–(6.33). As result, the secular terms appear among the inhomogeneous terms on the right sides of these equations. Elimination of the secular terms is necessary to obtain the solutions that are bounded. After detecting the secular terms, the following solvability conditions can be formulated

$$\frac{\partial B_1}{\partial \tau_1} = 0, \qquad \frac{\partial \bar{B}_1}{\partial \tau_1} = 0, \tag{6.38}$$

$$\frac{\partial B_2}{\partial \tau_1} = 0, \qquad \frac{\partial \bar{B}_2}{\partial \tau_1} = 0. \tag{6.39}$$

Making use of the solvability conditions, one can obtain the following equations of the second-order approximation that do not lead to non-physical solutions

$$\frac{\partial^2 \phi_2}{\partial \tau_0^2} + \phi_2 = \mathrm{i}w(j_3 - j_2)\,\hat{\omega}_z\left(B_2 e^{\mathrm{i}w\tau_0} - \bar{B}_2 e^{-\mathrm{i}w\tau_0}\right), \tag{6.40}$$

$$\frac{\partial^2 \theta_2}{\partial \tau_0^2} + w^2\theta_2 = \mathrm{i}\ (j_4 - 1)\,\hat{\omega}_z\left(B_1 e^{\mathrm{i}\tau_0} - \bar{B}_1 e^{-\mathrm{i}\tau_0}\right). \tag{6.41}$$

The particular solutions to equations (6.40)–(6.41) are as follows

$$\phi_2(\tau_0,\tau_1,\tau_2) = \frac{\mathrm{i}w(j_2 - j_3)\,\hat{\omega}_z\left(B_2 e^{\mathrm{i}w\tau_0} - \bar{B}_2 e^{-\mathrm{i}w\tau_0}\right)}{w^2 - 1}, \tag{6.42}$$

$$\theta_2(\tau_0,\tau_1,\tau_2) = \frac{\mathrm{i}\ (j_4 - 1)\,\hat{\omega}_z\left(B_1 e^{\mathrm{i}\tau_0} - \bar{B}_1 e^{-\mathrm{i}\tau_0}\right)}{w^2 - 1}. \tag{6.43}$$

Inserting of solutions (6.36)–(6.37) and (6.40)–(6.43) into equations (6.34)–(6.35) generates the secular terms. According to the expectation mentioned before, that the solutions are limited, the secular terms should be eliminated. So, we obtain the consecutive solvability conditions

$$-2\mathrm{i}\frac{\partial B_1}{\partial \tau_2} + \left(\hat{\omega}_z^2 \frac{j_1\left(1-w^2\right) + \left(j_2 - j_3\right)\left(j_4 - 1\right)}{w^2 - 1} + 3\alpha_1 B_1 \bar{B}_1 - \mathrm{i}\,\hat{c}_1\right) B_1 = 0, \tag{6.44}$$

$$2\mathrm{i}\frac{\partial \bar{B}_1}{\partial \tau_2} + \left(\hat{\omega}_z^2 \frac{j_1\left(1-w^2\right) + \left(j_2 - j_3\right)\left(j_4 - 1\right)}{w^2 - 1} + 3\alpha_1 B_1 \bar{B}_1 + \mathrm{i}\,\hat{c}_1\right) \bar{B}_1 = 0, \tag{6.45}$$

$$\begin{gathered} -2\mathrm{i}\,w\frac{\partial B_2}{\partial \tau_2} + \left(\hat{\omega}_z^2 \frac{j_4\left(1 - w^2\left(1 + j_2 - j_3\right)\right) + w^2\left(j_2 - j_3\right)}{w^2 - 1} + \right. \\ \left. 2 j_4 B_1 \bar{B}_1 + 3\alpha_2 B_2 \bar{B}_2 - \mathrm{i}\,w\hat{c}_2\right) B_2 = 0, \end{gathered} \tag{6.46}$$

$$\begin{gathered} 2\mathrm{i}\,w\frac{\partial \bar{B}_2}{\partial \tau_2} + \left(\hat{\omega}_z^2 \frac{j_4\left(1 - w^2\left(1 + j_2 - j_3\right)\right) + w^2\left(j_2 - j_3\right)}{w^2 - 1} + \right. \\ \left. 2 j_4 B_1 \bar{B}_1 + 3\alpha_2 B_2 \bar{B}_2 + \mathrm{i}\,w\hat{c}_2\right) \bar{B}_2 = 0. \end{gathered} \tag{6.47}$$

Taking advantage of the solvability conditions leads to the following equations of third-order approximation

$$\begin{gathered} \frac{\partial^2 \phi_3}{\partial \tau_0^2} + \phi_3 = \hat{f}_0 \sin\left(p\tau_0\right) + \alpha_1\left(e^{3\mathrm{i}\tau_0} B_1^3 + e^{-3\mathrm{i}\tau_0}\bar{B}_1^3\right) + 2w j_2\left(e^{\mathrm{i}(2w+1)\tau_0} B_1 B_2^2 + \right. \\ \left. e^{-\mathrm{i}(2w+1)\tau_0}\bar{B}_1 \bar{B}_2^2 - e^{\mathrm{i}(2w-1)\tau_0}\bar{B}_1 B_2^2 - e^{-\mathrm{i}(2w-1)\tau_0} B_1 \bar{B}_2^2\right), \end{gathered} \tag{6.48}$$

$$\begin{gathered} \frac{\partial^2 \theta_3}{\partial \tau_0^2} + w^2\theta_3 = \alpha_2\left(e^{3\mathrm{i}w\tau_0} B_2^3 + e^{-3\mathrm{i}w\tau_0}\bar{B}_2^3\right) - j_4\left(e^{\mathrm{i}(w+2)\tau_0} B_1^2 B_2 + \right. \\ \left. e^{-\mathrm{i}(w+2)\tau_0}\bar{B}_1^2 \bar{B}_2\right) - j_4\left(e^{\mathrm{i}(w-2)\tau_0}\bar{B}_1^2 B_2 + e^{-\mathrm{i}(w-2)\tau_0} B_1^2 \bar{B}_2\right) \end{gathered} \tag{6.49}$$

that do not contain any secular terms. The particular solutions to equations (6.48)–(6.49) can be written as

$$\begin{gathered} \phi_3\left(\tau_0, \tau_1, \tau_2\right) = \frac{\hat{f}_0 \sin\left(p\tau_0\right)}{1 - p^2} - \frac{\alpha_1}{8}\left(e^{3\mathrm{i}\tau_0} B_1^3 + e^{-3\mathrm{i}\tau_0}\bar{B}_1^3\right) - \\ j_2\frac{\left(e^{\mathrm{i}(2w+1)\tau_0} B_1 B_2^2 + e^{-\mathrm{i}(2w+1)\tau_0}\bar{B}_1 \bar{B}_2^2\right)}{2\left(w+1\right)} + \\ j_2\frac{\left(e^{\mathrm{i}(2w-1)\tau_0}\bar{B}_1 B_2^2 + e^{-\mathrm{i}(2w-1)\tau_0} B_1 \bar{B}_2^2\right)}{2\left(w-1\right)}, \end{gathered} \tag{6.50}$$

$$\theta_3(\tau_0,\tau_1,\tau_2) = -\frac{\alpha_2}{8w^2}\left(e^{3iw\tau_0}B_2^3 + e^{-3iw\tau_0}\bar{B}_2^3\right) + \tag{6.51}$$
$$j_4\frac{e^{i(w+2)\tau_0}B_1^2B_2 + e^{-i(w+2)\tau_0}\bar{B}_1^2\bar{B}_2}{4(w+1)} - j_4\frac{e^{i(w-2)\tau_0}\bar{B}_1^2B_2 + e^{-i(w-2)\tau_0}B_1^2\bar{B}_2}{4(w-1)}.$$

The unknown complex-valued functions $B_1(\tau_1,\tau_2)$, $B_2(\tau_1,\tau_2)$ and their complex conjugates $\bar{B}_1(\tau_1,\tau_2)$, $\bar{B}_2(\tau_1,\tau_2)$ occurring in solutions (6.36)–(6.37), (6.42)–(6.43) and (6.50)–(6.51) are restricted by the solvability conditions. The elimination of the secular terms does not only ensure that bounded solutions are possible to obtain, but also leads to an initial value problem which gives the possibility to determine the unknown functions $B_1(\tau_1,\tau_2)$, $B_2(\tau_1,\tau_2)$. Based on equations (6.38)–(6.39), one can conclude that all these functions do not depend on the variable τ_1.

It is reasonable to present the unknown functions in the exponential form

$$B_1(\tau_2) = \frac{1}{2}b_1(\tau_2)e^{i\psi_1(\tau_2)}, \qquad \bar{B}_1(\tau_2) = \frac{1}{2}b_1(\tau_2)e^{-i\psi_1(\tau_2)}, \tag{6.52}$$

$$B_2(\tau_2) = \frac{1}{2}b_2(\tau_2)e^{i\psi_2(\tau_2)}, \qquad \bar{B}_2(\tau_2) = \frac{1}{2}b_2(\tau_2)e^{-i\psi_2(\tau_2)}, \tag{6.53}$$

where the functions $b_1(\tau_2)$, $b_2(\tau_2)$, $\psi_1(\tau_2)$, $\psi_2(\tau_2)$ are real-valued.

After making use of equations (6.38)–(6.39), solvability conditions (6.44)–(6.45) remain the only restrictions imposed on the unknown functions $b_1(\tau_2)$, $b_2(\tau_2)$, $\psi_1(\tau_2)$, $\psi_2(\tau_2)$. Let us insert the relationships (6.52)–(6.53) into equations (6.44)–(6.47), and then solve the last ones with respect to the derivatives. As result, we obtain

$$\frac{db_1}{d\tau_2} = -\frac{\hat{c}_1}{2}b_1, \tag{6.54}$$

$$\frac{db_2}{d\tau_2} = -\frac{\hat{c}_2}{2}b_2, \tag{6.55}$$

$$\frac{d\psi_1}{d\tau_2} = -\frac{3}{8}\alpha_1 b_1^2 + \frac{\hat{\omega}_z^2}{2}\left(j_1 - \frac{(j_2-j_3)(j_4-1)}{w^2-1}\right), \tag{6.56}$$

$$\frac{d\psi_2}{d\tau_2} = -\frac{3}{8w}\alpha_2 b_2^2 - \frac{j_4}{2w}b_1^2 + \hat{\omega}_z^2\frac{w^2(j_3-j_2)(1-j_4)+j_4(w^2-1)}{2w(w^2-1)}. \tag{6.57}$$

We gain a system of four ordinary differential equations regarding the slowest time variable. However, a problem arises when we want to write the initial conditions for equations (6.54)–(6.57) that would be consistent with the initial conditions (6.23). Therefore, it is purposeful to transform these equations onto equivalent ones, but formulated with respect to the original time τ. Considering definition (6.25), one can notice that the following relationships

$$\frac{db_i}{d\tau} = \varepsilon^2\frac{\partial b_i}{\partial \tau_2}, \qquad \frac{d\psi_i}{d\tau} = \varepsilon^2\frac{\partial \psi_i}{\partial \tau_2}, \qquad i = 1,2, \tag{6.58}$$

are true in the case of any function depending only on the variable τ_2. So, substituting equalities (6.58) into equations (6.54)–(6.57), and taking into account relations (6.29), we gain

$$\frac{da_1}{d\tau} = -\frac{1}{2}c_1 a_1, \tag{6.59}$$

$$\frac{da_2}{d\tau} = -\frac{1}{2}c_2 a_2, \tag{6.60}$$

$$\frac{d\psi_1}{d\tau} = -\frac{3\alpha_1}{8}a_1^2 + \frac{\omega_z^2}{2}\left(j_1 - \frac{(j_2 - j_3)(j_4 - 1)}{w^2 - 1}\right), \tag{6.61}$$

$$\frac{d\psi_2}{d\tau} = -\frac{j_4}{4w}a_1^2 - \frac{3\alpha_2}{8}a_2^2 + \omega_z^2 \frac{w^2 (j_3 - j_2)(1 - j_4) + j_4 (w^2 - 1)}{2w(w^2 - 1)}, \tag{6.62}$$

where $a_i(\tau) = \varepsilon\, b_i(\tau),\ i = 1,\ 2, 3.$

The functions $a_1(\tau),\ a_2(\tau)$ and $\psi_1(\tau),\ \psi_2(\tau)$ stand for the amplitudes and the phases, respectively, of the main components of the approximate solution. By the term "main components" we mean solutions (6.36)–(6.37) to the first-order approximation equations. Since equations (6.59)–(6.62) are primarily formulated with respect to the slowest time scale, they describe the slow modulation of the amplitudes and phases in time. To formulate the initial value problem governing the modulation, equations (6.59)–(6.62) should be supplemented by the initial conditions which can be written as

$$a_1(0) = a_{10}, \qquad a_2(0) = a_{20}, \qquad \psi_1(0) = \psi_{10}, \qquad \psi_2(0) = \psi_{20}, \tag{6.63}$$

where the initial values of the amplitudes and the phases $a_{10},\ a_{20},\ \psi_{10},\ \psi_{20}$ are known and compatible with the primary introduced initial values $u_{01},\ u_{02},\ u_{03},\ u_{04}$.

The exact solution to the initial value problem (6.59)–(6.62) and (6.63) that is often termed the modulation problem has the form

$$a_1 = a_{10}e^{-\frac{c_1\tau}{2}}, \tag{6.64}$$

$$a_2 = a_{20}e^{-\frac{c_2\tau}{2}}, \tag{6.65}$$

$$\psi_1 = -\frac{3\alpha_1}{8c_1}a_{10}^2\left(1 - e^{-c_1\tau}\right) + \frac{\omega_z^2}{2}\left(j_1 - \frac{(j_2 - j_3)(j_4 - 1)}{w^2 - 1}\right)\tau + \psi_{10}, \tag{6.66}$$

$$\begin{aligned}\psi_2 = &-\frac{3\alpha_2}{8c_2 w}a_{20}^2\left(1 - e^{-c_2\tau}\right) - \frac{j_4 a_{10}^2}{4c_1 w}\left(1 - e^{-c_1\tau}\right) + \\ &\omega_z^2 \frac{w^2 (j_3 - j_2)(1 - j_4) + j_4 (w^2 - 1)}{2w(w^2 - 1)}\tau + \psi_{20}.\end{aligned} \tag{6.67}$$

Gathering the solutions of the subsequent orders according to equations (6.27)–(6.28), we obtain the approximate solution to the considered problem in the form

$$\varphi(\tau) = \frac{f_0 \sin(p_1 \tau)}{1-p^2} + a_1 \cos(\tau + \psi_1) - \frac{\alpha_1}{32} a_1^3 \cos(3\tau + 3\psi_1) + \frac{j_2 a_1 a_2^2}{8(w-1)} \cos((1-2w)\tau + \psi_1 - 2\psi_2) - \frac{j_2 a_1 a_2^2}{8(w+1)} \cos((1+2w)\tau + \psi_1 + 2\psi_2) + \frac{(j_3 - j_2) w \omega_z}{w^2-1} a_2 \sin(w\tau + \psi_2), \tag{6.68}$$

$$\vartheta(\tau) = a_2 \cos(w\tau + \psi_2) - \frac{\alpha_2 a_2^3}{32 w^2} \cos(3w\tau + 3\psi_2) - \frac{j_4 a_1^2 a_2}{16(w-1)} \cos((2-w)\tau + 2\psi_1 - \psi_2) + \frac{j_4 a_1^2 a_2}{16(w+1)} \cos((2+w)\tau + 2\psi_1 + \psi_2) - \frac{(j_4 - 1)\omega_z a_1}{w^2-1} \sin(\tau + \psi_1), \tag{6.69}$$

where the functions a_1, a_2 and ψ_1, ψ_2 are determined by equations (6.64)–(6.67).

One can notice that the function $a_1(\tau)$ stands for the amplitude of the first term of the asymptotic expansion that approximates according to equation (6.27), the generalized coordinate $\varphi(\tau)$, whereas the function $\psi_1(\tau)$ is the phase of this term. Similarly, $a_2(\tau)$ and $\psi_2(\tau)$ are the amplitude and the phase, respectively, of the first term of asymptotic expansion (6.28). All the functions depict the slow evolution of the amplitudes and phases in time.

Solutions (6.68)–(6.69) describe the forced, but non-resonant vibration of the micromechanical gyroscope caused by the harmonic torque acting about the drive axis. It is worth emphasizing that although they are approximations, however, they have the analytical form. Excluding the case when $w = 0$ that has no practical sense, the solution given by formulas (6.68)–(6.69) fails when $p = 1$ or $w = 1$. These cases correspond to the main and the internal resonance, respectively. Therefore, the solution, given by equations (6.68)–(6.69), is useful to the description of the gyroscope behavior in conditions being far from the resonance. The resonant cases should be considered separately based on some additionally formulated assumptions.

To present and test the approximate solution obtained, we set the following values of the parameters which determine the gyroscope motion: $w = 0.8$, $\alpha_1 = 2.5$, $\alpha_2 = 2.5$, $j_1 = 0.95$, $j_2 = 0.9$, $j_3 = 1$, $j_4 = 0.9$, $c_1 = 0.0045$, $c_2 = 0.0045$, $f_0 = 0.0015$, $p = 0.08$, $\omega_z = 0.01$.

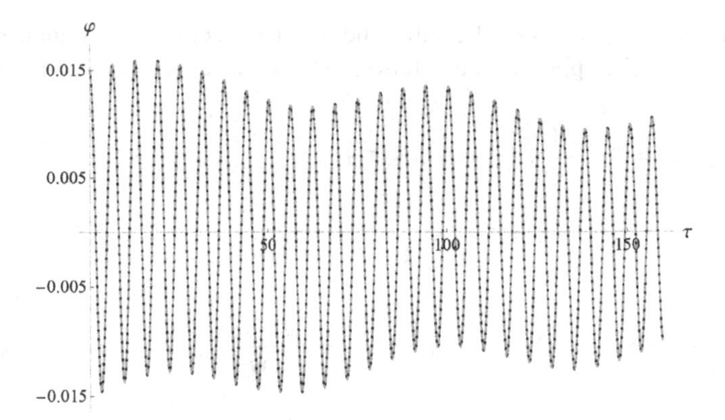

FIGURE 6.4 The oscillations $\varphi(\tau)$ of the intermediate frame with significantly decreasing amplitude at the beginning of motion estimated analytically (MSM) and numerically (dotted line).

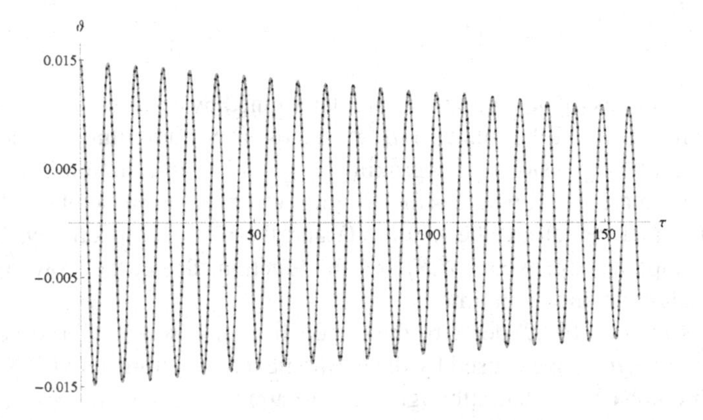

FIGURE 6.5 The oscillations $\vartheta(\tau)$ of the sensing plate at the beginning of motion estimated analytically (MSM) and numerically (dotted line).

The following initial values occurring in conditions (6.23) are assumed: $u_{01} = 0.015$, $u_{02} = 0$, $u_{03} = 0.015$, $u_{04} = 0$.

These initial values are compatible with the following initial values occurring in conditions (6.63): $a_{10} = 0.0150023$, $a_{20} = 0.0149998$, $\psi_{10} = 0.00402424$, $\psi_{20} = -0.00629011$.

Figures 6.4–6.7 present how the generalized coordinates $\varphi(\tau)$ and $\vartheta(\tau)$ change themselves in the motion caused by the harmonically changing torque and the static equilibrium disturbance. There are two curves depicted in each of these graphs. The solid black line presents the approximate analytical solution obtained using MSM whereas the numerical solution obtained using the standard *NDSolve* of the *Mathematica* software is drawn using the dotted gray line. One can observe the high compliance between the numerical and the approximate analytical solutions.

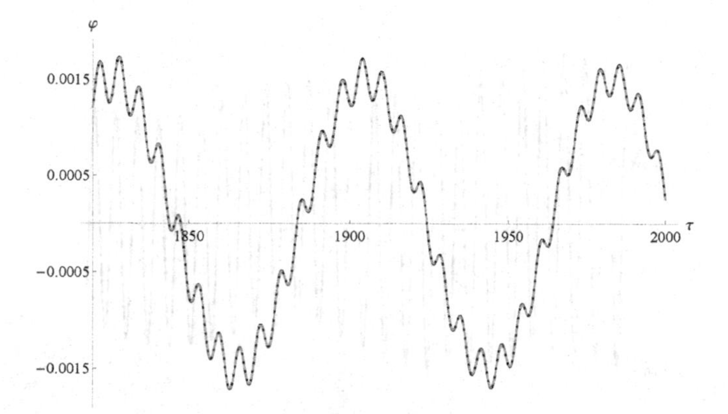

FIGURE 6.6 Almost stationary oscillations $\varphi(\tau)$ of the intermediate frame at the end of the motion simulation estimated analytically (MSM) and numerically (dotted line).

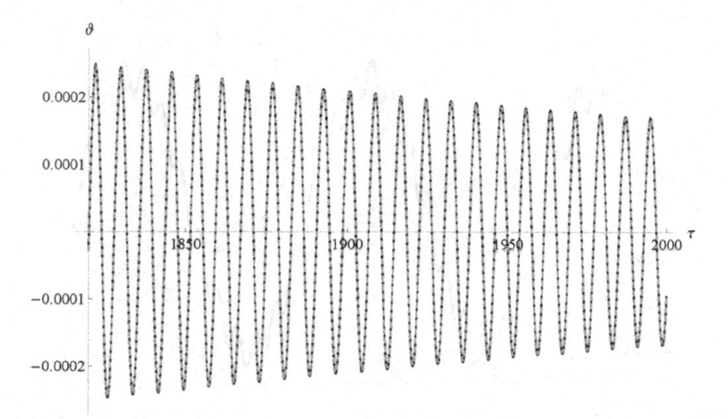

FIGURE 6.7 Almost stationary oscillations $\vartheta(\tau)$ of the sensing plate at the end of the motion simulation estimated analytically (MSM) and numerically (dotted line).

Figures 6.4–6.5 show the vibration course of the intermediate frame and the sensing plate, respectively, in the initial stage of the motion when the transient vibration occurs. The values of both amplitudes decrease significantly. The rate of this decline is exponential over time, according to relationships (6.64)–(6.65). Thus, the oscillations registered at the end of the simulation and presented in Figures 6.6–6.7 are characterized not only by significantly smaller values of the amplitudes but also almost stationary conditions.

The effect of the overlapping of the self-vibration and the fixed component of the forced vibration is noticeable in the case of the vibration of the intermediate frame. Although the character of this overlap changes itself diametrically with time. Let

$$\varphi_c(\tau) = \frac{f_0 \sin(p_1 \tau)}{1 - p^2} \tag{6.70}$$

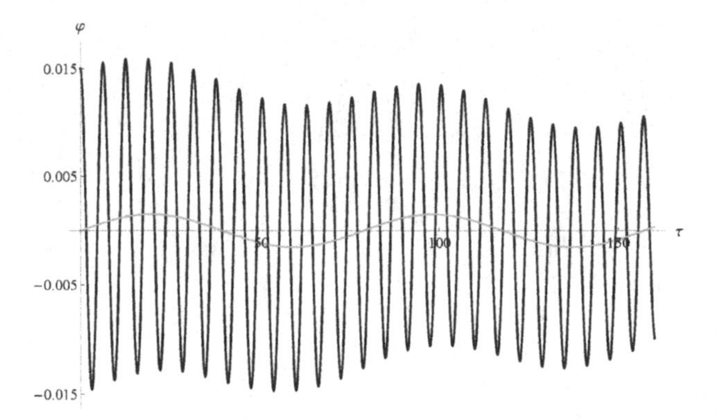

FIGURE 6.8 Transient vibration $\varphi(\tau)$ of the intermediate frame and its fixed component $\varphi_c(\tau)$ in transient motion.

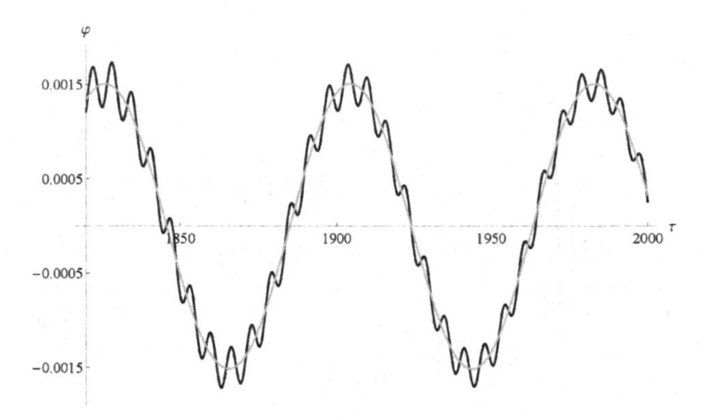

FIGURE 6.9 Almost stationary vibration $\varphi(\tau)$ of the intermediate frame and its fixed component $\varphi_c(\tau)$.

denotes the fixed component of the frame vibration. The approximate analytical solution obtained using MSM and the component $\varphi_c(\tau)$ are shown in Figures 6.8–6.9.

Initially, the amplitude of the slowly changing fixed component is significantly smaller than the amplitude of the fast self-vibrations, which is depicted in Figure 6.8. Therefore, we can observe the slow modulation of the intermediate frame oscillations due to the external forcing. The amplitude of the self vibration decreases over time. As a result, the fixed component becomes dominant in the image of the genaralized coordinate $\varphi(\tau)$ variability, as shown in Figure 6.9.

As in the previous chapters, the error satisfying the micro gyroscope motion equations (6.21)–(6.22) is taken as a measure of the accuracy of both the approximate analytical and the numerical solution. Let $G_1(\varphi_a(\tau), \vartheta_a(\tau))$ and $G_2(\varphi_a(\tau), \vartheta_a(\tau))$ stand for the differential operators occurring on the left sides of equations (6.21) and 6.22), respectively. The error satisfying the motion equations is defined as follows

$$\delta_i = \sqrt{\frac{1}{\tau_e - \tau_s} \int_{\tau_s}^{\tau_e} \left(G_i\left(\varphi_a(\tau),\, \vartheta_a(\tau)\right) - 0\right)^2 d\tau}, \quad i = 1,2, \tag{6.71}$$

where $\varphi_a(\tau)$, $\vartheta_a(\tau)$ are the approximate solutions obtained by means of the MSM or numerically, whereas τ_s and τ_e denote the chosen instants.

The values of the error δ_i for the analytical and the numerical approximate solutions which are presented in Figures 6.4–6.7 are gathered in Table 6.1.

TABLE 6.1
The values of the error satisfying the governing equations for the non-resonant vibration

	MSM solution	numerical solution
transient stage (Figures 6.4–6.5) $\tau_0 = 0,\ \tau_e = 160$	$\delta_1 = 6.63599 \cdot 10^{-7}$ $\delta_2 = 8.71245 \cdot 10^{-8}$	$\delta_1 = 7.31750 \cdot 10^{-7}$ $\delta_2 = 3.44559 \cdot 10^{-8}$
almost stationary stage (Figures 6.6–6.7) $\tau_0 = 1820,\ \tau_e = 2000$	$\delta_1 = 3.42183 \cdot 10^{-7}$ $\delta_2 = 7.45993 \cdot 10^{-8}$	$\delta_1 = 2.11391 \cdot 10^{-7}$ $\delta_2 = 4.87939 \cdot 10^{-8}$

6.6 RESONANT VIBRATION IN SIMULTANEOUSLY OCCURRING EXTERNAL AND INTERNAL RESONANCES

The approximate analytical solution derived in the previous section allows one to detect the conditions at which the resonant phenomena may occur during the motion of the gyroscope that is excited by harmonically changing torque acting around the drive axis. When any of the denominators in the terms occurring on the right sides of equations (6.68)–(6.69) is close to zero then the resonant vibration appears in the system. Excluding the case $w = 0$ having no practical sense, the approximate solution based on the approach with three time scales allows one to recognize the following resonant cases:

(i) the primary external resonances, when $p \approx 1$,
(ii) the internal resonance, when $w \approx 1$.

The non-dimensional frequencies around which the main and the internal resonances occur are determined by the same value. This characteristic feature of the considered type of MEMS gyroscope results from its structure. The sense and drive axes around which the sensing plate and the frame can rotate independently are mutually orthogonal.

The micromechanical gyroscope works as a torsional resonator. It means that its desirable work regime is the simultaneous resonance around both the sense and the drive axes. That intended state is achieved when designing involves the proper choice of the inertial properties of the MEMS parts and the elastic features of the torsional joints. Analyzing equations (6.15)–(6.16), it can be noted that the coincidence of two resonances is possible due to the tiny impact of the substrate angular velocity ω_z on the resonant frequencies range. This slight impact can be justified by recalling the assumptions (6.29) that are used in motion equations (6.21)–(6.22).

Assuming that the permanent equalization of the eigenfrequencies at designing and manufacturing is unobtainable, we take into account the following resonance conditions

$$p = 1 + \sigma_1, \qquad w \approx 1 + \sigma_2, \tag{6.72}$$

where the detuning parameters σ_1 and σ_2 are assumed to be small numbers of the order $O\left(\varepsilon^2\right)$, i.e.

$$\sigma_1 = \varepsilon \hat{\sigma}_1 \qquad \sigma_2 = \varepsilon \hat{\sigma}_2 \tag{6.73}$$

The coefficients $\hat{\sigma}_1$ and $\hat{\sigma}_2$ are understood as $O(1)$ when $\varepsilon \to 0$.

Substituting relationships (6.72)–(6.73) into equations (6.21)–(6.22) yields the mathematical model governing the vibration around the stable equilibrium position of the considered micro gyroscope in the case of the doubled resonance. The model consists of the following differential equations

$$\begin{aligned} &\ddot{\varphi} + \left(1 + j_1\omega_z^2\right)\varphi - \alpha_1\varphi^3 + \left(c_1 + 2j_2\vartheta(\tau)\dot{\vartheta}(\tau)\right)\dot{\varphi} + \\ &\left(j_2 - j_3\right)\omega_z\dot{\vartheta}(\tau) - f_0\sin\left(\left(1 + \varepsilon^2\hat{\sigma}_1\right)\tau\right) = 0, \end{aligned} \tag{6.74}$$

$$\ddot{\vartheta} + \left(\left(1 + \varepsilon^2\hat{\sigma}_2\right)^2 + j_4\omega_z^2 - j_4\dot{\varphi}(\tau)^2\right)\vartheta - \alpha_2\vartheta^3 + c_2\dot{\vartheta} + \left(1 - j_4\right)\omega_z\dot{\varphi} = 0 \tag{6.75}$$

and the initial conditions

$$\varphi(0) = u_{01}, \quad \dot{\varphi}(0) = u_{02}, \quad \vartheta(0) = u_{03}, \quad \dot{\vartheta}(0) = u_{04}, \tag{6.76}$$

where $u_{01}, u_{02}, u_{03}, u_{04}$ are known.

The assumptions (6.29) concerning the smallness of some parameters remain still valid. The approximate solution to the initial-value problem is determined using the multiple scales method. As previously, we introduce three variables τ_0, τ_1, τ_2 replacing the original time τ. The first and the second derivatives over time τ are substituted by the partial differential operators defined by equations (6.25)–(6.26). According to the MSM rules, we employ formulas (6.27)–(6.28), for the weakly nonlinear and weakly coupled dynamical system. As a result of all the substitutions carried out, the small parameter ε appears in some various powers. After rearranging the equations according to the powers of the small parameter and omitting the terms accompanied by ε in the powers higher than three, we require the motion equations should be satisfied for any value of the small parameter. Hereby, we get a system of six differential equations with partial derivatives. We organize the equations obtained into three groups. To the first group is a set of homogeneous equations that contain the terms standing at ε. They are as follows

$$\frac{\partial^2\phi_1}{\partial\tau_0^2}+\phi_1=0, \tag{6.77}$$

$$\frac{\partial^2\theta_1}{\partial\tau_0^2}+\theta_1=0. \tag{6.78}$$

The terms accompanied by ε^2 form the equations of the second-order approximation

$$\frac{\partial^2\phi_2}{\partial\tau_0^2}+\phi_2=-2\frac{\partial^2\phi_1}{\partial\tau_0\partial_1}+(j_3-j_2)\,\hat{\omega}_z\frac{\partial\theta_1}{\partial\tau_0}, \tag{6.79}$$

$$\frac{\partial^2\theta_2}{\partial\tau_0^2}+\theta_2=-2\hat{\sigma}_2\theta_1-2\frac{\partial^2\theta_1}{\partial\tau_0\partial\tau_1}+(j_4-1)\,\hat{\omega}_z\frac{\partial\phi_1}{\partial\tau_0}. \tag{6.80}$$

To the equations of the third-order approximation belong the following equations in which the terms accompanied by ε^3 are gathered

$$\begin{aligned}\frac{\partial^2\phi_3}{\partial\tau_0^2}+\phi_3&=\hat{f}_0\sin\left((1+\varepsilon\hat{\sigma}_1)\,\tau_0\right)-j_1\hat{\omega}_z^2\phi_1+\alpha_1\phi_1^3-\hat{c}_1\frac{\partial\phi_1}{\partial\tau_0}-\frac{\partial^2\phi_1}{\partial\tau_1^2}-\\ &2\left(\frac{\partial^2\phi_1}{\partial\tau_0\partial\tau_2}+\frac{\partial^2\phi_2}{\partial\tau_0\partial\tau_1}\right)+\hat{\omega}_z\,(j_3-j_2)\left(\frac{\partial\theta_1}{\partial\tau_1}+\frac{\partial\theta_2}{\partial\tau_0}\right)-2j_2\theta_1\frac{\partial\theta_1}{\partial\tau_0}\frac{\partial\phi_1}{\partial\tau_0},\end{aligned} \tag{6.81}$$

$$\begin{aligned}\frac{\partial^2\theta_3}{\partial\tau_0^2}+\theta_3&=-\left(j_4\hat{\omega}_z^2+\hat{\sigma}_2^2\right)\theta_1-2\hat{\sigma}_2\theta_2+\alpha_2\theta_1^3-\hat{c}_2\frac{\partial\theta_1}{\partial\tau_0}-\frac{\partial^2\theta_1}{\partial\tau_1^2}-\\ &2\left(\frac{\partial^2\theta_1}{\partial\tau_0\partial\tau_2}+\frac{\partial^2\theta_2}{\partial\tau_0\partial\tau_1}\right)+\hat{\omega}_z\,(j_4-1)\left(\frac{\partial\phi_1}{\partial\tau_1}+\frac{\partial\phi_2}{\partial\tau_0}\right)+j_4\theta_1\left(\frac{\partial\phi_1}{\partial\tau_0}\right)^2,\end{aligned} \tag{6.82}$$

where the functions $\phi_k(\tau_0,\ \tau_1,\ \tau_2)$, $\theta_k(\tau_0,\ \tau_1,\ \tau_2)$, $k=1,\ \ldots,3$, are to be sought.

The weak coupling assumed in the system manifests itself in such a way that the linear operators in each pair of the equations of the subsequent approximations are mutually independent.

The system of six partial differential equations is solved recursively starting with homogeneous equations (6.77)–(6.78). The general solutions to the equations are

$$\phi_1(\tau_0,\tau_1,\tau_2)=B_1(\tau_1,\tau_2)\,e^{i\tau_0}+\bar{B}_1(\tau_1,\tau_2)\,e^{-i\tau_0}, \tag{6.83}$$

$$\theta_1(\tau_0,\tau_1,\tau_2)=B_2(\tau_1,\tau_2)\,e^{i\tau_0}+\bar{B}_2(\tau_1,\tau_2)\,e^{-i\tau_0}, \tag{6.84}$$

where the functions $B_1(\tau_1,\tau_2)$, $B_2(\tau_1,\tau_2)$ and their complex conjugates $\bar{B}_1(\tau_1,\tau_2)$, $\bar{B}_2(\tau_1,\tau_2)$ are unknown.

Inserting general solutions (6.83)–(6.84) into equations (6.79)–(6.80) starts the solving procedure of the second-order approximation equations. The secular terms are generated, and elimination of the same is necessary to obtain the solutions that are bounded in time, which leads to the following solvability conditions

$$2\frac{\partial B_1}{\partial \tau_1} - \hat{\omega}_z (j_3 - j_2) B_2 = 0, \tag{6.85}$$

$$2\frac{\partial \bar{B}_1}{\partial \tau_1} - \hat{\omega}_z (j_3 - j_2) \bar{B}_2 = 0, \tag{6.86}$$

$$2\,\mathrm{i}\frac{\partial B_2}{\partial \tau_1} - \mathrm{i}\,\hat{\omega}_z (j_4 - 1) B_1 + 2\hat{\sigma}_2 B_2 = 0, \tag{6.87}$$

$$2\,\mathrm{i}\,\frac{\partial \bar{B}_2}{\partial \tau_1} - \mathrm{i}\,\hat{\omega}_z (j_4 - 1) \bar{B}_1 - 2\hat{\sigma}_2 \bar{B}_2 = 0. \tag{6.88}$$

After elimination of the secular terms, equations (6.79)–(6.80) become homogeneous ones of the form

$$\frac{\partial^2 \phi_2}{\partial \tau_0^2} + \phi_2 = 0, \tag{6.89}$$

$$\frac{\partial^2 \theta_2}{\partial \tau_0^2} + \theta_2 = 0. \tag{6.90}$$

The constant function equal to zero is taken as the particular solution to each of equations (6.89)–(6.90). Inserting solutions (6.83)–(6.84) into equations (6.81)–(6.82) generates the secular terms that should be detected and then eliminated by equalizing to zero. Thus, we obtain the subsequent solvability conditions

$$\begin{aligned} &-2\,\mathrm{i}\frac{\partial B_1}{\partial \tau_2} - \left(\mathrm{i}\,\hat{c}_1 + \left(j_1 - \frac{1}{4}(j_3 - j_2)(j_4 - 1)\right)\hat{\omega}_z^2\right) B_1 + \\ &\frac{\mathrm{i}}{2}(j_3 - j_2)\,\hat{\sigma}_2 \hat{\omega}_z B_2 + \left(3\alpha_1 B_1^2 - 2 j_2 B_2^2\right)\bar{B}_1 - \mathrm{i}\frac{\hat{f}_0}{2} e^{\mathrm{i}\,\varepsilon \hat{\sigma}_1 \tau_0} = 0, \end{aligned} \tag{6.91}$$

$$\begin{aligned} &2\,\mathrm{i}\frac{\partial \bar{B}_1}{\partial \tau_2} + \left(\mathrm{i}\,\hat{c}_1 - \left(j_1 - \frac{1}{4}(j_3 - j_2)(j_4 - 1)\right)\hat{\omega}_z^2\right)\bar{B}_1 - \\ &\frac{\mathrm{i}}{2}(j_3 - j_2)\,\hat{\sigma}_2 \hat{\omega}_z \bar{B}_2 + \left(3\alpha_1 \bar{B}_1^2 - 2 j_2 \bar{B}_2^2\right) B_1 + \mathrm{i}\frac{\hat{f}_0}{2} e^{-\mathrm{i}\,\varepsilon \hat{\sigma}_1 \tau_0} = 0, \end{aligned} \tag{6.92}$$

$$\begin{aligned} &-2\,\mathrm{i}\,\frac{\partial B_2}{\partial \tau_2} - \left(\mathrm{i}\,\hat{c}_2 + \left(j_4 - \frac{1}{4}(j_3 - j_2)(j_4 - 1)\right)\hat{\omega}_z^2 - 2 j_4 B_1 \bar{B}_1\right) B_2 + \\ &\left(3\alpha_2 B_2^2 - j_4 B_1^2\right)\bar{B}_2 - \frac{\mathrm{i}}{2}(j_4 - 1)\,\hat{\sigma}_2 \hat{\omega}_z B_1 = 0, \end{aligned} \tag{6.93}$$

$$2\,\mathrm{i}\,\frac{\partial \bar{B}_2}{\partial \tau_2} + \left(\mathrm{i}\,\hat{c}_2 - \left(j_4 - \frac{1}{4}(j_3 - j_2)(j_4 - 1)\right)\hat{\omega}_z^2 + 2 j_4 B_1 \bar{B}_1\right)\bar{B}_2 + \left(3\alpha_2 \bar{B}_2^2 - j_4 \bar{B}_1^2\right) B_2 + \frac{\mathrm{i}}{2}(j_4 - 1)\,\hat{\sigma}_2 \hat{\omega}_z \bar{B}_1 = 0. \tag{6.94}$$

Eliminating all the secular terms from the equations of the third-order approximation, we get

$$\frac{\partial^2 \phi_3}{\partial \tau_0^2} + \phi_3 = \left(\alpha_1 B_1^2 + 2 j_2 B_2^2\right) B_1 e^{3\mathrm{i}\tau_0} + \left(\alpha_1 \bar{B}_1^2 + 2 j_2 \bar{B}_2^2\right) \bar{B}_1 e^{-3\mathrm{i}\tau_0}, \tag{6.95}$$

$$\frac{\partial^2 \theta_3}{\partial \tau_0^2} + \theta_3 = \left(\alpha_2 B_2^2 - j_4 B_1^2\right) B_2 e^{3\mathrm{i}\tau_0} + \left(\alpha_2 \bar{B}_2^2 - j_4 \bar{B}_1^2\right) \bar{B}_2 e^{-3\mathrm{i}\tau_0}. \tag{6.96}$$

The particular solutions to equations (6.95)–(6.96) are

$$\phi_3(\tau_0, \tau_1, \tau_2) = -\frac{1}{8}\left(\alpha_1 B_1^2 + 2 j_2 B_2^2\right) B_1 e^{3\mathrm{i}\tau_0} - \frac{1}{8}\left(\alpha_1 \bar{B}_1^2 + 2 j_2 \bar{B}_2^2\right) \bar{B}_1 e^{-3\mathrm{i}\tau_0}, \tag{6.97}$$

$$\theta_3(\tau_0, \tau_1, \tau_2) = -\frac{1}{8}\left(\alpha_2 B_2^2 - j_4 B_1^2\right) B_2 e^{3\mathrm{i}\tau_0} - \frac{1}{8}\left(\alpha_2 \bar{B}_2^2 - j_4 \bar{B}_1^2\right) \bar{B}_2 e^{-3\mathrm{i}\tau_0}. \tag{6.98}$$

We present the unknown complex-valued functions $B_1(\tau_1, \tau_2)$, $B_2(\tau_1, \tau_2)$ and their complex conjugates $\bar{B}_1(\tau_1, \tau_2)$, $\bar{B}_2(\tau_1, \tau_2)$ in the exponential form

$$B_j(\tau_1, \tau_2) = \frac{1}{2} b_j(\tau_1, \tau_2) e^{\mathrm{i}\psi_j(\tau_1, \tau_2)}, \quad \bar{B}_j(\tau_1, \tau_2) = \frac{1}{2} b_j(\tau_1, \tau_2) e^{-\mathrm{i}\psi_j(\tau_1, \tau_2)}, \tag{6.99}$$

where the functions $b_j(\tau_1, \tau_2)$, $\psi_j(\tau_1, \tau_2)$ are real-valued, and $j = 1, 2$.

Elimination of the secular terms guarantees not only that solutions of the problem considered are bounded but also gives a way to determine the unknown functions $B_1(\tau_1, \tau_2)$, $B_2(\tau_1, \tau_2)$. Inserting relationships (6.99) into equations (6.85)–(6.88) and (6.91)–(6.94) yields the system of eight partial differential equations of the first-order with unknown functions $b_j(\tau_1, \tau_2)$, $\psi_j(\tau_1, \tau_2)$, $j = 1, 2$. The equations are as follows

$$\left(-\mathrm{i}\frac{\partial b_1}{\partial \tau_1} + b_1 \frac{\partial \psi_1}{\partial \tau_1}\right) e^{\mathrm{i}\,\psi_1} + \frac{1}{2}\mathrm{i}\,\hat{\omega}_z (j_3 - j_2)\, b_2 e^{\mathrm{i}\,\psi_2} = 0, \tag{6.100}$$

$$\left(\mathrm{i}\frac{\partial b_1}{\partial \tau_1} + b_1 \frac{\partial \psi_1}{\partial \tau_1}\right) e^{-\mathrm{i}\,\psi_1} - \frac{1}{2}\mathrm{i}\,\hat{\omega}_z (j_3 - j_2)\, b_2 e^{-\mathrm{i}\,\psi_2} = 0, \tag{6.101}$$

$$\left(-\mathrm{i}\frac{\partial b_2}{\partial \tau_1} + b_2 \left(\frac{\partial \psi_2}{\partial \tau_1} - \hat{\sigma}_2\right)\right) e^{\mathrm{i}\,\psi_2} + \frac{1}{2}\mathrm{i}\,\hat{\omega}_z (j_4 - 1)\, b_1 e^{\mathrm{i}\,\psi_1} = 0, \tag{6.102}$$

$$\left(\mathrm{i}\,\frac{\partial b_2}{\partial \tau_1}+b_2\left(\frac{\partial \psi_2}{\partial \tau_1}-\hat{\sigma}_2\right)\right)e^{-\mathrm{i}\,\psi_2}-\frac{1}{2}\mathrm{i}\,\hat{\omega}_z\,(j_4-1)\,b_1e^{-\mathrm{i}\,\psi_1}=0, \tag{6.103}$$

$$\begin{gathered}\left(-\mathrm{i}\,\frac{\partial b_1}{\partial \tau_2}+b_1\frac{\partial \psi_1}{\partial \tau_2}\right)e^{\mathrm{i}\,\psi_1}+\\ \left(\frac{3}{8}\alpha_1 b_1^2-\frac{1}{2}\mathrm{i}\,\hat{c}_1+\frac{1}{8}\hat{\omega}_z^2\,((j_3-j_2)\,(j_4-1)-4j_1)\right)b_1e^{\mathrm{i}\,\psi_1}-\\ \frac{1}{2}\,\mathrm{i}\,\hat{f}_0e^{\mathrm{i}\,\varepsilon\hat{\sigma}_1\tau_0}+\frac{1}{4}\left(\mathrm{i}\,(j_3-j_2)\,\hat{\sigma}_2\hat{\omega}_z-j_2b_1b_2e^{-\mathrm{i}\,(\psi_1-\psi_2)}\right)b_2e^{\mathrm{i}\,\psi_2}=0,\end{gathered} \tag{6.104}$$

$$\begin{gathered}\left(\mathrm{i}\,\frac{\partial b_1}{\partial \tau_2}+b_1\frac{\partial \psi_1}{\partial \tau_2}\right)e^{-\mathrm{i}\,\psi_1}+\\ \left(\frac{3}{8}\alpha_1 b_1^2+\frac{1}{2}\mathrm{i}\,\hat{c}_1+\frac{1}{8}\hat{\omega}_z^2\,((j_3-j_2)\,(j_4-1)-4j_1)\right)b_1e^{-\mathrm{i}\,\psi_1}+\\ \frac{1}{2}\,\mathrm{i}\,\hat{f}_0e^{-\mathrm{i}\,\varepsilon\hat{\sigma}_1\tau_0}-\frac{1}{4}\left(\mathrm{i}\,(j_3-j_2)\,\hat{\sigma}_2\hat{\omega}_z+j_2b_1b_2e^{\mathrm{i}\,(\psi_1-\psi_2)}\right)b_2e^{-\mathrm{i}\,\psi_2}=0,\end{gathered} \tag{6.105}$$

$$\begin{gathered}\left(-\mathrm{i}\,\frac{\partial b_2}{\partial \tau_2}+b_2\frac{\partial \psi_2}{\partial \tau_2}\right)e^{\mathrm{i}\,\psi_2}+\\ \left(\frac{3}{8}\alpha_2 b_2^2-\frac{1}{2}\mathrm{i}\,\hat{c}_2+\frac{1}{8}\hat{\omega}_z^2\,((j_3-j_2)\,(j_4-1)-4j_4)\right)b_2e^{\mathrm{i}\,\psi_2}-\\ \frac{1}{4}\mathrm{i}\,(j_4-1)\,\hat{\sigma}_2\hat{\omega}_zb_1e^{\mathrm{i}\,\psi_1}+\frac{j_4}{8}\left(2-e^{2\,\mathrm{i}\,(\psi_1-\psi_2)}\right)b_1^2b_2e^{\mathrm{i}\,\psi_2}=0,\end{gathered} \tag{6.106}$$

$$\begin{gathered}\left(\mathrm{i}\,\frac{\partial b_2}{\partial \tau_2}+b_2\frac{\partial \psi_2}{\partial \tau_2}\right)e^{-\mathrm{i}\,\psi_2}+\\ \left(\frac{3}{8}\alpha_2 b_2^2+\frac{1}{2}\mathrm{i}\,\hat{c}_2+\frac{1}{8}\hat{\omega}_z^2\,((j_3-j_2)\,(j_4-1)-4j_4)\right)b_2e^{-\mathrm{i}\,\psi_2}+\\ \frac{1}{4}\mathrm{i}\,(j_4-1)\,\hat{\sigma}_2\hat{\omega}_zb_1e^{-\mathrm{i}\,\psi_1}+\frac{j_4}{8}\left(2-e^{-2\,\mathrm{i}\,(\psi_1-\psi_2)}\right)b_1^2b_2e^{-\mathrm{i}\,\psi_2}=0.\end{gathered} \tag{6.107}$$

Let us solve the system of equations (6.100)–(6.107) with respect to the partial derivatives

$$\frac{\partial b_j}{\partial \tau_1},\quad \frac{\partial b_j}{\partial \tau_2},\quad \frac{\partial \psi_j}{\partial \tau_1},\quad \frac{\partial \psi_j}{\partial \tau_2},\quad j=1,2.$$

The solution is as follows

$$\frac{\partial b_1}{\partial \tau_1}=\frac{1}{2}\hat{\omega}_z\,(j_3-j_2)\,b_2\cos\left(\psi_1-\psi_2\right), \tag{6.108}$$

$$\frac{\partial b_1}{\partial \tau_2} = -\frac{1}{2}\hat{c}_1 b_1 - \frac{1}{2}\hat{f}_0 \cos(\varepsilon\hat{\sigma}_1\tau_0 - \psi_1) + \frac{1}{4} j_2 b_1 b_2^2 \sin(2(\psi_1 - \psi_2)) + \frac{1}{4}(j_3 - j_2)\hat{\sigma}_2\hat{\omega}_z b_2 \cos(\psi_1 - \psi_2), \tag{6.109}$$

$$\frac{\partial b_2}{\partial \tau_1} = \frac{1}{2}\hat{\omega}_z (j_4 - 1) b_1 \cos(\psi_1 - \psi_2), \tag{6.110}$$

$$\frac{\partial b_2}{\partial \tau_2} = -\frac{1}{2}\hat{c}_2 b_2 - \frac{1}{4}(j_4 - 1)\hat{\sigma}_2\hat{\omega}_z b_1 \cos(\psi_1 - \psi_2) - \frac{1}{8} j_4 b_1^2 b_2 \sin(2(\psi_1 - \psi_2)), \tag{6.111}$$

$$\frac{\partial \psi_1}{\partial \tau_1} = -\frac{1}{2}(j_3 - j_2)\hat{\omega}_z \frac{b_2}{b_1} \sin(\psi_1 - \psi_2), \tag{6.112}$$

$$\frac{\partial \psi_1}{\partial \tau_2} = -\frac{3}{8}\alpha_1 b_1^2 - \frac{\hat{f}_0}{2\, b_1} \sin(\varepsilon\hat{\sigma}_1\tau_0 - \psi_1) + \frac{1}{4} j_2 b_2^2 \cos(2(\psi_1 - \psi_2)) - \frac{(j_3 - j_2)\hat{\sigma}_2\hat{\omega}_z}{4b_1} b_2 \sin(\psi_1 - \psi_2) + \frac{1}{2}\left(j_1 - \frac{1}{4}(j_3 - j_2)(j_4 - 1)\right)\hat{\omega}_z^2 \tag{6.113}$$

$$\frac{\partial \psi_2}{\partial \tau_1} = \hat{\sigma}_2 + \frac{1}{2}(j_4 - 1)\hat{\omega}_z \frac{b_1}{b_2} \sin(\psi_1 - \psi_2), \tag{6.114}$$

$$\frac{\partial \psi_2}{\partial \tau_2} = -\frac{3}{8}\alpha_2 b_2^2 + \frac{1}{8} j_4 b_1^2 (\cos(2(\psi_1 - \psi_2)) - 2) - \frac{(j_4 - 1)\hat{\sigma}_2\hat{\omega}_z}{4b_2} b_1 \sin(\psi_1 - \psi_2) + \frac{1}{2}\left(j_4 - \frac{1}{4}(j_3 - j_2)(j_4 - 1)\right)\hat{\omega}_z^2. \tag{6.115}$$

It is convenient to transform equations (6.108)–(6.115) onto the problem formulated for the time variable τ. The sought functions $b_j(\tau_1, \tau_2)$, $\psi_j(\tau_1, \tau_2)$, $j = 1, 2$, depend only on both slow time variables. Referring to definition (6.25), one can write formally the derivatives of these functions with respect to the time variable τ in the form

$$\frac{db_j}{d\tau} = \varepsilon \frac{\partial b_i}{\partial \tau_1} + \varepsilon^2 \frac{\partial b_j}{\partial \tau_2}, \quad j = 1, 2, \tag{6.116}$$

$$\frac{d\psi_j}{d\tau} = \varepsilon \frac{\partial \psi_j}{\partial \tau_1} + \varepsilon^2 \frac{\partial \psi_j}{\partial \tau_2}, \quad j = 1, 2. \tag{6.117}$$

Inserting solutions (6.108)–(6.115) into equations (6.116)–(6.117) and applying inverse relationships (6.29) and (6.73), one gains four ordinary differential equations of the first-order, as follows

$$\begin{aligned}\frac{da_1}{d\tau} &= -\frac{c_1}{2}a_1(\tau) - \frac{f_0}{2}\cos(\sigma_1\tau - \psi_1(\tau)) + \\ &\frac{\omega_z}{4}(j_3 - j_2)(\sigma_2 + 2)a_2(\tau)\cos(\psi_1(\tau) - \psi_2(\tau)) + \\ &\frac{j_2}{4}a_1(\tau)a_2^2(\tau)\sin(2(\psi_1(\tau) - \psi_2(\tau))),\end{aligned} \tag{6.118}$$

$$\begin{aligned}\frac{da_2}{d\tau} = &-\frac{c_2}{2}a_2(\tau) - \frac{\omega_z}{4}(j_4 - 1)(\sigma_2 - 2)a_1(\tau)\cos(\psi_1(\tau) - \psi_2(\tau)) - \\ &\frac{j_4}{8}a_1^2(\tau)a_2(\tau)\sin(2(\psi_1(\tau) - \psi_2(\tau))),\end{aligned} \tag{6.119}$$

$$\begin{aligned}\frac{d\psi_1}{d\tau} &= -\frac{f_0}{2a_1(\tau)}\sin(\sigma_1\tau - \psi_1(\tau)) - \frac{3}{8}\alpha_1 a_1^2(\tau) + \\ &\frac{j_2}{4}a_2^2(\tau)\cos(2(\psi_1(\tau) - \psi_2(\tau))) - \\ &\frac{(j_3 - j_2)(\sigma_2 + 2)\omega_z}{4a_1(\tau)}a_2(\tau)\sin(\psi_1(\tau) - \psi_2(\tau)) + \\ &\frac{1}{2}\left(j_1 - \frac{1}{4}(j_3 - j_2)(j_4 - 1)\right)\hat{\omega}_z^2,\end{aligned} \tag{6.120}$$

$$\begin{aligned}\frac{d\psi_2}{d\tau} = \;&\sigma_2 - \frac{3}{8}\alpha_2 a_2^2(\tau) + \frac{j_4}{8}a_1^2(\tau)(\cos(2(\psi_1(\tau) - \psi_2(\tau))) - 2) - \\ &\frac{(j_4 - 1)(\sigma_2 - 2)\omega_z}{4a_2(\tau)}a_1(\tau)\sin(\psi_1(\tau) - \psi_2(\tau)) + \\ &\frac{1}{2}\left(j_4 - \frac{1}{4}(j_3 - j_2)(j_4 - 1)\right)\hat{\omega}_z^2,\end{aligned} \tag{6.121}$$

where $a_j(\tau) = \varepsilon b_j(\tau)$, $j = 1,2$.

The system of equations (6.118)–(6.121) is supplemented by the following initial conditions

$$a_1(0) = a_{10}, \quad a_2(0) = a_{20}, \quad \psi_1(0) = \psi_{10}, \quad \psi_2(0) = \psi_{20}, \tag{6.122}$$

where the quantities a_{10}, a_{20}, ψ_{10}, ψ_{20} are compatible with the initial values u_{01}, u_{02}, u_{03}, u_{04} occurring in conditions (6.23).

Although differential equations (6.118)–(6.121) are written using derivatives with respect to the time τ, they govern the evolution concerning the slow time variables because the sought functions depend a priori only on τ_1 and τ_2.

The essential difference between the modulation equations related to the resonant case and the previously discussed equations (6.59)–(6.62) lies in the occurrence of the trigonometric functions with the arguments containing the unknown phases $\psi_1(\tau)$, $\psi_2(\tau)$. That circumstance shows that the initial-value problem (6.118)–(6.121) cannot be solved analytically.

Substituting solutions (6.83)–(6.84) and (6.97)–(6.98) into asymptotic expansions (6.27)–(6.28) and then employing formulas (6.72)–(6.73) and (6.29), the last inversely, we obtain the approximate solution describing the problem of the micromechanical gyroscope vibration in the doubled resonance. This solution is as follows

$$\varphi(\tau) = a_1(\tau)\cos(\tau+\psi_1(\tau)) - \frac{\alpha_1}{32}a_1^3(\tau)\cos(3\tau+3\psi_1(\tau)) - \frac{j_2}{16}a_1(\tau)a_2^2\cos(3\tau+\psi_1(\tau)+2\psi_2(\tau)), \tag{6.123}$$

$$\vartheta(\tau) = a_2(\tau)\cos(\tau+\psi_2(\tau)) - \frac{\alpha_2}{32}a_2^3(\tau)\cos(3\tau+3\psi_2(\tau)) + \frac{j_4}{32}a_1^2(\tau)a_2(\tau)\cos(3\tau+2\psi_1(\tau)+\psi_2(\tau)), \tag{6.124}$$

where $a_1(\tau)$, $a_2(\tau)$ and $\psi_1(\tau)$, $\psi_2(\tau)$ are numerically obtained solutions to the slow modulation problem defined by equations (6.118)–(6.121) and (6.122).

The functions $a_1(\tau)$ and $\psi_1(\tau)$ stand for strictly the amplitude and the phase, respectively, of the first term of asymptotic expansion (6.123). Similarly, $a_2(\tau)$ and $\psi_2(\tau)$ depict the slowly changing amplitude and phase, respectively, of the first term of asymptotic expansion (6.124). Thus, the functions describe the slow evolution of the amplitudes and phases in time.

The simulation concerning the simultaneously occurring external and internal resonances for the micromechanical gyroscope was carried out for the following values of parameters: $\alpha_1 = 2.5$, $\alpha_2 = 2.5$, $j_1 = 0.95$, $j_2 = 0.9$, $j_3 = 1$, $j_4 = 0.9$, $c_1 = 0.001$, $c_2 = 0.001$, $f_0 = 0.0001$, $\sigma_1 = 0.015$, $\sigma_2 = 0.015$, $\omega_z = 0.03$.

The initial values for motion equations (6.74)–(6.75) are as follows: $u_{01} = 0.02$, $u_{02} = 0$, $u_{03} = 0.02$, $u_{04} = 0$.

In turn, the following initial values related to the modulation problem: $a_{10} = 2.00011\times10^{-2}$, $a_{20} = 2.00004\times 10{-}2$, $\psi_{10} = -0.00148889$, $\psi_{20} = -0.00195973$ remain in compliance with the initial values of the amplitudes and phases.

The approximate solutions obtained using MSM and numerically employing the standard procedure *NDSolve* of the *Mathematica* software have been compared. In Figures 6.10–6.13, the solid black line depicts the approximate analytical solution obtained using MSM whereas the numerical solution is drawn by the dotted light gray line. Figures 6.10–6.11 present the initial course of oscillations of both parts of the micro gyroscope. The graphs presented in Figures 6.12–6.13 show the oscillations after a certain time from the starting point when the values of the amplitudes decrease already but the periodic modulation of the amplitudes is still observable. High compliance between the numerical and the asymptotic solutions on the whole range is worth taking note of.

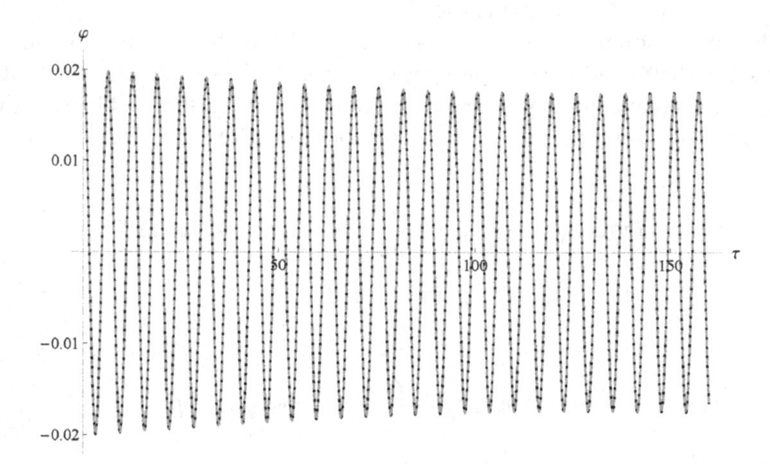

FIGURE 6.10 The oscillations $\varphi(\tau)$ of the intermediate frame at the beginning of motion estimated analytically (MSM) and numerically (dotted line).

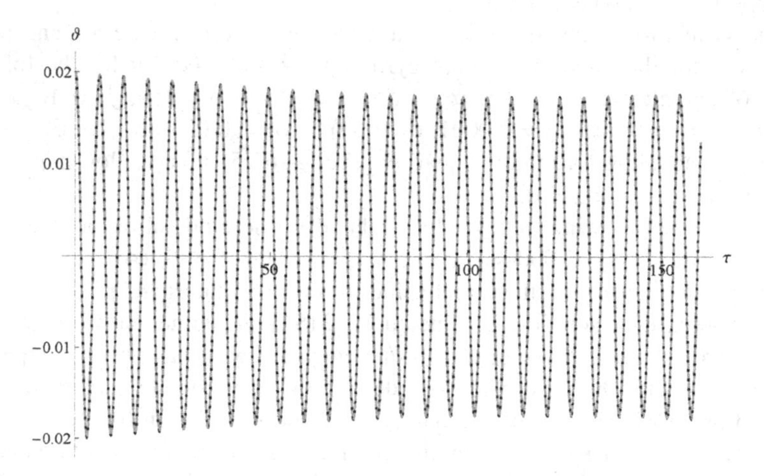

FIGURE 6.11 The oscillations $\vartheta(\tau)$ of the sensing plate at the beginning of motion estimated analytically (MSM) and numerically (dotted line).

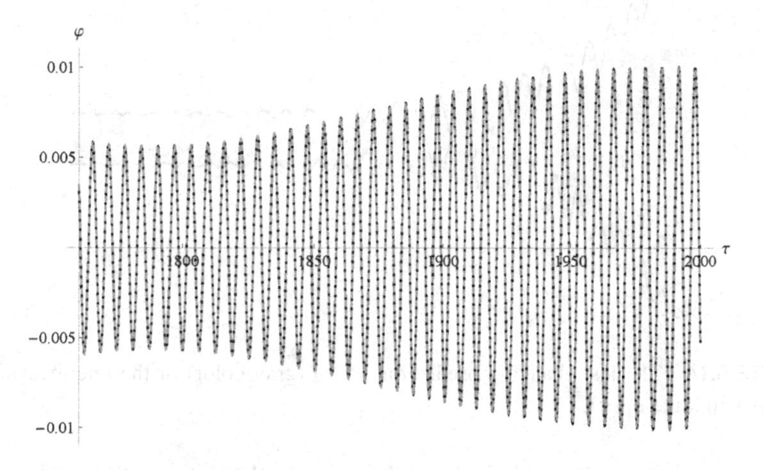

FIGURE 6.12 The resonant oscillations $\varphi(\tau)$ of the intermediate gimbal with clear amplitude modulation estimated analytically (MSM) and numerically (dotted line).

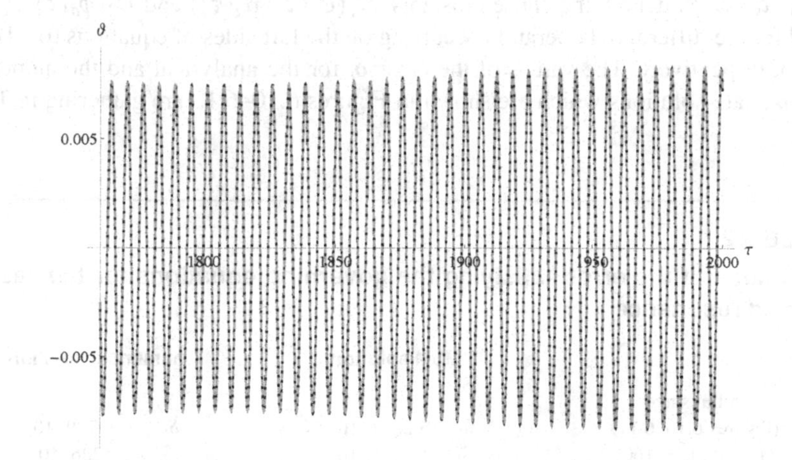

FIGURE 6.13 The resonant oscillations $\vartheta(\tau)$ of the sensing plate with amplitude modulation estimated analytically (MSM) and numerically (dotted line).

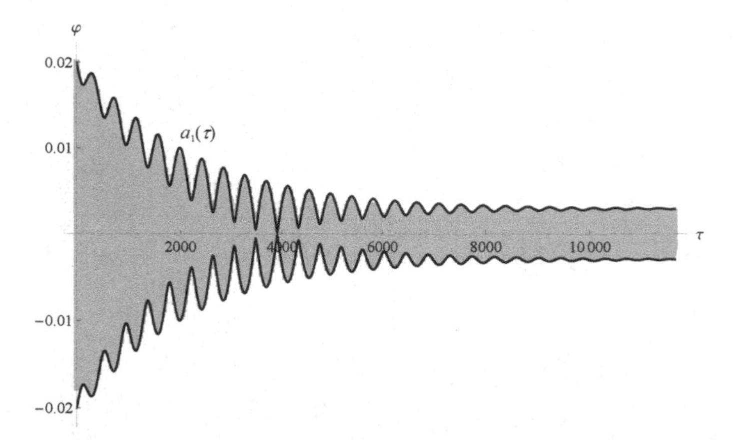

FIGURE 6.14 The fast-changing oscillations $\varphi(\tau)$ (gray color) of the intermediate frame and the amplitude $a_1(\tau)$.

The accuracy of the analytical and the numerical solutions was quantified employing the error satisfying equations (6.74)–(6.75). The error is defined as follows

$$\delta_i = \sqrt{\frac{1}{\tau_e - \tau_s}\int_{\tau_s}^{\tau_e} \left(H_i\left(\varphi_a(\tau), \vartheta_a(\tau)\right) - 0\right)^2 d\tau}, i = 1,2, \tag{6.125}$$

where $\varphi_a(\tau)$, $\vartheta_a(\tau)$ are the approximate solutions obtained using the MSM or numerically, τ_s and τ_e denote the chosen instants, $H_1(\varphi_a(\tau), \vartheta_a(\tau))$ and $H_2(\varphi_a(\tau), \vartheta_a(\tau))$ stand for the differential operators occurring on the left sides of equations (6.74) and (6.75), respectively. The values of the error δ_i for the analytical and the numerical approximate solutions which are shown in Figures 6.10–6.13 are gathering in Table 6.2.

TABLE 6.2

The values of the error satisfying the governing equations for the case of doubled resonance

	MSM solution	**numerical solution**
initial stage (Figures 6.10, 6.11) $\tau_0 = 0,\ \tau_e = 160$	$\delta_1 = 5.2887 \cdot 10^{-7}$ $\delta_2 = 1.0065 \cdot 10^{-7}$	$\delta_1 = 1.47799 \cdot 10^{-4}$ $\delta_2 = 1.5226 \cdot 10^{-4}$
end of the simulation (Figures 6.12, 6.13) $\tau_0 = 1760,\ \tau_e = 2000$	$\delta_1 = 5.41512 \cdot 10^{-7}$ $\delta_2 = 1.07535 \cdot 10^{-7}$	$\delta_1 = 2.08634 \cdot 10^{-7}$ $\delta_2 = 2.28243 \cdot 10^{-7}$

At the beginning of the micromechanical gyroscope motion, the error of the solution obtained using MSM is distinctly smaller than the error calculated for the numerical solution. At the end of the motion simulation, the tendency reverses itself and the numerical solution turns out to be slightly more accurate.

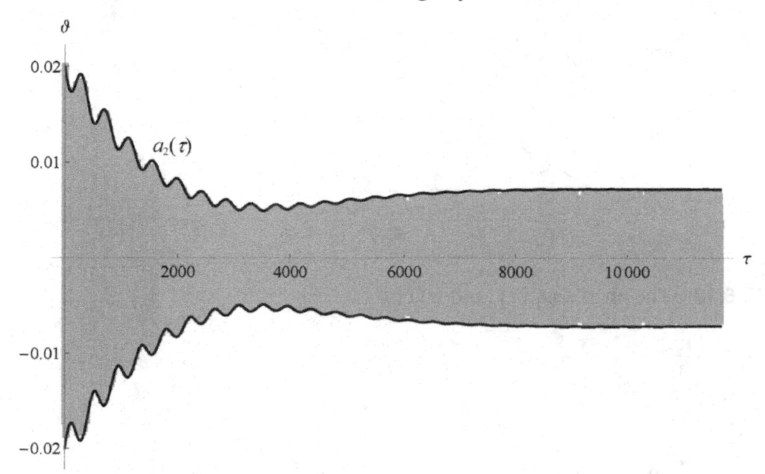

FIGURE 6.15 The fast-changing oscillations $\vartheta(\tau)$ (gray color) of the sensing plate and the amplitude $a_2(\tau)$ for transient modulation.

The time intervals on which the solutions are drawn in Figures 6.10–6.13 are relatively short, too short to exemplify the slow changes of the amplitudes. The time courses of the generalized coordinates $\varphi(\tau)$ and $\vartheta(\tau)$ on longer ranges are presented in Figures 6.14–6.15. The approximate analytical solutions obtained using MSM are drawn using the gray line. Because of the fast variability of the solutions, the line merges into one area. The graphs of the amplitudes $a_1(\tau), -a_1(\tau)$ and $a_2(\tau), -a_2(\tau)$ are drawn with the black line. Strictly, these lines present the amplitudes of the first terms of both solutions (6.123)–(6.124). Nevertheless, they can be treated in the approximate sense as the amplitudes of the whole approximate solutions. This circumstance confirms that the subsequent terms of the uniformly valid expansion are significantly smaller than its first term for each of the solutions given by (6.123)–(6.124).

The modulation of the amplitudes $a_1(\tau)$ and $a_2(\tau)$ presented in Figures 6.14, 6.15 are transient initially. The slow oscillations of both amplitudes fix themselves on a longer time range. On the whole time range, the amplitudes $a_1(\tau)$ and $a_2(\tau)$ perform with great accuracy the role of the amplitudes of the approximate solution.

In Figure 6.16 the time variability of the functions $\psi_1(\tau)$ and $\psi_2(\tau)$ is shown. The functions depict the phases of the first term of asymptotic expansion (6.123) and (6.124), respectively. On a sufficiently long time range, the rate of the growth of each phase fixes. At the transient oscillations, the increase of each of the phases is of a different course, but with time the plots become almost parallel to each other, so the phases have approximately the same rate of growth. The graph presenting the difference between the phases $\psi_1(\tau)$ and $\psi_2(\tau)$ when they vary almost linearly is

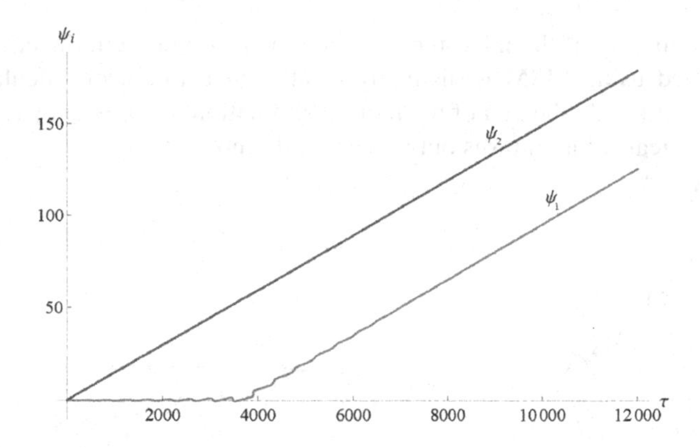

FIGURE 6.16 The phases $\psi_1(\tau)$ and $\psi_2(\tau)$.

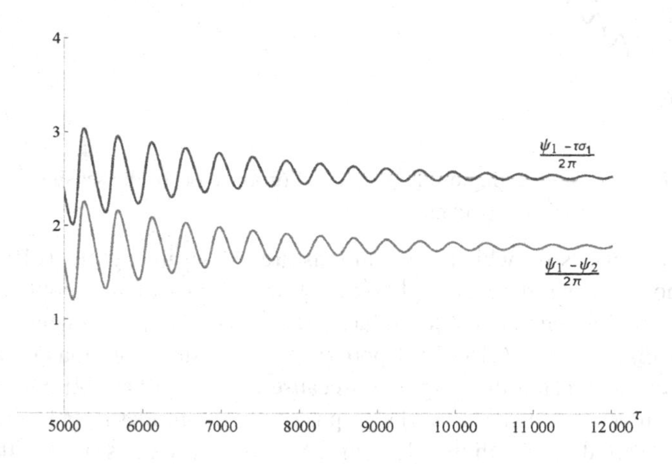

FIGURE 6.17 Convergence of the differences $\psi_1(\tau) - \psi_2(\tau)$ and $\psi_1(\tau) - \sigma_1\tau$ with time.

shown in Figure 6.17. The variability of the difference $\psi_1(\tau) - \sigma_1\tau$ is also shown in the figure. To eliminate the impact of the transient state, both differences have been divided by 2π.

6.7 STEADY-STATE RESPONSES IN SIMULTANEOUSLY OCCURRING RESONANCES

The initial value problem given by equations (6.118)–(6.121) and (6.122), arose in the process of solving equations of motion (6.74)–(6.79) using MSM, and it is a good basis for studying the steady-state forced vibration of the micromechanical gyroscope. For this purpose, it is convenient to introduce the modified phases. The quantities $\gamma_1(\tau)$ and $\gamma_2(\tau)$ defined as follows

$$\gamma_1(\tau) = \sigma_1\tau - \psi_1(\tau), \gamma_2(\tau) = \psi_1(\tau) - \psi_2(\tau) \tag{6.126}$$

converge with time to fixed values, which is shown in Figure 6.17. So, they are assumed as the modified phases. Applying equations (6.126) to modulation equations (6.118)–(6.121) yields

$$\begin{aligned} \frac{da_1}{d\tau} &= -\frac{c_1}{2}a_1(\tau) - \frac{f_0}{2}\cos(\gamma_1(\tau)) + \\ &\frac{\omega_z}{4}(j_3 - j_2)(\sigma_2 + 2)a_2(\tau)\cos(\gamma_2(\tau)) + \\ &\frac{j_2}{4}a_1(\tau)a_2^2(\tau)\sin(2\gamma_2(\tau)), \end{aligned} \tag{6.127}$$

$$\begin{aligned} \frac{da_2}{d\tau} &= -\frac{c_2}{2}a_2(\tau) - \frac{\omega_z}{4}(j_4 - 1)(\sigma_2 - 2)a_1(\tau)\cos(\gamma_2(\tau)) - \\ &\frac{j_4}{8}a_1^2(\tau)a_2(\tau)\sin(2\gamma_2(\tau)), \end{aligned} \tag{6.128}$$

$$\begin{aligned} \frac{d\gamma_1}{d\tau} &= \sigma_1 + \frac{3}{8}\alpha_1 a_1^2(\tau) + \frac{f_0}{2a_1(\tau)}\sin(\gamma_1(\tau)) - \frac{j_2}{4}a_2^2(\tau)\cos(2\gamma_2(\tau)) + \\ &\frac{(j_3 - j_2)(\sigma_2 + 2)\omega_z}{4a_1(\tau)}a_2(\tau)\sin(\gamma_2(\tau)) - \frac{1}{2}\left(j_1 - \frac{1}{4}(j_3 - j_2)(j_4 - 1)\right)\hat{\omega}_z^2, \end{aligned} \tag{6.129}$$

$$\begin{aligned} \frac{d\gamma_2}{d\tau} &= -\sigma_2 + \frac{j_1 - j_4}{2}\hat{\omega}_z^2 + \frac{3}{8}\left(\alpha_2 a_2^2(\tau) - \alpha_1 a_1^2(\tau)\right) - \frac{f_0}{2a_1(\tau)}\sin(\gamma_1(\tau)) + \\ &\frac{j_2}{4}a_2^2(\tau)\cos(2\gamma_2(\tau)) + \frac{j_4}{8}a_1^2(\tau)(2 - \cos(2\gamma_2(\tau))) + \\ &\frac{\omega_z}{4}\left(\frac{(j_4 - 1)(\sigma_2 - 2)}{a_2(\tau)}a_1(\tau) - \frac{(j_3 - j_2)(\sigma_2 + 2)}{a_1(\tau)}a_2(\tau)\right)\sin(\gamma_2(\tau)). \end{aligned} \tag{6.130}$$

The autonomous form of modulation equations (6.127)–(6.130) is suitable to analyze the steady-state motion. When all transient processes disappear the oscillations of the forced system can achieve the steady-state. Unchanging values of the amplitudes and the modified phases are the symptoms of this state. Therefore, formulating the conditions for the steady-state we require the derivatives of the amplitudes and modified phases to be equal to zero. This way we obtain the following system of four equations

$$2c_1a_1 + 2f_0\cos\gamma_1 - \omega_z(j_3 - j_2)(\sigma_2 + 2)a_2\cos\gamma_2 - j_2a_1a_2^2\sin(2\gamma_2) = 0, \tag{6.131}$$

$$4c_2a_2+2\omega_z(j_4-1)(\sigma_2-2)a_1\cos\gamma_2+j_4a_1^2a_2\sin(2\gamma_2)=0, \tag{6.132}$$

$$\begin{gathered}8\sigma_1a_1+3\alpha_1a_1^3+4f_0\sin\gamma_1-2j_2a_1a_2^2\cos(2\gamma_2)+\\ 2(j_3-j_2)(\sigma_2+2)\omega_za_2\sin\gamma_2-(4j_1-(j_3-j_2)(j_4-1))a_1\omega_z^2=0,\end{gathered} \tag{6.133}$$

$$\begin{gathered}-8\sigma_2a_1a_2+4(j_1-j_4)\omega_z^2a_1a_2+3\left(\alpha_2a_2^2-\alpha_1a_1^2\right)a_1a_2-4f_0a_2\sin\gamma_1+\\ 2j_2a_1a_2^3\cos(2\gamma_2)+j_4a_2a_1^3(2-\cos(2\gamma_2))+\\ 2\omega_z\left((j_4-1)(\sigma_2-2)a_1^2-(j_3-j_2)(\sigma_2+2)a_2^2\right)\sin\gamma_2=0,\end{gathered} \tag{6.134}$$

where symbols a_1, a_2 and θ_1,θ_2 denote amplitudes and phases corresponding to the steady-state.

The values satisfying the system (6.127)–(6.130) are the fixed amplitudes and modified phases in the case of the main and internal resonances occurring simultaneously.

There are both algebraic and trigonometric nonlinearities in equations (6.131)–(6.134). The following substitutions

$$\sin\gamma_i=\frac{2t_i}{1+t_i^2},\quad \cos\gamma_i=\frac{1-t_i^2}{1+t_i^2},\quad i=1,2, \tag{6.135}$$

where $t_i=\tan\left(\frac{\gamma_i}{2}\right)$, allow one to transform equations (6.131)–(6.134) to equivalent ones of the pure algebraic form. The system of four algebraic equations describing the steady state is as follows

$$\begin{gathered}2\left(2j_2a_2^2\left(t_2^2-1\right)t_2+c_1\left(1+t_2^2\right)^2\right)a_1\left(1+t_1^2\right)-\\ \left(1+t_2^2\right)\left(2f_0\left(t_1^2-1\right)\left(1+t_2^2\right)-(j_3-j_2)(\sigma_2+2)\omega_za_2\left(1+t_1^2\right)\left(t_2^2-1\right)\right)=0,\end{gathered} \tag{6.136}$$

$$\begin{gathered}2\left(1+t_1^2\right)\left(-2j_4t_2\left(t_2^2-1\right)a_1^2a_2+2c_2\left(1+t_2^2\right)^2a_2-\right.\\ \left.(j_4-1)(\sigma_2-2)\omega_z\left(t_2^4-1\right)a_1\right)=0,\end{gathered} \tag{6.137}$$

$$\begin{aligned}&3\alpha_1\left(1+t_1^2\right)\left(1+t_2^2\right)^2a_1^3+8f_0t_1\left(1+t_2^2\right)^2+\\ &4(j_3-j_2)(\sigma_2+2)\omega_z\left(1+t_1^2\right)\left(t_2+t_2^3\right)a_2+\\ &\left(8\sigma_1-(4j_1-(j_3-j_2)(j_4-1))\omega_z^2\right)\left(1+t_1^2\right)\left(1+t_2^2\right)^2a_1-\\ &2j_2\left(1+t_1^2\right)\left(1-6t_2^2+t_2^4\right)a_1a_2^2=0,\end{aligned} \tag{6.138}$$

$$\begin{aligned}&\left(4\left(j_1-j_4\right)\omega_z^2-8\sigma_2+3\left(\alpha_2a_2^2-\alpha_1a_1^2\right)\right)\left(1+t_1^2\right)\left(1+t_2^2\right)^2a_1a_2-\\&\quad 8f_0t_1\left(1+t_2^2\right)^2a_2+2j_4\left(1+t_1^2\right)\left(1+t_2^2\right)^2a_2a_1^3+\\&\quad\left(2j_2a_2^2-j_4a_1^2\right)\left(1+t_1^2\right)\left(1-6t_2^2+t_2^4\right)a_1a_2+\\&4\omega_z\left(\left(j_4-1\right)\left(\sigma_2-2\right)a_1^2-\left(j_3-j_2\right)\left(\sigma_2+2\right)a_2^2\right)\left(1+t_1^2\right)\left(t_2+t_2^3\right)=0.\end{aligned} \tag{6.139}$$

The solution to the system of nonlinear equations (6.136)–(6.139) determines the values of the amplitudes and the modified phases when the main and the internal resonances occur simultaneously.

Let us analyze the resonant response of the micromechanical gyroscope in the general case, i.e. without any additional assumptions about its properties. The following values of the parameters are fixed: $\alpha_1=1.38$, $\alpha_2=2.19$, $j_1=0.17436$, $j_2=0$, $j_3=0.845$, $j_4=0$, $c_1=3.21\cdot10^{-5}$, $c_2=2.44\cdot10^{-5}$, $f_0=4.36\cdot10^{-7}$, $\sigma_2=0.000032$, $\omega_z=0.000062606$.

The value of the detuning parameter σ_1 is increased regularly by 0.000001, starting from $\sigma_1=-0.0001$. The resonant response curves are obtained solving equations (6.136)–(6.139) with help of the procedure *NSolve* offered in *Mathematica* 12.0. The results are presented in Figure 6.18.

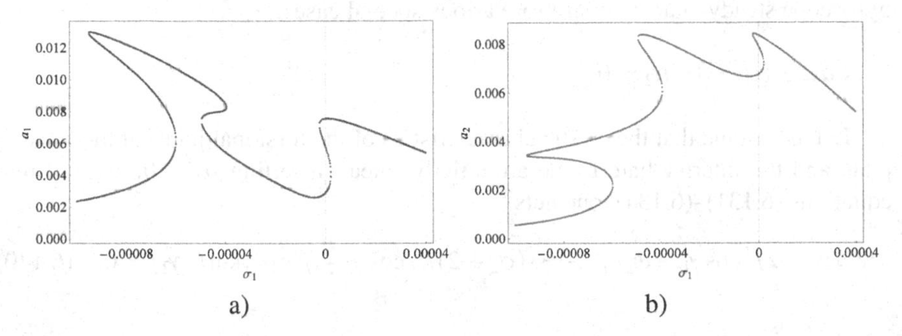

FIGURE 6.18 Resonance curves: (a) for the intermediate frame vibration; (b) for the sensing element vibration.

The resonant responses of the system exhibit high coincidence with the numerical solution of the equations of motion. The simulation was carried out assuming the same values of the parameters. Additionally, it is assumed: $\sigma_1=0$, $a_{10}=0.01$, $a_{20}=0.01$, $\psi_{10}=0$, $\psi_{20}=0$.

Time histories of the generalized coordinates for this data are given in Figure 6.19. Both graphs present the vibration just before the end of simulation which was realized for $\tau\in(0,0.7\times10^5)$. The amplitudes read from the resonance curves (Fig. 6.18) are $a_1=0.0028877$ and $a_2=0.0067565$, whereas the amplitude obtained from the time histories (Fig. 6.19) are $a_1=0.0028858$ and $a_2=0.0067554$ which confirms the good agreement between the results obtained analytically and numerically.

The values of error (6.125) for the approximate solution obtained using MSM are as follows

$$\delta_1=5.2955\times10^{-12},\qquad \delta_2=4.3024\times10^{-12},$$

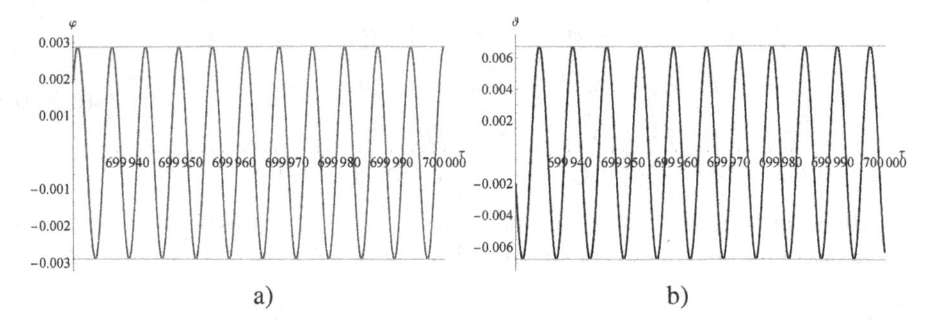

FIGURE 6.19 Time histories of the generalized coordinates of $\varphi(\tau)$ (a) and $\vartheta(\tau)$ (b); gray lines indicate the amplitude.

while the values of the error for the numerical results obtained using *Mathematica* software are

$$\delta_1 = 1.6802 \times 10^{-7}, \qquad \delta_2 = 2.9094 \times 10^{-7}.$$

Equations (6.131)–(6.134) allow for a complete analysis of the micromechanical gyroscope steady-state response for various special cases.

Case 1 $(\alpha_1 = 0,\ \alpha_2 = 0)$

Let us assume that the elastic characteristics of the torsional joints at the sensing plate and the intermediate frame are strictly linear. Inserting $\alpha_1 = 0,\ \alpha_2 = 0$ into equations (6.131)–(6.134), one gets

$$2c_1a_1 + 2f_0\cos\gamma_1 - \omega_z(j_3 - j_2)(\sigma_2 + 2)a_2\cos\gamma_2 - j_2a_1a_2^2\sin(2\gamma_2) = 0, \tag{6.140}$$

$$4c_2a_2 + 2\omega_z(j_4 - 1)(\sigma_2 - 2)a_1\cos\gamma_2 + j_4a_1^2a_2\sin(2\gamma_2) = 0, \tag{6.141}$$

$$\begin{gathered} 8\sigma_1a_1 + 4f_0\sin\gamma_1 - 2j_2a_1a_2^2\cos(2\gamma_2) + \\ 2(j_3 - j_2)(\sigma_2 + 2)\omega_za_2\sin\gamma_2 - (4j_1 - (j_3 - j_2)(j_4 - 1))a_1\omega_z^2 = 0, \end{gathered} \tag{6.142}$$

$$\begin{gathered} -8\sigma_2a_1a_2 + 4(j_1 - j_4)\omega_z^2a_1a_2 - 4f_0a_2\sin\gamma_1 + \\ 2j_2a_1a_2^3\cos(2\gamma_2) + j_4a_2a_1^3(2 - \cos(2\gamma_2)) + \\ 2\omega_z\left((j_4 - 1)(\sigma_2 - 2)a_1^2 - (j_3 - j_2)(\sigma_2 + 2)a_2^2\right)\sin\gamma_2 = 0. \end{gathered} \tag{6.143}$$

The assumption about vanishing the nonlinear terms does not cause any significant simplifications. The modulation equations are still nonlinear and a majority of the nonlinear terms are conditioned by the inertial properties of the sensing plate.

Case 2 $(\alpha_1 = 0,\ \alpha_2 = 0,\ j_2 = 0,\ j_4 = 0)$

Observe that the steady-state equations become much simpler when additionally $j_2 = j_4 = 0$. These relations are satisfied if the components J_x and J_y of the diagonal inertia tensor $\hat{\mathbf{J}}$ of the sensing plate are equal to each other. The modulation equations take then, the following form

$$2c_1 a_1 + 2f_0 \cos\gamma_1 - \omega_z j_3 (\sigma_2 + 2) a_2 \cos\gamma_2 = 0, \tag{6.144}$$

$$4c_2 a_2 - 2\omega_z (\sigma_2 - 2) a_1 \cos\gamma_2 = 0, \tag{6.145}$$

$$8\sigma_1 a_1 + 4f_0 \sin\gamma_1 + 2j_3 (\sigma_2 + 2)\,\omega_z a_2 \sin\gamma_2 - (4j_1 + j_3) a_1 \omega_z^2 = 0, \tag{6.146}$$

$$\begin{gathered} -8\sigma_2 a_1 a_2 + 4(j_1)\,\omega_z^2 a_1 a_2 - 4f_0 a_2 \sin\gamma_1 - \\ 2\omega_z \left((\sigma_2 - 2) a_1^2 + j_3 (\sigma_2 + 2) a_2^2\right) \sin\gamma_2 = 0, \end{gathered} \tag{6.147}$$

The modified phases can be eliminated from equations (6.144)–(6.147) using trigonometric identities which allows one to express explicitly, the amplitude-frequency dependencies

$$\frac{a_2^2\left(16c_2^2 + (8\sigma_2 - 8\sigma_1 + j_3\omega_z^2)^2\right)}{4a_1^2(\sigma_2 - 2)^2\omega_z^2} = 1, \tag{6.148}$$

$$\begin{gathered} \left(4a_1 j_1 \omega_z^2 - 8\sigma_1 a_1 + \omega_z^2 a_1 j_3 + \frac{a_2^2 j_3 (2 + \sigma_2)(8\sigma_2 - 8\sigma_1 + j_3\omega_z^2)}{a_1(\sigma_2 - 2)}\right)^2 + \\ \frac{16\left(c_1(\sigma_2 - 2)a_1^2 + c_2(2 + \sigma_2) j_3 a_2^2\right)^2}{a_1^2(\sigma_2 - 2)^2} = 16f_0^2. \end{gathered} \tag{6.149}$$

We get the system of two algebraic equations of the 8-th order with respect to the unknown amplitudes a_1 and a_2. The resonance curves obtained as a result of solving system (6.148)–(6.149) for the following data of parameters $f_0 = 9\cdot 10^{-7}$, $c_1 = 10^{-5}$, $c_2 = 10^{-5}$, $j_1 = 0.00826$, $j_3 = 0.9917$ and for several values of ω_z are presented in Figure 6.20.

The full symmetry of the graphs depicted in Figure 6.20 is typical behavior of the considered two degrees-of-freedom linear system.

The mentioned symmetry is disturbed when the system is not perfectly tuned to the internal resonance, i.e. when $\sigma_2 \neq 0$. The angular velocity of the substrate has also a crucial influence on the resonant response. The influence on the resonant response of the detuning parameter $\sigma_2 \neq 0$, which means that the system is not perfectly tuned, and the influence of the angular velocity ω_z is presented in Figure 6.21.

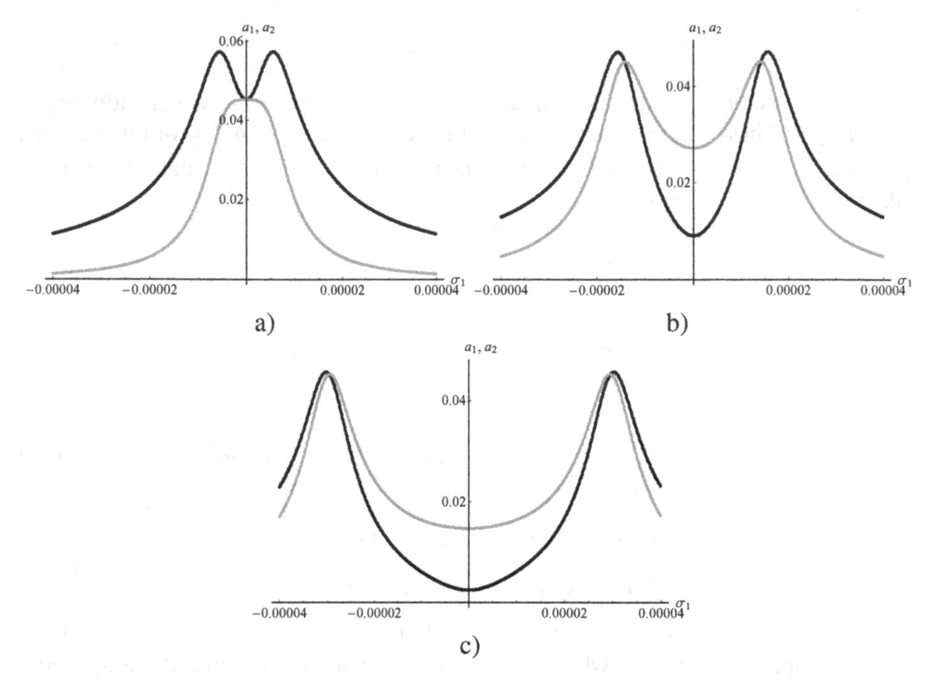

FIGURE 6.20 Resonance curves including: a_1 – amplitude of the frame (black), and a_2 amplitude of the sensing plate (gray): a) $\sigma_2 = 0,\ \omega_z = 1 \cdot 10^{-5}$; b) $\sigma_2 = 0,\ \omega_z = 3 \cdot 10^{-5}$; c) $\sigma_2 = 0,\ \omega_z = 6 \cdot 10^{-5}$.

Comparing several cases presented in Figure 6.21 we can observe that resonant peaks are moving away from each other when ω_z increases. However, the increase of σ_2 disturbs the symmetry of the graphs. The line of the amplitude a_2 moves right, whereas the line of the amplitude a_1 loses its symmetry.

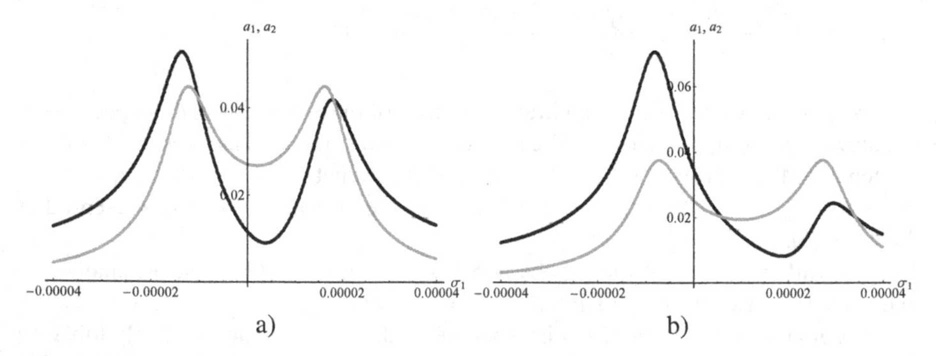

FIGURE 6.21 Resonance curves including: a_1 – amplitude of the frame (black), and a_2 amplitude of the sensing plate (gray): a) $\sigma_2 = 4 \cdot 10^{-6}$, $\omega_z = 3 \cdot 10^{-5}$; b) $\sigma_2 = 2 \cdot 10^{-5}$, $\omega_z = 3 \cdot 10^{-5}$.

Case 3 $(j_1 = j_2 = j_4 = 0,\ \ j_3 = 1)$

Let us analyze the case when the moments of inertia of the gimbal are assumed as negligible and the tensor of inertia of the sensor element is isotropic. Inserting the assumptions $j_1 = j_2 = j_4 = 0,\ j_3 = 1$ into modulation equations (6.131)–(6.134), we get

$$2c_1 a_1 + 2f_0 \cos\gamma_1 - \omega_z(\sigma_2 + 2)\, a_2 \cos\gamma_2 = 0, \tag{6.150}$$

$$4c_2 a_2 - 2\omega_z(\sigma_2 - 2)\, a_1 \cos\gamma_2 = 0, \tag{6.151}$$

$$8\sigma_1 a_1 + 3\alpha_1 a_1^3 + 4f_0 \sin\gamma_1 + 2(\sigma_2 + 2)\,\omega_z a_2 \sin\gamma_2 + a_1\omega_z^2 = 0, \tag{6.152}$$

$$\begin{aligned} &-8\sigma_2 a_1 a_2 + 3\left(\alpha_2 a_2^2 - \alpha_1 a_1^2\right) a_1 a_2 - 4f_0 a_2 \sin\gamma_1 - \\ &2\omega_z\left((\sigma_2 - 2)\, a_1^2 + (\sigma_2 + 2)\, a_2^2\right) \sin\gamma_2 = 0. \end{aligned} \tag{6.153}$$

The considered assumptions cause the vanishing of trigonometric functions whose argument is the modified phase γ_2 multiplied by two. This circumstance allows one to eliminate from equations (6.150)–(6.153) the modified phases. In this manner, the following implicit dependence between the amplitudes and frequencies in the resonance zone is obtained

$$\frac{a_2^2\left(16c_2^2 + \left(-3\alpha_2 a_2^2 + 8\sigma_2 - 8\sigma_1 + \omega_z^2\right)^2\right)}{4a_1^2(\sigma_2 - 2)^2\omega_z^2} = 1, \tag{6.154}$$

$$\begin{aligned} &\left(-3\alpha_1 a_1^3 - 8\sigma_1 a_1 + \omega_z^2 a_1 + \frac{a_2^2\,(2 + \sigma_2)\left(-3\alpha_2 a_2^3 + 8\sigma_2 - 8\sigma_1 + \omega_z^2\right)}{a_1\,(\sigma_2 - 2)}\right)^2 + \\ &\frac{16\left(c_1\,(\sigma_2 - 2)\, a_1^2 + c_2\,(2 + \sigma_2)\, a_2^2\right)^2}{a_1^2(\sigma_2 - 2)^2} = 16 f_0^2. \end{aligned} \tag{6.155}$$

Equations (6.154)–(6.155) form the algebraic set of 6th order with respect to the unknown amplitudes a_1 and a_2, where the amplitudes appear only in even powers. These equations allow one to perform the qualitative analysis of the resonance steady-state amplitudes versus detuning parameter σ_1 for various parameters. The influence of the various parameters on the resonance response is presented in Figures 6.22–6.26.

The analysis of the influence of the damping coefficients on the amplitudes is presented in Figure 6.22. The assumed values of the system parameters are: $f_0 = 4 \cdot 10^{-7}$, $\alpha_1 = 1$, $\alpha_2 = 1$, $\omega_z = 6 \cdot 10^{-5}$, $\sigma_2 = 0$.

The influence of the value of the amplitude f_0 on the resonance curves shape, for the same data and additionally $c_1 = c_2 = 3 \cdot 10^{-5}$, is presented in Figure 6.23.

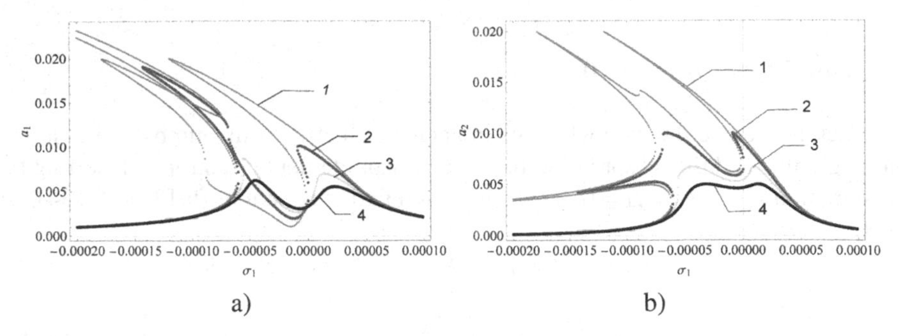

FIGURE 6.22 Resonance curves in the plane $a_1(\sigma_1)$ (a) and $a_2(\sigma_1)$ (b) for: 1) $c_1 = c_2 = 1 \cdot 10^{-5}$, 2) $c_1 = c_2 = 2 \cdot 10^{-5}$, 3) $c_1 = c_2 = 3 \cdot 10^{-5}$, 4) $c_1 = c_2 = 4 \cdot 10^{-5}$.

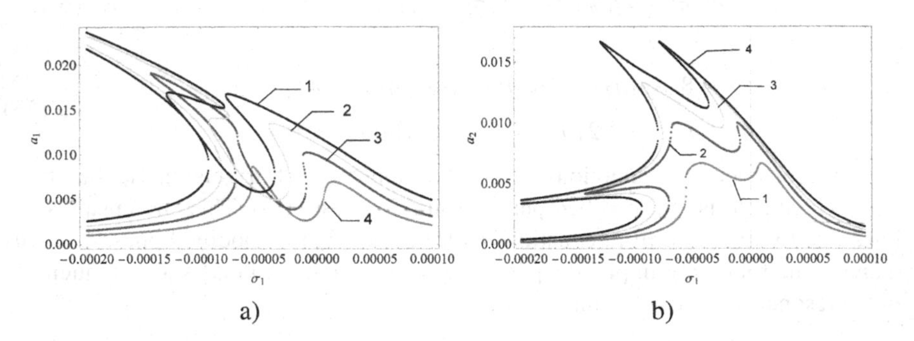

FIGURE 6.23 Resonance curves in the plane $a_1(\sigma_1)$ (a) and $a_2(\sigma_1)$ (b) for: 1) $f_0 = 4 \cdot 10^{-7}$, 2) $f_0 = 6 \cdot 10^{-7}$, 3) $f_0 = 8 \cdot 10^{-7}$, 4) $f_0 = 10 \cdot 10^{-7}$.

The parameters α_1 and α_2 describing nonlinear features the torsional joints have also an essential impact on the dynamic response. Their influence on the resonant curves shape for the same values of parameters uses previously, and $c_1 = c_2 = 2 \cdot 10^{-5}$, is presented in Figure 6.24.

The resonance response for various values of the angular velocity of the support ω_z is visualized in Figure 6.25.

The vibrating system is well-tuned if $\sigma_2 = 0$. The impact of the variation of σ_2 on the resonance curves is presented in Figure 6.26. The data are assumed as previously with $c_1 = c_2 = 3 \cdot 10^{-5}$.

The disproportion between the position of the local maxima in the amplitude resonance curves depends strongly on the value of σ_2.

Many parameters influence the unambiguity of the resonance curves, which is visible in Figures 6.22–6.26. The values of the parameter for which the curves become unique can be estimated in the way of numerical simulations, for the given microdevice. For example, for data values assumed in this section, the coefficients $c_1 = c_2 = 0.000031$ fulfill this criterion.

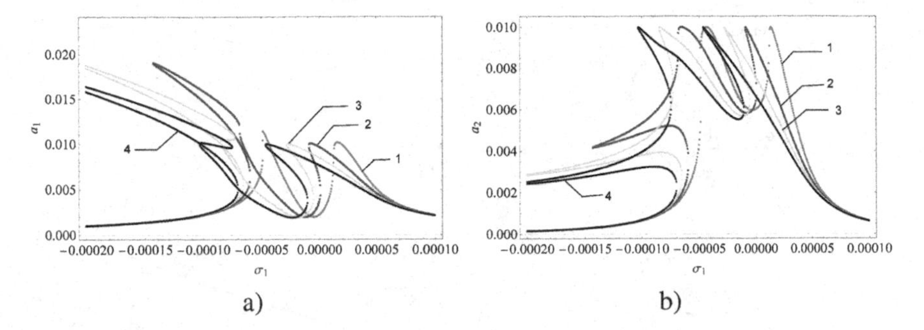

FIGURE 6.24 Resonance curves in the plane $a_1\,(\sigma_1)$ (a) and $a_2\,(\sigma_1)$ (b) for: 1) $\alpha_1 = \alpha_2 = 0.4$, 2) $\alpha_1 = \alpha_2 = 1$, 3) $\alpha_1 = \alpha_2 = 1.5$, 4) $\alpha_1 = \alpha_2 = 2$.

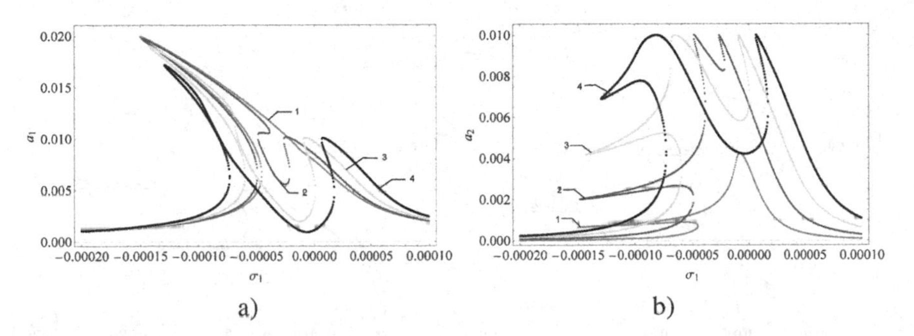

FIGURE 6.25 Resonance curves in the plane $a_1\,(\sigma_1)$ (a) and $a_2\,(\sigma_1)$ (b) for: 1) $\omega_z = 1 \cdot 10^{-5}$, 2) $\omega_z = 3 \cdot 10^{-5}$, 3) $\omega_z = 6 \cdot 10^{-5}$, 4) $\omega_z = 9 \cdot 10^{-5}$.

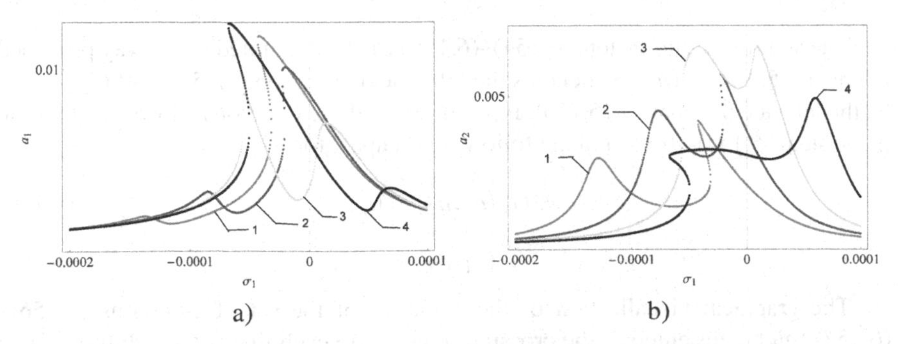

FIGURE 6.26 Resonance curves in the plane $a_1\,(\sigma_1)$ (a) and $a_2\,(\sigma_1)$ (b) for: 1) $\sigma_2 = -1.2 \cdot 10^{-4}$, 2) $\sigma_2 = -6 \cdot 10^{-5}$, 3) $\sigma_2 = 0$, 4) $\sigma_2 = 6 \cdot 10^{-5}$.

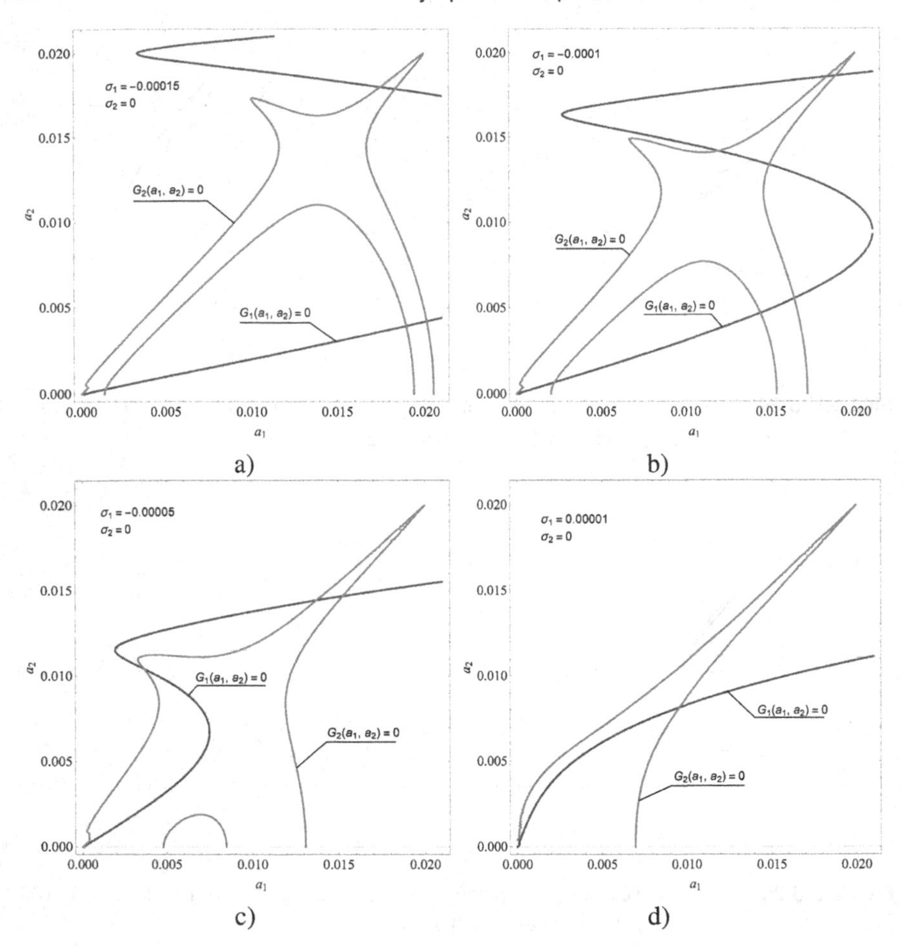

FIGURE 6.27 Influence of the detuning parameter σ_1 on the number of possible steady states: a) $\sigma_1 = -0.00005$, $\sigma_2 = 0$; b) $\sigma_1 = 0.00001$, $\sigma_2 = 0$.

The solution of the system (6.154)–(6.155) can be illustrated in the way proposed in chapter 5. Let $G_1(a_1, a_2)$ denotes the left side of equation (6.154), and $G_2(a_1, a_2)$ be the left side of equation (6.155). Introducing of these functions allows one to write the system (6.154)–(6.152) in the following compact form

$$G_1(a_1, a_2) = 0, \tag{6.156}$$

$$G_2(a_1, a_2) = 0. \tag{6.157}$$

The graphical visualization of the solutions of the set of equations (6.156)–(6.157) can be presented as the crossing lines where each line is the solution of one of equations (6.156)–(6.157), similarly as in chapter 5. Let us assume the data values and $c_1 = c_2 = 1 \cdot 10^{-5}$ (the same as for the line "1" in Figure 6.22). In Figure 6.27 the curves drawn in gray and black represent the solution to equations (6.156) and

(6.157) respectively, so the crossing points of these two curves indicate the steady-state vibration amplitudes of the intermediate frame (a_1) and the sensing plate (a_2).

The graphs collected in Fig. 6.27 indicate that the number of real positive solutions to the system (6.156)–(6.157) can be equal to five, seven, three, or one. Note that the values and number of possible steady-state amplitudes for the assumed data are the same in Figures 6.22 and 6.25. The presented method of visualization allows analyzing the number and values of the possible amplitudes in the steady state for various combinations of parameters.

Case 4 ($\omega_z = 0$)

In the case when the substrate is immovable the form of the modulation equations given by equations (6.131)–(6.134) is simplified to the following one

$$2c_1a_1 + 2f_0\cos\gamma_1 - j_2a_1a_2^2\sin(2\gamma_2) = 0, \tag{6.158}$$

$$4c_2a_2 + j_4a_1^2a_2\sin(2\gamma_2) = 0, \tag{6.159}$$

$$8\sigma_1a_1 + 3\alpha_1a_1^3 + 4f_0\sin\gamma_1 - 2j_2a_1a_2^2\cos(2\gamma_2) = 0, \tag{6.160}$$

$$\begin{aligned}&-8\sigma_2a_1a_2 + 3\left(\alpha_2a_2^2 - \alpha_1a_1^2\right)a_1a_2 - 4f_0a_2\sin\gamma_1 + \\ &2j_2a_1a_2^3\cos(2\gamma_2) + j_4a_2a_1^3\left(2 - \cos(2\gamma_2)\right) = 0,\end{aligned} \tag{6.161}$$

The inertial parameters j_1 and j_3 do not appear in equations (6.158)–(6.161). The trigonometric identities allow one again to eliminate the functions γ_1 and γ_2.

Case 5 ($\omega_z = 0,\ j_2 = 0,\ j_4 = 0$)

Let us assume that the substrate is immovable and additionally $J_x = J_z$, so $j_2 = j_4 = 0$. Substituting these assumptions into equations (6.158)–(6.161) we obtain

$$c_1a_1 + f_0\cos\gamma_1 = 0, \tag{6.162}$$

$$c_2a_2 = 0, \tag{6.163}$$

$$8\sigma_1a_1 + 3\alpha_1a_1^3 + 4f_0\sin\gamma_1 = 0, \tag{6.164}$$

$$-8\sigma_2a_1a_2 + 3\left(\alpha_2a_2^2 - \alpha_1a_1^2\right)a_1a_2 - 4f_0a_2\sin\gamma_1 = 0, \tag{6.165}$$

The assumptions considered here lead to the conclusion that $a_2 = 0$. When the substrate is at rest, then the sensing plate does not oscillate. It is justified because the Coriolis torque is the reason causing the sense plate to vibrate. In turn, equation (6.162) gives the possibility to estimate the amplitude a_1 as follows

$$a_1 \leq \frac{f_0}{c_1}.$$

Due to the position of the substrate and the sensor plate at rest, the amplitude-frequency relationship for the intermediate frame vibration is completely uncoupled and has the following form

$$a_1^2\left(16c_1^2 + \left(3\alpha_1 a_1^2 + 8\sigma_1\right)^2\right) = 16f_0^2. \tag{6.166}$$

The family of resonance curves obtained using equation (6.166), for various values of the damping coefficient c_1 is presented in Figure 6.28. The values of parameters assumed in this simulation are: $\sigma_2 = 0$, $f_0 = 4 \cdot 10^{-7}$, $\alpha_1 = 2$, $j_2 = 0$, $j_4 = 0$.

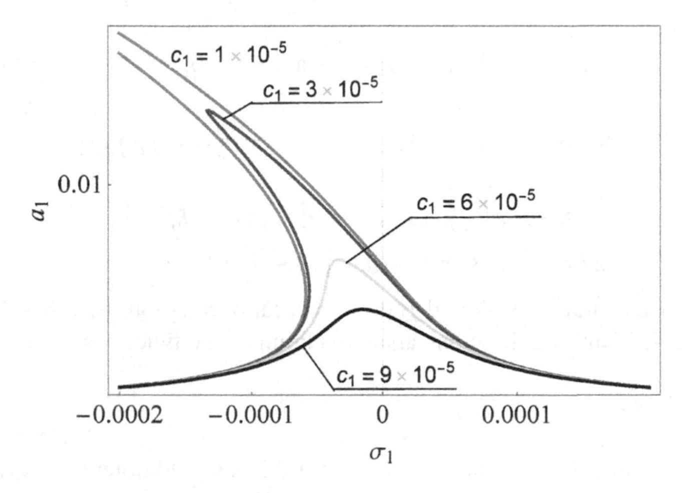

FIGURE 6.28 Resonance curves $a_1(\sigma_1)$ for the intermediate frame when the substrate is immovable for different damping coefficient.

The coordinates of the top of the bend for each resonance curve presented in Figure 6.28 are

$$\left(-\frac{3}{8}\frac{\alpha_1 f_0}{c_1^2}, \frac{f_0}{c_1}\right)$$

which results from equations (6.162) and (6.164).

Case 6 (identification of the angular velocity)

The identification problem of the substrate angular velocity requires an unambiguous character of each of both resonance response curves. So, the determination of the value of the damping coefficient which guarantees this unambiguity for a given micromechanical gyroscope is of grat importance. It is also important to reduce the number of measurement system parameters that have influence on the identification. One of the earlier discussed cases leads to the essential simplification of the equations of steady-state. It is the case when the gimbal moments of inertia are sufficiently slight and the tensor of inertia of the sense plate is isotropic. These assumptions that simplify the steady-state equations to the form (6.150)–(6.153) can be written as follows $j_1 = j_2 = j_4 = 0$, $j_3 = 1$. Equation (6.151) is relatively simple and contains only a few numbers of the parameters of micromechanical gyroscope. Let us solve equation (6.151) for the angular velocity

$$\omega_z = \frac{2a_2c_2}{(\sigma_2 - 2)\, a_1 \cos\gamma_2}. \tag{6.167}$$

When the system is perfectly designed, made, and tuned, the detuning parameter σ_2 is equal to zero. Thus it is enough to know the value of damping coefficient c_2 and to measure the values of amplitudes a_1 and a_2 in the resonance. In the steady-state, the difference of the modified phases should be set at value π. Assuming all these circumstances, we can write that

$$\omega_z = \frac{a_2c_2}{a_1}. \tag{6.168}$$

The following simulation is carried out to apply this identification method. We assume some values of parameters of the micromechanical gyroscope including the value of the substrate angular velocity in the case of the strict resonance ($\sigma_1 = 0, \sigma_2 = 0$), namely $p = 1$, $w = 1$, $\omega_z = 6 \cdot 10^{-5}$, $c_1 = 8 \cdot 10^{-5}$, $c_2 = 8 \cdot 10^{-5}$, $\alpha_1 = 1$, $\alpha_2 = 1$, $f_1 = 4 \cdot 10^{-7}$, $j_1 = 0$, $j_2 = 0$, $j_3 = 1$, $j_4 = 0$, $a_{10} = 0.002$, $a_{20} = 0.002$, $\gamma_{10} = 0$, $\gamma_{20} = 0$.

Then, we find the approximate solution of the initial value problem (6.74)–(6.75) and determine the values of amplitudes, by reading them directly from the graphs. This way the real measurement is simulated. The variability over time of both generalized coordinates for the steady state vibration is presented in Figure 6.29. Using graphs, we determine $a_1 = 0.0032005$, $a_2 = 0.0023969$, and hence from equation (6.168) we obtain $\omega_z = 5.991 \cdot 10^{-5}$ which yields the approximation of the assumed value with the relative error equal to about 0.015.

Although this method of determining the angular velocity ω_z of the substrate is very simple and efficient but it fails when the system is not tuned perfectly, i.e. when $\sigma_2 \neq 0$. In that situation the coincidence of both resonances does not appear. The measurement precision of the angular velocity depends strongly on the detuning parameter σ_2. The higher value of $|\sigma_2|$ causes the less accurate measurement. The influence of σ_2 on the steady state vibration amplitudes, both for sensing plate and frame is shown in Figure 6.30. The graphs present a family of curves for various damping coefficients. The influence of damping is the most spectacular especially for small values of c_1 and c_2 when the curves are ambiguous. However, even for appropriately high damping when the curves are unique, the influence of σ_2 on the amplitudes is significant. The numerical simulations were carried out assuming the same values of the parameters i.e. $\sigma_1 = 0$, $\omega_z = 6 \cdot 10^{-5}$, $\alpha_1 = 1$, $\alpha_2 = 1$, $f_1 = 4 \cdot 10^{-7}$.

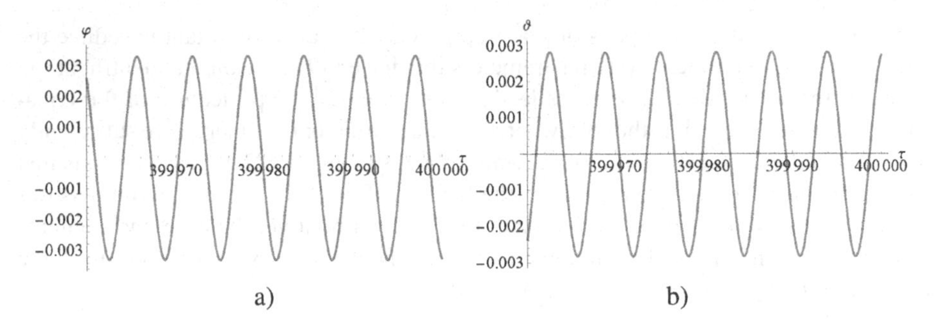

FIGURE 6.29 The steady state vibration for the frame (a) and the sensing plate (b).

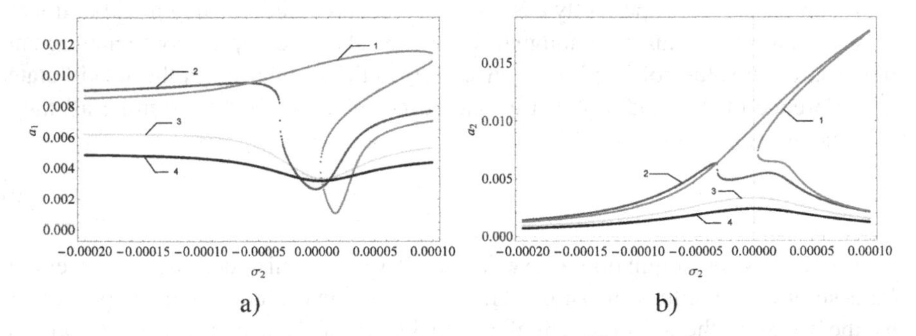

FIGURE 6.30 The amplitudes of the frame a_1 (a) and the sensing plate a_2 (b) versus σ_2 for various values of the damping coefficients: 1) $c_1 = c_2 = 1 \cdot 10^{-5}$, 2) $c_1 = c_2 = 3 \cdot 10^{-5}$, 3) $c_1 = c_2 = 6 \cdot 10^{-5}$, 4) $c_1 = c_2 = 8 \cdot 10^{-5}$.

6.8 STABILITY ANALYSIS

The procedure presented in the previous section is aimed of obtaining the amplitude-frequency dependence. Equations (6.131)–(6.134) that are the most general form of the conditions under which the periodic steady oscillations occur are of trigonometric-algebraic type. Among the solutions, there are both real as well as complex values. Only real and positive values of the amplitudes and real values of the modified phases have physical meaning. In the case of the unambiguous responses, like in *Case 2*, each branch of the resonance curve is stable. The situation becomes much more complicated when ambiguity occurs. To examine which of the found steady states realize a stable objectives, stability analysis is needed. To do that we have to analyze the stability of the fixed points of the modulation equations.

Let us test the stability concerning the general case of the steady state vibration in the doubled resonance. We introduce the denotation $(a_{1s}, a_{2s}, \gamma_{1s}, \gamma_{2s})$ for any fixed point of the dynamical system (6.127)–(6.130). To examine the stability of the point in the sense of Lyapunov, we focus on the solutions of the system (6.127)–(6.130) that are close to the state $(a_{1s}, a_{2s}, \gamma_{1s}, \gamma_{2s})$. Introducing the functions $\tilde{a}_1(\tau)$, $\tilde{a}_2(\tau)$,

$\tilde{\gamma}_1(\tau)$, $\tilde{\gamma}_2(\tau)$, that are understood as small perturbations, one can write these solutions in the form

$$\begin{aligned} a_1(\tau) &= a_{1s} + \tilde{a}_1(\tau), \qquad & a_2(\tau) &= a_{2s} + \tilde{a}_2(\tau), \\ \gamma_1(\tau) &= \gamma_{1s} + \tilde{\gamma}_1(\tau), \qquad & \gamma_2(\tau) &= \gamma_{2s} + \tilde{\gamma}_2(\tau), \end{aligned} \tag{6.169}$$

where $a_{1s}, a_{2s}, \gamma_{1s}, \gamma_{2s}$ are the solutions of equations (6.131)–(6.134).

Substituting equations (6.169) into modulation equations (6.127)–(6.130), we obtain

$$\begin{aligned} \frac{d\tilde{a}_1}{d\tau} &= -\frac{c_1}{2}(a_{1s}+\tilde{a}_1) - \frac{f_0}{2}\cos(\gamma_{1s}+\tilde{\gamma}_1) + \\ &\frac{\omega_z}{4}(j_3-j_2)(\sigma_2+2)(a_{2s}+\tilde{a}_2)\cos(\gamma_{2s}+\tilde{\gamma}_2) + \\ &\frac{j_2}{4}(a_{1s}+\tilde{a}_1)(a_{2s}+\tilde{a}_2)^2\sin(2(\gamma_{2s}+\tilde{\gamma}_2)), \end{aligned} \tag{6.170}$$

$$\begin{aligned} \frac{d\tilde{a}_2}{d\tau} = -\frac{c_2}{2}(a_{2s}+\tilde{a}_2) - \frac{\omega_z}{4}(j_4-1)(\sigma_2-2)(a_{1s}+\tilde{a}_1)\cos(\gamma_{2s}+\tilde{\gamma}_2) - \\ \frac{j_4}{8}a_1^2(\tau)(a_{2s}+\tilde{a}_2)\sin(2(\gamma_{2s}+\tilde{\gamma}_2)), \end{aligned} \tag{6.171}$$

$$\begin{aligned} \frac{d\tilde{\gamma}_1}{d\tau} &= \sigma_1 + \frac{3}{8}\alpha_1(a_{1s}+\tilde{a}_1)^2 + \frac{f_0}{2(a_{1s}+\tilde{a}_1)}\sin(\gamma_{1s}+\tilde{\gamma}_1) - \\ &\frac{j_2}{4}(a_{2s}+\tilde{a}_2)^2\cos(2(\gamma_{2s}+\tilde{\gamma}_2)) + \\ &\frac{(j_3-j_2)(\sigma_2+2)\,\omega_z}{4\,(a_{1s}+\tilde{a}_1)}(a_{2s}+\tilde{a}_2)\sin(\gamma_{2s}+\tilde{\gamma}_2) - \\ &\frac{1}{2}\left(j_1 - \frac{1}{4}(j_3-j_2)(j_4-1)\right)\hat{\omega}_z^2, \end{aligned} \tag{6.172}$$

$$\begin{aligned} \frac{d\tilde{\gamma}_2}{d\tau} &= -\sigma_2 + \frac{j_1-j_4}{2}\hat{\omega}_z^2 + \frac{3}{8}\left(\alpha_2(a_{2s}+\tilde{a}_2)^2 - \alpha_1(a_{1s}+\tilde{a}_1)^2\right) - \\ &\frac{f_0}{2(a_{1s}+\tilde{a}_1)}\sin(\gamma_{1s}+\tilde{\gamma}_1) + \frac{j_2}{4}(a_{2s}+\tilde{a}_2)^2\cos(2(\gamma_{2s}+\tilde{\gamma}_2)) + \\ &\frac{j_4}{8}(a_{1s}+\tilde{a}_1)^2(2-\cos(2(\gamma_{2s}+\tilde{\gamma}_2))) + \\ &\frac{\omega_z}{4}\left(\frac{(j_4-1)(\sigma_2-2)}{a_{2s}+\tilde{a}_2}(a_{1s}+\tilde{a}_1) - \frac{(j_3-j_2)(\sigma_2+2)}{(a_{1s}+\tilde{a}_1)}(a_{2s}+\tilde{a}_2)\right)\sin(\gamma_{2s}+\tilde{\gamma}_2). \end{aligned} \tag{6.173}$$

The right-hand sides of equations (6.170)–(6.173) are expanded in the Taylor series around the fixed point $(a_{1s}, a_{2s}, \gamma_{1s}, \gamma_{2s})$. The terms of order one are kept and the higher ones are rejected. Then, taking into account that values $a_{1s}, a_{2s}, \gamma_{1s}, \gamma_{2s}$, which fulfill the steady-state equations (6.131)–(6.134), we obtain the set of the linear ordinary differential equations with small perturbations $\tilde{a}_1$, $\tilde{a}_2$, $\tilde{\theta}_1$, $\tilde{\theta}_2$ as unknown functions. By introducing some symbols to shorten the form of the notation, we can write the equations as follows

$$\frac{d\tilde{a}_1}{d\tau} = A_{11}\tilde{a}_1 + A_{12}\tilde{a}_2 + A_{13}\tilde{\gamma}_1 + A_{14}\tilde{\gamma}_2, \tag{6.174}$$

$$\frac{d\tilde{a}_2}{d\tau} = A_{21}\tilde{a}_1 + A_{22}\tilde{a}_2 + A_{23}\tilde{\gamma}_1 + A_{24}\tilde{\gamma}_2, \tag{6.175}$$

$$\frac{d\tilde{\gamma}_1}{d\tau} = A_{31}\tilde{a}_1 + A_{32}\tilde{a}_2 + A_{33}\tilde{\gamma}_1 + A_{34}\tilde{\gamma}_2, \tag{6.176}$$

$$\frac{d\tilde{\gamma}_2}{d\tau} = A_{41}\tilde{a}_1 + A_{42}\tilde{a}_2 + A_{43}\tilde{\gamma}_1 + A_{44}\tilde{\gamma}_2, \tag{6.177}$$

where

$$A_{11} = -\frac{c_1}{2} + \frac{1}{4}a_{2s}^2 \sin(2\gamma_{2s}),$$

$$A_{12} = \frac{1}{4}(j_3 - j_2)(\sigma_2 + 2)\,\omega_z \cos(\gamma_{2s}) + \frac{1}{2}j_2 a_{1s} a_{2s} \sin(2\gamma_{2s}),$$

$$A_{13} = \frac{1}{2}f_1 \sin\gamma_1,$$

$$A_{14} = \frac{1}{2}j_2 a_{1s} a_{2s}^2 \cos(2\gamma_{2s}) - \frac{(j_3 - j_2)}{4}(\sigma_2 + 2)\,\omega_z a_{2s} \sin(\gamma_{2s}),$$

$$A_{21} = -\frac{1}{4}(j_4 - 1)(\sigma_2 - 2)\,\omega_z \cos(\gamma_{2s}) - \frac{j_4}{4}a_{1s} a_{2s} \sin(2\gamma_{2s}),$$

$$A_{22} = -\frac{1}{2}c_2 - \frac{1}{8}j_4 a_{1s}^2 \sin(2\gamma_{2s}),$$

$$A_{23} = 0,$$

$$A_{24} = -\frac{1}{4}j_4 a_{1s}^2 a_{2s} \cos(2\gamma_{2s}) + \frac{1}{4}(j_4 - 1)(\sigma_2 - 2)\,\omega_z a_{1s} \sin(\gamma_{2s}),$$

$$A_{31} = \frac{3}{4}\alpha_1 a_{1s} - \frac{f_0}{2a_{1s}^2}\sin\gamma_{1s} - \frac{(j_3 - j_2)(\sigma_2 + 2)\,\omega_z}{4a_{1s}^2}a_{2s}\sin(\gamma_{2s}),$$

$$A_{32} = -\frac{1}{2}j_2 a_{2s} \cos(2\gamma_{2s}) + \frac{(j_3 - j_2)(\sigma_2 + 2)\,\omega_z}{4a_{1s}}\sin(\gamma_{2s}),$$

$$A_{33} = \frac{f_0}{2a_{1s}}\cos(\gamma_{1s}),$$

$$A_{34} = \frac{1}{2}j_2 a_{2s}^2 \sin(2\gamma_{2s}) + \frac{(j_3 - j_2)(\sigma_2 + 2)\,\omega_z}{4a_{1s}}a_{2s}\cos(\gamma_{2s}),$$

$$A_{41} = -\frac{3}{4}\alpha_1 a_{1s} + \frac{j_4}{4}\left(2 - \cos\left(2\gamma_{2s}\right)\right) a_{1s} + \frac{f_0}{2a_{1s}^2}\sin(\gamma_{1s}) + \frac{(j_4 - 1)(\sigma_2 - 2)\,\omega_z}{4a_{2s}}\sin\left(\gamma_{2s}\right) + \frac{(j_3 - j_2)(\sigma_2 + 2)\,\omega_z}{4a_{1s}^2}\sin\left(\gamma_{2s}\right),$$

$$A_{42} = \frac{3}{4}\alpha_2 a_{2s} + \frac{j_2}{2}a_{2s}\cos\left(2\gamma_{2s}\right) - \frac{(j_4 - 1)(\sigma_2 - 2)\,\omega_z}{4a_{2s}^2}a_{1s}\sin\left(\gamma_{2s}\right) - \frac{(j_3 - j_2)(\sigma_2 + 2)\,\omega_z}{4\,a_{1s}}\sin\left(\gamma_{2s}\right),$$

$$A_{43} = -\frac{f_0}{2a_{1s}}\cos\left(\gamma_{1s}\right),$$

$$A_{44} = \left(\frac{(j_4 - 1)(\sigma_2 - 2)\,\omega_z}{4a_{2s}}a_{1s} - \frac{(j_3 - j_2)(\sigma_2 + 2)\,\omega_z}{4a_{1s}}a_{2s}\right)\cos\left(\gamma_{2s}\right) + \left(\frac{1}{4}j_4 a_{1s}^2 - \frac{j_2}{2}a_{2s}^2\right)\sin\left(2\gamma_{2s}\right).$$

The fixed point is stable in the Lyapunov sense if and only if the real parts of all eigenvalues of the characteristic matrix

$$\mathbf{A} = \begin{bmatrix} A_{11} & A_{12} & A_{13} & A_{14} \\ A_{21} & A_{22} & A_{23} & A_{24} \\ A_{31} & A_{32} & A_{33} & A_{34} \\ A_{41} & A_{42} & A_{43} & A_{44} \end{bmatrix} \tag{6.178}$$

are negative.

The stability evaluation of the resonance response curve presented in Figure 6.31 has been made using the procedure described above. The eigenvalues of matrix **A** were calculated using *Eigenvalues* standard procedure in the *Wolfram Mathematica* software. Results of the examination are presented in Figure 6.31, where the black and gray colours depict unstable and stable branches of the resonance curves respectively. The calculations were made for the following values of the parameters: $\omega_z = 6\cdot 10^{-5}$, $c_1 = 2\cdot 10^{-5}$, $c_2 = 2\cdot 10^{-5}$, $\alpha_1 = 1$, $\alpha_2 = 1$, $f_1 = 5\cdot 10^{-7}$, $j_1 = 0$, $j_2 = 0$, $j_3 = 1$, $j_4 = 0$.

In Figure 6.31 we can identify an interval $\sigma_1 \in (-0.000066, -0.00003)$ for which no stable steady state vibration is possible. For example for $\sigma_1 = -0.00009$ two stationary amplitudes are possible: $a_1 = 0.003250117$ or $a_1 = 0.015470713$ and $a_2 = 0.001081953$ or $a_2 = 0.006024170$, for $\sigma_1 = 0$ there is only one stable branch and stationary amplitudes are $a_1 = 0.0104829841$ and $a_2 = 0.009287976$, whereas for $\sigma_1 = -0.00004$ there are no stable branches of the resonance curve, so the steady-state amplitude does not exist. The confirmation of the above discussion can be the time histories of amplitudes a_1 and a_2 presented in Figures 6.32–6.34 obtained for the same values of the parameters.

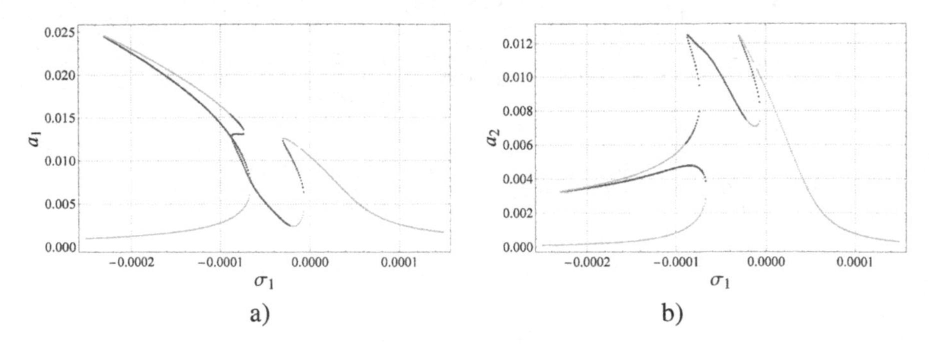

FIGURE 6.31 The resonance response curves θ_1 (a) and θ_2 (b) versus the detuning parameter σ_1: stable branches (gray), unstable (black).

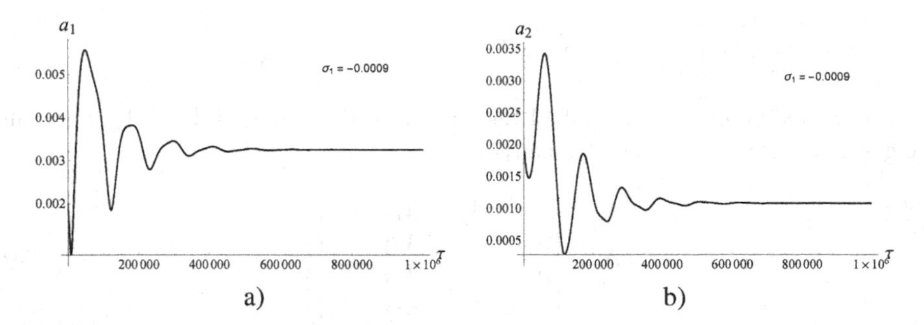

FIGURE 6.32 Time courses of the amplitudes a_1 (a) and a_2 (b) in the case of the steady-state vibration for $\sigma_1 = -0.0009$.

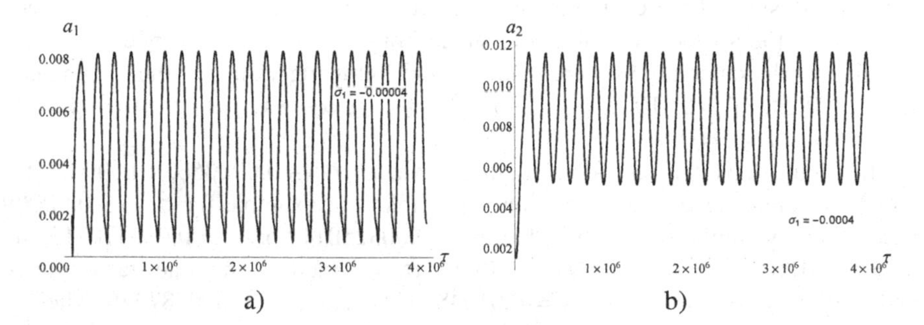

FIGURE 6.33 Time courses of the amplitudes a_1 (a) and a_2 (b) when no steady-state vibration for $\sigma_1 = -0.0004$.

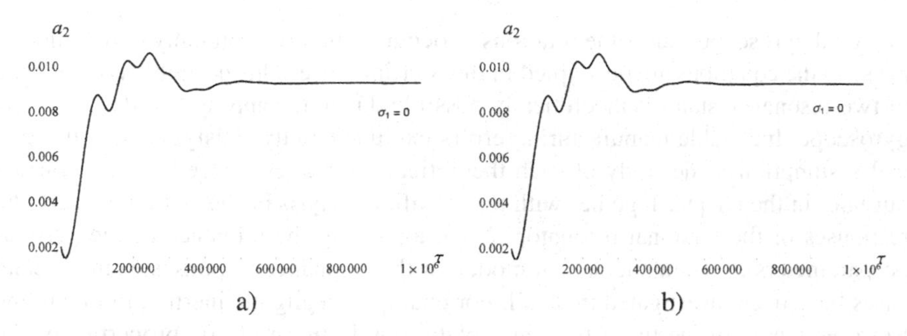

FIGURE 6.34 Time courses of the amplitude a_1 (a) and a_2 (b) in the case of the steady-state vibration for $\sigma_1 = 0$.

6.9 CLOSING REMARKS

Generally, linear and torsional gyroscopes can be distinguished in the class of vibratory micromechanical devices serving as angular velocity sensors. This distinction is based on some model assumptions, according to which parts of a linear micromechanical gyroscope can only do the translational movement. Their linear displacements are negligible while only the spherical motion is allowed in the case of rotational gyroscopes. The subject of this chapter is the dynamics of a mechanical system of the vibratory MEMS gyroscope of the second type.

The mathematical model presented here takes into account the nonlinear properties of the beams acting as torsional connections between the gyroscope's parts. The cubic type relationship between the twisting moment and the angle of twisting has been assumed. As in the previous chapters, the equations of motion were derived employing the Lagrange formalism. The viscous damping and the external torque acting around the drive axis and changing harmonically, have been introduced as the generalized forces. The simplified equations describing the motion at small angular displacements were derived from the primary equations valid for any angles. Due to the smallness of the angular displacements, these equations are of practical importance for rotational micromechanical gyroscopes. Although they concern the motion for small angles, they are still nonlinear equations with preserved the cubic terms and the coupling via the generalized velocities.

The approximate asymptotic solutions to the initial value problems concerning both the non-resonant and resonant vibration were obtained employing the method of multiple scales with three variables in the time domain. In the case of the non-resonant vibration, the initial value problem describing the modulation of the amplitudes and phases was solved analytically. The approximate analytical solutions have been validated using the proposed measure of the error satisfying the governing equations and also by comparison to the numerical simulation results.

By reflecting the operating conditions of the micromechanical torsional gyroscopes, the case with both external and internal resonances occuring simultaneously, was considered. Such a state is potentially possible because the structure of the torsional gyroscope causes the rotations to occur around two mutually perpendicular axes, so the equations are decoupled in the inertial sense. This desirable coincidence of two resonance states is theoretically possible due to the appropriate design of the gyroscope. Inevitable manufacturing errors exclude strictly satisfying such theoretical assumptions. The study of such theoretical double resonance has been carried out later in the chapter together with the stability analysis of the stationary periodic responses of the torsional resonator. When analyzing the influence of the particular parameters of the mathematical model on the resonance responses, some special cases have been investigated in detail. For example, negligible inertia of gimbal, the isotropic tensor of inertia of the sense plate, purely linear elastic properties of the torsional links, the resonance when the substrate is immovable have been studied as separate special cases.

7 Torsional Oscillations of a Two-Disk Rotating System

In this chapter, the problem of coupled and periodically driven torsional oscillators is analyzed. The system consists of a harmonic oscillator of a large mass to which a nonlinear oscillator of much lower mass is connected. From the mathematical model, the internal motion is separated and then the approximate effective equation of motion is derived. This equation is transformed to a first-order differential equation for complex functions and then efficiently solved in the vicinity of main resonance using the multiple scale method. A detailed analysis of transient vibrations is carried out, which allows us to determine the critical values of the nonlinearity parameter for which there is a qualitative change in the dynamics of the system. For the steady motion, the amplitude-frequency dependencies are determined, the curves of the resonance response were plotted and their stability tested.

7.1 INTRODUCTION

The exchange of energy is a subject of great interest to many researchers. Especially, nonlinear properties of elasticity or damping have big application potentials ([184], Gendelman et al. [76], Vakakis and Gendelman [216], Gendelman [74, 75], Kerschen et al. [96], Manevitch et al. [134]). It could be crucial in the tasks of control or designing of the passive damping systems. Such a method has been firstly proposed in the paper (Haxton and Barr [82]). Many less or more similar approaches are widely discussed in the review article (Oueini et al. [162]). Determining the conditions in which this phenomenon occurs allows for the appropriate selection of the parameters of the system in order to transfer energy from one part to another. Coupled oscillators play an important role in many fields of science, including mechanics, medicine and biology (Awrejcewicz [17], Kozłowski et al. [105]).

This chapter examines the motion of coupled oscillators with two degrees of freedom under periodic excitation. The dynamics of such a system is very complicated (Awrejcewicz [17]). The method used here consists of two major steps. The first is to simplify the problem by extracting internal motion in the system and describing it with an appropriate, one effective equation of motion. The effective equation for a similar system was investigated by Kyziol and Okniński [111], Okniński and Kyziol [159]. The second step is to analyze the dynamics of the effective equation. For this purpose, the main idea contained in the work (Okniński and Kyziol [160]) was adapted. In the latter work the authors used an innovative approach allowing for qualitative and quantitative study of the behavior of nonlinear systems with one degree of freedom, exchanging energy with external excitation.

It is obvious that in general, reduction of 2 DoF systems to 1 DoF systems is impossible. Besides the mentioned work, we have found the earlier investigation of

DOI: 10.1201/9781003270706-7

Hough (1988) where a 2 DoF conservative mechanical system was reduced to 1 DoF kinematic system under the change of an independent variable. With the help of Hamiltonian integral, the problem was simplified to study a second-order ODE for function specifying the orbital curve.

7.2 HARMONIC OSCILLATOR WITH ADDED NONLINEAR OSCILLATOR

In many cases, the displacement of real mechanical system is relatively small during operations, so their dynamical behavior can be described using linear differential equations with good approximation. The elimination of undesirable vibrations can be achieved by attaching to such a device an oscillator with a low mass and appropriately selected nonlinear characteristics. Therefore, an important issue from the point of view of potential practical applications is the dynamic analysis of two coupled oscillators, one of which the higher mass is linear.

In such structures, energy could be transferred from one part of the system to the other one. The added subsystem can be used as a damper.

Let us analyze the dynamics of two coupled oscillators where one has big inertia and linear elastic property, and the second has small inertia and nonlinear elastic characteristics. The in-depth analysis of this type of the coupled oscillators is investigated in the paper (Okniński and Kyziol [160]).

The equations of motion of such system in the general case can be written as follows:

$$L_1(q_1) + L_n(q_2 - q_1) = F_1, \tag{7.1}$$

$$L_2(q_2) - L_n(q_2 - q_1) = F_2 \tag{7.2}$$

where, q_1 and q_2 are generalized coordinates, L_1 and L_2 are linear differential operators, L_n is a nonlinear differential operator, F_1 and F_2 are external loadings.

Introducing new denotations $x = q_1$ and $y = q_2 - q_1$, the system (7.1)–(7.2) can be written

$$L_1(x) + L_n(y) = F_1, \tag{7.3}$$

$$L_2(y) + L_2(x) - L_n(y) = F_2. \tag{7.4}$$

Adding (7.3) and (7.4) we obtain an important linear relation between variables x and y.

$$L_2(y) + L_1(x) + L_2(x) = F_1 + F_2. \tag{7.5}$$

Moreover, it is possible to separate variables in (7.3) and (7.4) due to their structure. By acting on equation (7.3) with the L_2 operator, and on equation (7.4) with the L_1 operator, one can obtain

$$L_2(L_1(x)) + L_2(L_n(y)) = L_2(F_1), \tag{7.6}$$

$$L_1(L_2(y)) + L_1(L_2(x)) - L_1(L_n(y)) = L_1(F_2). \tag{7.7}$$

There is a possibility of obtaining an equation for the variable y alone, which describe the internal motion. The differential operators L_1 and L_2 are linear, so $L_1(L_2(x)) = L_2(L_1(x))$. Taking advantage of this property and subtracting equation (7.6) from equation (7.7) we obtain

$$L_1(L_2(y)) - L_2(L_n(y)) - L_1(L_n(y)) = L_1(F_2) - L_2(F_1) \tag{7.8}$$

or after simple transformation

$$(L_1 + L_2)(-L_n(y)) + L_1(L_2(y)) = L_1(F_2) - L_2(F_1). \tag{7.9}$$

As a result of the above mathematical manipulations, the one equation for the unknown function y has been formally derived. Equation (7.9) describes the internal motion in the system, so we can investigate the original dynamical problem of two degrees of freedom by analyzing only one differential equation.

7.3 FORMULATION OF THE PROBLEM

Later in this chapter we will analyze a system as an example to study coupled oscillators. The method presented in this example is general and can be used to test other dynamically coupled systems. As the example of coupled oscillators analyzed here serves as a mechanical system exhibiting torsional oscillations, the system has two degrees of freedom.

Torsional oscillations belong to major problems in the design of power transmission systems. The dynamic stresses caused by torsional oscillations, especially near resonance when the amplitudes grow significantly, may be very large and can lead to failure of the whole structure.

Both discrete and continuous models are commonly applied in literature to investigate the torsional vibration of the power transmission systems (Ying et al. [239], Porter [173], Wenming et al. [233]).

The problem of oscillations in nonlinear systems has to be treated with due care. The numerical methods commonly used in solving nonlinear problems do not provide the possibility of qualitative analysis of various dynamic behavior. Some surprising types of oscillation motion cannot be detected when only numerical methods are used. The multiple scale method (MSM) in the time domain allows one not only to obtain approximate solutions, but also (thanks to their analytical form) it makes it possible to accurately examine the influence of selected parameters on its motion. This method is applied in solving the problem of torsional oscillation of the shaft on which the rigid disks are mounted. The relationships obtained in the analytical form, between the amplitudes and phases of vibration at resonance are examined. Combination of the MSM with an interesting approach proposed by Manevich and Musienko [136] known as Limiting Phase Trajectory (LPT) method regarding the system with two disks allows predicting the sudden change of the shape of amplitude modulation which appears in resonance at a certain value of the nonlinear parameter. A similar approach, but based completely on the proposed method by Manevich formulation in the field of complex functions, was applied to investigate the torsional oscillation in the system with one degree of freedom (Starosta [196]).

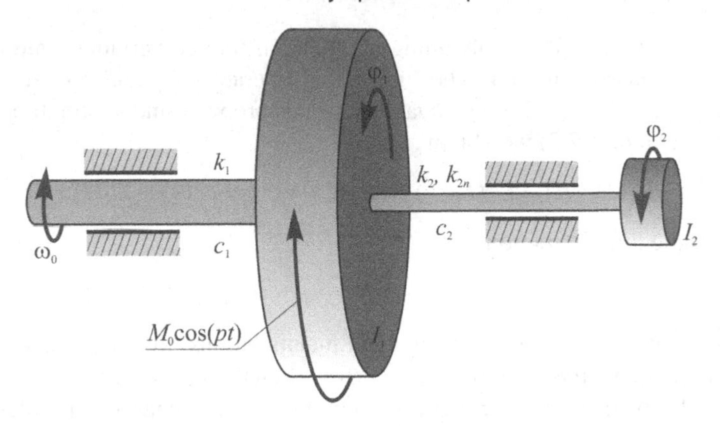

FIGURE 7.1 Model of the rotating 2 DoF system.

Let us consider a rotating system, consisting of two disks mounted on a shaft. The system studied is shown in Figure 7.1. The left side of the shaft rotates with a constant angular velocity ω_0. The disks are considered rigid. Their moments of inertia around the axis of rotation are denoted by I_1 and I_2, respectively. We assume that $I_1 \gg I_2$. Since the shaft is relatively thin and light, its inertia may be neglected. The shaft provides torsional stiffness only. The left part of the shaft is linearly elastic with the stiffness coefficient k_1, whereas two coefficients of stiffness, marked by k_2 and k_{2n}, describe the linear and nonlinear properties of the elasticity of the shaft between wheels. Also viscous dampings, represented by the damping coefficients c_1 and c_2 are taken into account. The whole system is mounted on frictionless bearings which are also ideal in the geometric sense. Moreover, one of the disks is under the action of the harmonically changing torque $M(t) = M_0 \cos(p_0 t)$. The system has two degrees of freedom and the angles of rotation of both wheels φ_1 and φ_2 are chosen as the generalized coordinates.

The similar like problem has been earlier solved in (Awrejcewicz and Starosta [27], Awrejcewicz et al. [29, 30]).

7.4 MATHEMATICAL MODEL

The governing equations are derived using Lagrange formalism. The kinetic energy of the system takes the form

$$T = \frac{1}{2}\left(I_1(\omega_0 + \dot{\varphi}_1(t))^2 + I_2(\omega_0 + \dot{\varphi}_2(t))^2\right), \tag{7.10}$$

and the potential energy of the elastic conservative forces reads

$$V = \frac{1}{2}k_1\varphi_1(t)^2 + \frac{1}{2}k_2(\varphi_2(t) - \varphi_1(t))^2 + \frac{1}{4}k_{2n}(\varphi_2(t) - \varphi_1(t))^4. \tag{7.11}$$

The external excitation and damping effects are included in the model as the generalized forces corresponding to the generalized coordinates

$$Q_1 = M_0 \cos(p_0 t) - c_1 \dot{\varphi}_1(t) - c_2(\dot{\varphi}_1(t) - \dot{\varphi}_2(t)), \tag{7.12}$$

$$Q_2 = -c_2(\dot{\varphi}_2(t) - \dot{\varphi}_1(t)). \tag{7.13}$$

The Lagrange equations of motion take the form

$$\begin{aligned} &k_1 \varphi_1(t) + k_2(\varphi_1(t) - \varphi_2(t)) + k_{2n}(\varphi_1(t) - \varphi_2(t))^3 + \\ &c_1 \dot{\varphi}_1(t) + c_2(\dot{\varphi}_1(t) - \dot{\varphi}_2(t)) + I_1 \ddot{\varphi}_1(t) = M_0 \cos(p_0 t), \end{aligned} \tag{7.14}$$

$$\begin{aligned} &k_2(\varphi_2(t) - \varphi_1(t)) + k_{2n}(\varphi_2(t) - \varphi_1(t))^3 + \\ &c_2(\dot{\varphi}_2(t) - \dot{\varphi}_1(t)) + I_2 \ddot{\varphi}_2(\tau) = 0. \end{aligned} \tag{7.15}$$

Note that the constant angular velocity ω_0 of the shaft does not appear in the equations of motion (7.14)–(7.15). It means that the stiff rotation has been eliminated. In other words, equations (7.14)–(7.15) describe only the vibration processes emerging in the system due to the elasticity of the shaft.

Let us introduce the new coordinates

$$x(t) = \varphi_1(t), \qquad y(t) = \varphi_2(t) - \varphi_1(t). \tag{7.16}$$

Now equations (7.14)–(7.15) take the form

$$k_1 x(t) - k_2 y(t) - k_{2n}(y(t))^3 + c_1 \dot{x}(t) - c_2 \dot{y}(t) + I_1 \ddot{x}(t) = M_0 \cos(p_0 t), \tag{7.17}$$

$$k_2 y(t) + k_{2n}(y(t))^3 + c_2 \dot{y}(t) + I_2 \ddot{y}(\tau) + I_2 \ddot{x}(\tau) = 0. \tag{7.18}$$

Let us transform equations (7.17)–(7.18) into a more convenient dimensionless form. In this way, the number of parameters defining the system will be reduced. The dimensionless time is defined as $\tau = \omega_2 t$, where $\omega_2 = \sqrt{k_2 \frac{I_1 + I_2}{I_1 I_2}}$ is a characteristic frequency suitable for further analysis. The new functions of dimensionless time are defined $\phi(\tau) = y\left(\frac{\tau}{\omega_2}\right)$ and $\phi_1(\tau) = x\left(\frac{\tau}{\omega_2}\right)$. In further analysis, the "dot" over the function denotes the derivative with regard to the dimensionless time τ.

The dimensionless form of the governing equations is as follows

$$\begin{aligned} &(1+\mu)\ddot{\phi}_1(\tau) + \gamma \dot{\phi}_1(\tau) + \alpha \phi_1(\tau) - \phi(\tau) - \\ &\eta_e(\phi(\tau))^3 - \gamma_e \dot{\phi}(\tau) = m_0 \cos(p\,\tau), \end{aligned} \tag{7.19}$$

$$\phi(\tau) + \eta_e(\phi(\tau))^3 + \gamma_e \dot{\phi}(\tau) + \frac{\mu+1}{\mu} \ddot{\phi}(\tau) + \frac{\mu+1}{\mu} \ddot{\phi}_1(\tau) = 0, \tag{7.20}$$

where

$$\gamma = \frac{c_1}{k_2}\omega_2,\ \alpha = \frac{k_1}{k_2},\ \eta_e = \frac{k_{2n}}{k_2},\ \gamma_e = \frac{c_2}{k_2}\omega_2,\ \mu = \frac{I_1}{I_2},\ m_0 = \frac{M_0}{k_2},\ p = \frac{p_0}{\omega_2}.$$

Equations (7.19)–(7.20) are supplemented by the initial conditions

$$\phi(0) = \phi_0,\ \dot{\phi}(0) = \omega_0,\ \phi_1(0) = \phi_{10},\ \dot{\phi}_1(0) = \omega_{10}, \tag{7.21}$$

where $\phi_0,\ \omega_0,\ \phi_{10},\ \omega_{10}$ are known.

The three differential operators L_1, L_2 and L_n can be distinguished in equations (7.19)–(7.20) according to the formulas (7.3)–(7.4), and they have the following form

$$L_1(\chi) = (1+\mu)\frac{d^2\chi}{d\tau^2} + \gamma\frac{d\chi}{d\tau} + \alpha\chi, \tag{7.22}$$

$$L_2(\chi) = \frac{1+\mu}{\mu}\frac{d^2\chi}{d\tau^2}, \tag{7.23}$$

$$L_n(\chi) = -\gamma_e\frac{d\chi}{d\tau} - \chi - \eta_e\chi^3. \tag{7.24}$$

Therefore taking advantage of the definitions (7.22)–(7.24), the equations of motion could be written in its counterpart compact form

$$L_1(\phi_1(\tau)) + L_n(\phi(\tau)) = m_0\cos(p\,\tau), \tag{7.25}$$

$$L_2(\phi_1(\tau)) + L_2(\phi(\tau)) - L_n(\phi(\tau)) = 0. \tag{7.26}$$

Addition of equations (7.25)–(7.26) one to another, according to (7.5), leads to the linear relation between the generalized coordinates ϕ and ϕ_1, i.e. we have

$$L_1(\phi_1(\tau)) + L_2(\phi_1(\tau)) + L_2(\phi(\tau)) = m_0\cos(p\,\tau), \tag{7.27}$$

or in explicit form

$$\begin{aligned}&(1+\mu)\ddot{\phi}_1(\tau) + \gamma\dot{\phi}_1(\tau) + \alpha\phi_1(\tau) + \frac{\mu+1}{\mu}\ddot{\phi}_1(\tau) + \\ &\frac{\mu+1}{\mu}\ddot{\phi}(\tau) = m_0\cos(p\tau).\end{aligned} \tag{7.28}$$

The relation (7.28) allows determining ϕ_1 assuming that ϕ is already known.

Then, taking advantage of the procedure described in section 7.2 of this chapter, there is a possibility to write one equation containing the function $\phi(\tau)$ alone. This equation in the form of the formula (7.9) is as follows

$$\begin{aligned}&\left(\frac{(1+\mu)^2}{\mu}\frac{d^2}{d\tau^2} + \gamma\frac{d}{d\tau} + \alpha\right)\left(\gamma_e\frac{d\phi}{d\tau} + \phi + \eta_e\phi^3\right) + \\ &\left((1+\mu)\frac{d^2}{d\tau^2} + \gamma\frac{d}{d\tau} + \alpha\right)\left(\frac{1+\mu}{\mu}\frac{d^2\phi}{d\tau^2}\right) = \frac{1+\mu}{\mu}m_0p^2\cos(pt),\end{aligned} \tag{7.29}$$

or, after rearranging it to a more suitable form for further analysis

$$\left(\frac{(1+\mu)^2}{\mu}\frac{d^2}{d\tau^2}+\gamma\frac{d}{d\tau}+\alpha\right)\left(\frac{d^2\phi}{d\tau^2}+\gamma_e\frac{d\phi}{d\tau}+\phi+\eta_e\phi^3\right)+ \frac{1}{\mu}\left(\gamma\frac{d}{d\tau}+\alpha\right)\frac{d^2\phi}{d\tau^2}=\frac{1+\mu}{\mu}m_0p^2\cos(p\,t). \tag{7.30}$$

In this way, one equation of the fourth order has been derived. Observe that equation (7.30) is equivalent, concerning the function $\phi(\tau)$, to the previous set of equations (7.19)–(7.20).

The assumption adopted at the beginning of section 7.3 that $I_1 \gg I_2$ causes the quantity $\frac{1}{\mu}$ to be small. Therefore, the second component on the left-hand side in equation (7.30), is small compared to the other ones, and will be omitted in further analysis. It allows writing the approximate equation for the internal motion in the following form

$$\frac{(1+\mu)^2}{\mu}\ddot{g}(\tau)+\gamma\dot{g}(\tau)+\alpha g(\tau)=\frac{1+\mu}{\mu}m_0p^2\cos(p\,t), \tag{7.31}$$

where

$$g(\tau)=\ddot{\phi}(\tau)+\gamma_e\dot{\phi}(\tau)+\phi(\tau)+\eta_e\phi(\tau)^3. \tag{7.32}$$

After normalization, equation (7.31) may be presented in its simpler form

$$\left(\frac{d^2}{d\tau^2}+2\tilde{\gamma}\frac{d}{d\tau}+\tilde{\alpha}^2\right)g(\tau)=\tilde{m}_0p^2\cos(pt), \tag{7.33}$$

where: $2\tilde{\gamma}=\gamma\frac{\mu}{(1+\mu)^2}$, $\tilde{\alpha}^2=\alpha\frac{\mu}{(1+\mu)^2}$, $\tilde{m}_0=\frac{m_0}{1+\mu}$.

Since equation (7.33) governs the forced vibration of a damped linear oscillator, it has the following analytical solution

$$g(\tau)=\frac{\tilde{m}_0p^2}{\sqrt{\Delta}}\left(\left(\tilde{\alpha}^2-p^2\right)\cos(p\tau)+\tilde{\gamma}p\sin(p\tau)\right)+ e^{-\frac{\tilde{\gamma}\tau}{2}}\left(d_1\cos(\beta\tau)+d_2\sin(\beta\tau)\right), \tag{7.34}$$

where $\beta=\sqrt{\tilde{\alpha}^2-\tilde{\gamma}^2}$, $\Delta=\left(\tilde{\alpha}^2-p^2\right)^2+\tilde{\gamma}^2p^2$, d_1 and d_2 are the integration constants.

The second component of the solution (7.34) exponentially tends to zero. We take into further consideration only its non-decaying part. Using some shortening denotations, and then introducing (7.34) into equation (7.32), the following equation is obtained

$$\ddot{\phi}(\tau)+\gamma_e\dot{\phi}(\tau)+\phi(\tau)+\eta_e\phi(\tau)^3=P\cos(p\tau-\Phi) \tag{7.35}$$

where: $\tan(\Phi)=\frac{p\tilde{\gamma}}{\tilde{\alpha}^2-p^2}$, $P=\frac{\tilde{m}_0p^2}{\sqrt{\Delta}}$.

Owing to the transformations presented above, the original problem of the motion of the rotating and vibrating shaft has been reduced to the one equation describing its

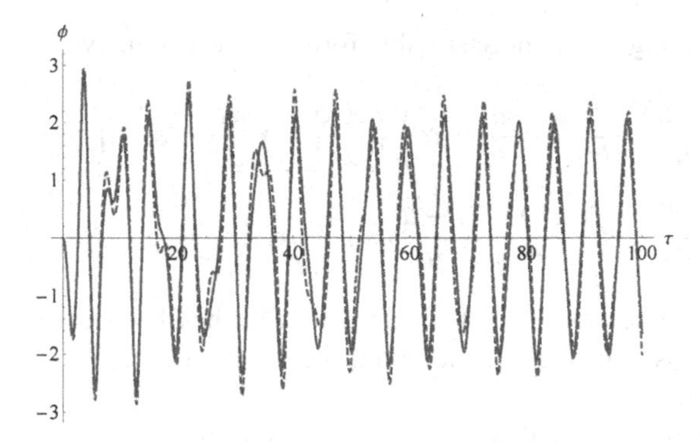

FIGURE 7.2 Time history (transient vibration) for the internal coordinate ϕ obtained from (7.19)–(7.20), and from effective equation (7.35) (dashed).

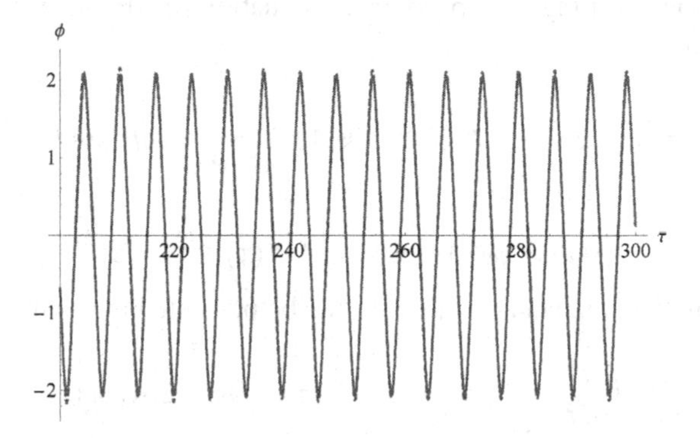

FIGURE 7.3 Time history (steady-state vibration) for the internal coordinate ϕ obtained from (7.19)–(7.20), and from effective equation (7.35) (dashed).

internal motion. The effective equation (7.35) has the form of the Duffing equation widely discussed in the literature (Kovacic [101]).

In order to compare the solutions yielded by the exact original equations (7.19)–(7.20) and by the approximate equation (7.35) the time histories obtained in these two ways are presented in Figure 7.2 (7.3) for transient (steady-state) oscillations for the following fixed parameters: $m_0 = 2.5, p = 1, \mu = 0.01, \eta_e = 0.4,\ \alpha = 0.1, \gamma_e = 0.03, \gamma = 0.04$, and the following initial conditions: $\phi(0) = 0,\ \dot{\phi}(0) = 0,\ \phi_1(0) = 0, \dot{\phi}_1(0) = 0$.

Some transient differences are visible in Figure 7.2 for the early stage of motion. In the steady-stage motion, the dynamical response obtained from the original equations of motion and equation (7.35) is very similar. That confirms the proper description of the internal motion by the effective equation (7.35).

7.5 COMPLEX REPRESENTATION

The further analysis concerns effective equation (7.35). In this section, the modulation equations for amplitude and phase are derived in a similar way as in chapter 4. The special complex functions are introduced into the governing equation.

Let us assume the homogeneous initial conditions of the form

$$\phi(0) = 0, \qquad \dot{\phi}(0) = 0. \tag{7.36}$$

The original problem (7.35)–(7.36) can be transformed into the equivalent counterpart form in the phase space, where

$$\dot{\phi}(\tau) = \omega(\tau), \qquad \ddot{\phi}(\tau) = \dot{\omega}(\tau). \tag{7.37}$$

In this way, we obtain the set of two ODEs of the first-order

$$\dot{\phi}(\tau) - \omega(\tau) = 0, \tag{7.38}$$

$$\dot{\omega}(\tau) + \gamma_e \omega(\tau) + \phi(\tau) + \eta_e \phi(\tau)^3 = P\cos(p\tau - \Phi). \tag{7.39}$$

Crucial meaning in the further procedure is the introduction into above equations (7.38)–(7.39) the new complex function $\psi(\tau)$ and its complex conjugate of the following forms

$$\psi(\tau) = \omega(\tau) + \mathrm{i}\,\phi(\tau), \quad \bar{\psi}(\tau) = \omega(\tau) - \mathrm{i}\,\phi(\tau). \tag{7.40}$$

Now, the phase space coordinates are

$$\phi(\tau) = -\frac{1}{2}\mathrm{i}(\psi(\tau) - \bar{\psi}(\tau)), \quad \omega(\tau) = \frac{1}{2}(\psi(\tau) + \bar{\psi}(\tau)). \tag{7.41}$$

Introduction of equations (7.41) into system (7.38)–(7.39) yields the essential simplification of the studied problem because the latter is reduced to complex first-order ODE:

$$\begin{aligned} &\frac{d\psi(\tau)}{d\tau} + \frac{1}{2}\gamma_e(\psi(\tau) + \bar{\psi}(\tau)) - \mathrm{i}\,\psi(\tau) + \\ &\frac{1}{8}\mathrm{i}\eta_e(\psi(\tau) - \bar{\psi}(\tau))^3 = P\cos(p\tau - \Phi), \end{aligned} \tag{7.42}$$

with the initial condition $\psi(0) = 0$ and the ODE conjugated to it of the following form

$$\begin{aligned} &\frac{d\bar{\psi}(\tau)}{d\tau} + \frac{1}{2}\gamma_e(\psi(\tau) + \bar{\psi}(\tau)) + \mathrm{i}\,\bar{\psi}(\tau) + \\ &\frac{1}{8}\mathrm{i}\eta_e(\psi(\tau) - \bar{\psi}(\tau))^3 = P\cos(p\tau - \Phi), \end{aligned} \tag{7.43}$$

with the initial condition $\bar{\psi}(0) = 0$.

The new function $\Psi(\tau)$ introduced in the following way

$$\psi(\tau) = \Psi(\tau)e^{\mathrm{i}\,\tau}, \quad \bar{\psi}(\tau) = \bar{\Psi}(\tau)e^{-\mathrm{i}\tau}, \tag{7.44}$$

obeys the following equation

$$\begin{aligned}\frac{d\Psi(\tau)}{d\tau} + \frac{1}{8}\eta_e\left(\mathrm{i}e^{2\mathrm{i}\tau}\Psi(\tau)^3 - 3\mathrm{i}|\Psi(\tau)|^2\Psi(\tau) + 3\mathrm{i}e^{-2\mathrm{i}\tau}|\Psi(\tau)|^2\bar{\Psi}(\tau) - \right. \\ \left. \mathrm{i}e^{-4\mathrm{i}\tau}\bar{\Psi}(\tau)^3\right) + \frac{1}{2}\gamma_e\left(\Psi(\tau) + e^{-2\mathrm{i}\tau}\bar{\Psi}(\tau)\right) = e^{-\mathrm{i}\tau}P\cos(p\tau - \Phi),\end{aligned} \tag{7.45}$$

with the initial condition $\Psi(0) = 0$ while the function $\bar{\Psi}(\tau)$ should satisfy the conjugated equation

$$\begin{aligned}\frac{d\bar{\Psi}(\tau)}{d\tau} + \frac{1}{8}\eta_e\left(-\mathrm{i}e^{-2\mathrm{i}\tau}\bar{\Psi}(\tau)^3 + 3\mathrm{i}|\Psi(\tau)|^2\bar{\Psi}(\tau) - 3\mathrm{i}e^{2\mathrm{i}\tau}|\Psi(\tau)|^2\Psi(\tau) + \right. \\ \left. \mathrm{i}e^{4\mathrm{i}\tau}\Psi(\tau)^3\right) + \frac{1}{2}\gamma_e\left(\bar{\Psi}(\tau) + e^{2\mathrm{i}\tau}\Psi(\tau)\right) = e^{\mathrm{i}\tau}P\cos(p\tau - \Phi),\end{aligned} \tag{7.46}$$

with the initial condition $\bar{\Psi}(0) = 0$.

7.6 SOLUTION METHOD

The asymptotic analysis of the initial value problems (7.45) and (7.46) is performed using the MSM. Two time scales τ_0 and τ_1 are adopted in the analysis and defined with the help of the small parameter $0 < \varepsilon \ll 1$, i.e. we assume

$$\tau_0 = \tau, \qquad \tau_1 = \varepsilon\tau. \tag{7.47}$$

Due to the above definition, the time scale τ_0 is called fast, while τ_1 is called slow.

The time derivative appearing in equations (7.45) and (7.46) should be redefined according to the definitions of time scales

$$\frac{d}{d\tau} = \frac{\partial}{\partial\tau_0} + \varepsilon\frac{\partial}{\partial\tau_1}. \tag{7.48}$$

The assumed solution has the form

$$\Psi(\tau;\varepsilon) = \sum_{k=0}^{1}\varepsilon^k\Psi_k(\tau_0,\tau_1) + O(\varepsilon^3), \tag{7.49}$$

where $\Psi_0(\tau_0,\tau_1)$ and $\Psi_1(\tau_0,\tau_1)$ are the unknown complex functions.

We assume that the system is weakly damped and that the nonlinear characteristics of elasticity and the amplitude of the external excitation are small. Therefore, some parameters can be expressed using the small parameter. Let us introduce the new notations as follows

$$\gamma_e = 2\varepsilon\widehat{\gamma}_e, \quad \eta_e = 8\varepsilon\widehat{\eta}_e, \quad P = 2\varepsilon\widehat{F}. \tag{7.50}$$

7.7 RESONANT OSCILLATION

Since the further analysis concerns the case of the main resonance, we assume that $p \approx 1$. The small detuning parameter $\sigma = \varepsilon\widehat{\sigma}$ is adopted as the measure to the distance of the pure (strict) resonance. Therefore the substitution

$$p = 1 + \sigma = 1 + \varepsilon\widehat{\sigma} \tag{7.51}$$

into equations (7.45)–(7.46) allows investigation of the behavior of the system in some neighbourhood of the main resonance.

According to the MSM, the expansion (7.49) is introduced into equations (7.45) and (7.46) taking into account substitutions (7.50) and (7.51). The time derivative is calculated according to the definition (7.48). In this way, we obtain the equations containing small parameter ε. These equations are ordered for the powers of ε. The coefficients standing at ε of the same power yield the equations with unknown functions $\Psi_0(\tau_0,\tau_1)$ and $\Psi_1(\tau_0,\tau_1)$.

The equations of order ε^0 have the simple homogeneous form

$$\frac{\partial\Psi_0(\tau_0,\tau_1)}{\partial\tau_0} = 0, \tag{7.52}$$

$$\frac{\partial\bar{\Psi}_0(\tau_0,\tau_1)}{\partial\tau_0} = 0. \tag{7.53}$$

Consequently, it follows that functions Ψ_0 and $\bar{\Psi}_0$ do not depend on the fast time τ_0, so $\Psi_0(\tau_0,\tau_1) \equiv \Psi_0(\tau_1)$ and $\bar{\Psi}_0(\tau_0,\tau_1) \equiv \bar{\Psi}_0(\tau_1)$.

The equation of order ε^1 is

$$\begin{aligned}\frac{\partial\Psi_1(\tau_0,\tau_1)}{\partial\tau_0} &= -\frac{\partial\Psi_0(\tau_1)}{\partial\tau_1} - \hat{\gamma}_e\left(\Psi_0(\tau_1) + e^{-2i\tau_0}\bar{\Psi}_0(\tau_1)\right) - \\ &\hat{\eta}_e i\left(e^{2i\tau_0}\Psi_0(\tau_1)^3 - 3|\Psi_0(\tau_1)|^2\Psi_0(\tau_1) + 3e^{-2i\tau_0}|\Psi_0(\tau_1)|^2\bar{\Psi}_0(\tau_1) - \right. \\ &\left. e^{-4i\tau_0}\bar{\Psi}_0(\tau_1)^3\right) + \hat{F}e^{i(\hat{\sigma}\tau_1-\Phi)} + \hat{F}e^{-i(2\tau_0+\hat{\sigma}\tau_1-\Phi)},\end{aligned} \tag{7.54}$$

and its counter part conjugated equation take the form

$$\begin{aligned}\frac{\partial\bar{\Psi}_1(\tau_0,\tau_1)}{\partial\tau_0} &= -\frac{\partial\bar{\Psi}_0(\tau_1)}{\partial\tau_1} - \hat{\gamma}_e\left(e^{2i\tau_0}\Psi_0(\tau_1) + \bar{\Psi}_0(\tau_1)\right) - \\ &\hat{\eta}_e i\left(e^{4i\tau_0}\Psi_0(\tau_1)^3 - 3e^{2i\tau_0}|\Psi_0(\tau_1)|^2\Psi_0(\tau_1) + 3|\Psi_0(\tau_1)|^2\bar{\Psi}_0(\tau_1) - \right. \\ &\left. e^{-2i\tau_0}\bar{\Psi}_0(\tau_1)^3\right) + \hat{F}e^{-i(\hat{\sigma}\tau_1-\Phi)} + \hat{F}e^{i(2\tau_0+\hat{\sigma}\tau_1-\Phi)}.\end{aligned} \tag{7.55}$$

The solution should be bounded, therefore the components which generate the secular terms should be eliminated. The structure of the first-order equations (7.54)–(7.55) allow recognizing these terms as those independent of τ_0, which implies the following condition for elimination of secular terms in equation (7.54)

$$\frac{\partial\Psi_0(\tau_1)}{\partial\tau_1} + \hat{\gamma}_e\Psi_0(\tau_1) - 3\hat{\eta}_e\, i|\Psi_0(\tau_1)|^2\Psi_0(\tau_1) - \hat{F}e^{i(\hat{\sigma}\tau_1-\Phi)} = 0, \tag{7.56}$$

and in equation (7.55) when

$$\frac{\partial \bar{\Psi}_0(\tau_1)}{\partial \tau_1} + \hat{\gamma}_e \bar{\Psi}_0(\tau_1) + 3\hat{\eta}_e \mathrm{i} |\Psi_0(\tau_1)|^2 \bar{\Psi}_0(\tau_1) - \hat{F} e^{-\mathrm{i}(\hat{\sigma}\tau_1 - \Phi)} = 0. \quad (7.57)$$

Let us introduce the polar representation of the complex functions

$$\Psi_0 = \hat{a}(\tau_1) e^{\mathrm{i}\hat{\delta}(\tau_1)}, \qquad \bar{\Psi}_0 = \hat{a}(\tau_1) e^{-\mathrm{i}\hat{\delta}(\tau_1)}, \quad (7.58)$$

where $\hat{a}(\tau_1)$ and $\hat{\delta}(\tau_1)$ are real-valued.

The real-valued representation (7.58) of the complex functions Ψ_0 and $\bar{\Psi}_0$, introduced into equations (7.56) gives the new form of the solvability conditions

$$\frac{d\hat{a}(\tau_1)}{d\tau_1} + \mathrm{i}\hat{a}(\tau_1) \frac{d\hat{\delta}(\tau_1)}{d\tau_1} + \hat{\gamma}_e \hat{a}(\tau_1) - 3\hat{\eta}_e \mathrm{i}(\hat{a}(\tau_1))^3 - \hat{F} e^{\mathrm{i}(\hat{\sigma}\tau_1 - \Phi - \hat{\delta}(\tau_1))} = 0, \quad (7.59)$$

and similarly, from equation (7.57) we get

$$\frac{d\hat{a}(\tau_1)}{d\tau_1} - \mathrm{i}\hat{a}(\tau_1) \frac{d\hat{\delta}(\tau_1)}{d\tau_1} + \hat{\gamma}_e \hat{a}(\tau_1) + 3\hat{\eta}_e \mathrm{i}(\hat{a}(\tau_1))^3 - \hat{F} e^{-\mathrm{i}(\hat{\sigma}\tau_1 - \Phi - \hat{\delta}(\tau_1))} = 0. \quad (7.60)$$

Now, multiplying equations (7.59)–(7.60) by ε and returning to the original denotations (7.50) and (7.51) and using the definition of the derivative (7.48) we can transform equation (7.59) to the form

$$\frac{da(\tau)}{d\tau} + \mathrm{i}a(\tau) \frac{d\delta(\tau)}{d\tau} + \frac{1}{2}\gamma_e a(\tau) - \frac{3}{8}\eta_e \mathrm{i}(a(\tau))^3 - \frac{P}{2} e^{\mathrm{i}(\sigma\tau - \Phi - \delta(\tau))} = 0, \quad (7.61)$$

and similarly, from equation (7.60) we get

$$\frac{da(\tau)}{d\tau} - \mathrm{i}a(\tau) \frac{d\delta(\tau)}{d\tau} + \frac{1}{2}\gamma_e a(\tau) + \frac{3}{8}\eta_e \mathrm{i}(a(\tau))^3 - \frac{P}{2} e^{-\mathrm{i}(\sigma\tau - \Phi - \delta(\tau))} = 0. \quad (7.62)$$

where the function $a(\tau) = \hat{a}(\tau_1)$ and $\delta(\tau) = \hat{\delta}(\tau_1)$.

Separation of the real and imaginary part of equations (7.61)–(7.62) and using the trigonometric representation of the exponent functions gives the following first-order ODEs

$$\frac{da(\tau)}{d\tau} + \frac{1}{2}\gamma_e a(\tau) = \frac{P}{2}\cos(\sigma\tau - \Phi - \delta(\tau)), \quad (7.63)$$

$$a(\tau)\frac{d\delta(\tau)}{d\tau} - \frac{3}{8}\eta_e (a(\tau))^3 = \frac{P}{2}\sin(\sigma\tau - \Phi - \delta(\tau)), \quad (7.64)$$

Now using substitutions (7.44) in the first of definitions (7.41) and using real representations of the complex functions (7.58) we get

$$\begin{aligned} \phi(\tau) &= -\frac{1}{2}i(\psi(\tau) - \bar{\psi}(\tau)) = -\frac{1}{2}i\left(\Psi(\tau)e^{\mathrm{i}\tau} - \bar{\Psi}(\tau)e^{-\mathrm{i}\tau}\right) = \\ &\hat{a}(\tau_1)\sin\left(\hat{\delta}(\tau_1) + \tau\right) = a(\tau)\sin(\delta(\tau) + \tau). \end{aligned} \quad (7.65)$$

The above relation (7.65) indicates that the functions $a(\tau)$ and $\delta(\tau)$ represent modulations of amplitude and phase of the vibration of the variable $\phi(\tau)$ in the first-order of the asymptotic approximation.

It is convenient in further analysis to introduce the modified phase into equations (7.63)–(7.64) of the form

$$\theta(\tau) = \sigma\tau - \Phi - \delta(\tau), \tag{7.66}$$

Substituting (7.66) into (7.63)–(7.64), the modulation equations take the following autonomous form

$$\frac{da(\tau)}{d\tau} + \frac{1}{2}\gamma_e a(\tau) = \frac{P}{2}\cos(\theta(\tau)), \tag{7.67}$$

$$-a(\tau)\frac{d\theta(\tau)}{d\tau} + a(\tau)\sigma - \frac{3}{8}\eta_e(a(\tau))^3 = \frac{P}{2}\sin(\theta(\tau)). \tag{7.68}$$

The approximate solution (7.65) can be recast to the following form

$$\phi(\tau) = a(\tau)\sin(\tau(\sigma+1) - \theta(\tau) + \Phi). \tag{7.69}$$

7.8 STEADY-STATE MOTION

Equations (7.67)–(7.68) allow the investigation of both steady and non-steady state motion. The steady-state occurs when time derivatives $\frac{da}{d\tau} = 0$ and $\frac{d\theta}{d\tau} = 0$. This assumption leads to the system of the algebraic equations for the constant values a and θ:

$$\frac{1}{2}\gamma_e a = \frac{P}{2}\cos(\theta), \tag{7.70}$$

$$a\sigma - \frac{3}{8}\eta_e a^3 = \frac{P}{2}\sin(\theta). \tag{7.71}$$

The system (7.70)–(7.71) has nonlinearities due to trigonometric and power functions, but one can easily reduce them to one equation. Thanks to the autonomous form of these equations we can eliminate the modified phase $\theta(\tau)$. Taking advantage of the trigonometric identity, after elementary transformations we get the amplitude-frequency function

$$\gamma_e^2(a)^2 + \left(2a\sigma - \frac{3}{4}\eta_e a^3\right)^2 = P^2. \tag{7.72}$$

The above relation (7.72) describes the dependence between amplitude and frequency of the external loading and gives the possibility to draw the amplitude back-bone response curve in resonance. The exemplary amplitude-frequency dependence is illustrated in Figure 7.4. The curve bends to the right that reflects the assumed hard characteristics of the shaft elasticity.

The back-bone curve presented in Figure 7.4 shows that for some values of detuning parameter more than one value of steady-state amplitude can appear. The stability analysis is needed to recognize which part of the curve is stable, i.e. in order to get physically reliable solution.

7.8.1 STABILITY

Though equation (7.72) is very useful in obtaining the amplitude curve but it does not contain the phase. The stability analysis of the stationary solution of the modulation equations (7.67)–(7.68) demand to determine both amplitude and phase. It is convenient to introduce the following substitutions

$$\sin(\theta) = \frac{2t_g}{1+t_g^2}, \qquad \cos(\theta) = \frac{1-t_g^2}{1+t_g^2}, \tag{7.73}$$

into equations (7.70)–(7.71), where $t_g = \tan(\theta/2)$. In this way, we obtain the following set of algebraic equations with unknowns a and t_g:

$$\gamma_e a = \frac{1-t_g^2}{1+t_g^2} P, \tag{7.74}$$

$$a\sigma - \frac{3}{8}\eta_e a^3 = \frac{t_g}{1+t_g^2} P. \tag{7.75}$$

The system (7.74)–(7.75) can be solved with regard to variables a and t_g. However, observe that t_g uniquely define the value of modified phase $\theta = 2\arctan t_g$. The amplitude response obtained from the solution of (7.74)–(7.75) gives the same graph as presented in Figure 7.4, while the resonance response of the modified phase generated for the same parameters ($P = 0.001, \eta_e = .002, \gamma_e = 0.0004$) is presented in Figure 7.5.

Let us assume that a_s and θ_s are fixed points of the modulation equations (7.67)–(7.68). That means a_s and θ_s satisfies equations (7.70)–(7.71). To check the stability in sense of Lyapunov of these solutions, the new functions $\tilde{a}(\tau)$ and $\tilde{\theta}(\tau)$ assumed

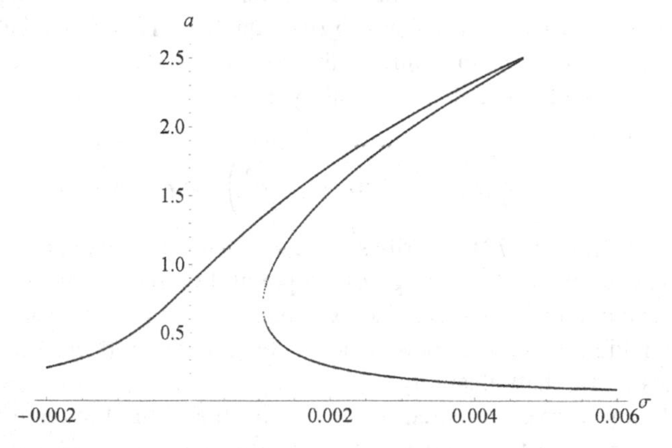

FIGURE 7.4 Amplitude as a function of the detuning parameter σ for $P = 0.001, \eta_e = 0.002, \gamma_e = 0.0004$.

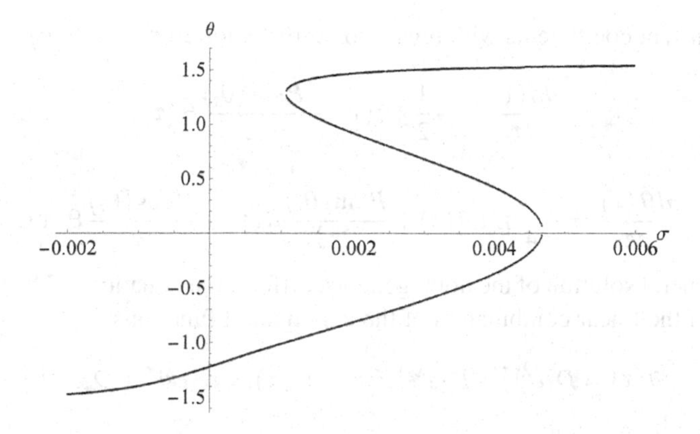

FIGURE 7.5 Modified phases for $P = 0.001, \eta_e = 0.002, \gamma_e = 0.0004$.

as small perturbations, are added to the stationary solutions in the following way

$$a(\tau) = a_s + \tilde{a}(\tau), \qquad \theta(\tau) = \theta_s + \tilde{\theta}(\tau). \tag{7.76}$$

The definitions (7.76) are introduced into equations (7.67)–(7.68) which take the form

$$\frac{d\tilde{a}(\tau)}{d\tau} = -\frac{1}{2}\gamma_e(a_s + \tilde{a}(\tau)) + \frac{P}{2}\cos\left(\theta_s + \tilde{\theta}(\tau)\right), \tag{7.77}$$

$$\frac{d\tilde{\theta}(\tau)}{d\tau} = \sigma - \frac{3}{8}\eta_e(a_s + \tilde{a}(\tau))^2 - \frac{P}{2(a_s + \tilde{a}(\tau))}\sin\left(\theta_s + \tilde{\theta}(\tau)\right). \tag{7.78}$$

Then, the right hand sides of equations (7.77)–(7.78) are expanded in power series around the fixed point (a_s, θ_s). In further analysis, only first-order expansion components are taken into account and products of functions $\tilde{a}(\tau)$ and $\tilde{\theta}(\tau)$ are omitted as a small quantity of higher order. The linearized perturbed equations take the following form

$$\frac{d\tilde{a}(\tau)}{d\tau} = -\frac{1}{2}\gamma_e(a_s + \tilde{a}(\tau)) + \frac{P}{2}\left(\cos(\theta_s) - \sin(\theta_s)\tilde{\theta}(\tau)\right), \tag{7.79}$$

$$\begin{aligned}\frac{d\tilde{\theta}(\tau)}{d\tau} &= \sigma - \frac{3}{8}\eta_e\left(a_s^2 + 2a_s\tilde{a}(\tau)\right) - \\ \frac{P}{2a_s^2}&\left(a_s\sin(\theta_s) - \tilde{a}(\tau)\sin(\theta_s) + a_s\tilde{\theta}(\tau)\cos(\theta_s)\right).\end{aligned} \tag{7.80}$$

Taking into account that a_s and θ_s satisfy the steady-state equations (7.70)–(7.71), the above set (7.79)–(7.80) is simplified to the form of linear first-order ODEs having constant coefficients with regard to perturbation $\tilde{a}(\tau)$ and $\tilde{\theta}(\tau)$

$$\frac{d\tilde{a}(\tau)}{d\tau} = -\frac{1}{2}\gamma_e\tilde{a}(\tau) - \frac{P\sin(\theta_s)}{2}\tilde{\theta}(\tau), \tag{7.81}$$

$$\frac{d\tilde{\theta}(\tau)}{d\tau} = -\frac{3}{4}\eta_e a_s\tilde{a}(\tau) + \frac{P\sin(\theta_s)}{2a_s^2}\tilde{a}(\tau) - \frac{P\cos(\theta_s)}{2a_s}\tilde{\theta}(\tau). \tag{7.82}$$

The general solution of the homogeneous differential equations (7.81)–(7.82) has the form of the linear combination of the exponential functions

$$\tilde{a}(\tau) = D_1 e^{\lambda_1\tau} + D_2 e^{\lambda_2\tau}, \qquad \tilde{\theta}(\tau) = D_3 e^{\lambda_1\tau} + D_4 e^{\lambda_2\tau}, \tag{7.83}$$

where $D_i^{(j)}$ are constants dependent on the initial conditions and λ_i are the roots of the characteristic equation

$$|\mathbf{A} - \lambda\mathbf{I}| = 0. \tag{7.84}$$

Here **I** is the identity matrix, and **A** is the characteristic matrix of the differential equations (7.81)–(7.82)

$$\mathbf{A} = \begin{bmatrix} -\frac{\gamma_e}{2} & -\frac{P}{2}\sin(\theta_s) \\ -\frac{3}{4}a_s\eta_e + \frac{P\sin(\theta_s)}{2a_s^2} & -\frac{P\cos(\theta_s)}{2\,a_s} \end{bmatrix}. \tag{7.85}$$

The fixed point $(a_s,\ \theta_s)$ is stable in the sense of Lyapunov, when the eigenvalues λ_1, λ_2 of matrix **A** have negative real parts.

Taking advantage of the above described procedure we can check the stability of the resonant curves. Let us test the stability of the curves presented in Figures 7.4 and 7.5. The stability is estimated by calculating the eigenvalues of matrix **A** for the subsequent points along the curves. The result of the procedure is presented in Figures 7.6 and 7.7. The light gray colour indicates the points for which both eigenvalues λ_1, λ_2 have negative real parts that means stable amplitude and phase.

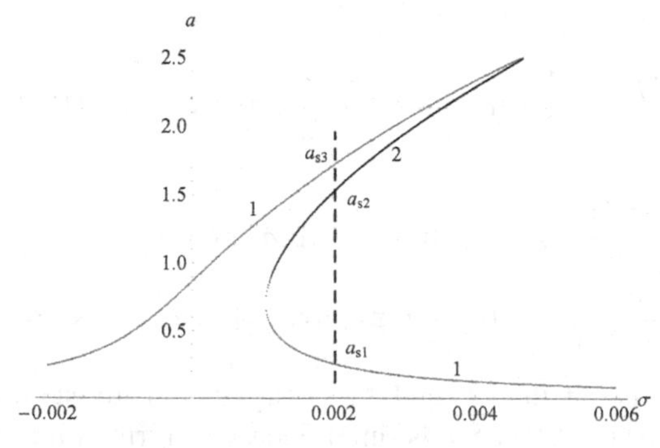

FIGURE 7.6 Amplitude vs. detuning parameter; 1 and 2 curves denote the stable and unstable branches respectively for $P = 0.001, \eta_e = 0.002, \gamma_e = 0.0004$.

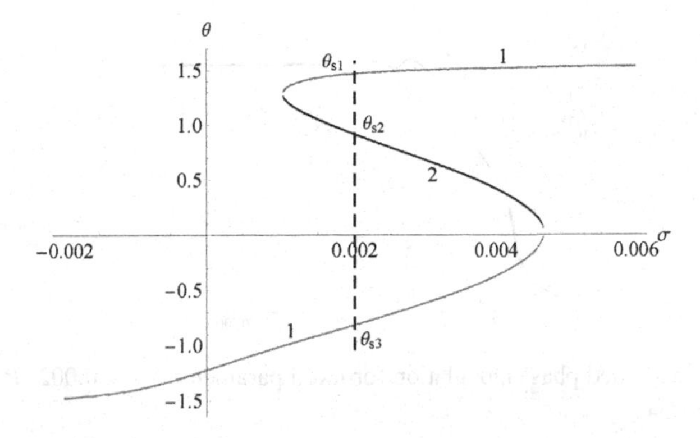

FIGURE 7.7 Modified phase vs. detuning parameter; 1 and 2 curves denote the stable and unstable branches respectively $P = 0.001, \eta_e = 0.002, \gamma_e = 0.0004$.

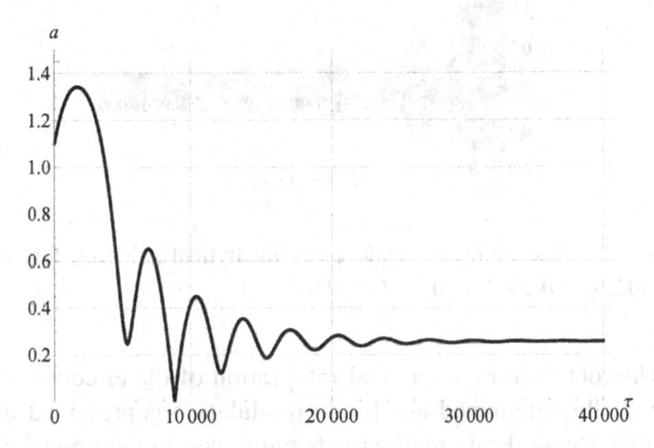

FIGURE 7.8 Amplitude modulation for fixed parameters: $\sigma = 0.002$, $P = 0.001, \eta_e = 0.002, \gamma_e = 0.0004$.

The dashed vertical line in Figure 7.6 allows one to recognize three different steady-state values of the amplitude for $\sigma = 0.002$, where two of them are stable $a_{s1} = 0.255$ and $a_{s3} = 1.717$ and one unstable $a_{s2} = 1.523$. The oscillation can be realized with amplitudes either a_{s1} or a_{s2} depending on the initial conditions or just on the prior state of motion. Similarly, the counterpart to the stable amplitudes are phases $\theta_{s1} = 1.469$, $\theta_{s2} = 0.916$ and $\theta_{s3} = -0.814$, marked in Figure 7.7. Time histories of the amplitude $a(\tau)$ and phase $\theta(\tau)$ obtained by integration of the modulation equations (7.67)–(7.68) are presented in Figures 7.8 and 7.9, respectively. The initial conditions are $a(0) = 1.1, \theta(0) = 0$, and other parameters are the same as earlier.

The steady-state phase is determined by equations (7.70)–(7.71) with accuracy to the additive constant 2π. Time history obtained through the approximate analytical solution (7.69) is presented in Figure 7.10. The amplitude of this vibration is governed by the set (7.67)–(7.68), so the amplitude modulation presented in Figure 7.8 fits excellently to the analytically obtained time history.

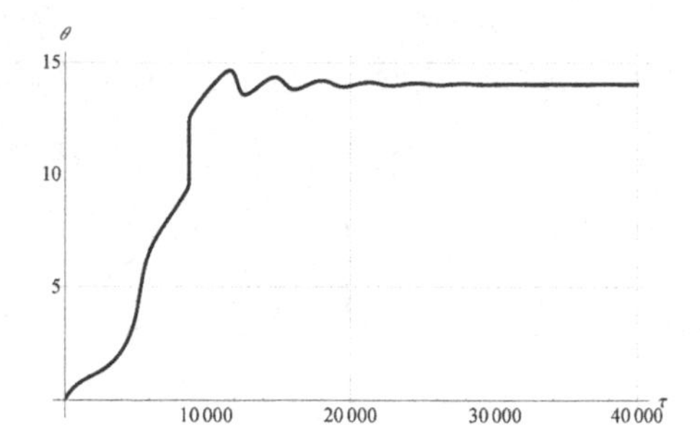

FIGURE 7.9 Modified phase modulation for fixed parameters: $\sigma = 0.002,\ P = 0.001, \eta_e = 0.002, \gamma_e = 0.0004$.

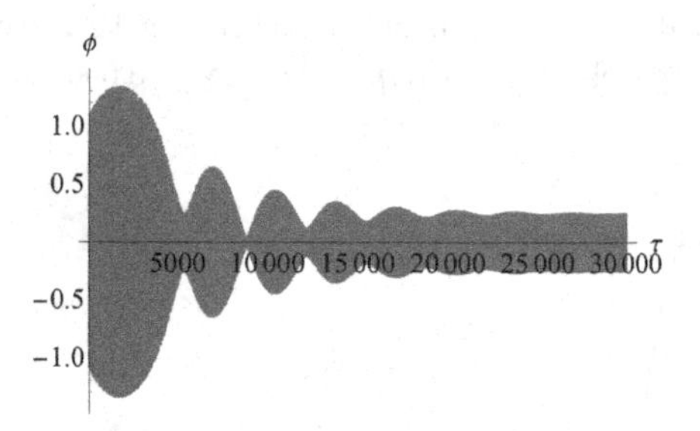

FIGURE 7.10 Time history of the angle $\phi(\tau)$ (analytical solution) for $\sigma = 0.002,\ P = 0.001, \eta_e = 0.002, \gamma_e = 0.0004,\ a(0) = 1.1,\ \theta(0) = 0$.

The solution obtained by numerical integration of the effective equation (7.35) and the analytically determined amplitude modulation is presented in Figure 7.11. The fourth-order Runge-Kutta method was employed in the procedure *NDSolve* in *Mathematica* while carrying out the numerical calculations. The amplitude modulation curve (black line) perfectly envelopes the fast-changing oscillations (gray colour) which validates the correctness and high accuracy of the asymptotic solution.

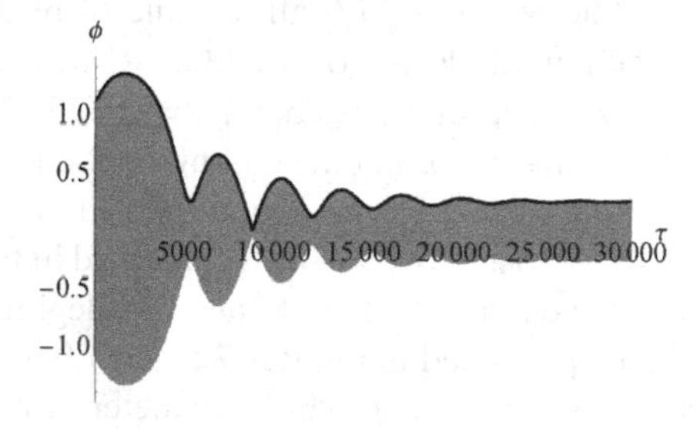

FIGURE 7.11 Time history of the angle $\phi(\tau)$ (numerical solution) for $\sigma = 0.002,\ P = 0.001, \eta_e = 0.002, \gamma_e = 0.0004,\ a(0) = 1.1,\ \theta(0) = 0$.

The amplitude of vibration stabilizes at the value 0.255, which corresponds to the above mentioned lower steady-state amplitude a_{s1}.

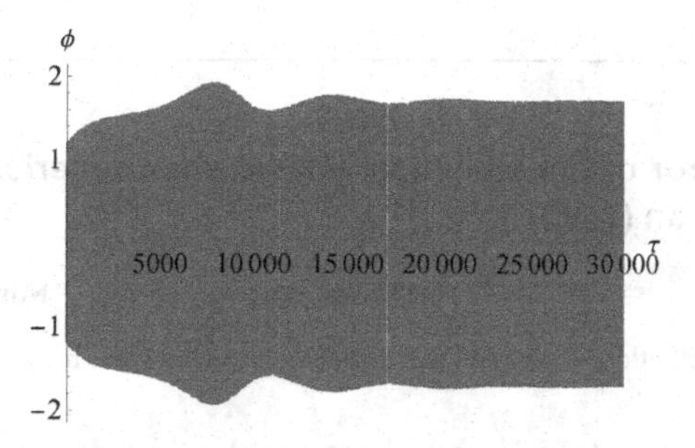

FIGURE 7.12 Time history of the angle $\phi(\tau)$ (analytical solution) for $\sigma = 0.002, P = 0.001,$ $\eta_e = 0.002,\ \gamma_e = 0.0004,\ a(0) = 1.2,\ \theta(0) = 0.$

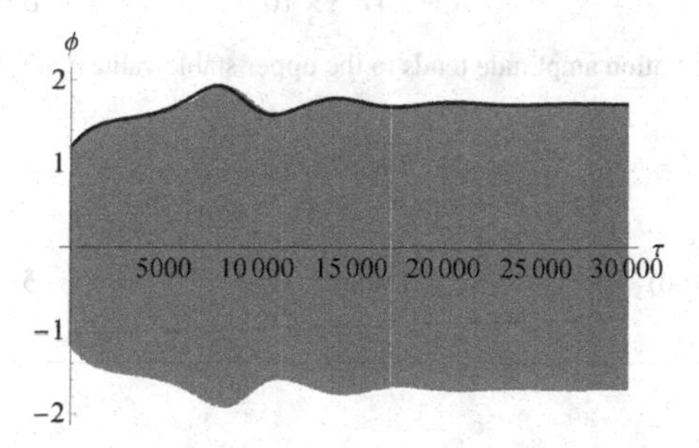

FIGURE 7.13 Time history of the angle $\phi(\tau)$ (numerical solution) for $\sigma = 0.002,\ P = 0.001, \eta_e = 0.002, \gamma_e = 0.0004,\ a(0) = 1.2,\ \theta(0) = 0.$

Let us assume slightly greater initial amplitude $a(0) = 1.2,\ \theta(0) = 0$. The companion asymptotic solution is presented in Figure 7.12. In this case, the amplitude tends to the second stable solution of the set (7.70)–(7.71). This stands for the value a_{s3} indicated in Figure 7.6.

The numerically obtained time history (gray colour) and amplitude modulation curve (black line) are reported in Figure 7.13.

The correctness and high accuracy of the asymptotic solutions based on numerical validation are visible in Figures 7.10–7.13.

The objective way for measuring of the error of the numerical and analytical solution is proposed by the formula

$$\delta = \frac{1}{\tau_e - \tau_s} \int_{\tau_s}^{\tau_e} \left| \ddot{\phi}_a(\tau) + \gamma_e \dot{\phi}_a(\tau) + \phi_a(\tau) + \eta_e \phi_a(\tau)^3 - P\cos(p\tau - \Phi) \right| d\tau, \quad (7.86)$$

where τ_s and τ_e are the instants of the beginning and end of the tested time history. The error δ is the measure of the fulfillment of equation (7.35) by the analytical solution ϕ_a. The same measure for the numerical solution ϕ_n is used.

Values of the error (7.86) for the analytical and numerical approximate solutions discussed above are collected in Table 7.1.

TABLE 7.1
The values of error of the analytical (MSM) and numerical solutions of the governing equation (7.35)

	MSM solution	Numerical solution
Oscillation amplitude tends to the lower stable value $a_{s1} = 0.255$		
Transient stage $\tau \in (0, 100)$	$\delta = 4.4022 \times 10^{-4}$	$\delta = 5.0713 \times 10^{-7}$
Stationary stage $\tau \in (10000, 10100)$	$\delta = 5.4775 \times 10^{-6}$	$\delta = 2.5223 \times 10^{-7}$
Oscillation amplitude tends to the upper stable value $a_{s3} = 1.717$		
Transient stage $\tau \in (0, 100)$	$\delta = 5.6837 \times 10^{-4}$	$\delta = 4.7348 \times 10^{-7}$
Stationary stage $\tau \in (10000, 10100)$	$\delta = 1.4024 \times 10^{-3}$	$\delta = 2.9334 \times 10^{-7}$

The error (7.86) of the fulfilment of the governing equation (7.35) is smaller in the case of numerical than analytical solution, however, in all cases, it is very small comparing to the values of the vibration amplitudes.

Case study: Undamped vibrations

Some analytical in-depth study of the system behavior is possible in the case when dissipative effects are neglected. Let us discuss the case of no damping motion γ_e=0. From equations (7.70)–(7.71), one can find that the modified phase in the steady-state motion $= \pm\frac{\pi}{2} + n\pi$, where $n \in C$. Limiting modified phase to the range $-\pi < \theta < \pi$, the modified phase $\theta = -\frac{\pi}{2}$ corresponds to the algebraic equation

$$2a\sigma - \frac{3}{4}\eta_e a^3 = -P, \tag{7.87}$$

while $\theta = \frac{\pi}{2}$ corresponds to the equation

$$2a\sigma - \frac{3}{4}\eta_e a^3 = P. \tag{7.88}$$

The roots of equations (7.87)–(7.88) are the amplitudes of the steady-state vibration near resonance. The number of roots depends on the sign of the discriminant

$$D_1 = -\frac{512\sigma^3}{729\eta_e^3} + \frac{4P^2}{9\eta_e^2}, \tag{7.89}$$

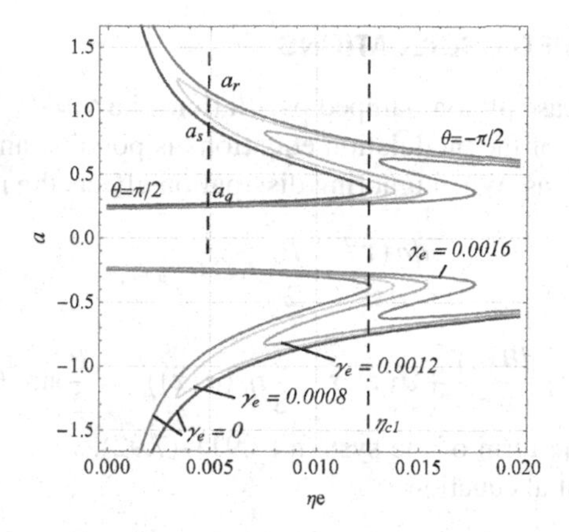

FIGURE 7.14 Position of the stationary points of the system (7.67)–(7.68) versus η_e for various damping coefficients γ_c.

whereby D_1>0 for

$$\eta_e > \eta_{c1} = \frac{128\,\sigma^3}{81\,P^2}. \tag{7.90}$$

In the latter case, both equations (7.87) and (7.88) have one real root. For $\theta = -\frac{\pi}{2}$ its value is positive and indicates the amplitude of the resonance center "a_r" in the phase plane (a, θ) presented in Figure 7.15. Whereas for $\theta = \frac{\pi}{2}$ the root is negative (nonphysical). In the case when $\eta_e < \eta_{c1}$, each of equations (7.87)–(7.88) has three real roots. For $\theta = \frac{\pi}{2}$ one of them is negative which is nonphysical, and the other two represent the quasilinear centre "a_q" and the saddle point "a_s". For $\theta = -\frac{\pi}{2}$ there is one positive root denoting the resonance center "a_r", where $a = a_r \to \infty$ when $\eta_e \to 0$. The above discussion illustrates the graph shown in Figure 7.14 for the fixed parameters: $\sigma = 0.002$, $P = 0.001$. The critical value in this figure is marked as η_{c1}. The gray lines denote the position of the steady state amplitudes in the case of the damped oscillation. Only the positive values of a have physical meaning. The position of stationary amplitudes $a_q = 0.268$, $a_s = 0.872$, $a_r = 1.140$ for $\eta_e = 0.005$ are marked in Figure 7.14.

7.9 NON-STATIONARY OSCILLATIONS

Most researchers emphasize the analysis of the steady-state motion as the most important and practical. However, in some cases of strong resonance and small damping, a transient state can be localized for a long time. Therefore, there is a need to investigate also non-stationary oscillations, where intensive energy exchange takes place.

7.9.1 NON-DAMPED OSCILLATIONS

Let us analyze the case of non-damped oscillations, i.e $\gamma_e = 0$. With this assumption the analytical study of the modulation equations is possible and we can draw some qualitative conclusions. When ignoring dissipation effects the modulation equations take the form

$$\frac{da(\tau)}{d\tau} = \frac{P}{2}\cos(\theta(\tau)), \tag{7.91}$$

$$-a(\tau)\frac{d\theta(\tau)}{d\tau} + a(\tau)\sigma - \frac{3}{8}\eta_e(a(\tau))^3 = \frac{P}{2}\sin(\theta(\tau)). \tag{7.92}$$

According to the form of the system (7.91)–(7.92), we can write the following companion differential equation

$$\begin{gathered} a(\tau)\frac{P}{2}\cos(\theta(\tau))d\theta(\tau) + \\ \left(\frac{P}{2}\sin(\theta(\tau)) + \frac{3}{8}\eta_e a(\tau)^3 - a(\tau)\sigma\right)da(\tau) = 0, \end{gathered} \tag{7.93}$$

which has the form of the exact differential equation, and has the first integral

$$H = -a\frac{P}{2}\sin(\theta) + \sigma\frac{a^2}{2} - \frac{3}{32}\eta_e a^4 = const. \tag{7.94}$$

The constant on the right side depends on the initial condition for the system (7.91)–(7.92). Equation (7.94) allows us to draw phase trajectories in the plane (a, θ). The example of such trajectories calculated for various initial conditions are presented in Figure 7.15 for the same data as above, so the position of stationary amplitudes a_q, a_s, a_r correspond to their location marked in Figure 7.14.

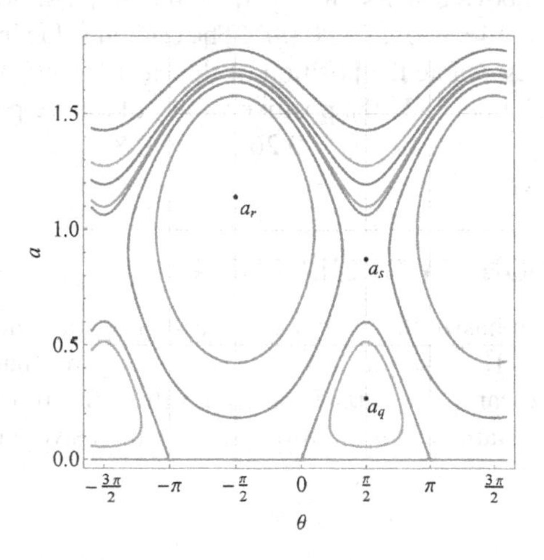

FIGURE 7.15 Trajectories in the plane (a, θ) for the following fixed parameters: $\sigma = 0.002$, $P = 0.001$, $\eta_e = 0.005$.

Let us analyze the important case when the maximum transfer of energy between the system and the external loading appears. This situation takes place for $H = 0$ in equation (7.94). In that case the first integral of equation (7.93) takes the form

$$H_0 = -16aP\sin(\theta) + 16\sigma a^2 - 3\eta_e a^3 = 0. \tag{7.95}$$

The trajectory corresponding to the maximum energy transfer $H_0 = 0$ is called Limiting Phase Trajectory (LPT) (Manevitch and Musienko [136]). The total differential $dH_0 = 0$, which implies

$$\left(16\sigma - 9\eta_e a^2\right) da - 16P\cos(\theta)\, d\theta = 0, \tag{7.96}$$

and consequently

$$\frac{da}{d\theta} = \frac{16P\cos(\theta)}{16\sigma - 9\eta_e a^2}. \tag{7.97}$$

Therefore, the curve H_0 has extremum for $\theta = -\frac{\pi}{2} + k\pi$ and $\theta = \frac{\pi}{2} + k\pi$, where $k \in C$. Let us limit our further analysis to the interval $-\pi < \theta < \pi$.

Case study 1: $\theta = -\frac{\pi}{2}$.

The number of roots of equation (7.95) for $\sigma > 0$ and $P > 0$ depends on the sign of the expression

$$D = \frac{64P^2}{9\,\eta_e^2} - \frac{4096\sigma^3}{729\eta_e^3}, \tag{7.98}$$

whereas $D = 0$, when

$$\eta_e = \eta_c = \frac{64\sigma^3}{81P^2} = \frac{1}{2}\eta_{c1}. \tag{7.99}$$

The maximum value of the amplitude a_{max} of non-damped resonant vibration is presented in Figure 7.16, which were drawn on the basis of equation (7.95) for $\sigma = 0.002$ and $P = 0.001$. Now for the critical nonlinearity parameter $\eta_c = 0.00632$, the jump appears. The modified phase changes from $\frac{\pi}{2}$ to $-\frac{\pi}{2}$ and the maximum value of amplitude significantly increases.

For $\eta_e < \eta_c$ also $D < 0$, and equation (7.95) has three real roots, where one of them is positive (denoted by "1" in Figures 7.16 and 7.17) and corresponds to the open phase trajectory on the plane (θ, a) (see Fig. 7.17). In this situation, phase θ grows unlimitedly in time. Two other roots are negative and therefore nonphysical ($a > 0$ by definition).

For $\eta_e > \eta_c$ also $D > 0$, and equation (7.95) has two complex roots and one real positive denoted by "2" in Figure 7.16 and in Figure 7.17. This root corresponds to LPT and indicates the maximum amplitude on the closed trajectory which encircles the resonance center a_r on the plane (θ, a) describing motion with maximum energy exchange.

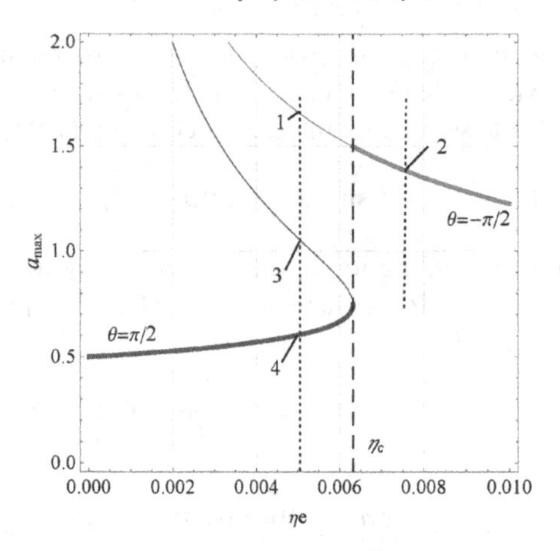

FIGURE 7.16 Dependence of a_{max} vs. η_e estimated based on equation (7.95); thick line identifies the maximum amplitude of vibration; thin line corresponds to the open curves on the phase plane (θ, a) and does not describe oscillations.

Case study 2: $\theta = \frac{\pi}{2}$.

The number of roots depends on the sign of D, similarly as in the latter case. For $\eta_e < \eta_c$, equation (7.95) has three real roots, but now one is negative (nonphysical). The second one corresponds to the open trajectory on the phase plane (point "3" in Figures 7.16 and 7.17). In this case, phase θ grows unlimitedly in time. The third one (point "4" in Figures 7.16 and 7.17) corresponds to the closed trajectory which encircles the quasi-linear center a_q (see also Fig. 7.15). For $\eta_e > \eta_c$, equation (7.95)

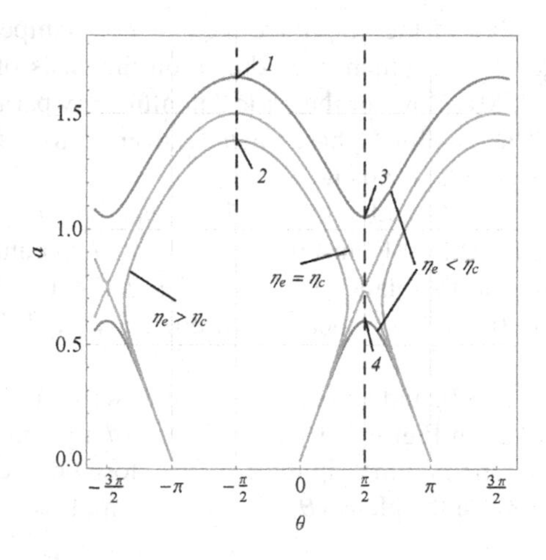

FIGURE 7.17 Limiting phase trajectories for various values of the nonlinearity η_e (points 1, 2, 3, 4 correspond to the real roots of equation (7.95) for $\theta = -\frac{\pi}{2}$ and $\theta = \frac{\pi}{2}$).

has two complex and one real negative root, so in this case, there are no physical solutions and no one trajectory on the analyzed phase plane corresponds to it. The visualization of the positions of the roots of equation (7.95) and trajectories connected to them are presented in Figure 7.17 for $\sigma = 0.002$ and $P = 0.001$.

The above presented discussion of equation (7.95) and the trajectories presented in Figure 7.17 emphasizes the important qualitative transition in the dynamical behavior of the system when the value of the parameter η_e responsible for the nonlinearity crosses the critical value η_c.

In Figure 7.18 the phase portraits in the plane (θ, a) are presented for the parameter η_e slightly less (Fig. 7.18a) and slightly greater (Fig. 7.18b) with regard to its critical value η_c. Though the trajectories correspond to various values of H, LPT (for $H = 0$) are indicated by the black thick line. The numerical data assumed for these graphs are the same as earlier ($\sigma = 0.002$, $P = 0.001$, so $\eta_c = 0.00632$). Such comparison allows for detection of the qualitative change in the limiting phase trajectory corresponding to the maximum energy exchange, i.e. for $H = 0$. One can observe the transformation of LPT from a trajectory encircling quasilinear center for $\eta_e < \eta_c$ at $\theta = \pi/2$ to a curve encircling strongly nonlinear center for $\eta_e > \eta_c$ at $\theta = -\pi/2$. It means that for $\eta_e = \eta_c$ the qualitative threshold change appears in the non-steady vibration of the system. The shape of the trajectory changes and the maximum value of the amplitude of the vibration significantly increases when η_e crosses η_c which corresponds to the structural transition in the system.

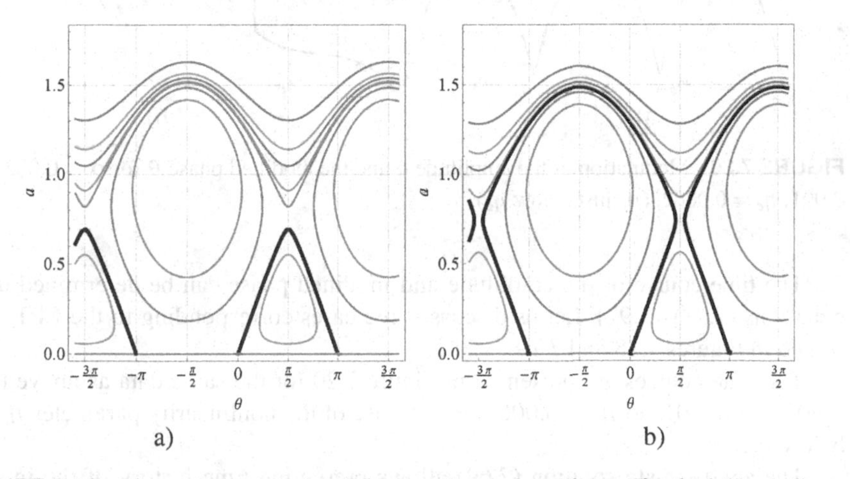

FIGURE 7.18 Phase-plane portraits of the non-damped oscillator. The thick line indicates LPT: a) for $\eta_e = 0.0062$ (just below η_c), b) for $\eta_e = 0.0064$ (just above η_c).

If the parameter η_e increases further, one more qualitative change takes place in the phase plane (a, θ). For $\eta_e = \eta_{c1} = 2\eta_c$ the discriminant $D = 0$ and two positive real roots of equation (7.95) annihilate. The saddle point and quasilinear center coincide, so the topological transformation appears. For $\eta_e > 2\eta_c$ there are no trajectories for $\theta = \pi/2$. The situation for η_e just above $2\eta_c$ is illustrated in Figure 7.19. The second structural transition appears for $\eta_e = 2\eta_c$.

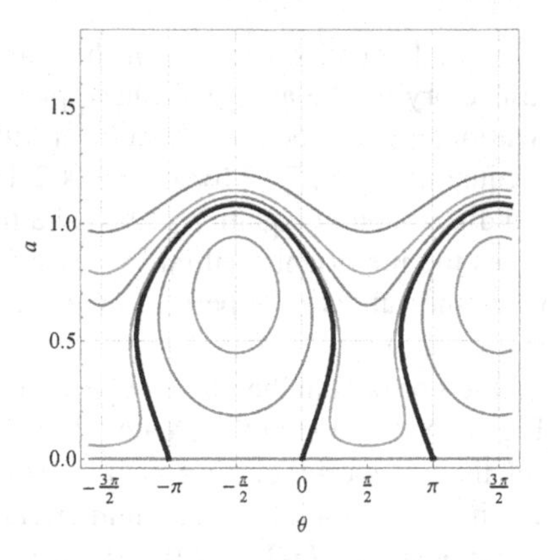

FIGURE 7.19 Phase-plane portraits of the non-damped oscillator. The thick line indicates LPT for $\eta_e = 0.013$ (just above $2\eta_c$).

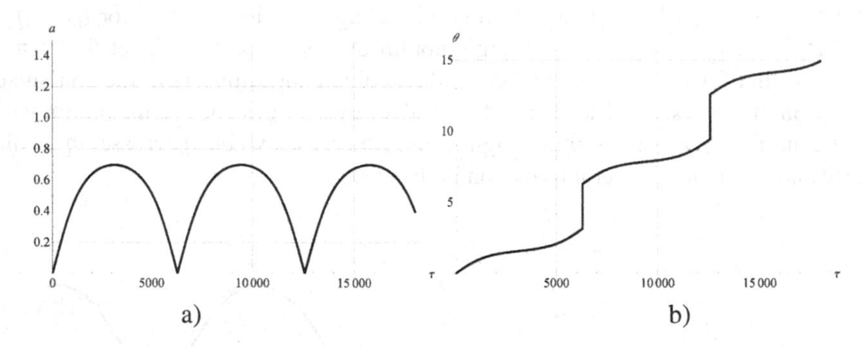

FIGURE 7.20 Modulation of the amplitude a and the modified phase θ for $\sigma = 0.002$, $P = 0.001$, $\eta_e = 0.0062$, (η_e just below η_c).

The time course of the amplitude and modified phase can be determined using equations (7.91)–(7.92). Let us discuss some cases corresponding to the LPTs presented in Figures 7.18 and 7.19.

The time courses are presented in Figure 7.20 for the same data as above ($\sigma = 0.002$, $P = 0.001$, so $\eta_e = 0.0062$) in the case of the nonlinearity parameter η_e just below η_c.

The approximate solution (7.69) allows presenting time history of the internal oscillation $\phi(\tau)$. The analytical solution (7.69) and the solution obtained by numerical integration of equation (7.35) with initial conditions (7.36) are reported in Figure 7.21. The graphs obtained analytically and numerically presented in Figure 7.21 are very similar, and confirm the correctness and high accuracy of the analytical approach.

The time courses of amplitude and phase for $\sigma = 0.002$, $P = 0.001$ and the nonlinearity $\eta_e = 0.0064$ (slightly greater than η_c) are presented in Figure 7.22, whereas time histories of the function $\phi(\tau)$ obtained analytically and numerically are shown

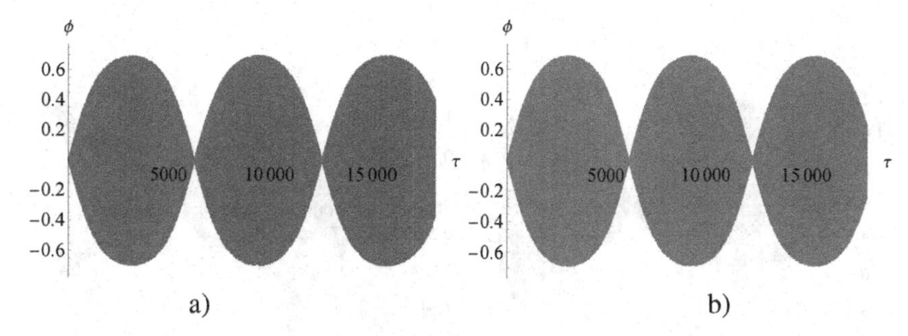

FIGURE 7.21 Time history of the function $\phi(\tau)$ for $\sigma = 0.002$, $P = 0.001$, $\eta_e = 0.0062$ (η_e just below η_c): a) analytical solution equation (7.69), b) obtained by the numerical integration of the initial problem (7.35)–(7.36).

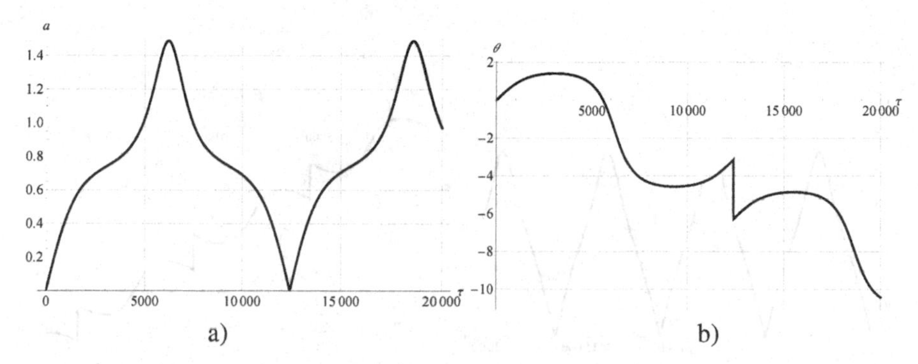

FIGURE 7.22 Modulation of the amplitude a and the modified phase θ for $\sigma = 0.002$, $P = 0.001$, $\eta_e = 0.0064$ (η_e just above η_c).

in Figure 7.23. The change in modulation is substantial, which confirms the qualitative transformation in the behavior of the system. The nonlinearity $\eta_e = \eta_c$ causes changes of the shape of the modulation curves and significantly increases the maximum amplitude. The maximum value of the amplitude of oscillations is compatible with that presented in Figure 7.16.

As mentioned above, the second qualitative transformation of the system appears for $\eta_e = 2\eta_c$. The modulation curves and time history of the oscillation for $\sigma = 0.002$, $P = 0.001$ and $\eta_e = 0.013$ (η_e just above $2\eta_c$) are presented in Figure 7.24 and Figure 7.25, respectively. Also in this case the analytical and numerical solutions coincide perfectly.

It is visible that with the increase of the nonlinearity parameter η_e the amplitude modulation function $a(\tau)$ tends to the saw-tooth form. In this regime, the oscillation could be studied as a vibro-impact processes (Vakakis et al. [217]).

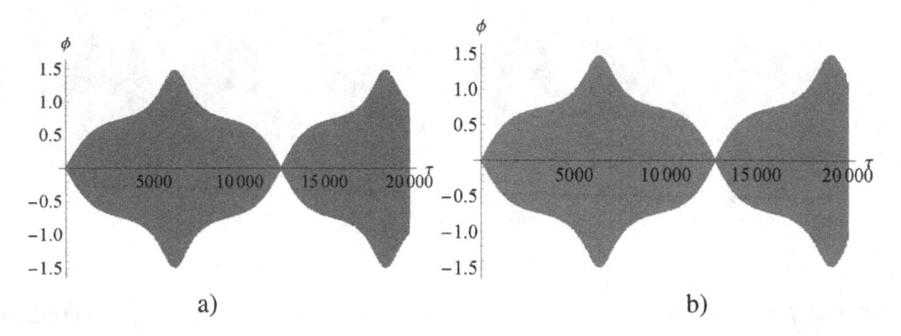

FIGURE 7.23 Time history of the function $\phi(\tau)$ for $\sigma = 0.002$, $P = 0.001$, $\eta_e = 0.0064$ (η_e just above η_c): a) analytical solution equation (7.69); b) obtained by the numerical integration of the initial problem (7.35)–(7.36).

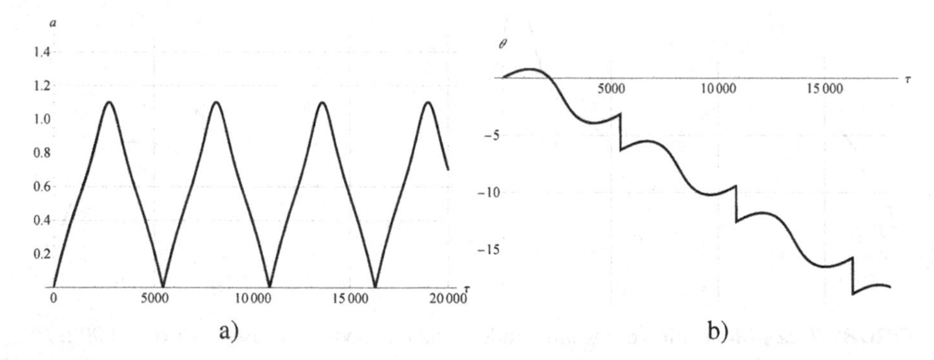

FIGURE 7.24 Modulation of the amplitude a and the modified phase θ for $\sigma = 0.002$, $P = 0.001$, $\eta_e = 0.0128$ (η_e just above $2\eta_c$).

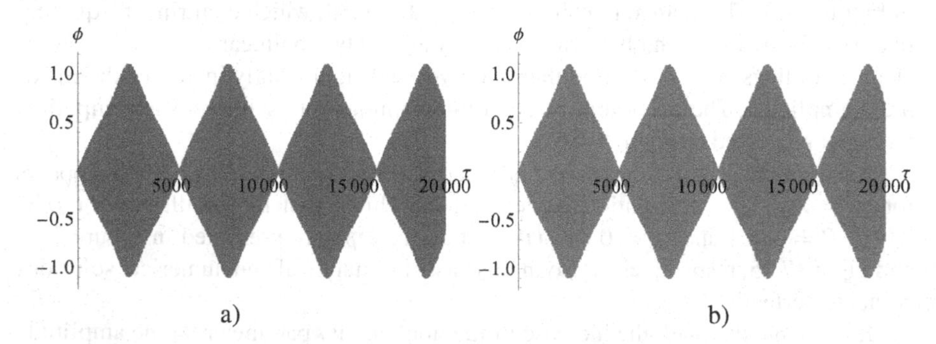

FIGURE 7.25 Time history of the function $\phi(\tau)$ for $\sigma = 0.002$, $P = 0.001$, $\eta_e = 0.0128$ (η_e just above $2\eta_c$): a) analytical solution equation (7.69); b) numerical solution of the initial problem (7.35)–(7.36).

All graphs presented in this section have been made for the same parameters P and σ to give the possibility to compare results on various graphs.

7.9.2 DAMPED OSCILLATIONS

The previous subsection deals with the vibration of the non-damped system, which allows getting some information, mainly qualitative, about the behavior of the system in the limiting case when $\gamma_e \to 0$. It is obvious that damping is a permanent feature in real systems, however, the dynamics analyzed in subsection 7.9.1 can fairly approximate the transient behavior of the damped system in the case when the dissipative effects are appropriately small (especially in the beginning phase of motion). Let us discuss the case when the initial conditions are: $a(0) = 0$ and $\theta(0) = 0$. The damped oscillation in the analyzed problem tends to the steady-state, so the amplitude of the vibration after some time tends to the value close to the quasi-linear center a_q and $\theta \approx \pi/2$ when $\eta_e < \eta_c$ or tends to the strongly nonlinear resonant center a_r and $\theta \approx -\pi/2$ when $\eta_e > \eta_c$.

The phase trajectories on the plane (a, θ) for the small damping coefficient $\gamma_e = 0.00001$ and the nonlinearity η_e close to the critical value η_c (in this case $\eta_c \approx 0.00685$) are presented in Figure 7.26. The last parameters are the same as previously ($\sigma = 0.002$, $P = 0.001$).

Modulations of the amplitude and phases for $\sigma = 0.002$, $P = 0.001$, $\gamma_e = 0.00001$ as the function of the dimensionless time are presented in Figures 7.27 and 7.28, assuming the nonlinearity coefficient η_e is smaller and larger than its critical value.

The limiting steady amplitudes and phases can be computed from equations (7.67)–(7.68) assuming $\frac{da}{d\tau} = 0$ and $\frac{d\theta}{d\tau} = 0$. The steady-state amplitude and the modified phase are $a = 0.2770$ and $\theta = 1.5431$ in the case presented in Figure 7.27, while $a = 0.9841$ and $\theta = -1.4722$ refer to the case presented in Figure 7.28.

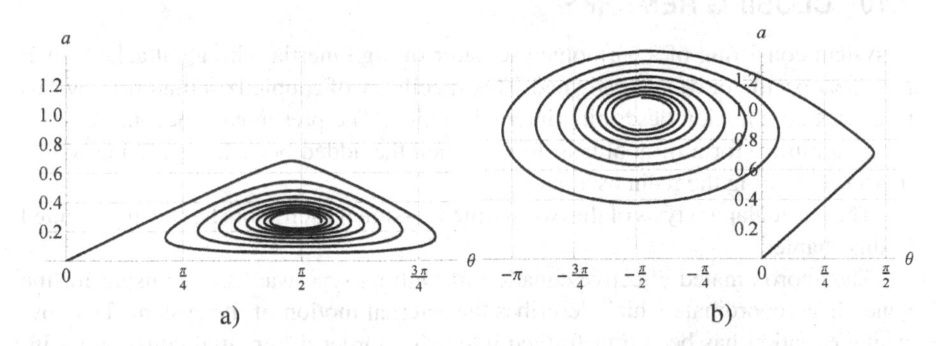

FIGURE 7.26 Phase trajectory for $\sigma = 0.002$, $P = 0.001$: a) $\eta_e = 0.0068$ (η_e just below η_c); b) $\eta_e = 0.0069$ (η_e just above η_c) obtained by the numerical integration of the initial problem (7.67)–(7.68).

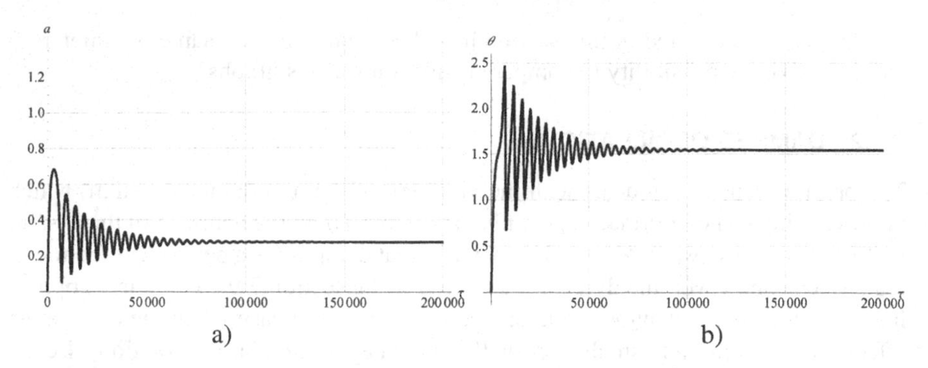

FIGURE 7.27 Modulation of the amplitude a and the modified phase θ for $\sigma = 0.002$, $P = 0.001$, $\gamma_e = 0.00001$, $\eta_e = 0.0062$ (η_e below η_c).

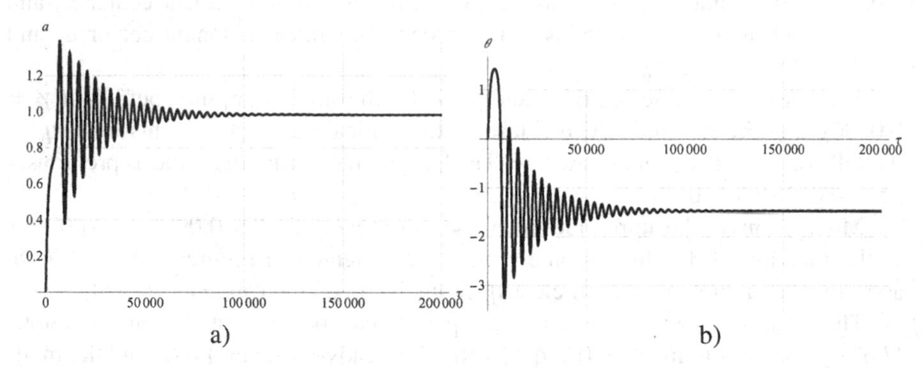

FIGURE 7.28 Modulation of the amplitude a and the modified phase θ for $\sigma = 0.002$, $P = 0.001$, $\gamma_e = 0.00001$, $\eta_e = 0.0069$ (η_e over η_c).

7.10 CLOSING REMARKS

A system consisting of a harmonic oscillator of high inertia with an attached small nonlinear oscillator was investigated. This specificity of coupled oscillators allows us to extract an equation describing internal motion. The presented procedure is valid for an arbitrary form of nonlinear forces when the added body has sufficiently low inertia comparing the main oscillator.

The particular analysis of the two coupled torsional oscillators has been presented in this chapter.

The approximated effective equation of Duffing type was derived using for the generalized coordinate which describes the internal motion of the system. The governing equation has been transformed into a first-order differential equation by introducing a complex representation of the phase variables. This equation was the subject of the asymptotic analysis performed using the multiple scales method. The study concerns a case of main resonance.

Two-time scales occurred to be enough in this analysis. The modulation equations derived from MSM allowed for a detailed discussion of the steady and nonstationary motion of the system. The fixed points of the modulation equations have been detected and their number versus nonlinearity parameter has been determined analytically in the case of non-damped vibrations. Moreover, the amplitude-frequency curves were obtained and its stability in sense of Lyapunov has been tested. It has been shown that the steady-state vibration amplitude can reach a different value for different initial conditions. The objective measure of the error of the solution confirms its good quality.

Afterward, the analysis of nonstationary vibrations has been discussed in detail. This analysis allowed us to determine the critical value of the nonlinearity parameter for which qualitative change in the dynamics of the system appears. For the nonlinearity parameter lower than its critical value, quasi-linear oscillations occur, while when the critical value is exceeded, the vibrations are strongly nonlinear. Moreover, the equation describing limiting phase trajectories has been derived which allows analyzing motion in the phase space of amplitudes and phases. In addition, the maximum value of amplitude and its step change in the non-steady-state versus nonlinearity parameter has been determined.

8 Oscillator with a Springs-in-Series

In this chapter, the differential equations governing dynamics of an oscillator with a springs-in-series is solved analytically, or partly analytically with the support of numerical solutions to nonlinear modulation equations, through the use of the multiple scales method (MSM). The variant of the MSM with three variables describing the motion of the oscillator in time is used. The non-resonant vibrations and vibrations at the main resonance are studied separately. The procedures serving for the construction of the approximate solutions for both cases are described in detail. The accuracy of the solutions is assessed by comparison with the solutions obtained numerically using procedure *NDSolve* of *Mathematica* software. The periodic stationary solutions at resonances together with their stability in the Lyapunov sense are also analyzed. The system of two algebraic equations that govern the steady-state is derived from the differential equations governing the slow modulation of the amplitude and phase. The resonance response curves obtained numerically, as a result of an iterative procedure, are presented in the form of a few dependencies between the amplitude, modified phase, and detuning parameter. The discussion about the influence of the parameters of the mathematical model on the resonance response curves has been carried out.

8.1 INTRODUCTION

Elastic elements arranged in various types of connection (in serial, in parallel, or in branching) are widely applied in many kinds of constructions, mechanisms, mechatronic devices, and micromechanical systems. Modeling of the springs as massless elastic links leads to the mathematical equations among which there are algebraic equations beside the differential ones. As the massless springs are connected in series or ramification, the algebraic equations are the equilibrium equations of the node at which the springs join together. In the linear case, when the superposition principle applies, the connections in series create no difficulties. Depending on the degree of complexity of connection of the springs, the equivalent spring constant can be introduced or one can retain the linear algebraic equations in the model. The last approach results in dealing with a positive semi-definite mass matrix.

In nonlinear systems, the principle of superposition does not apply, which is a source of model-related and computational difficulties. It should be also emphasized that the connections of springs in series increase the system compliance, thus they manifest even more nonlinear character of the problem.

The mechanical systems which contain parallel or serially connected massless springs are widely investigated and discussed in the theoretical and applied mechanics. They have found applications in mechanical and civil engineering, mechatronic

DOI: 10.1201/9781003270706-8

devices, and more recently in micromechanical systems. Various configurations of the connections between the springs, including also their spatial orientation, can lead to the complex dynamical behavior of those systems, especially when the elastic elements have nonlinear characteristics. Such systems could exhibit a variety of interesting behaviors, sometimes even surprising which especially concern the resonance states.

Models of many real systems demand to introduce rigid body approximation where some springs and dampers are connected in various configurations. The car suspension containing systems of the parallel and serially connected springs is investigated in [12, 206]. The authors show that such connections have a great impact on the vibration transmissibility from the rough road to the car body.

Telli and Kopmaz [211] studied a one-dimensional oscillator mounted via two springs wherein one of them is linear and the second one has nonlinear features. They proposed two mathematical models for the system considered. The differential-algebraic equations on which the first approach is based have been solved numerically. The second model based on a single differential equation obtained using the relative displacement variables allows for getting the approximate analytical solution. Comparing the results obtained using the two approaches they determined the values of the parameters describing elastic properties of the springs for which the high agreement is observed.

The system similar to the one analyzed by us is the subject of the paper [221], where tethered satellites are modeled by two serially connected springs. Another application of the similar system one can find in the paper [231], where the springs assembled both in the serial and parallel configurations are used by the authors as a model of the structural equivalent stiffness. In that way, they produce a powerful procedure to study structural behavior. A one-dimensional mass-spring system of two degrees of freedom is studied in [136]. Its mathematical model consists of two strongly coupled differential equations. The analytical method of nonlinear normal modes was successively applied to qualitative analysis of the behavior of that system.

In paper [198], two systems with one- and two degrees-of-freedom having the serially connected springs were investigated. The equations governing the systems contain both differential and algebraic equations. The appropriately adopted multiple scales method (MSM) has been presented and successfully applied by the authors in the case of the undamped free vibration.

8.2 MATHEMATICAL MODEL

In this section the nonlinear nature of the studied problem results from the nonlinear properties of the springs and the damper. The constitutive relationships for elastic forces are postulated in the form of the third-order power law, and the damping is described by the Rayleigh model. The equations of motion were derived from the Lagrange equations, but the starting point was a one-dimensional oscillator with two bodies connected by a spring, i.e. a system with two degrees of freedom. By assuming that one of the bodies, the intermediate one, is massless, one of the differential equations of motion becomes the algebraic equation. It depicts the equilibrium condition of the massless point linking two springs. The equivalent spring constant of the analogical, but linear, oscillator is assumed as the reference quantity, necessary to form dimensionless equations.

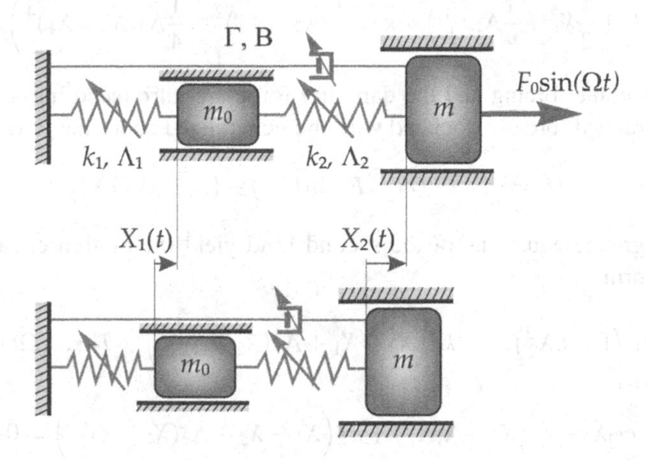

FIGURE 8.1 System of coupled oscillators.

Let us assume a one-dimensional system of two coupled oscillators of mass m_0 and m, as shown in Figure 8.1. The bodies are connected by a spring with the non-linear elastic features of the cubic type. The second spring has similar properties. The relationship between the elastic force and the elongation for each spring can be written as follows

$$F_i = k_i(\Delta_i + \Lambda_i \Delta_i^3), \qquad i = 1, 2, \tag{8.1}$$

where k_i, Λ_i – the elastic coefficients of the i-th spring, Δ_i – the total elongation of the spring.

The Rayleigh damping model [179] is admitted, so the magnitude of the damping force depending on the first and third power of the velocity can be written as

$$R = \Gamma v_2 + B v_2^3, \tag{8.2}$$

where v_2 stands for the velocity of the body with the mass m; B, Γ – the damping coefficients. The problem of nonlinear damping and its modeling is described among others by Elliot et al. [65].

Both springs and the damper are assumed to be massless. The known force F acting horizontally changes according to $F = F_0 \sin(\Omega t)$. In Figure 8.1, there is also depicted, the static equilibrium position of the coupled oscillators when the external forcing does not act. The system has two degrees of freedom. In order to apply the Lagrange formalism, we introduce the generalized coordinates $X_1(t)$ and $X_2(t)$ which describe the position of both bodies in relation to the static equilibrium state. The kinetic energy relative to the reference position has the form

$$T = \frac{1}{2} m_0 \dot{X}_1^2 + \frac{1}{2} m \dot{X}_2^2. \tag{8.3}$$

The potential energy of the conservative forces expresses in terms of the generalized coordinates is given by

$$V = k_1\left(\frac{1}{2}X_1^2 + \frac{1}{4}\Lambda_1 X_1^4\right) + k_2\left(\frac{1}{2}(X_2 - X_1)^2 + \frac{1}{4}\Lambda_2 (X_2 - X_1)^4\right). \tag{8.4}$$

The harmonic forcing and the damping force are introduced into consideration as the generalized forces associated with the generalized coordinates as follows

$$Q_1 = 0, \qquad Q_2 = F_0 \sin(\Omega t) - \left(\Gamma + B\dot{X}_2^2\right)\dot{X}_2. \tag{8.5}$$

The Lagrange equations of the second kind yield the motion equations of the following form

$$m\ddot{X}_2 + \left(\Gamma + B\dot{X}_2^2\right)\dot{X}_2 + k_2\left(X_2 - X_1 + \Lambda_2 (X_2 - X_1)^3\right) = F_0 \sin(\Omega t), \tag{8.6}$$

$$m_0\ddot{X}_1 + k_1\left(X_1 + \Lambda_1 X_1^3\right) + k_2\left(X_1 - X_2 + \Lambda_2 (X_1 - X_2)^3\right) = 0. \tag{8.7}$$

Bishop et al. [42] discussed the special cases of linear systems leading to the formulation with positively semi-definite matrices of mass or stiffness. The case with the positively semi-definite matrix of mass is characterized by the fact that at least one of the Lagrange equations becomes an algebraic equation. Repeating this approach, we assume that

$$m_0 = 0 \tag{8.8}$$

in the system presented in Figure 8.1. As a result of this assumption, we obtain a system with one slider connected to the immovable wall using two springs serially linked. In Figure 8.2, showing this system, the point connecting the springs is marked with the symbol S. The point S can be treated as the massless counterpart of the slider the mass of which tends to zero.

Taking assumption (8.8) into account, we can derive the following equations

$$m\ddot{X}_2 + \left(\Gamma + B\dot{X}_2^2\right)\dot{X}_2 + k_2\left(X_2 - X_1 + \Lambda_2 (X_2 - X_1)^3\right) = F_0 \sin(\Omega t), \tag{8.9}$$

$$k_1\left(X_1 + \Lambda_1 X_1^3\right) + k_2\left(X_1 - X_2 + \Lambda_2 (X_1 - X_2)^3\right) = 0, \tag{8.10}$$

from equations (8.6)–(8.7). Equation (8.10) is algebraic and describes the equilibrium condition concerning the forces of both springs. The mathematical model, given by equations (8.9)–(8.10), belongs to the class of dynamical systems governed by the differential-algebraic equations (DAEs).

In the linear case, when the superposition principle applies, the connections of springs in series create no difficulties. Depending on the degree of complexity of the connection, the equivalent spring constant can be introduced or one can retain in the model the algebraic equations, which results in the positive semi-definite matrix of mass. However, also then it is easy to eliminate one of the coordinates. So, if the

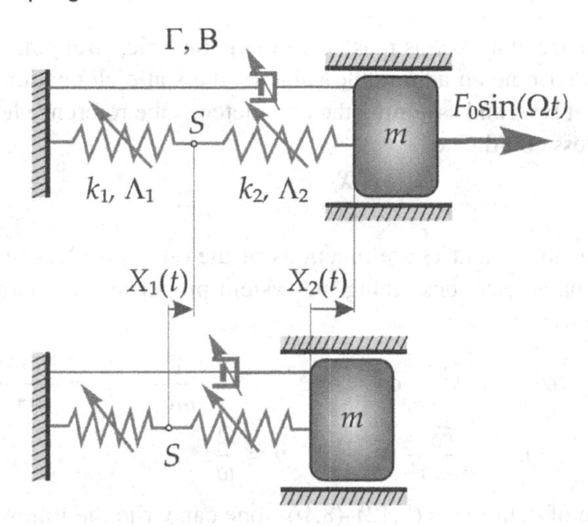

FIGURE 8.2 Slider with two springs connected serially.

springs are linear, the system similar to the considered one can be treated as a system with one degree of freedom. In nonlinear systems, the principle of superposition does not apply, which is a source of model-related and computational difficulties. It is not possible to introduce the equivalent stiffness in the form of a close relationship. Also, the elimination in the strict sense of the X_1 coordinate is not an option, so both coordinates are necessary to define unambiguously the state of the system. Assuming that equation (8.10) is an equation of constraints, the X_1 coordinate can be regarded as complementary and dependent one, and the system shown in Figure 8.2 – as the system with one degree of freedom.

The initial conditions of the form

$$X_2(0) = X_0, \qquad \dot{X}_2(0) = V_0, \tag{8.11}$$

where X_0, V_0 are known quantities, are necessary for an unambiguous solving of equations (8.9)–(8.10).

In order to transform equations (8.9)–(8.10) into the dimensionless form, we introduce the reference mechanical system with the analogous structure as the considered one but with the springs whose elastic properties are linear, i.e. with $\Lambda_1 = 0$ and $\Lambda_2 = 0$. The effective stiffness of the reference system is

$$k_e = \frac{k_1 k_2}{k_1 + k_2}. \tag{8.12}$$

The eigenfrequency of the reference system

$$\omega = \sqrt{\frac{k_e}{m}} \tag{8.13}$$

is used for the definition of the dimensionless time

$$\tau = \omega t. \tag{8.14}$$

Taking into account various reasons of the geometric, strength, or structural nature one can determine an admissible value of the static elongation of the springs. Having denote it by Δ and assuming the parameter as the reference length, we define the dimensionless coordinates

$$x_i = \frac{X_i}{\Delta}, \qquad i = 1,2. \tag{8.15}$$

The coordinates x_1 and x_2 are functions of the dimensionless time τ. The other dimensionless parameters describing the system properties and harmonic force are defined as follows:

$$\lambda = \frac{k_2}{k_1}, \quad \alpha_1 = \Lambda_1 \Delta^2, \quad \alpha_2 = \Lambda_2 \Delta^2, \quad \gamma = \frac{\Gamma}{m\omega}, \quad \beta = \frac{B\omega\Delta^2}{m},$$
$$f_0 = \frac{F_0}{\Delta m \omega^2} = \frac{F_0}{\Delta k_e}, \quad p = \frac{\Omega}{\omega}. \tag{8.16}$$

Making use of definitions (8.12)–(8.16), one can write the following dimensionless equations of the mathematical model

$$\ddot{x}_2 + \gamma \dot{x}_2 + \beta \dot{x}_2^3 + (1+\lambda)(x_2 - x_1) + \alpha_2 (1+\lambda)(x_2 - x_1)^3 - f_0 \sin(p\tau) = 0, \tag{8.17}$$

$$x_1 \left(1 + \alpha_1 x_1^2\right) + \lambda (x_1 - x_2)\left(1 + \alpha_2 (x_1 - x_2)^2\right) = 0, \tag{8.18}$$

$$x_2(0) = x_0, \qquad \dot{x}_2(0) = v_0, \tag{8.19}$$

where $x_0 = \frac{X_0}{\Delta}$, $v_0 = \frac{V_0}{\Delta\omega}$.

8.3 SOLUTION METHOD

The approximate analytical solution in the asymptotic sense to the problem given by (8.17)–(8.19) is obtained using the multiple scales method (MSM). Some special MSM approach is needed to solve the problem due to the occurrence of both differential and algebraic equations.

Following MSM in the time domain, we introduce the small parameter ε that should satisfy a priori the inequalities $0 < \varepsilon \ll 1$. The system evolution is described using n variables instead of the dimensionless time τ. We decide to choose the variant of the MSM with three variables: $\tau_0,\ \tau_1,\ \tau_2$ that are related to the time τ as follows

$$\tau_i = \varepsilon^i \tau, \qquad i = 0,1,2. \tag{8.20}$$

The functions $x_1(\tau)$ and $x_2(\tau)$ are approximated by the following asymptotic expansions

$$x_1(\tau;\varepsilon) = \sum_{k=0}^{2} \varepsilon^k \xi_{1k}(\tau_0,\ \tau_1,\ \tau_2) + O\left(\varepsilon^3\right), \tag{8.21}$$

$$x_2(\tau;\varepsilon) = \sum_{k=0}^{2} \varepsilon^k \xi_{2k}(\tau_0,\ \tau_1,\ \tau_2) + O\left(\varepsilon^3\right), \tag{8.22}$$

where functions $\xi_{1k}(\tau_0,\ \tau_1,\ \tau_2),\ \xi_{2k}(\tau_0,\ \tau_1,\ \tau_2),\ k = 1,2,3$ are sought.

The differential operators relating to the time τ are replaced by operators with partial derivatives. According to the chain rule, they take the form

$$\frac{d}{d\tau}=\sum_{k=0}^{2}\varepsilon^k\frac{\partial}{\partial\tau_k}=\frac{\partial}{\partial\tau_0}+\varepsilon\frac{\partial}{\partial\tau_1}+\varepsilon^2\frac{\partial}{\partial\tau_2}, \tag{8.23}$$

$$\frac{d^2}{d\tau^2}=\frac{\partial^2}{\partial\tau_0^2}+2\varepsilon\frac{\partial^2}{\partial\tau_0\partial\tau_1}+\varepsilon^2\left(\frac{\partial^2}{\partial\tau_1^2}+2\frac{\partial^2}{\partial\tau_0\partial\tau_2}\right)+O\left(\varepsilon^3\right). \tag{8.24}$$

Limiting the considerations only to weakly nonlinear systems, a few parameters describing the system and its loading are assumed to be small which can be expressed using the small parameter

$$\alpha_1=\varepsilon\hat{\alpha}_1,\quad \alpha_2=\varepsilon\hat{\alpha}_2,\quad \gamma=\varepsilon\hat{\gamma},\quad \beta=\varepsilon\hat{\beta},\quad f_0=\varepsilon^2\hat{f}_0. \tag{8.25}$$

The coefficients $\hat{\alpha}_1$, $\hat{\alpha}_2$, $\hat{\gamma}$, $\hat{\beta}$, $\hat{f}_0$ can be understood as $O(1)$ when $\varepsilon\to 0$.

8.4 NON-RESONANT VIBRATION

Inserting equations (8.20)–(8.25) into the governing equations (8.17)–(8.18), we obtain two equations in which the small parameter appears in a few different powers. Both equations should be satisfied for any value of ε. After ordering the terms of each equation according to the powers of the small parameter, omitting all terms of the order $O\left(\varepsilon^3\right)$ and higher ones, and using the method of undetermined coefficients, we obtain the set of six equations with unknown functions $\xi_{1k}(\tau_0,\ \tau_1,\ \tau_2)$, $\xi_{2k}(\tau_0,\ \tau_1,\ \tau_2)$, $k=0,\ \ldots,2$. The equations are organized into three groups. To the first group belongs the equations that contain the terms standing at ε^0

$$\frac{\partial^2\xi_{20}}{\partial\tau_0^2}+(1+\lambda)(\xi_{20}-\xi_{10})=0, \tag{8.26}$$

$$\lambda\xi_{20}-(1+\lambda)\xi_{10}=0. \tag{8.27}$$

Equations (8.26)–(8.27) are called the first-order approximation equations. The terms standing at ε^1 create the equations of the second-order approximation

$$\begin{aligned}\frac{\partial^2\xi_{21}}{\partial\tau_0^2}&+(1+\lambda)(\xi_{21}-\xi_{11})=-(1+\lambda)\hat{\alpha}_2(\xi_{20}-\xi_{10})^3-\\ &\hat{\gamma}\frac{\partial\xi_{20}}{\partial\tau_0}-\hat{\beta}\left(\frac{\partial\xi_{20}}{\partial\tau_0}\right)^3-2\frac{\partial^2\xi_{20}}{\partial\tau_0\partial\tau_1},\end{aligned} \tag{8.28}$$

$$\lambda\xi_{21}-(1+\lambda)\xi_{11}=\hat{\alpha}_1\xi_{10}^3-\lambda\hat{\alpha}_2(\xi_{20}-\xi_{10})^3. \tag{8.29}$$

The coefficients that are accompanied by ε^2 form the equations of the third-order approximation

$$\frac{\partial^2 \xi_{22}}{\partial \tau_0^2} + (1+\lambda)(\xi_{22} - \xi_{12}) = \hat{f}_0 \sin(p\tau_0) -$$

$$3(1+\lambda)\hat{\alpha}_2(\xi_{20} - \xi_{10})^2(\xi_{21} - \xi_{11}) - \hat{\gamma}\left(\frac{\partial \xi_{20}}{\partial \tau_1} + \frac{\partial \xi_{21}}{\partial \tau_0}\right) - \tag{8.30}$$

$$3\hat{\beta}\left(\frac{\partial \xi_{20}}{\partial \tau_0}\right)^2\left(\frac{\partial \xi_{20}}{\partial \tau_1} + \frac{\partial \xi_{21}}{\partial \tau_0}\right) - \frac{\partial^2 \xi_{20}}{\partial \tau_1^2} - 2\frac{\partial^2 \xi_{20}}{\partial \tau_0 \partial \tau_2} - 2\frac{\partial^2 \xi_{21}}{\partial \tau_0 \partial \tau_1},$$

$$\lambda \xi_{22} - (1+\lambda)\xi_{12} = -3\lambda \hat{\alpha}_2(\xi_{20} - \xi_{10})^2(\xi_{21} - \xi_{11}) + 3\,\hat{\alpha}_1 \xi_{10}^2 \xi_{11}. \tag{8.31}$$

The system of equations (8.26)–(8.31) is solved recursively. Notice that among the equations of each order approximation one is algebraic and the second one differential. Solving the approximate problem, we start always with the algebraic equation. The algebraic equations enable one to express the functions $\xi_{1k}(\tau_0, \tau_1, \tau_2)$ in terms of the functions $\xi_{2k}(\tau_0, \tau_1, \tau_2)$, where $k = 0, 1, 2$.

Solving equation (8.27) with respect to the function $\xi_{10}(\tau_0, \tau_1, \tau_2)$, we obtain

$$\xi_{10} = \frac{\lambda}{(1+\lambda)}\xi_{20}. \tag{8.32}$$

Substituting (8.32) into equation (8.26) gives the following homogeneous differential equation

$$\frac{\partial^2 \xi_{20}}{\partial \tau_0^2} + \xi_{20} = 0, \tag{8.33}$$

the general solution of which is

$$\xi_{20} = D(\tau_1, \tau_2)e^{i\tau_0} + \bar{D}(\tau_1, \tau_2)e^{-i\tau_0}, \tag{8.34}$$

where i denotes the imaginary unit, D and its complex conjugate $\bar{D}$ are unknown complex-valued functions of both slower time scales.

Taking into account relationship (8.32), we can write

$$\xi_{10} = \frac{\lambda}{(1+\lambda)}D(\tau_1, \tau_2)e^{i\tau_0} + \frac{\lambda}{(1+\lambda)}\bar{D}(\tau_1, \tau_2)e^{-i\tau_0}. \tag{8.35}$$

We begin now the procedure for solving equations of the second-order approximation. Employing of the MSM shows that the nonlinear constraint equation (8.18) can be approximated by a linear equation with respect to the functions $\xi_{11}(\tau_0, \tau_1, \tau_2)$ and $\xi_{21}(\tau_0, \tau_1, \tau_2)$. Solving equation (8.29) with respect to $\xi_{11}(\tau_0, \tau_1, \tau_2)$ yields

$$\xi_{11} = \frac{\lambda}{(1+\lambda)}\xi_{21} - \frac{\lambda(\lambda^2\hat{\alpha}_1 - \hat{\alpha}_2)}{(1+\lambda)^4}\xi_{20}^3. \tag{8.36}$$

Inserting relationship (8.36), containing the periodic solution (8.34) with the period of 2π, and (8.35) into differential equation (8.28) implies the appearance of the secular terms. The secular terms must be eliminated because they exclude obtaining the bounded solution to equation (8.28). The conditions based on which we eliminate the secular terms become the solvability conditions. They have the following form

$$2\mathrm{i}\frac{\partial D}{\partial \tau_1}+\mathrm{i}\hat{\gamma}D+3\left(\frac{\lambda^3\hat{\alpha}_1+\hat{\alpha}_2}{(1+\lambda)^3}+\mathrm{i}\hat{\beta}\right)D^2\bar{D}, \tag{8.37}$$

$$2\mathrm{i}\frac{\partial \bar{D}}{\partial \tau_1}+\mathrm{i}\,\hat{\gamma}\bar{D}-3\left(\frac{\lambda^3\hat{\alpha}_1+\hat{\alpha}_2}{(1+\lambda)^3}-\mathrm{i}\hat{\beta}\right)D\bar{D}^2=0. \tag{8.38}$$

After eliminating the secular terms, accordingly to (8.37)–(8.38), differential equation (8.28) becomes

$$\begin{aligned}\frac{\partial^2\xi_{21}}{\partial\tau_0^2}+\xi_{21}=&-\frac{\lambda^3\hat{\alpha}_1+\hat{\alpha}_2-\mathrm{i}(1+\lambda)^3\hat{\beta}}{(1+\lambda)^3}e^{3\mathrm{i}\tau_0}D^3-\\&\frac{\lambda^3\hat{\alpha}_1+\hat{\alpha}_2+\mathrm{i}(1+\lambda)^3\hat{\beta}}{(1+\lambda)^3}e^{-3\mathrm{i}\tau_0}\bar{D}^3.\end{aligned} \tag{8.39}$$

The particular solution to equation (8.39) has the following form

$$\begin{aligned}\xi_{21}(\tau_0,\,\tau_1,\,\tau_2)=&\frac{\lambda^3\hat{\alpha}_1+\hat{\alpha}_2-\mathrm{i}(1+\lambda)^3\hat{\beta}}{8(1+\lambda)^3}e^{3\mathrm{i}\tau_0}D^3+\\&\frac{\lambda^3\hat{\alpha}_1+\hat{\alpha}_2+\mathrm{i}(1+\lambda)^3\hat{\beta}}{8(1+\lambda)^3}e^{-3\mathrm{i}\tau_0}\bar{D}^3.\end{aligned} \tag{8.40}$$

Making use of relationship (8.36), we can determine the function $\xi_{11}(\tau_0,\tau_1,\tau_2)$ and write that

$$\begin{aligned}\xi_{11}(\tau_0,\,\tau_1,\,\tau_2)=&-\frac{3\lambda\left(\lambda^2\hat{\alpha}_1-\hat{\alpha}_2\right)}{(1+\lambda)^4}\left(e^{3\mathrm{i}\tau_0}D^2\bar{D}+e^{-3\mathrm{i}\tau_0}\bar{D}^2D\right)+\\&\frac{\lambda\left(\lambda^2(\lambda-8)\hat{\alpha}_1+9\hat{\alpha}_2-\mathrm{i}(1+\lambda)^3\hat{\beta}\right)(\hat{\alpha}_1-\hat{\alpha}_2)}{8(1+\lambda)^4}e^{3\mathrm{i}\tau_0}D^3+\\&\frac{\lambda\left(\lambda^2(\lambda-8)\hat{\alpha}_1+9\hat{\alpha}_2+\mathrm{i}(1+\lambda)^3\hat{\beta}\right)(\hat{\alpha}_1-\hat{\alpha}_2)}{8(1+\lambda)^4}e^{-3\mathrm{i}\tau_0}\bar{D}^3.\end{aligned} \tag{8.41}$$

Let us note that when inserting the solutions of the first and second-order approximation into equation (8.30) we should know the second derivatives $\frac{\partial^2 D}{\partial\tau_1^2}$ and $\frac{\partial^2 \bar{D}}{\partial\tau_1^2}$ due to the occurrence of the term $\frac{\partial^2 \xi_{20}}{\partial \tau_1^2}$. From equations (8.37)–(8.38), it follows that

$$\frac{\partial D}{\partial \tau_1} = \frac{1}{2} D \left(\frac{3\mathrm{i}\,\left(\lambda^3 \hat{\alpha}_1 + \hat{\alpha}_2\right)}{(1+\lambda)^3} D\bar{D} - 3\hat{\beta} D\bar{D} - \hat{\gamma} \right), \tag{8.42}$$

$$\frac{\partial \bar{D}}{\partial \tau_1} = -\frac{1}{2} \bar{D} \left(\frac{3\mathrm{i}\,\left(\lambda^3 \hat{\alpha}_1 + \hat{\alpha}_2\right)}{(1+\lambda)^3} D\bar{D} + 3\hat{\beta} D\bar{D} + \hat{\gamma} \right). \tag{8.43}$$

Differentiating (8.42)–(8.43) with respect to τ_1 gives

$$\begin{aligned} \frac{\partial^2 D}{\partial \tau_1^2} &= \frac{\hat{\gamma}^2}{4} D + 3D^2 \bar{D} \hat{\gamma} \left(\hat{\beta} - \mathrm{i} \frac{\lambda^3 \hat{\alpha}_1 + \hat{\alpha}_2}{(1+\lambda)^3} \right) - \\ & \frac{9}{4} D^3 \bar{D}^2 \left(\frac{\left(\left(\lambda^3 \hat{\alpha}_1 + \hat{\alpha}_2\right)^2 + 4\mathrm{i}(1+\lambda)^3 \left(\lambda^3 \hat{\alpha}_1 + \hat{\alpha}_2\right) \hat{\beta} \right)}{(1+\lambda)^6} - 3\hat{\beta}^2 \right), \end{aligned} \tag{8.44}$$

$$\begin{aligned} \frac{\partial^2 \bar{D}}{\partial \tau_1^2} &= \frac{\hat{\gamma}^2}{4} \bar{D} + 3D\bar{D}^2 \hat{\gamma} \left(\hat{\beta} + \mathrm{i} \frac{\lambda^3 \hat{\alpha}_1 + \hat{\alpha}_2}{(1+\lambda)^3} \right) + \\ & \frac{9}{4} D^2 \bar{D}^3 \left(\frac{\left(-\left(\lambda^3 \hat{\alpha}_1 + \hat{\alpha}_2\right)^2 + 4\mathrm{i}(1+\lambda)^3 \left(\lambda^3 \hat{\alpha}_1 + \hat{\alpha}_2\right) \hat{\beta} \right)}{(1+\lambda)^6} - 3\hat{\beta}^2 \right). \end{aligned} \tag{8.45}$$

Similarly, as before, we first determine $\xi_{12}(\tau_0, \tau_1, \tau_2)$ in terms of $\xi_{22}(\tau_0, \tau_1, \tau_2)$. Employing equation (8.31), we obtain

$$\begin{aligned} \xi_{12} = & \frac{\lambda}{(1+\lambda)} \xi_{22} + \frac{3\lambda^2 \left(\lambda^2 \hat{\alpha}_1 - \hat{\alpha}_2\right) \left(\lambda \hat{\alpha}_1 + \hat{\alpha}_2\right)}{(1+\lambda)^7} \xi_{20}^5 - \\ & \frac{3\lambda \left(\lambda^2 \hat{\alpha}_1 - \hat{\alpha}_2\right)}{(1+\lambda)^4} \xi_{20}^2 \xi_{21}. \end{aligned} \tag{8.46}$$

The function $\xi_{20}(\tau_0, \tau_1, \tau_2)$ has the period which is equal to 2π. Substituting equations (8.34), (8.36), (8.40)–(8.41), and (8.46) into differential equation (8.30) causes that its solution becomes unbounded. Necessary elimination of the secular terms leads to the next solvability conditions in the form

$$\begin{aligned} & -2\mathrm{i} \frac{\partial D}{\partial \tau_2} + \frac{1}{4} \hat{\gamma}^2 D + 3\mathrm{i} \hat{\gamma} \frac{\lambda^3 \hat{\alpha}_1 + \hat{\alpha}_2}{2(1+\lambda)^3} D^2 \bar{D} - \\ & \frac{9}{8} \hat{\beta}^2 D^3 \bar{D}^2 - 3\mathrm{i} \hat{\beta} \frac{\lambda^3 \hat{\alpha}_1 + \hat{\alpha}_2}{(1+\lambda)^3} D^3 \bar{D}^2 + \\ & 15 \frac{\lambda^5 (\lambda + 16) \hat{\alpha}_1^2 - 30\lambda^3 \hat{\alpha}_1 \hat{\alpha}_2 + (16\lambda + 1) \hat{\alpha}_2^2}{8(1+\lambda)^6} D^3 \bar{D}^2 = 0, \end{aligned} \tag{8.47}$$

$$
\begin{aligned}
&2\mathrm{i}\frac{\partial \bar{D}}{\partial \tau_2}+\frac{1}{4}\hat{\gamma}^2\bar{D}-3\mathrm{i}\hat{\gamma}\frac{\lambda^3\hat{\alpha}_1+\hat{\alpha}_2}{2(1+\lambda)^3}D\bar{D}^2-\\
&\frac{9}{8}\hat{\beta}^2D^2\bar{D}^3+3\mathrm{i}\hat{\beta}\frac{\lambda^3\hat{\alpha}_1+\hat{\alpha}_2}{(1+\lambda)^3}D^2\bar{D}^3+\\
&15\frac{\lambda^5(\lambda+16)\hat{\alpha}_1^2-30\lambda^3\hat{\alpha}_1\hat{\alpha}_2+(16\lambda+1)\hat{\alpha}_2^2}{8(1+\lambda)^6}D^2\bar{D}^3=0.
\end{aligned}
\tag{8.48}
$$

Eliminating the secular terms from equation (8.30), according to (8.47)–(8.48), yields the following differential equation

$$
\begin{aligned}
&\frac{\partial^2\xi_{22}}{\partial\tau_0^2}+\xi_{22}=\\
&\frac{3}{8}\left(3\hat{\beta}^2-\frac{\lambda^5(\lambda-8)\hat{\alpha}_1^2+18\lambda^3\hat{\alpha}_1\hat{\alpha}_2+(1-8\lambda)\hat{\alpha}_2^2}{(1+\lambda)^6}\right)\left(e^{5\mathrm{i}\tau_0}D^5+e^{-5\mathrm{i}\tau_0}\bar{D}^5\right)+\\
&\frac{3}{8}\left(\frac{\lambda^5(7\lambda+40)\hat{\alpha}_1^2-66\lambda^3\hat{\alpha}_1\hat{\alpha}_2+(40+7)\hat{\alpha}_2^2}{(1+\lambda)^6}-9\hat{\beta}^2\right)D\bar{D}\left(e^{3\mathrm{i}\tau_0}D^3+e^{-3\mathrm{i}\tau_0}\bar{D}^3\right)+\\
&3\mathrm{i}\frac{(\lambda^3\hat{\alpha}_1+\hat{\alpha}_2)\hat{\beta}}{2(1+\lambda)^3}\left(e^{5\mathrm{i}\tau_0}D^5-e^{-5\mathrm{i}\tau_0}\bar{D}^5+2D\bar{D}\left(e^{3\mathrm{i}\tau_0}D^3-e^{-3\mathrm{i}\tau_0}\bar{D}^3\right)\right)+\\
&\frac{3}{4}\left(\mathrm{i}\frac{\lambda^2\hat{\alpha}_1+\hat{\alpha}_2}{(1+\lambda)^3}-\hat{\beta}\right)\hat{\gamma}e^{3\mathrm{i}\tau_0}D^3-\frac{3}{4}\left(\mathrm{i}\frac{\lambda^2\hat{\alpha}_1+\hat{\alpha}_2}{(1+\lambda)^3}+\hat{\beta}\right)\hat{\gamma}e^{-3\mathrm{i}\tau_0}\bar{D}^3+\hat{f}_0\sin(p\tau_0).
\end{aligned}
\tag{8.49}
$$

The particular solution to equation (8.49) is given by

$$
\begin{aligned}
&\xi_{22}(\tau_0,\ \tau_1,\ \tau_2)=\\
&\left(\frac{\lambda^5(\lambda-8)\hat{\alpha}_1^2+18\lambda^3\hat{\alpha}_1\hat{\alpha}_2+(1-8\lambda)\hat{\alpha}_2^2}{64(1+\lambda)^6}-\frac{3}{64}\hat{\beta}^2\right)\left(e^{5\mathrm{i}\tau_0}D^5+e^{-5\mathrm{i}\tau_0}\bar{D}^5\right)-\\
&\frac{3}{64}\left(\frac{\lambda^5(7\lambda+40)\hat{\alpha}_1^2-66\lambda^3\hat{\alpha}_1\hat{\alpha}_2+(40\lambda+7)\hat{\alpha}_2^2}{(1+\lambda)^6}-9\hat{\beta}^2\right)D\bar{D}\left(e^{3\mathrm{i}\tau_0}D^3+e^{-3\mathrm{i}\tau_0}\bar{D}^3\right)-\\
&3\mathrm{i}\frac{(\lambda^3\hat{\alpha}_1+\hat{\alpha}_2)\hat{\beta}}{8(1+\lambda)^3}\left(\frac{1}{6}\left(e^{5\mathrm{i}\tau_0}D^5-e^{-5\mathrm{i}\tau_0}\bar{D}^5\right)+D\bar{D}\left(e^{3\mathrm{i}\tau_0}D^3-e^{-3\mathrm{i}\tau_0}\bar{D}^3\right)\right)-\\
&\frac{3}{32}\left(\mathrm{i}\frac{\lambda^2\hat{\alpha}_1+\hat{\alpha}_2}{(1+\lambda)^3}-\hat{\beta}\right)\hat{\gamma}e^{3\mathrm{i}\tau_0}D^3-\\
&\frac{3}{32}\left(\mathrm{i}\frac{\lambda^2\hat{\alpha}_1+\hat{\alpha}_2}{(1+\lambda)^3}+\hat{\beta}\right)\hat{\gamma}e^{-3\mathrm{i}\tau_0}\bar{D}^3+\frac{\hat{f}_0}{1-p^2}\sin(p\tau_0).
\end{aligned}
\tag{8.50}
$$

Taking into account equations (8.34), (8.40), (8.46), and (8.50), we can determinate the function $\xi_{12}((\tau_0, \tau_1, \tau_2)$, i.e. the last one which is necessary to construct the approximate solution. It is as follows

$$\xi_{12} = \left(\frac{\lambda^6 (\lambda - 8)\hat{\alpha}_1^2 + 18\lambda^4 \hat{\alpha}_1 \hat{\alpha}_2 + \lambda (1 - 8\lambda)\hat{\alpha}_2^2}{64(1+\lambda)^7} - \right.$$

$$\left. \frac{3\lambda \hat{\beta}^2}{64(1+\lambda)} \right) \left(e^{5i\tau_0} D^5 + e^{-5i\tau_0} \bar{D}^5 \right) -$$

$$\frac{3}{64} \left(\frac{\lambda^6 (7\lambda + 40)\hat{\alpha}_1^2 - 66\lambda^4 \hat{\alpha}_1 \hat{\alpha}_2 + \lambda (40\lambda + 7)\hat{\alpha}_2^2}{(1+\lambda)^7} - \right.$$

$$\left. \frac{9\lambda \hat{\beta}^2}{1+\lambda} \right) D\bar{D} \left(e^{3i\tau_0} D^3 + e^{-3i\tau_0} \bar{D}^3 \right) -$$

$$3i \frac{(\lambda^3 \hat{\alpha}_1 + \hat{\alpha}_2)\lambda \hat{\beta}}{8(1+\lambda)^4} \left(\frac{1}{6} \left(e^{5i\tau_0} D^5 - e^{-5i\tau_0} \bar{D}^5 \right) + D\bar{D} \left(e^{3i\tau_0} D^3 - e^{-3i\tau_0} \bar{D}^3 \right) \right) -$$

$$\frac{3\lambda}{32} i \left(\frac{\lambda^2 \hat{\alpha}_1 + \hat{\alpha}_2}{(1+\lambda)^4} + \frac{i\hat{\beta}}{1+\lambda} \right) \hat{\gamma} e^{3i\tau_0} D^3 - \frac{3\lambda}{32} i \left(\frac{\lambda^2 \hat{\alpha}_1 + \hat{\alpha}_2}{(1+\lambda)^4} - \frac{i\hat{\beta}}{1+\lambda} \right) \hat{\gamma} e^{-3i\tau_0} \bar{D}^3 +$$

$$\frac{3\lambda^2 (\lambda^2 \hat{\alpha}_1 - \hat{\alpha}_2)(\lambda \hat{\alpha}_1 + \hat{\alpha}_2)}{(1+\lambda)^7} \left(D(\tau_1, \tau_2) e^{i\tau_0} + \bar{D}(\tau_1, \tau_2) e^{-i\tau_0} \right)^5 - \tag{8.51}$$

$$\frac{3\lambda (\lambda^2 \hat{\alpha}_1 - \hat{\alpha}_2)}{(1+\lambda)^4} \left(D(\tau_1, \tau_2) e^{i} \tau_0 + \right.$$

$$\left. \bar{D}(\tau_1, \tau_2) e^{-i\tau_0} \right)^2 \frac{\lambda^3 \hat{\alpha}_1 - i\left((1+\lambda)^3 \hat{\beta} + i\hat{\alpha}_2 \right)}{8(1+\lambda)^3} e^{3i\tau_0} D^3 -$$

$$\frac{3\lambda (\lambda^2 \hat{\alpha}_1 - \hat{\alpha}_2)}{(1+\lambda)^4} \left(D(\tau_1, \tau_2) e^{i\tau_0} + \right.$$

$$\left. \bar{D}(\tau_1, \tau_2) e^{-i\tau_0} \right)^2 \frac{\lambda^3 \hat{\alpha}_1 + i\left((1+\lambda)^3 \hat{\beta} - i\hat{\alpha}_2 \right)}{8(1+\lambda)^3} e^{-3i\tau_0} \bar{D}^3 +$$

$$\frac{\lambda \hat{f}_0}{(1+\lambda)(1-p^2)} \sin(p\tau_0).$$

The unknown complex-valued functions $D(\tau_1, \tau_2)$ and $\bar{D}(\tau_1, \tau_2)$ are restricted by the solvability conditions. We depict both functions in the exponential form

$$D(\tau_1, \tau_2) = \frac{1}{2} a(\tau_1, \tau_2) e^{i\psi(\tau_1, \tau_2)}, \quad \bar{D}(\tau_1, \tau_2) = \frac{1}{2} a(\tau_1, \tau_2) e^{-i\psi(\tau_1, \tau_2)}, \tag{8.52}$$

where the functions $a(\tau_1, \tau_2)$ and $\psi(\tau_1, \tau_2)$ are real-valued.

Inserting relationships (8.52) into the solvability conditions, i.e. into equations (8.37)–(8.38) and (8.47)–(8.48) yields the following set of four partial differential equations with unknown functions $a(\tau_1,\tau_2)$ and $\psi(\tau_1,\tau_2)$

$$-\mathrm{i}\frac{\partial a}{\partial \tau_1}+\frac{1}{2}a\left(2\frac{\partial \psi}{\partial \tau_1}-\mathrm{i}\hat{\gamma}\right)-\frac{3a^3\left(\lambda^3\hat{\alpha}_1+\hat{\alpha}_2+\mathrm{i}(1+\lambda)^3\hat{\beta}\right)}{8(1+\lambda)^3}=0, \tag{8.53}$$

$$\mathrm{i}\frac{\partial a}{\partial \tau_1}+\frac{1}{2}a\left(2\frac{\partial \psi}{\partial \tau_1}+\mathrm{i}\hat{\gamma}\right)-\frac{3a^3\left(\lambda^3\hat{\alpha}_1+\hat{\alpha}_2-\mathrm{i}(1+\lambda)^3\hat{\beta}\right)}{8(1+\lambda)^3}=0, \tag{8.54}$$

$$\begin{gathered}-\mathrm{i}\frac{\partial a}{\partial \tau_2}+a\left(\frac{\partial \psi}{\partial \tau_2}+\frac{\hat{\gamma}^2}{8}\right)+3\mathrm{i}\frac{\left(\lambda^3\hat{\alpha}_1+\hat{\alpha}_2\right)}{16(1+\lambda)^3}\hat{\gamma}a^3-\frac{9}{256}\hat{\beta}^2a^5+\\ 15\frac{\lambda^5(16+\lambda)\hat{\alpha}_1^2-30\lambda^3\hat{\alpha}_1\hat{\alpha}_2+(1+16\lambda)\hat{\alpha}_2^2}{256(1+\lambda)^6}a^5+\\ 3\mathrm{i}\frac{\lambda^3\hat{\alpha}_1+\hat{\alpha}_2}{32(1+\lambda)^3}\hat{\beta}a^5=0,\end{gathered} \tag{8.55}$$

$$\begin{gathered}\mathrm{i}\frac{\partial a}{\partial \tau_2}+a\left(\frac{\partial \psi}{\partial \tau_2}+\frac{\hat{\gamma}^2}{8}\right)-3\mathrm{i}\frac{\left(\lambda^3\hat{\alpha}_1+\hat{\alpha}_2\right)}{16(1+\lambda)^3}\hat{\gamma}a^3-\frac{9}{256}\hat{\beta}^2a^5+\\ 15\frac{\lambda^5(16+\lambda)\hat{\alpha}_1^2-30\lambda^3\hat{\alpha}_1\hat{\alpha}_2+(1+16\lambda)\hat{\alpha}_2^2}{256(1+\lambda)^6}a^5+\\ 3\mathrm{i}\frac{\lambda^3\hat{\alpha}_1+\hat{\alpha}_2}{32(1+\lambda)^3}\hat{\beta}a^5=0.\end{gathered} \tag{8.56}$$

Solving equations (8.53)–(8.56) with respect to the partial derivatives gives

$$\frac{\partial a}{\partial \tau_1}=-\frac{1}{2}\hat{\gamma}a-\frac{3}{8}\hat{\beta}a^3, \tag{8.57}$$

$$\frac{\partial a}{\partial \tau_2}=-\frac{3a^3\left(\lambda^3\hat{\alpha}_1+\hat{\alpha}_2\right)\left(\hat{\beta}a^2-2\hat{\gamma}\right)}{32(1+\lambda)^3}, \tag{8.58}$$

$$\frac{\partial \psi}{\partial \tau_1}=3\frac{\left(\lambda^3\hat{\alpha}_1+\hat{\alpha}_2\right)}{8(1+\lambda)^3}a^2, \tag{8.59}$$

$$\begin{gathered}\frac{\partial \psi}{\partial \tau_2}=-\frac{15\lambda^5(16+\lambda)\hat{\alpha}_1^2}{256(1+\lambda)^6}a^4+\frac{225\lambda^3\hat{\alpha}_1\hat{\alpha}_2}{128(1+\lambda)^6}a^4-\\ \frac{15(1+16\lambda)\hat{\alpha}_2^2}{256(1+\lambda)^6}a^4+\frac{9}{256}\hat{\beta}^2a^4-\frac{\hat{\gamma}^2}{8}.\end{gathered} \tag{8.60}$$

Equations (8.57)–(8.60) govern the variability of the functions $a(\tau_1, \tau_2)$ and $\psi(\tau_1, \tau_2)$ which is described using only the slow variables. Reverse use of definition (8.23) and assumptions (8.25) allows one to express this variability in terms of the dimensionless time τ, as follows

$$\frac{da}{d\tau} = -\frac{1}{2}\gamma a + \frac{3}{16}\left(\gamma\frac{\lambda^3\alpha_1+\alpha_2}{(1+\lambda)^3} - 2\beta\right)a^3 - 3\beta\frac{\lambda^3\alpha_1+\alpha_2}{32(1+\lambda)^3}a^5, \tag{8.61}$$

$$\begin{aligned}\frac{d\psi}{d\tau} &= -\frac{\gamma^2}{8} + 3\frac{\lambda^3\alpha_1+\alpha_2}{8(1+\lambda)^3}a^2 + \\ &3\left(5\frac{-\lambda^5(16+\lambda)\alpha_1^2 + 30\lambda^3\alpha_1\alpha_2 - (1+16\lambda)\alpha_2^2}{256(1+\lambda)^6} + \frac{3}{256}\beta^2\right)a^4.\end{aligned} \tag{8.62}$$

According to equations (8.34) and (8.52), the functions $a(\tau)$ and $\psi(\tau)$ depict the amplitude and phase of the first term of the asymptotic expansion given by (8.22). Modulation equations (8.61)–(8.62) are supplemented by the initial conditions

$$a(0) = a_0, \quad \psi(0) = \psi_0, \tag{8.63}$$

where the quantities a_0 and ψ_0 are known and compatible with the initial values x_0, v_0.

The initial-value problem governing the slow modulation of the amplitude and phase requires numerical solving.

Assembling the partial solutions obtained using the recursive procedure, according to (8.21)–(8.22), we obtain the approximate solution to the problem considered

$$\begin{aligned}x_2(\tau) &= a(\tau)\cos(\tau+\psi(\tau)) + \beta\frac{\lambda^3\alpha_1+\alpha_2}{256(\lambda+1)^3}a^5(\tau)\sin(5\tau+5\psi(\tau)) + \\ &\left(\frac{\lambda^5(\lambda-8)\alpha_1^2 + 18\lambda^3\alpha_1\alpha_2 - (8\lambda-1)\alpha_2^2}{1024(1+\lambda)^6} - \frac{3\beta^2}{1024}\right)a^5(\tau)\cos(5\tau+5\psi(\tau)) + \\ &\frac{3}{1024}\left(9\beta^2 - \frac{\lambda^5(7\lambda+40)\alpha_1^2 - 66\lambda^3\alpha_1\alpha_2 + \alpha_2^2(40\lambda+7)}{(1+\lambda)^6}\right)a^5(\tau)\cos(3\tau+3\psi(\tau)) + \\ &\frac{3\beta(\lambda^3\alpha_1+\alpha_2)}{128(1+\lambda)^3}a^5(\tau)\sin(3\tau+3\psi(\tau)) + \\ &\left(\frac{\beta}{32} + 3\gamma\frac{\lambda^3\alpha_1+\alpha_2}{128(1+\lambda)^3}\right)a^3(\tau)\sin(3\tau+3\psi(\tau)) + \\ &\frac{4\lambda^3\alpha_1 + 4\alpha_2 + 3\beta\gamma(1+\lambda)^3}{128(1+\lambda)^3}a^3(\tau)\cos(3\tau+3\psi(\tau)) - \frac{f_0\sin(p\tau)}{p^2-1},\end{aligned} \tag{8.64}$$

$$
\begin{aligned}
x_1(\tau) = {} & \frac{\lambda}{1+\lambda} a(\tau)\cos(\tau+\psi(\tau)) + \\
& \beta\lambda \frac{\left(\lambda^2(\lambda-6)\alpha_1+7\alpha_2\right)}{256(\lambda+1)^4} a^5(\tau)\sin(5\tau+5\psi(\tau)) + \\
& \left(\frac{18\lambda^3(13\lambda-12)\alpha_1\alpha_2}{1024(1+\lambda)^7} - \frac{3\beta^2\lambda}{1024(1+\lambda)}\right) a^5(\tau)\cos(5\tau+5\psi(\tau)) + \\
& \frac{\lambda^5(\lambda^2-32\lambda+192)\alpha_1^2-25\lambda(8\lambda-1)\alpha_2^2}{1024(1+\lambda)^7} a^5(\tau)\cos(5\tau+5\psi(\tau)) + \\
& 3\beta\lambda \frac{\left(\lambda^2(\lambda-2)\alpha_1+3\alpha_2\right)}{128(\lambda+1)^4} a^5(\tau)\sin(3\tau+3\psi(\tau)) + \\
& \left(\frac{27\beta^2\lambda}{1024(1+\lambda)} + \frac{18\lambda^3(67\lambda-56)\alpha_1\alpha_2}{1024(1+\lambda)^7}\right) a^5(\tau)\cos(3\tau+3\psi(\tau)) - \\
& 3\lambda \frac{9(40\lambda-1)\alpha_2^2+\lambda^4(7\lambda^2+56\lambda-320)\alpha_1^2}{1024(1+\lambda)^7} a^5(\tau)\cos(3\tau+3\psi(\tau)) + \\
& \frac{3\beta\lambda(\alpha_2-\lambda^2\alpha_1)}{128(1+\lambda)^4} a^5(\tau)\sin(\tau+\psi(\tau)) - \qquad (8.65) \\
& 3\lambda \frac{\lambda^4(\lambda-80)\alpha_1^2-81\lambda^2(\lambda-1)\alpha_1\alpha_2+(80\lambda-1)\alpha_2^2}{128(1+\lambda)^7} a^5(\tau)\cos(\tau+\psi(\tau)) + \\
& \lambda \frac{4\beta(1+\lambda)^3+3\gamma(\lambda^3\alpha_1+\alpha_2)}{128(1+\lambda)^4} a^3(\tau)\sin(3\tau+3\psi(\tau)) + \\
& \lambda \frac{36\alpha_2+4\lambda^2(\lambda-8)\alpha_1+3(1+\lambda)^3\beta\gamma}{128(1+\lambda)^4} a^3(\tau)\cos(3\tau+3\psi(\tau)) + \\
& \frac{3\lambda(\alpha_2-\lambda^2\alpha_1)}{4(1+\lambda)^4} a^3(\tau)\cos(\tau+\psi(\tau)) - \frac{\lambda}{1+\lambda}\frac{f_0\sin(p\tau)}{p^2-1},
\end{aligned}
$$

where the functions $a(\tau), \psi(\tau)$ are the solutions to modulation problem (8.61)–(8.63).

The approximate asymptotic solution given by equations (8.64)–(8.65) describes the vibration of the body and variability of the strain of both springs with excluding the case when the dimensionless frequency of the harmonic force is close to 1.

Let us assume the following values of the dimensionless parameters having an impact on the vibration of the body:

$$
\lambda = 0.75, \quad \alpha_1 = 0.53, \quad \alpha_2 = 0.35, \quad \gamma = 0.01,
$$

$$
\beta = 0.01, \quad f_0 = 0.008, \quad p = 0.191.
$$

The initial values for the problem (8.17)–(8.19) let be:

$$x_0 = 0.9, \quad v_0 = 0.4.$$

The relationship between initial conditions (8.19) and (8.63) has been determined using the analytical form of the solution (8.64). Calculating the time derivative of solution (8.64), substituting $\tau = 0$ into the solution and its derivative, and requiring that the approximate solution should satisfy primary initial conditions (8.19), we get a set of algebraic equations with unknown a_0 and ψ_0. In this case, only a numerical solution is possible to obtain. For the data considered, it gives the following results:

$$a_0 \approx 0.977359, \quad \psi_0 \approx -0.402409.$$

The time histories of the body displacement obtained using the MSM are presented in Figures 8.3–8.5. The first of them shows how the vibration proceeds, just after the motion is initiated. The dominant effect observed here is the attenuation of the quite strong oscillations caused by the initial conditions. The transient process takes a relatively long time, see Figure 8.4, and the fixed vibration can be observed only around $\tau = 1500$, which is shown in Figure 8.5. The solution obtained by the numerical integration of the governing equations (8.17)–(8.19) is shown in each graph to verify the correctness and accuracy of the approximate analytical solution. The approximate asymptotic solution is drawn using a solid line. A dotted line depicts the numerical solution obtained using the standard *NDSolve* procedure of the *Mathematica* software.

One can note the high compatibility of the curves presenting the numerical and asymptotic solutions in each of the figures, which confirms the correctness of the

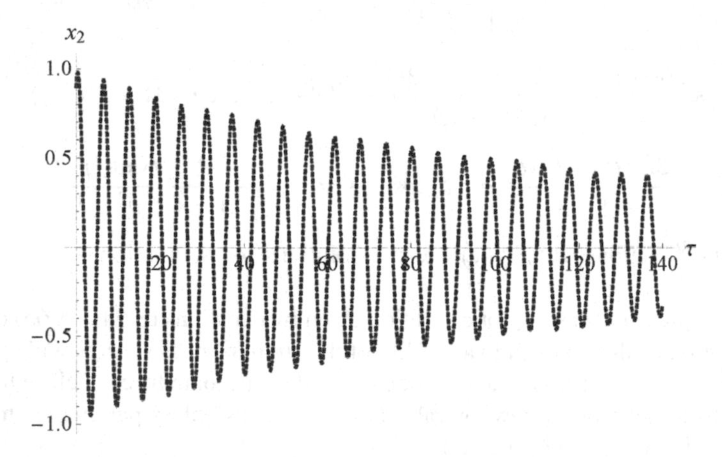

FIGURE 8.3 The body displacement $x_2(\tau)$ for transient non-resonant vibration; solid line – asymptotic solution, dotted line – numerical solution.

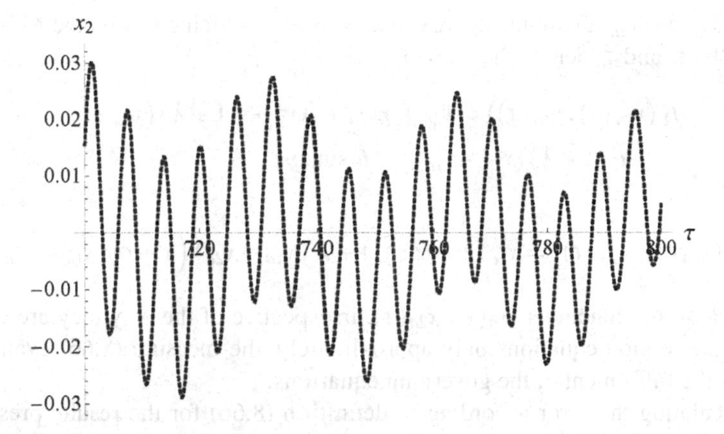

FIGURE 8.4 The body displacement $x_2(\tau)$ for still transient non-resonant vibration; solid line – asymptotic solution, dotted line – numerical solution.

approximate analytical solutions. The quantitative evaluation of the approximate solution (8.64)–(8.65) is made using the following measure

$$e_i = \sqrt{\frac{1}{\tau_e - \tau_s}\int_{\tau_s}^{\tau_e} (H_i(x_{1a}(\tau), x_{2a}(\tau)\,) - 0)^2 d\tau}, \quad i = 1, 2, \tag{8.66}$$

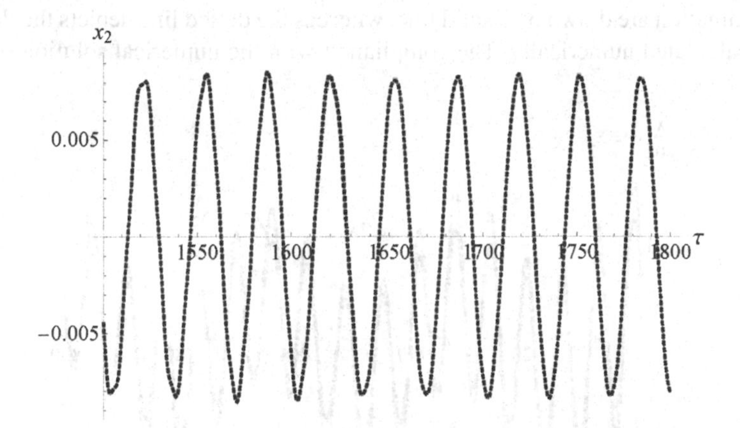

FIGURE 8.5 The body displacement $x_2(\tau)$ for almost stationary non-resonant vibration; solid line – asymptotic solution, dotted line – numerical solution.

where: $x_{1a}(\tau)$, $x_{2a}(\tau)$ are the approximate solutions obtained using the MSM or numerically, τ_s and τ_e denote the chosen instants,

$$H_1\left(x_{1a}(\tau),x_{2a}(\tau)\right)=\ddot{x}_{2a}+\gamma\dot{x}_{2a}+\beta\dot{x}_{2a}^3+(1+\lambda)\left(x_{2a}-x_{1a}\right)+ \\ \alpha_2(1+\lambda)\left(x_{2a}-x_{1a}\right)^3-f_0\sin(p\tau),$$

$$H_2\left(x_{1a}(\tau),x_{2a}(\tau)\right)=x_{1a}\left(1+\alpha_1x_{1a}^2\right)+\lambda\left(x_{1a}-x_{2a}\right)\left(1+\alpha_2(x_{1a}-x_{2a})^2\right),$$

Because the functions $x_{1a}(\tau)$, $x_{2a}(\tau)$, irrespective of the way they are obtained, satisfy the motion equations only approximately, the measure (8.66) evaluates the error of the fulfillment of the governing equations.

Calculating the error according to definition (8.66) for the results presented in Figures 8.3–8.5 we assumed:

$$\tau_s=0 \quad \text{and} \quad \tau_e=2000.$$

The values of the error are: $e_1 \approx 1.561\cdot10^{-4}$, $e_2 \approx 2.063\cdot10^{-5}$. For comparison, the numerical solution gives $e_1 \approx 1.785\cdot10^{-6}$, $e_2 \approx 1.513\cdot10^{-9}$ on the same time range. The accuracy of the results obtained using the MSM is lower, but they have the form of an analytical dependence that enables prediction and inference.

The time-varying deformations of both springs for the transient process and the stationary vibration are shown in Figures 8.6–8.7, respectively.

The elongation Δ_1 of the spring with one end immovable is equal to the coordinate $x_1(\tau)$ while the strain Δ_2 of the intermediate spring is calculated as the difference $x_2(\tau)-x_1(\tau)$. The deformations determined on the ground of the asymptotic approximation are drawn by a solid line, whereas the dotted line depicts the deformations calculated numerically. The compliance with the numerical solution observed

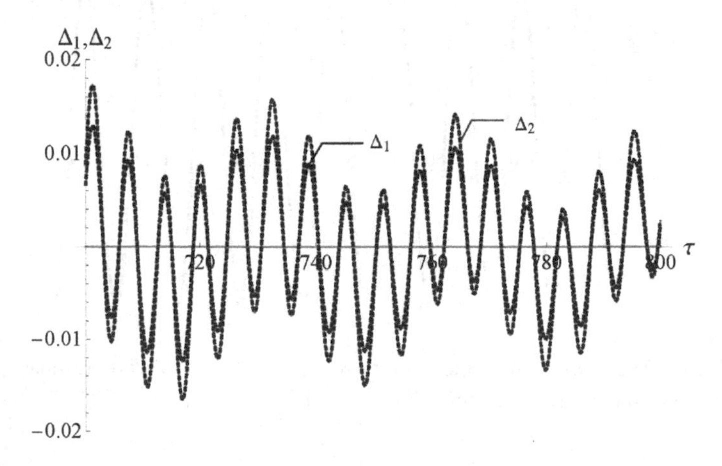

FIGURE 8.6 The strain of both springs for transient non-resonant vibration; solid line – asymptotic solution, dotted line – numerical solution.

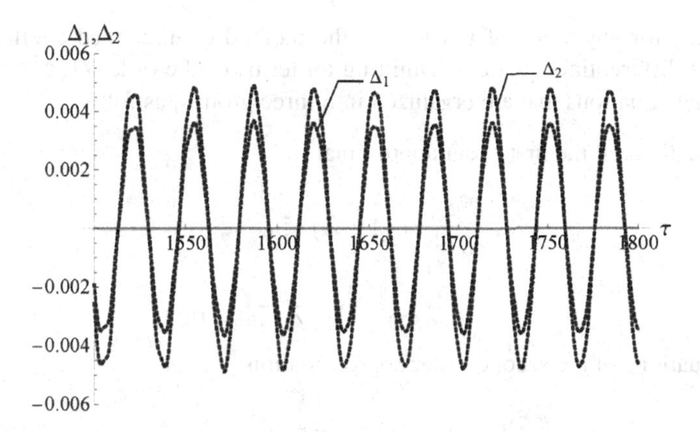

FIGURE 8.7 The strain of both springs for almost stationary non-resonant vibration; solid line – asymptotic solution, dotted line – numerical solution.

in these graphs and the small value of the error satisfying the algebraic equation confirm that the results obtained with the MSM can be considered reliable for the range of parameters consistent with the assumptions (8.25).

8.5 RESONANT VIBRATION

The functions given by (8.64)–(8.65) are not valid when the dimensionless frequency of the harmonic excitation is close to the one. In terms of dimensional quantities, this means that the solution fails when the harmonic force frequency is equal to the eigenfrequency of the reference system which is linear. Due to the assumption about the weak character of the nonlinearities and the damping, some neighborhood of the natural frequency of the reference system determines the zone of the resonance of the nonlinear system considered. This case corresponding to the external main resonance requires a separate approach based on additionally formulated assumptions.

Following the MSM, to express the nearness between the frequencies we introduce the detuning parameter σ, as follows

$$p = 1 + \sigma. \tag{8.67}$$

The detuning parameter is any but, rather, a small number. We assume that it is of the order $O(\varepsilon)$, so

$$\sigma = \varepsilon\hat{\sigma}. \tag{8.68}$$

The coefficient $\hat{\sigma}$ is understood as $O(1)$ when $\varepsilon \to 0$.

While maintaining the assumptions (8.25), inserting relations (8.20)–(8.24), being the main concept of the MSM, into the motion equations, and taking into account the resonance conditions (8.67)–(8.68), we get the equations containing the small parameter ε in various powers. The form of these equations is too long to be quoted here. To proceed further, these equations need to be rearranged according to the powers of a small parameter. The requirement that each of these equations should be satisfied for any value of ε leads via the method of undefined coefficients to a system of differential equations. Omitting all terms of the order $O(\varepsilon^3)$, we get six differential equations that are organized into three groups, as follows

(i) equations of the first-order approximation

$$\frac{\partial^2 \xi_{20}}{\partial \tau_0^2} + (1+\lambda)(\xi_{20} - \xi_{10}) = 0, \tag{8.69}$$

$$\lambda \xi_{20} - (1+\lambda)\xi_{10} = 0, \tag{8.70}$$

(ii) equations of the second-order approximation

$$\begin{aligned} &\frac{\partial^2 \xi_{21}}{\partial \tau_0^2} + (1+\lambda)(\xi_{21} - \xi_{11}) = \\ &-(1+\lambda)\hat{\alpha}_2(\xi_{20} - \xi_{10})^3 - \hat{\gamma}\frac{\partial \xi_{20}}{\partial \tau_0} - \hat{\beta}\left(\frac{\partial \xi_{20}}{\partial \tau_0}\right)^3 - 2\frac{\partial^2 \xi_{20}}{\partial \tau_0 \partial \tau_1}, \end{aligned} \tag{8.71}$$

$$\lambda \xi_{21} - (1+\lambda)\xi_{11} = \hat{\alpha}_1 \xi_{10}^3 - \lambda \hat{\alpha}_2 (\xi_{20} - \xi_{10})^3, \tag{8.72}$$

(iii) equations of the third-order approximation

$$\begin{aligned} &\frac{\partial^2 \xi_{22}}{\partial \tau_0^2} + (1+\lambda)(\xi_{22} - \xi_{12}) = \hat{f}_0 \sin(\tau_0 + \varepsilon\hat{\sigma}\tau_0) - \\ &3(1+\lambda)\hat{\alpha}_2(\xi_{20} - \xi_{10})^2(\xi_{21} - \xi_{11}) - \hat{\gamma}\left(\frac{\partial \xi_{20}}{\partial \tau_1} + \frac{\partial \xi_{21}}{\partial \tau_0}\right) - \\ &3\hat{\beta}\left(\frac{\partial \xi_{20}}{\partial \tau_0}\right)^2\left(\frac{\partial \xi_{20}}{\partial \tau_1} + \frac{\partial \xi_{21}}{\partial \tau_0}\right) - \frac{\partial^2 \xi_{20}}{\partial \tau_1^2} - 2\frac{\partial^2 \xi_{20}}{\partial \tau_0 \partial \tau_2} - 2\frac{\partial^2 \xi_{21}}{\partial \tau_0 \partial \tau_1}, \end{aligned} \tag{8.73}$$

$$\lambda \xi_{22} - (1+\lambda)\xi_{12} = -3\lambda \hat{\alpha}_2 (\xi_{20} - \xi_{10})^2 (\xi_{21} - \xi_{11}) + 3\,\hat{\alpha}_1 \xi_{10}^2 \xi_{11}. \tag{8.74}$$

The functions $\xi_{1k}(\tau_0, \tau_1, \tau_2)$, $\xi_{2k}(\tau_0, \tau_1, \tau_2)$, $k = 0, 1, 2$, being the terms of the asymptotic expansions (8.21)–(8.22), are to be determined. Similarly, as in the non-resonant case, the set of equations (8.69)–(8.74) is solved recursively starting from equations of the first-order approximation.

From equation (8.70) we get

$$\xi_{10} = \frac{\lambda}{(1+\lambda)}\xi_{20}. \tag{8.75}$$

Equation (8.69) after substituting relationship (8.75) becomes the following homogeneous differential equation

$$\frac{\partial^2 \xi_{20}}{\partial \tau_0^2} + \xi_{20} = 0, \tag{8.76}$$

the general solution of which is

$$\xi_{20} = D(\tau_1,\ \tau_2)\, e^{\mathrm{i}\tau_0} + \bar{D}(\tau_1,\ \tau_2)\, e^{-\mathrm{i}\tau_0}, \tag{8.77}$$

where D and its complex conjugate $\bar{D}$ are unknown complex-valued functions of both slower time scales.

Connection of relationships (8.75) and (8.77) implies that

$$\xi_{10} = \frac{\lambda}{(1+\lambda)} D(\tau_1,\ \tau_2)\, e^{\mathrm{i}\tau_0} + \frac{\lambda}{(1+\lambda)} \bar{D}(\tau_1,\ \tau_2) e^{-\mathrm{i}\tau_0}. \tag{8.78}$$

From equation (8.72) one can derive the relationship

$$\xi_{11} = \frac{\lambda}{(1+\lambda)} \xi_{21} - \frac{\lambda\left(\lambda^2 \hat{\alpha}_1 - \hat{\alpha}_2\right)}{(1+\lambda)^4} \xi_{20}^3. \tag{8.79}$$

Substitution of equation (8.79) and solutions (8.77)–(8.78) into equation (8.71) gives the following differential equation with one unknown function $\xi_{21}(\tau_0, \tau_1, \tau_2)$

$$\begin{aligned}
\frac{\partial^2 \xi_{21}}{\partial \tau_0^2} + \xi_{21} = &\left(\mathrm{i}\hat{\beta} - \frac{(\lambda^3 \hat{\alpha}_1 + \hat{\alpha}_2)}{(1+\lambda)^3}\right) e^{3\mathrm{i}\tau_0} D^3 - \\
&\left(\mathrm{i}\hat{\beta} + \frac{(\lambda^3 \hat{\alpha}_1 + \hat{\alpha}_2)}{(1+\lambda)^3}\right) e^{-3\mathrm{i}\tau_0} \bar{D}^3 - \\
&\left(2\mathrm{i}\frac{\partial D}{\partial \tau_1} + \mathrm{i}\hat{\gamma} D + 3\left(\frac{\lambda^3 \hat{\alpha}_1 + \hat{\alpha}_2}{(1+\lambda)^3} + \mathrm{i}\hat{\beta}\right) D^2 \bar{D}\right) e^{\mathrm{i}\tau_0} + \\
&\left(2\mathrm{i}\frac{\partial \bar{D}}{\partial \tau_1} + \mathrm{i}\hat{\gamma} \bar{D} - 3\left(\frac{\lambda^3 \hat{\alpha}_1 + \hat{\alpha}_2}{(1+\lambda)^3} - \mathrm{i}\hat{\beta}\right) D \bar{D}^2\right) e^{-\mathrm{i}\tau_0}.
\end{aligned} \tag{8.80}$$

The two last terms on the right side of equation (8.80) are the secular terms leading to an unbounded solution of the differential equation (8.80). It is necessary to eliminate them. This can be achieved by demanding that the coefficients accompanying the exponential functions $\exp(\mathrm{i}\tau_0)$ and $\exp(-\mathrm{i}\tau_0)$ be equal to zero. This way, we can write the following solvability conditions

$$2\mathrm{i}\frac{\partial D}{\partial \tau_1} + \mathrm{i}\hat{\gamma} D + 3\left(\frac{\lambda^3 \hat{\alpha}_1 + \hat{\alpha}_2}{(1+\lambda)^3} + \mathrm{i}\hat{\beta}\right) D^2 \bar{D}, \tag{8.81}$$

$$2\mathrm{i}\frac{\partial \bar{D}}{\partial \tau_1} + \mathrm{i}\hat{\gamma} \bar{D} - 3\left(\frac{\lambda^3 \hat{\alpha}_1 + \hat{\alpha}_2}{(1+\lambda)^3} - \mathrm{i}\hat{\beta}\right) D \bar{D}^2 = 0. \tag{8.82}$$

The following conclusions result from the solvability conditions

$$\frac{\partial D}{\partial \tau_1}=\frac{1}{2}D\left(\frac{3\mathrm{i}\,\left(\lambda^3\hat{\alpha}_1+\hat{\alpha}_2\right)}{(1+\lambda)^3}D\bar{D}-3\hat{\beta}D\bar{D}-\hat{\gamma}\right), \tag{8.83}$$

$$\frac{\partial \bar{D}}{\partial \tau_1}=-\frac{1}{2}\bar{D}\left(\frac{3\mathrm{i}\,\left(\lambda^3\hat{\alpha}_1+\hat{\alpha}_2\right)}{(1+\lambda)^3}D\bar{D}+3\hat{\beta}D\bar{D}+\hat{\gamma}\right). \tag{8.84}$$

Differentiating (8.83)–(8.84) with respect to τ_1 gives

$$\begin{aligned}\frac{\partial^2 D}{\partial \tau_1^2}&=\frac{\hat{\gamma}^2}{4}D+3D^2\bar{D}\hat{\gamma}\left(\hat{\beta}-\mathrm{i}\frac{\lambda^3\hat{\alpha}_1+\hat{\alpha}_2}{(1+\lambda)^3}\right)-\\ &\frac{9}{4}D^3\bar{D}^2\left(\frac{\left(\left(\lambda^3\hat{\alpha}_1+\hat{\alpha}_2\right)^2+4\mathrm{i}(1+\lambda)^3\left(\lambda^3\hat{\alpha}_1+\hat{\alpha}_2\right)\hat{\beta}\right)}{(1+\lambda)^6}-3\hat{\beta}^2\right),\end{aligned} \tag{8.85}$$

$$\begin{aligned}\frac{\partial^2 \bar{D}}{\partial \tau_1^2}&=\frac{\hat{\gamma}^2}{4}\bar{D}+3D\bar{D}^2\hat{\gamma}\left(\hat{\beta}+\mathrm{i}\frac{\lambda^3\hat{\alpha}_1+\hat{\alpha}_2}{(1+\lambda)^3}\right)+\\ &\frac{9}{4}D^2\bar{D}^3\left(\frac{\left(-\left(\lambda^3\hat{\alpha}_1+\hat{\alpha}_2\right)^2+4\mathrm{i}(1+\lambda)^3\left(\lambda^3\hat{\alpha}_1+\hat{\alpha}_2\right)\hat{\beta}\right)}{(1+\lambda)^6}-3\hat{\beta}^2\right).\end{aligned} \tag{8.86}$$

The particular solution to equation (8.80) obtained after eliminating the secular terms according to (8.81)–(8.82) is as follows

$$\begin{aligned}\xi_{21}\left(\tau_0,\ \tau_1,\ \tau_2\right)=&\frac{\lambda^3\hat{\alpha}_1+\hat{\alpha}_2-\mathrm{i}(1+\lambda)^3\hat{\beta}}{8(1+\lambda)^3}e^{3\mathrm{i}\tau_0}D^3+\\ &\frac{\lambda^3\hat{\alpha}_1+\hat{\alpha}_2+\mathrm{i}(1+\lambda)^3\hat{\beta}}{8(1+\lambda)^3}e^{-3\mathrm{i}\tau_0}\bar{D}^3.\end{aligned} \tag{8.87}$$

Making use of equations (8.77), (8.79), and (8.87) we can write that

$$\begin{aligned}\xi_{11}\left(\tau_0,\tau_1,\tau_2\right)=&-\frac{3\lambda\left(\lambda^2\hat{\alpha}_1-\hat{\alpha}_2\right)}{(1+\lambda)^4}\left(e^{3\mathrm{i}\tau_0}D^2\bar{D}+e^{-3\mathrm{i}\tau_0}\bar{D}^2D\right)+\\ &\frac{\lambda\left(\lambda^2(\lambda-8)\hat{\alpha}_1+9\hat{\alpha}_2-\mathrm{i}(1+\lambda)^3\hat{\beta}\right)\left(\hat{\alpha}_1-\hat{\alpha}_2\right)}{8(1+\lambda)^4}e^{3\mathrm{i}\tau_0}D^3+\\ &\frac{\lambda\left(\lambda^2(\lambda-8)\hat{\alpha}_1+9\hat{\alpha}_2+\mathrm{i}(1+\lambda)^3\hat{\beta}\right)\left(\hat{\alpha}_1-\hat{\alpha}_2\right)}{8(1+\lambda)^4}e^{-3\mathrm{i}\tau_0}\bar{D}^3.\end{aligned} \tag{8.88}$$

Using equation (8.74), we can derive the following relationship

$$\xi_{12} = \frac{\lambda}{(1+\lambda)}\xi_{22} + \frac{3\lambda^2\left(\lambda^2\hat{\alpha}_1 - \hat{\alpha}_2\right)\left(\lambda\hat{\alpha}_1 + \hat{\alpha}_2\right)}{(1+\lambda)^7}\xi_{20}^5 - \frac{3\lambda\left(\lambda^2\hat{\alpha}_1 - \hat{\alpha}_2\right)}{(1+\lambda)^4}\xi_{20}^2\xi_{21}. \tag{8.89}$$

Substituting equations (8.77), (8.79), and (8.87)–(8.88) into differential equation (8.73) generates subsequent secular terms. They should be eliminated according to the solvability conditions

$$-\frac{1}{2}\mathrm{i}\hat{f}_0 e^{\mathrm{i}\varepsilon\hat{\sigma}\tau_0} - 2\mathrm{i}\frac{\partial D}{\partial\tau_2} + \frac{1}{4}\hat{\gamma}^2 D + 3\mathrm{i}\hat{\gamma}\frac{\lambda^3\hat{\alpha}_1 + \hat{\alpha}_2}{2(1+\lambda)^3}D^2\bar{D} - \frac{9}{8}\hat{\beta}^2 D^3\bar{D}^2 - 3\mathrm{i}\hat{\beta}\frac{\lambda^3\hat{\alpha}_1 + \hat{\alpha}_2}{(1+\lambda)^3}D^3\bar{D}^2 + 15\frac{\lambda^5(\lambda+16)\hat{\alpha}_1^2 - 30\lambda^3\hat{\alpha}_1\hat{\alpha}_2 + (16\lambda+1)\hat{\alpha}_2^2}{8(1+\lambda)^6}D^3\bar{D}^2 = 0, \tag{8.90}$$

$$\frac{1}{2}\mathrm{i}\hat{f}_0 e^{-\mathrm{i}\varepsilon\hat{\sigma}\tau_0} + 2\mathrm{i}\frac{\partial \bar{D}}{\partial\tau_2} + \frac{1}{4}\hat{\gamma}^2 \bar{D} - 3\mathrm{i}\hat{\gamma}\frac{\lambda^3\hat{\alpha}_1 + \hat{\alpha}_2}{2(1+\lambda)^3}D\bar{D}^2 - \frac{9}{8}\hat{\beta}^2 D^2\bar{D}^3 + 3\mathrm{i}\hat{\beta}\frac{\lambda^3\hat{\alpha}_1 + \hat{\alpha}_2}{(1+\lambda)^3}D^2\bar{D}^3 + 15\frac{\lambda^5(\lambda+16)\hat{\alpha}_1^2 - 30\lambda^3\hat{\alpha}_1\hat{\alpha}_2 + (16\lambda+1)\hat{\alpha}_2^2}{8(1+\lambda)^6}D^2\bar{D}^3 = 0. \tag{8.91}$$

Eliminating the secular terms from equation (8.73), according to (8.90)–(8.91), gives the following differential equation

$$\begin{aligned}
\frac{\partial^2\xi_{22}}{\partial\tau_0^2} + \xi_{22} = {} & \\
& \frac{3}{8}\left(3\hat{\beta}^2 - \frac{\lambda^5(\lambda-8)\hat{\alpha}_1^2 + 18\lambda^3\hat{\alpha}_1\hat{\alpha}_2 + (1-8\lambda)\hat{\alpha}_2^2}{(1+\lambda)^6}\right)\left(e^{5\mathrm{i}\tau_0}D^5 + e^{-5\mathrm{i}\tau_0}\bar{D}^5\right) + \\
& \frac{3}{8}\left(\frac{\lambda^5(7\lambda+40)\hat{\alpha}_1^2 - 66\lambda^3\hat{\alpha}_1\hat{\alpha}_2 + (40\lambda+7)\hat{\alpha}_2^2}{(1+\lambda)^6} - 9\hat{\beta}^2\right)D\bar{D}\left(e^{3\mathrm{i}\tau_0}D^3 + e^{-3\mathrm{i}\tau_0}\bar{D}^3\right) + \\
& 3\mathrm{i}\frac{\left(\lambda^3\hat{\alpha}_1 + \hat{\alpha}_2\right)\hat{\beta}}{2(1+\lambda)^3}\left(e^{5\mathrm{i}\tau_0}D^5 - e^{-5\mathrm{i}\tau_0}\bar{D}^5 + 2D\bar{D}\left(e^{3\mathrm{i}\tau_0}D^3 - e^{-3\mathrm{i}\tau_0}\bar{D}^3\right)\right) + \\
& \frac{3}{4}\left(\mathrm{i}\frac{\lambda^2\hat{\alpha}_1 + \hat{\alpha}_2}{(1+\lambda)^3} - \hat{\beta}\right)\hat{\gamma}e^{3\mathrm{i}\tau_0}D^3 - \frac{3}{4}\left(\mathrm{i}\frac{\lambda^2\hat{\alpha}_1 + \hat{\alpha}_2}{(1+\lambda)^3} + \hat{\beta}\right)\hat{\gamma}e^{-3\mathrm{i}\tau_0}\bar{D}^3.
\end{aligned} \tag{8.92}$$

The particular solution to equation (8.92) is given by

$$\xi_{22}(\tau_0,\tau_1,\tau_2) = \left(\frac{\lambda^5(\lambda-8)\hat{\alpha}_1^2 + 18\lambda^3\hat{\alpha}_1\hat{\alpha}_2 + (1-8\lambda)\hat{\alpha}_2^2}{64(1+\lambda)^6} - \frac{3}{64}\hat{\beta}^2\right)\left(e^{5i\tau_0}D^5 + e^{-5i\tau_0}\bar{D}^5\right) - \frac{3}{64}\left(\frac{\lambda^5(7\lambda+40)\hat{\alpha}_1^2 - 66\lambda^3\hat{\alpha}_1\hat{\alpha}_2 + (40\lambda+7)\hat{\alpha}_2^2}{(1+\lambda)^6} - 9\hat{\beta}^2\right)D\bar{D}\left(e^{3i\tau_0}D^3 + e^{-3i\tau_0}\bar{D}^3\right) - 3i\frac{(\lambda^3\hat{\alpha}_1+\hat{\alpha}_2)\hat{\beta}}{8(1+\lambda)^3}\left(\frac{1}{6}\left(e^{5i\tau_0}D^5 - e^{-5i\tau_0}\bar{D}^5\right) + D\bar{D}\left(e^{3i\tau_0}D^3 - e^{-3i\tau_0}\bar{D}^3\right)\right) - \frac{3}{32}\left(i\frac{\lambda^2\hat{\alpha}_1+\hat{\alpha}_2}{(1+\lambda)^3} - \hat{\beta}\right)\hat{\gamma}e^{3i\tau_0}D^3 - \frac{3}{32}\left(i\frac{\lambda^2\hat{\alpha}_1+\hat{\alpha}_2}{(1+\lambda)^3} + \hat{\beta}\right)\hat{\gamma}e^{-3i\tau_0}\bar{D}^3. \tag{8.93}$$

Taking into account equations (8.77), (8.87), (8.89), and (8.93), we obtain

$$\xi_{12} = \left(\frac{\lambda^6(\lambda-8)\hat{\alpha}_1^2 + 18\lambda^4\hat{\alpha}_1\hat{\alpha}_2 + \lambda(1-8\lambda)\hat{\alpha}_2^2}{64(1+\lambda)^7} - \frac{3\lambda\hat{\beta}^2}{64(1+\lambda)}\right)\left(e^{5i\tau_0}D^5 + e^{-5i\tau_0}\bar{D}^5\right) - \frac{3}{64}\left(\frac{\lambda^6(7\lambda+40)\hat{\alpha}_1^2 - 66\lambda^4\hat{\alpha}_1\hat{\alpha}_2 + \lambda(40\lambda+7)\hat{\alpha}_2^2}{(1+\lambda)^7} - \frac{9\lambda\hat{\beta}^2}{1+\lambda}\right)D\bar{D}\left(e^{3i\tau_0}D^3 + e^{-3i\tau_0}\bar{D}^3\right) - 3i\frac{(\lambda^3\hat{\alpha}_1+\hat{\alpha}_2)\lambda\hat{\beta}}{8(1+\lambda)^4}\left(\frac{1}{6}\left(e^{5i\tau_0}D^5 - e^{-5i\tau_0}\bar{D}^5\right) + D\bar{D}\left(e^{3i\tau_0}D^3 - e^{-3i\tau_0}\bar{D}^3\right)\right) - \frac{3\lambda}{32}\left(i\frac{\lambda^2\hat{\alpha}_1+\hat{\alpha}_2}{(1+\lambda)^4} - \frac{\hat{\beta}}{1+\lambda}\right)\hat{\gamma}e^{3i\tau_0}D^3 - \frac{3\lambda}{32}\left(i\frac{\lambda^2\hat{\alpha}_1+\hat{\alpha}_2}{(1+\lambda)^4} + \frac{\hat{\beta}}{1+\lambda}\right)\hat{\gamma}e^{-3i\tau_0}\bar{D}^3 + \frac{3\lambda^2(\lambda^2\hat{\alpha}_1-\hat{\alpha}_2)(\lambda\hat{\alpha}_1+\hat{\alpha}_2)}{(1+\lambda)^7}\left(D(\tau_1,\tau_2)e^{i\tau_0} + \bar{D}(\tau_1,\tau_2)e^{-i\tau_0}\right)^5 - \tag{8.94}$$

$$\frac{3\lambda\left(\lambda^2\hat{\alpha}_1-\hat{\alpha}_2\right)}{(1+\lambda)^4}\left(D(\tau_1,\tau_2)e^{i\tau_0}+\right.$$
$$\left.\bar{D}(\tau_1,\tau_2)e^{-i\tau_0}\right)^2\frac{\lambda^3\hat{\alpha}_1-i\left((1+\lambda)^3\hat{\beta}+i\hat{\alpha}_2\right)}{8(1+\lambda)^3}e^{3i\tau_0}D^3-$$
$$\frac{3\lambda\left(\lambda^2\hat{\alpha}_1-\hat{\alpha}_2\right)}{(1+\lambda)^4}\left(D(\tau_1,\ \tau_2)e^{i\tau_0}+\right.$$
$$\left.\bar{D}(\tau_1,\ \tau_2)e^{-i\tau_0}\right)^2\frac{\lambda^3\hat{\alpha}_1+i\left((1+\lambda)^3\hat{\beta}-i\hat{\alpha}_2\right)}{8(1+\lambda)^3}e^{-3i\tau_0}\bar{D}^3.$$

The unknown complex-valued functions $D(\tau_1,\tau_2)$ and $\bar{D}(\tau_1,\tau_2)$ restricted by the solvability conditions (8.81)–(8.82) and (8.90)–(8.91) we depict in the exponential form using relationships (8.52). Inserting relations (8.52) into the aforementioned solvability conditions, we obtain the following set of four partial differential equations with unknown functions $a(\tau_1,\tau_2)$ and $\psi(\tau_1,\tau_2)$

$$-i\frac{\partial a}{\partial\tau_1}+\frac{1}{2}a\left(2\frac{\partial\psi}{\partial\tau_1}-i\hat{\gamma}\right)-\frac{3a^3\left(\lambda^3\hat{\alpha}_1+\hat{\alpha}_2+i(1+\lambda)^3\hat{\beta}\right)}{8(1+\lambda)^3}=0, \tag{8.95}$$

$$i\frac{\partial a}{\partial\tau_1}+\frac{1}{2}a\left(2\frac{\partial\psi}{\partial\tau_1}+i\hat{\gamma}\right)-\frac{3a^3\left(\lambda^3\hat{\alpha}_1+\hat{\alpha}_2-i(1+\lambda)^3\hat{\beta}\right)}{8(1+\lambda)^3}=0, \tag{8.96}$$

$$\begin{aligned}&-i\frac{\partial a}{\partial\tau_2}+a\left(\frac{\partial\psi}{\partial\tau_2}+\frac{\hat{\gamma}^2}{8}\right)-\frac{1}{2}i\hat{f}_0e^{i(\varepsilon\hat{\sigma}\tau_0-\psi)}+\\&3i\frac{\left(\lambda^3\hat{\alpha}_1+\hat{\alpha}_2\right)}{16(1+\lambda)^3}\hat{\gamma}a^3-\frac{9}{256}\hat{\beta}^2a^5+\\&15\frac{\lambda^5(16+\lambda)\hat{\alpha}_1^2-30\lambda^3\hat{\alpha}_1\hat{\alpha}_2+(1+16\lambda)\hat{\alpha}_2^2}{256(1+\lambda)^6}a^5-\\&3i\frac{\lambda^3\hat{\alpha}_1+\hat{\alpha}_2}{32(1+\lambda)^3}\hat{\beta}a^5=0,\end{aligned} \tag{8.97}$$

$$\begin{aligned}&i\frac{\partial a}{\partial\tau_2}+a\left(\frac{\partial\psi}{\partial\tau_2}+\frac{\hat{\gamma}^2}{8}\right)+\frac{1}{2}i\hat{f}_0e^{-i(\varepsilon\hat{\sigma}\tau_0-\psi)}-\\&3i\frac{\left(\lambda^3\hat{\alpha}_1+\hat{\alpha}_2\right)}{16(1+\lambda)^3}\hat{\gamma}a^3-\frac{9}{256}\hat{\beta}^2a^5+\\&15\frac{\lambda^5(16+\lambda)\hat{\alpha}_1^2-30\lambda^3\hat{\alpha}_1\hat{\alpha}_2+(1+16\lambda)\hat{\alpha}_2^2}{256(1+\lambda)^6}a^5+\\&3i\frac{\lambda^3\hat{\alpha}_1+\hat{\alpha}_2}{32(1+\lambda)^3}\hat{\beta}a^5=0.\end{aligned} \tag{8.98}$$

Solving equations (8.95)–(8.98) with respect to the partial derivatives yields

$$\frac{\partial a}{\partial \tau_1} = -\frac{1}{2}\hat{\gamma} a - \frac{3}{8}\hat{\beta} a^3, \tag{8.99}$$

$$\frac{\partial a}{\partial \tau_2} = -\frac{1}{2}\hat{f}_0 \cos(\varepsilon\hat{\sigma}\tau_0 - \psi) - \frac{3\left(\lambda^3\hat{\alpha}_1 + \hat{\alpha}_2\right)\left(\hat{\beta}a^2 - 2\hat{\gamma}\right)}{32(1+\lambda)^3} a^3, \tag{8.100}$$

$$\frac{\partial \psi}{\partial \tau_1} = 3\frac{\left(\lambda^3\hat{\alpha}_1 + \hat{\alpha}_2\right)}{8(1+\lambda)^3} a^2, \tag{8.101}$$

$$\begin{aligned}\frac{\partial \psi}{\partial \tau_2} &= -\hat{f}_0 \frac{\sin(\varepsilon\hat{\sigma}\tau_0 - \psi)}{2a} - \frac{15\lambda^5(16+\lambda)\hat{\alpha}_1^2}{256(1+\lambda)^6} a^4 + \\ &\frac{225\lambda^3\hat{\alpha}_1\hat{\alpha}_2}{128(1+\lambda)^6} a^4 - \frac{15(1+16\lambda)\hat{\alpha}_2^2}{256(1+\lambda)^6} a^4 + \frac{9}{256}\hat{\beta}^2 a^4 - \frac{\hat{\gamma}^2}{8}.\end{aligned} \tag{8.102}$$

Reverse use of definition (8.23) and assumptions (8.25) allows one to transform partial differential equations (8.99)–(8.102) into a system of two ordinary differential equations formulated in terms of the primal time variable τ. They are as written below

$$\begin{aligned}\frac{da}{d\tau} &= -\frac{f_0}{2}\cos(\sigma\tau_0 - \psi(\tau)) - \frac{1}{2}\gamma\, a(\tau) + \\ &\frac{3}{16}\left(\gamma\frac{\lambda^3\alpha_1 + \alpha_2}{(1+\lambda)^3} - 2\beta\right) a^3(\tau) - 3\beta\frac{\lambda^3\alpha_1 + \alpha_2}{32(1+\lambda)^3} a^5(\tau),\end{aligned} \tag{8.103}$$

$$\begin{aligned}\frac{d\psi}{d\tau} &= -\frac{f_0}{2\,a(\tau)}\sin(\sigma\tau_0 - \psi(\tau)) - \frac{\gamma^2}{8} + 3\frac{\lambda^3\alpha_1 + \alpha_2}{8(1+\lambda)^3} a^2(\tau) + \\ &3\left(5\frac{-\lambda^5(16+\lambda)\alpha_1^2 + 30\lambda^3\alpha_1\alpha_2 - (1+16\lambda)\alpha_2^2}{256(1+\lambda)^6} + \frac{3}{256}\beta^2\right) a^4(\tau).\end{aligned} \tag{8.104}$$

Modulation equations (8.103)–(8.104) are supplemented by the following initial conditions

$$a(0) = a_0, \quad \psi(0) = \psi_0, \tag{8.105}$$

where known quantities a_0 and ψ_0 are agreed with the initial values x_0, v_0.

The functions $a(\tau)$ and $\psi(\tau)$ present the amplitude and the phase of the first term of the asymptotic expansion given by (8.22) at the main resonance conditions. The initial-value problem governing the slow modulation of the amplitude and the phase needs to be solved numerically.

The partial solutions obtained using the recursive procedure are assembled according to equations (8.21)–(8.22), and this way we obtain the approximate solution in the form

$$
\begin{aligned}
x_2(\tau) &= a(\tau)\cos(\tau+\psi(\tau)) + \beta\frac{\lambda^3\alpha_1+\alpha_2}{256(\lambda+1)^3}a^5(\tau)\sin(5\tau+5\psi(\tau)) + \\
&\left(\frac{\lambda^5(\lambda-8)\alpha_1^2+18\lambda^3\alpha_1\alpha_2-(8\lambda-1)\alpha_2^2}{1024(1+\lambda)^6}-\frac{3\beta^2}{1024}\right)a^5(\tau)\cos(5\tau+5\psi(\tau)) + \\
&\frac{3}{1024}\left(9\beta^2-\frac{\lambda^5(7\lambda+40)\alpha_1^2-66\lambda^3\alpha_1\alpha_2}{(1+\lambda)^6}-\right. \\
&\left.\frac{\alpha_2^2(40\lambda+7)}{(1+\lambda)^6}\right)a^5(\tau)\cos(3\tau+3\psi(\tau)) + \\
&\frac{3\beta(\lambda^3\alpha_1+\alpha_2)}{128(1+\lambda)^3}a^5(\tau)\sin(3\tau+3\psi(\tau)) + \qquad (8.106) \\
&\left(\frac{\beta}{32}+3\gamma\frac{\lambda^3\alpha_1+\alpha_2}{128(1+\lambda)^3}\right)a^3(\tau)\sin(3\tau+3\psi(\tau)) + \\
&\frac{4\lambda^3\alpha_1+4\alpha_2+3\beta\gamma(1+\lambda)^3}{128(1+\lambda)^3}a^3(\tau)\cos(3\tau+3\psi(\tau)), \\
x_1(\tau) &= \frac{\lambda}{1+\lambda}a(\tau)\cos(\tau+\psi(\tau)) + \\
&\beta\lambda\frac{(\lambda^2(\lambda-6)\alpha_1+7\alpha_2)}{256(\lambda+1)^4}a^5(\tau)\sin(5\tau+5\psi(\tau)) + \\
&\left(\frac{18\lambda^3(13\lambda-12)\alpha_1\alpha_2}{1024(1+\lambda)^7}-\frac{3\beta^2\lambda}{1024(1+\lambda)}\right)a^5(\tau)\cos(5\tau+5\psi(\tau)) + \\
&\frac{\lambda^5(\lambda^2-32\lambda+192)\alpha_1^2-25\lambda(8\lambda-1)\alpha_2^2}{1024(1+\lambda)^7}a^5(\tau)\cos(5\tau+5\psi(\tau)) + \\
&3\beta\lambda\frac{(\lambda^2(\lambda-2)\alpha_1+3\alpha_2)}{128(\lambda+1)^4}a^5(\tau)\sin(3\tau+3\psi(\tau)) + \\
&\left(\frac{27\beta^2\lambda}{1024(1+\lambda)}+\frac{18\lambda^3(67\lambda-56)\alpha_1\alpha_2}{1024(1+\lambda)^7}\right)a^5(\tau)\cos(3\tau+3\psi(\tau)) - \\
&3\lambda\frac{9(40\lambda-1)\alpha_2^2+\lambda^4(7\lambda^2+56\lambda-320)\alpha_1^2}{1024(1+\lambda)^7}a^5(\tau)\cos(3\tau+3\psi(\tau)) +
\end{aligned}
$$

$$\frac{3\beta\lambda\left(\alpha_2-\lambda^2\alpha_1\right)}{128(1+\lambda)^4}a^5(\tau)\sin(\tau+\psi(\tau))- \tag{8.107}$$

$$3\lambda\frac{\lambda^4(\lambda-80)\alpha_1^2-81\lambda^2(\lambda-1)\alpha_1\alpha_2+(80\lambda-1)\alpha_2^2}{128(1+\lambda)^7}a^5(\tau)\cos(\tau+\psi(\tau))+$$

$$\lambda\frac{4\beta(1+\lambda)^3+3\gamma(\lambda^3\alpha_1+\alpha_2)}{128(1+\lambda)^4}a^3(\tau)\sin(3\tau+3\psi(\tau))+$$

$$\lambda\frac{36\alpha_2+4\lambda^2(\lambda-8)\alpha_1+3(1+\lambda)^3\beta\gamma}{128(1+\lambda)^4}a^3(\tau)\cos(3\tau+3\psi(\tau))+$$

$$\frac{3\lambda\left(\alpha_2-\lambda^2\alpha_1\right)}{4(1+\lambda)^4}a^3(\tau)\cos(\tau+\psi(\tau)),$$

where the functions $a(\tau)$ and $\psi(\tau)$ are the solutions to modulation problem (8.103)–(8.105). The approximate solution given by (8.106)–(8.107) can be applied at the conditions of the main external resonance.

Maintaining the same values of the parameters which define the system, i.e.

$$\lambda=0.75,\ \ \alpha_1=0.53,\ \ \alpha_2=0.35,\ \ \gamma=0.01,\ \ \beta=0.01,$$

and assuming the following values concerning the harmonic force and the initial

$$f_0=0.008,\ \ \sigma=0.004,\ \ x_0=0.2,\ \ v_0=0.0,$$

we carried out calculations the results of which are presented below.

The relationship between initial conditions (8.19) and (8.105) has been determined using the analytical form of the solution (8.106). Calculating, as before, the initial values for the modulation problem, we obtain

$$a_0\approx 0.200036,\qquad \psi_0\approx -0.025039.$$

The time-courses of the body displacement are shown in Figures 8.8–8.9. The transient vibration with a clear upward trend in the amplitude is presented in Figure 8.8. The oscillations of a fixed nature start at about τ = 600 which is presented in Figure 8.9. The approximate solution obtained by the numerical integration of the governing equations (8.17)–(8.19) is shown in both graphs. The solid lined is used to depict the solution determined by equation (8.106), whereas the numerical solution obtained using the standard *NDSolve* procedure of the *Mathematica* software is drawn by the dotted line.

One can observe high compliance between both solutions. The values of error for the asymptotic and numerical solutions which are presented in Figures 8.8–8.9 calculated according to formula (8.66) over the interval (0,2000) are collected in Table 8.1.

The accuracy of the error satisfying the algebraic equation (8.18) is higher mainly due to its mathematical nature and assumption about the weak nonlinearity of the

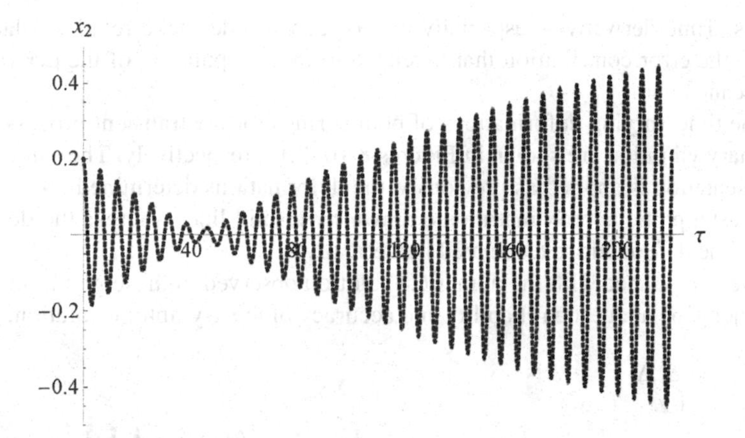

FIGURE 8.8 The body displacement $x_2(\tau)$ for transient vibration at the main resonance conditions; solid line – asymptotic solution, dotted line – numerical solution.

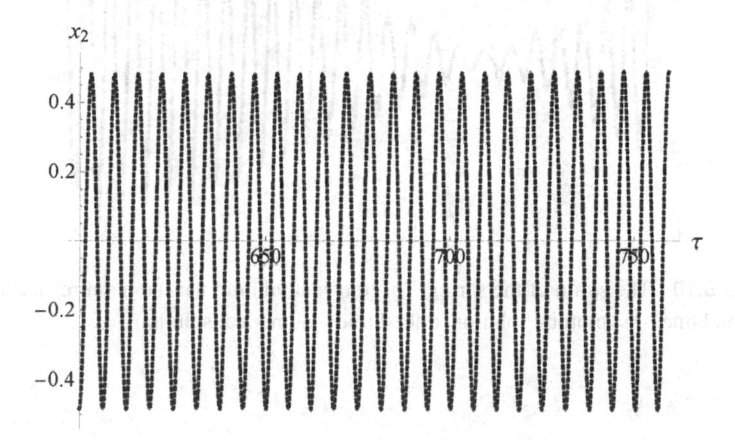

FIGURE 8.9 The body displacement $x_2(\tau)$ for stationary vibration at the main resonance; solid line – asymptotic solution, dotted line – numerical solution.

TABLE 8.1
The values of error of the fulfillment of the governing equations for the resonant vibration

	MSM solution	Numerical solution
$\tau_s = 0, \ \tau_e = 2000$	$e_1 = 7.4924 \cdot 10^{-5}$ $e_2 = 2.2315 \cdot 10^{-6}$	$e_1 = 3.2926 \cdot 10^{-6}$ $e_2 = 3.4733 \cdot 10^{-9}$

springs. Time derivatives, especially of the second-order, take relatively large values, so the error contribution that is related to the computation of the derivatives is significant.

The time-varying deformations of both springs for the transient process and the stationary vibration are shown in Figures 8.10–8.11, respectively. The same manner of presentation is applied as previously, the deformations determined on the ground of the asymptotic approximation are drawn by a solid line, whereas the dotted line depicts the deformations calculated numerically.

The compliance with the numerical solution observed in these graphs completes the conclusion about the reliability and accuracy of the asymptotic solution.

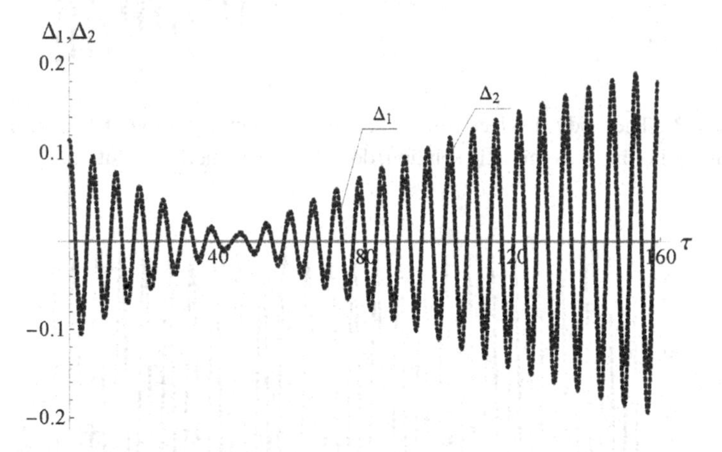

FIGURE 8.10 The strain of the springs for transient vibration at the main resonance conditions; solid line – asymptotic solution, dotted line – numerical solution.

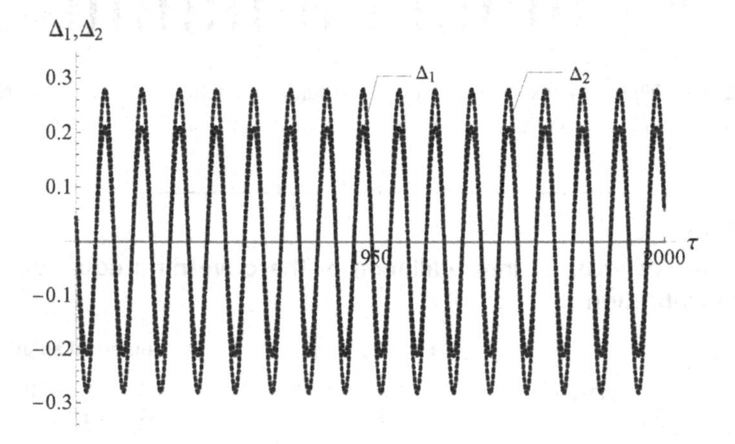

FIGURE 8.11 The strain of the springs for stationary vibration at the main resonance; solid line – asymptotic solution, dotted line – numerical solution.

8.6 STEADY-STATE VIBRATION AT MAIN RESONANCE

Following the MSM we define the modified phase

$$\theta(\tau) = \sigma\tau - \psi(\tau). \tag{8.108}$$

Introducing the modified phase into equations (8.103)–(8.104) enables us to transform them into an autonomous dynamical system given by

$$\begin{gathered}\frac{da}{d\tau} = -\frac{f_0}{2}\cos(\theta(\tau)) - \frac{1}{2}\gamma a(\tau) + \\ \frac{3}{16}\left(\gamma\frac{\lambda^3\alpha_1+\alpha_2}{(1+\lambda)^3} - 2\beta\right)a^3(\tau) - 3\beta\frac{\lambda^3\alpha_1+\alpha_2}{32(1+\lambda)^3}a^5(\tau),\end{gathered} \tag{8.109}$$

$$\begin{gathered}\frac{d\theta}{d\tau} = \frac{f_0}{2\,a(\tau)}\sin(\theta(\tau)) + \frac{\gamma^2}{8} + \sigma - 3\frac{\lambda^3\alpha_1+\alpha_2}{8(1+\lambda)^3}a^2(\tau) - \\ 3\left(5\frac{-\lambda^5(16+\lambda)\alpha_1^2 + 30\lambda^3\alpha_1\alpha_2 - (1+16\lambda)\alpha_2^2}{256(1+\lambda)^6} + \frac{3}{256}\beta^2\right)a^4(\tau).\end{gathered} \tag{8.110}$$

According to the assumptions about the non-stationary vibration, we postulate that the time derivatives of the amplitude and the modified phase become equal to zero. The amplitude and modified phase are therefore constant quantities. To avoid introducing additional symbols we denote them as a and θ without the argument τ. The steady-state vibration is governed by the following set of two equations

$$-\frac{f_0}{2}\cos\theta - \frac{1}{2}\gamma a + \frac{3}{16}\left(\gamma\frac{\lambda^3\alpha_1+\alpha_2}{(1+\lambda)^3} - 2\beta\right)a^3 - 3\beta\frac{\lambda^3\alpha_1+\alpha_2}{32(1+\lambda)^3}a^5 = 0, \tag{8.111}$$

$$\begin{gathered}\frac{f_0}{2a}\sin\theta + \frac{\gamma^2}{8} + \sigma - 3\frac{\lambda^3\alpha_1+\alpha_2}{8(1+\lambda)^3}a^2 - \\ 3\left(5\frac{-\lambda^5(16+\lambda)\alpha_1^2 + 30\lambda^3\alpha_1\alpha_2 - (1+16\lambda)\alpha_2^2}{256(1+\lambda)^6} + \frac{3}{256}\beta^2\right)a^4 = 0.\end{gathered} \tag{8.112}$$

The values satisfying the system (8.111)–(8.112) represent the fixed point of the dynamical system (8.109)–(8.110) at conditions of the main resonance.

Equations (8.111)–(8.112) are of the algebraic-trigonometric type. Employing the trigonometric identities, the modified phases can be eliminated from them which allows one to express the dependence between the stationary vibration amplitude and the frequency of the harmonic force in the form of the following algebraic equation

$$\begin{gathered}\left(\frac{3A_5}{128}a^5 + 3\frac{\lambda^3\alpha_1+\alpha_2}{4(1+\lambda)^3}a^3 - \frac{\gamma^2+8\sigma}{4}a\right)^2 + \\ \left(3\beta\frac{\lambda^3\alpha_1+\alpha_2}{16(1+\lambda)^3}a^5 + \frac{3}{8}\left(2\beta - \gamma\frac{\lambda^3\alpha_1+\alpha_2}{(1+\lambda)^3}\right)a^3 + \gamma a\right)^2 - f_0^2 = 0,\end{gathered} \tag{8.113}$$

where

$$A_5 = 3\beta^2 - 5\frac{\lambda^5(16+\lambda)\alpha_1^2 - 30\alpha_1\alpha_2\lambda^3 + (1+16\lambda)\alpha_2^2}{(1+\lambda)^6}.$$

The polynomial occurring on the left side of equation (8.113) is of tenth degree with respect to the variable a, hence the number of roots of the equation equals ten, but not all of them are real. Due to the fact that the polynomial is an even function of a, it is purposeful to introduce a new variable

$$u = a^2. \tag{8.114}$$

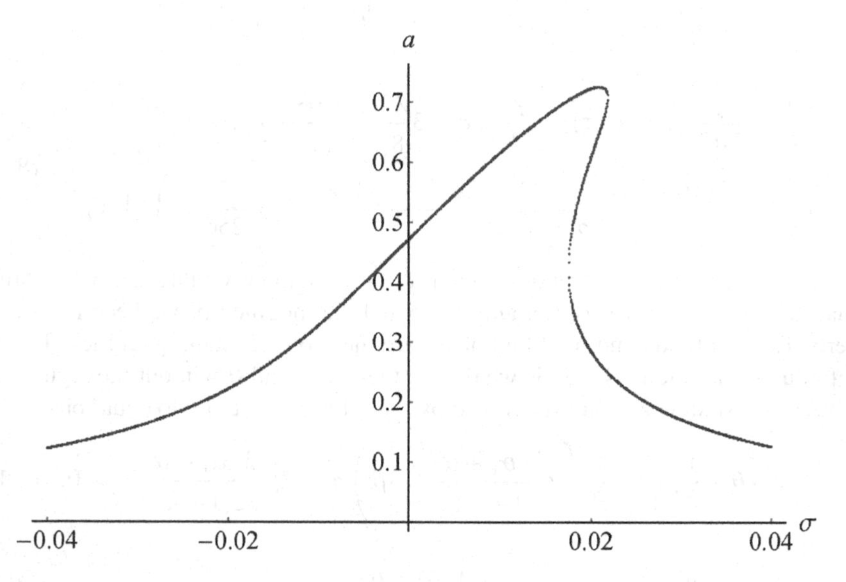

FIGURE 8.12 The stationary vibration amplitude a versus the detuning parameter σ at the main resonance.

The equation of the fifth degree with respect to the square of the amplitude can be written as

$$\left(\frac{3A_5}{128}u^2 + 3\frac{\lambda^3\alpha_1+\alpha_2}{4(1+\lambda)^3}u - \frac{\gamma^2+8\sigma}{4}\right)^2 u +$$
$$\left(3\beta\frac{\lambda^3\alpha_1+\alpha_2}{16(1+\lambda)^3}u^2 + \frac{3}{8}\left(2\beta - \gamma\frac{\lambda^3\alpha_1+\alpha_2}{(1+\lambda)^3}\right)u + \gamma\right)^2 u - f_0^2 = 0. \tag{8.115}$$

The roots of equation (8.115) are determined numerically using the standard procedure *NSolve* of the *Mathematica*. The complex solutions are then rejected.

The resonance response (backbone) curve for the oscillator with the same parameters as those adopted in the above-mentioned simulations is shown in Figure 8.12. The magnitude of the dimensionless amplitude of harmonic force is assumed to be

0.01, and the dimensionless frequency of the force is close to unity. The detuning parameter varies from -0.04 to 0.04 with the apriori assumed step equal to $5 \cdot 10^{-5}$.

It is very important to know the resonance response in the form of the dependency shown in Figure 8.12. However, in order to fully describe the stationary states of nonlinear mechanical systems, it is also necessary to recognize which of the states are stable. The stability analysis is the subject of the next subchapter. The influence of the values of the oscillator parameters on the shape of the backbone curves will be also discussed in the subchapter.

8.7 STABILITY ANALYSIS

The stability study requires knowledge of values of the amplitude as well as modified phase. Therefore, it is necessary to solve the system of equations (8.111)–(8.112). Let $g_1(a,\theta)$ and $g_2(a,\theta)$ stand for the left sides of equations (8.111)–(8.112), i.e.

$$
\begin{aligned}
g_1(a,\theta) = & -\frac{f_0}{2}\cos\theta - \frac{1}{2}\gamma a + \\
& \frac{3}{16}\left(\gamma\frac{\lambda^3\alpha_1+\alpha_2}{(1+\lambda)^3} - 2\beta\right)a^3 - 3\beta\frac{\lambda^3\alpha_1+\alpha_2}{32(1+\lambda)^3}a^5,
\end{aligned}
\tag{8.116}
$$

$$
\begin{aligned}
g_2(a,\theta) = & \frac{f_0}{2}\sin\theta + \left(\frac{\gamma^2}{8}+\sigma\right)a - 3\frac{\lambda^3\alpha_1+\alpha_2}{8(1+\lambda)^3}a^3 - \\
& 3\left(5\frac{-\lambda^5(16+\lambda)\alpha_1^2+30\lambda^3\alpha_1\alpha_2-(1+16\lambda)\alpha_2^2}{256(1+\lambda)^6} + \frac{3}{256}\beta^2\right)a^5.
\end{aligned}
\tag{8.117}
$$

Due to the periodicity of the functions $g_1(a,\theta)$ and $g_2(a,\theta)$, the unbounded two-dimensional area in which the roots are sought can be narrowed down to the band

$$
\Omega_1 = \left\{(a,\theta) : -\frac{\pi}{2} \le \theta < \frac{3\pi}{2}\right\}. \tag{8.118}
$$

Let us note that the following properties of the functions $g_1(a,\theta)$ and $g_2(a,\theta)$

$$
g_1(-a,\theta+\pi) = -g_1(a,\theta), \tag{8.119}
$$

$$
g_2(-a,\theta+\pi) = -g_2(a,\theta), \tag{8.120}
$$

allow one to limit the area Ω_1 to its part

$$
\Omega_0 = \left\{(a,\theta) : a > 0 \wedge \frac{\pi}{2} \le \theta < \frac{3\pi}{2}\right\}. \tag{8.121}
$$

Changing in the sign of the values of the functions in relations (8.119)–(8.120) is irrelevant because we are dealing with the equations of the form $g_i(\ldots)=0,\ i=1,2$.

The substitutions

$$\sin\theta = \frac{2w}{1+w^2}, \quad \cos\theta = \frac{1-w^2}{1+w^2}, \tag{8.122}$$

where

$$w = \tan\left(\frac{\theta}{2}\right),$$

allow one to transform equations (8.111)–(8.112) into a form with only one trigonometric function and avoid the difficulties related to the ambiguity in the sign of the root when the Pythagorean identity is used to express the cosine by sine or vice versa. Employing substitutions (8.122), we obtain the system of two algebraic equations of the form

$$\begin{aligned}&\frac{f_0}{2}\left(w^2-1\right)-\frac{1}{2}\gamma\left(w^2+1\right)a+\frac{3}{16}\left(w^2+1\right)\left(\gamma\frac{\lambda^3\alpha_1+\alpha_2}{(1+\lambda)^3}-2\beta\right)a^3-\\&3\beta\left(w^2+1\right)\frac{\lambda^3\alpha_1+\alpha_2}{32(1+\lambda)^3}a^5=0,\end{aligned} \tag{8.123}$$

$$\begin{aligned}&f_0w\left(\frac{\gamma^2}{8}+\sigma\right)a-3\left(w^2+1\right)\frac{\lambda^3\alpha_1+\alpha_2}{8(1+\lambda)^3}a^3-\\&3\left(w^2+1\right)\left(5\frac{-\lambda^5(16+\lambda)\alpha_1^2+30\lambda^3\alpha_1\alpha_2-(1+16\lambda)\alpha_2^2}{256(1+\lambda)^6}+\frac{3}{256}\beta^2\right)a^5=0,\end{aligned} \tag{8.124}$$

The system of equations (8.123)–(8.124) is solved numerically using the standard procedure *NSolve* of the *Mathematica*. Determining the roots of equations (8.123)–(8.124) in the iterative procedure enables the drawing of response curves while this term includes also the dependency between the modified phase and detuning parameter.

Continuing the example described in the previous sections, we determine both resonance response curves. As previously, the detuning parameter varies from -0.04 to 0.04 with the apriori assumed step equal to $5\cdot 10^{-5}$. The results are depicted in Figures 8.13–8.14. Making use of the symmetry properties (8.119)–(8.120), there are plotted only the single branch of each of these curves.

Let (a_s,θ_s) be any fixed point of the dynamical system governed by equations (8.109)–(8.110). In other words the values a_s,θ_s satisfy equations (8.111)–(8.112). To examine the stability of this point in the sense of Lyapunov, we introduce the functions $\tilde{a}(\tau)$ and $\tilde{\theta}(\tau)$ that can be treated as small perturbations, and depict and the solutions of equations (8.109)–(8.110) that are close to the point (a_s,θ_s) as follows

$$a(\tau)=a_s+\tilde{a}(\tau), \quad \theta(\tau)=\theta_s+\tilde{\theta}(\tau). \tag{8.125}$$

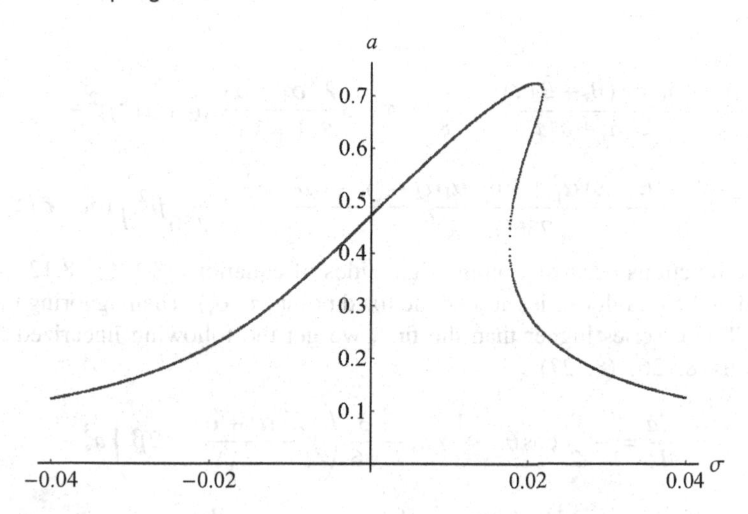

FIGURE 8.13 The amplitude a versus the detuning parameter σ at the main resonance.

Inserting functions (8.125) into equations (8.109)–(8.110) yields

$$\begin{aligned}\frac{d\tilde{a}}{d\tau} &= -\frac{f_0}{2}\cos\left(\theta_s+\tilde{\theta}(\tau)\right)-\frac{1}{2}\gamma\left(a_s+\tilde{a}(\tau)\right)+ \\ &\frac{3}{16}\left(\gamma\frac{\lambda^3\alpha_1+\alpha_2}{(1+\lambda)^3}-2\beta\right)\left(a_s+\tilde{a}(\tau)\right)^3-3\beta\frac{\lambda^3\alpha_1+\alpha_2}{32(1+\lambda)^3}\left(a_s+\tilde{a}(\tau)\right)^5,\end{aligned} \tag{8.126}$$

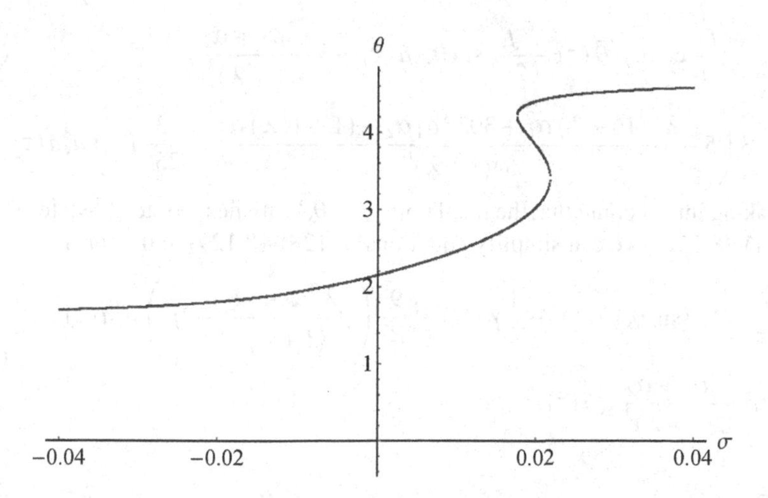

FIGURE 8.14 The modified phase θ versus the detuning parameter σ at the main resonance.

$$\frac{d\tilde{\theta}}{d\tau} = \frac{f_0 \sin\left(\theta_s + \tilde{\theta}(\tau)\right)}{2\left(a_s + \tilde{a}(\tau)\right)} + \frac{\gamma^2}{8} + \sigma - 3\frac{\lambda^3\alpha_1 + \alpha_2}{8(1+\lambda)^3}\left(a_s + \tilde{a}(\tau)\right)^2 - \left(5\frac{-\lambda^5(16+\lambda)\alpha_1^2 + 30\lambda^3\alpha_1\alpha_2 - (1+16\lambda)\alpha_2^2}{256(1+\lambda)^6} + \frac{3}{256}\beta^2\right)\left(a_s + \tilde{a}(\tau)\right)^4. \tag{8.127}$$

The functions occurring on the right sides of equations (8.126)–(8.127) are expanded in their Taylor series around the fixed point (a_s, θ_s). Then, ignoring the terms of the Taylor series higher than the first, we get the following linearized form of equations (8.126)–(8.127)

$$\frac{d\tilde{a}}{d\tau} = -\frac{f_0}{2}\cos\theta_s - \frac{1}{2}\gamma a_s + \frac{3}{16}\left(\gamma\frac{\lambda^3\alpha_1 + \alpha_2}{(1+\lambda)^3} - 2\beta\right)a_s^3 - 3\beta\frac{\lambda^3\alpha_1 + \alpha_2}{32(1+\lambda)^3}a_s^5 + \frac{f_0}{2}(\sin\theta_s)\tilde{\theta}(\tau) - \frac{1}{2}\gamma\tilde{a}(\tau) + \frac{9}{16}\left(\gamma\frac{\lambda^3\alpha_1 + \alpha_2}{(1+\lambda)^3} - 2\beta\right)a_s^2\tilde{a}(\tau) - 15\beta\frac{\lambda^3\alpha_1 + \alpha_2}{32(1+\lambda)^3}a_s^4\tilde{a}(\tau), \tag{8.128}$$

$$\frac{d\tilde{\theta}}{d\tau} = \frac{f_0}{2a_s}\sin\theta_s + \frac{\gamma^2}{8} + \sigma - 3\frac{\lambda^3\alpha_1 + \alpha_2}{8(1+\lambda)^3}a_s^2 - 3\left(5\frac{-\lambda^5(16+\lambda)\alpha_1^2 + 30\lambda^3\alpha_1\alpha_2 - (1+16\lambda)\alpha_2^2}{256(1+\lambda)^6} + \frac{3}{256}\beta^2\right)a_s^4 + \frac{f_0}{2a_s}\cos\theta_s \cdot \tilde{\theta}(\tau) - \frac{f_0}{2a_s^2}\sin\theta_s \cdot \tilde{a}(\tau) - 3\frac{\lambda^3\alpha_1 + \alpha_2}{4(1+\lambda)^3}a_s\tilde{a}(\tau) + 3\left(5\frac{-\lambda^5(16+\lambda)\alpha_1^2 + 30\lambda^3\alpha_1\alpha_2 - (1+16\lambda)\alpha_2^2}{64(1+\lambda)^6} + \frac{3}{256}\beta^2\right)a_s^3\tilde{a}(\tau). \tag{8.129}$$

Taking into account that the fixed point (a_s, θ_s) satisfies the steady-state equations (8.111)–(8.112), we can simplify equations (8.128)–(8.129) to the form

$$\frac{d\tilde{a}}{d\tau} = \frac{f_0}{2}(\sin\theta_s)\tilde{\theta}(\tau) - \frac{1}{2}\gamma\tilde{a}(\tau) + \frac{9}{16}\left(\gamma\frac{\lambda^3\alpha_1 + \alpha_2}{(1+\lambda)^3} - 2\beta\right)a_s^2\tilde{a}(\tau) - 15\beta\frac{\lambda^3\alpha_1 + \alpha_2}{32(1+\lambda)^3}a_s^4\tilde{a}(\tau), \tag{8.130}$$

$$\frac{d\tilde{\theta}}{d\tau} = \frac{f_0}{2a_s}\cos\theta_s \cdot \tilde{\theta}(\tau) - \frac{f_0}{2a_s^2}\sin\theta_s \cdot \tilde{a}(\tau) - 3\frac{\lambda^3\alpha_1 + \alpha_2}{4(1+\lambda)^3}a_s\tilde{a}(\tau) + 3\left(5\frac{-\lambda^5(\lambda+16)\alpha_1^2 + 30\lambda^3\alpha_1\alpha_2 - (16\lambda+1)\alpha_2^2}{64(1+\lambda)^6} + \frac{3}{256}\beta^2\right)a_s^3\tilde{a}(\tau). \tag{8.131}$$

The dynamical system governed by equations (8.130)–(8.131) is linear with respect to the small perturbations $\tilde{a}(\tau)$ and $\tilde{\theta}(\tau)$. Thus, the applied procedure ensures that the problem concerning the study of the stability of the stationary resonant states has been replaced with the much simpler, but approximate problem of the stability of the linear dynamical system.

Equations (8.130)–(8.131) are homogeneous with constant coefficients. The general solution of the system can be presented in the form of the linear combination of the exponential functions

$$\begin{aligned} \tilde{a}(\tau) &= C_1 e^{r_1\tau} + C_2 e^{r_2\tau}, \\ \tilde{\theta}(\tau) &= C_3 e^{r_1\tau} + C_4 e^{r_2\tau}, \end{aligned} \tag{8.132}$$

where r_j, $j = 1,2$, are the roots of the characteristic equation

$$|\mathbf{B} - r\mathbf{I}| = 0. \tag{8.133}$$

The symbol $\mathbf{I}$ denotes the identity 2×2 matrix, and $\mathbf{B}$ is the 2×2 matrix of the form

$$\mathbf{B} = \begin{bmatrix} b_{11} & b_{12} \\ b_{21} & b_{22} \end{bmatrix}, \tag{8.134}$$

where

$$\begin{aligned} b_{11} &= -\frac{15\beta\left(\lambda^3\alpha_1 + \alpha_2\right)a_s^4 + 18\left(2\beta(1+\lambda)^3 - \gamma(\lambda^3\alpha_1 + \alpha_2)\right)a_s^2 + 16\gamma(1+\lambda)^3}{32(1+\lambda)^3}, \\ b_{12} &= \frac{1}{2}f_0 \sin\theta_s, \\ b_{21} &= 3\left(5\frac{\lambda^5(\lambda+16)\alpha_1^2 - 30\lambda^3\alpha_1\alpha_2 + (16\lambda+1)\alpha_2^2}{64(1+\lambda)^6} - \frac{3\beta^2}{64}\right)a_s^3 - \\ &\quad 3\frac{\lambda^3\alpha_1 + \alpha_2}{4(1+\lambda)^3}a_s - \frac{f_0}{2a_s^2}\sin\theta_s, \\ b_{22} &= \frac{f_0}{2a_s}\cos\theta_s. \end{aligned} \tag{8.135}$$

The roots r_1, r_2 of equation (8.133) are the eigenvalues of the matrix $\mathbf{B}$. If the real parts of all eigenvalues of the matrix $\mathbf{B}$ are negative, then the fixed point (a_s, θ_s) is asymptotically stable in the sense of Lyapunov. The eigenvalues of the matrix $\mathbf{B}$ are determined using the standard procedure *Eigenvalues* of the *Mathematica*.

Making use of the procedure described above, the stability of the resonance response curves shown in Figures 8.13–8.14 has been analyzed. In Figures 8.15–8.17, the resonance responses of the oscillator at main resonance are presented both as the standard dependency between the amplitude and detuning parameter, as the dependency between the modified phase and detuning parameter, and as the relationship

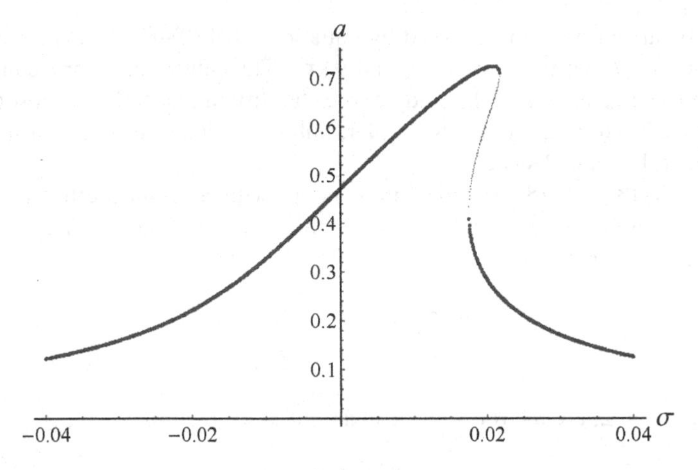

FIGURE 8.15 The amplitude a versus the detuning parameter σ at the main resonance; stable branches are drawn with larger dot size.

between the amplitude and modified phase. The stable branches were drawn using points of larger size.

The resonance response of the system can be also depicted as a three-dimensional curves in the space $(\sigma,\ a,\theta)$. It is a very illustrative way of presentation, which additionally allows us to avoid misinterpretation of the apparent points of intersection of curves, which may appear as a result of projection of the spatial curve on individual planes of space. Such a three-dimensional resonance response curve is shown in Figure 8.18. The grid lines showing the directions of the projection the three-dimensional curve onto the planes (σ,a) and (σ,θ) are also drawn. On the back

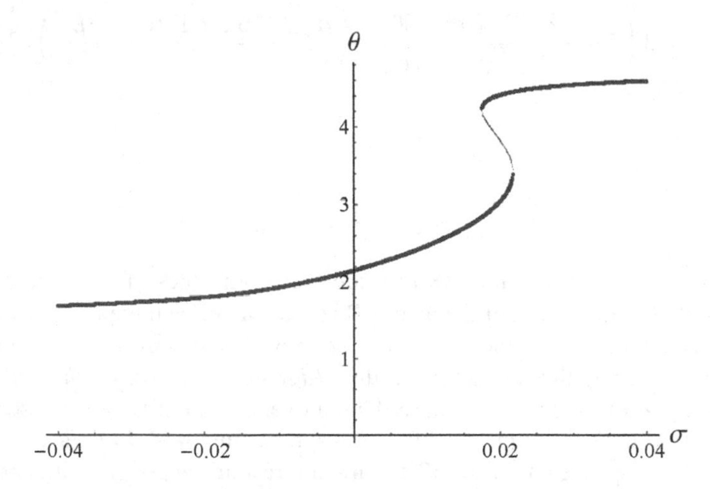

FIGURE 8.16 The modified phase θ versus the detuning parameter σ at the main resonance; stable branches are drawn with larger dot size.

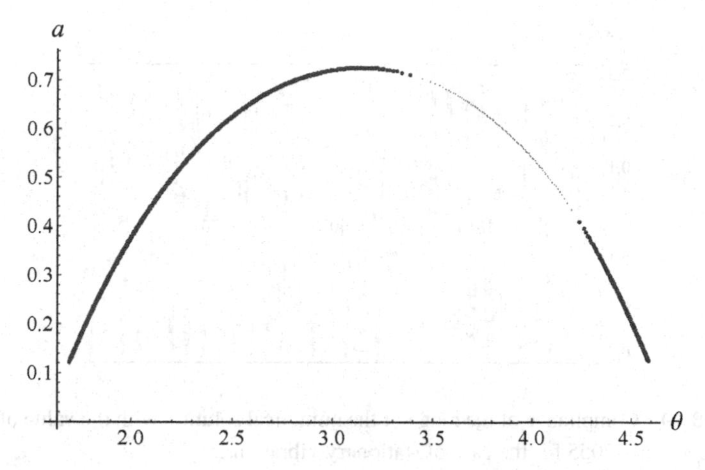

FIGURE 8.17 The amplitude a versus the modified phase θ at the main resonance; stable branches are drawn with larger dot size.

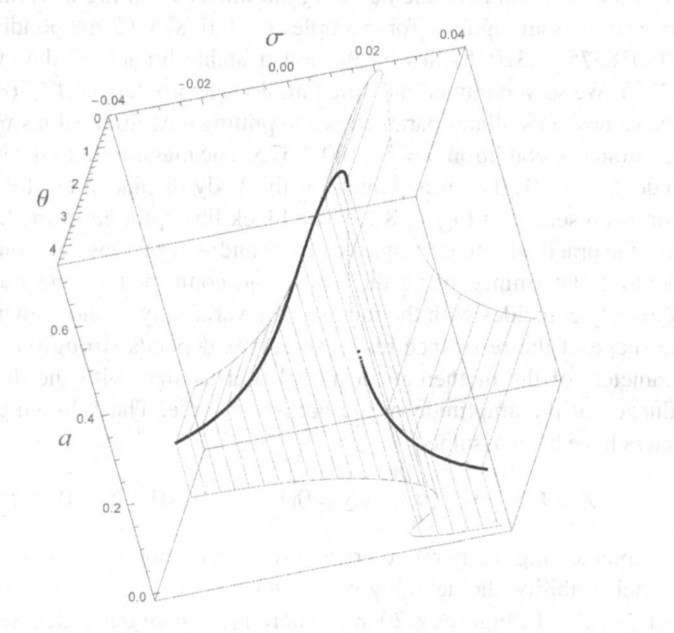

FIGURE 8.18 The three-dimensional resonance response curve for the main resonance and its projection onto the planes (σ, a) and (σ, θ); stable branches are drawn with larger dots size.

wall of the box, there is a projection corresponding to the curve shown in Figure 8.15, and on the bottom base of the box, there is a projection corresponding to the curve shown in Figure 8.16. For the sake of clarity of the drawing, the projection on the plane (θ, a) is skipped.

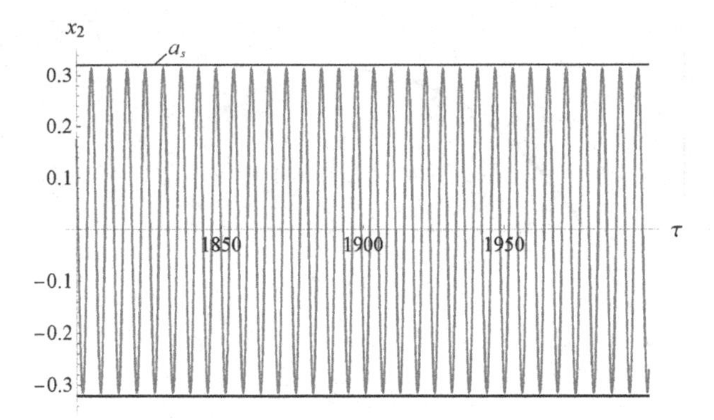

FIGURE 8.19 Compliance of the range of the numerical solution with the value of the stable amplitude $a_s \approx 0.32055$ for the case of stationary vibrations.

Let us take any value of the detuning parameter form the interval on which the backbone curve is ambiguous, for example $\sigma = 0.01875$. Corresponding to the value point (0.018575, 0.32055) lies on the lower stable branch of the curve shown in Figure 8.15. We solved numerically the initial value problem (8.17)–(8.19) assuming the same values of oscillator parameters and putting the initial values $x_0 = 0.1$, $v_0 = 0$ and the resonance condition $p = 1 + 0.018575$. The magnitude of the harmonic force amplitude $f_0 = 0.01$. The time-course of the body displacement for the stationary vibration is presented in Figure 8.19. The black lines plotted in the figure show the graphs of the practically fixed amplitude $a(\tau)$ and $-a(\tau)$. One can note that the value $a_s \approx 0.32055$ determined using the MSM and confirmed by the stability analysis almost exactly coincides with the range of the variability of the numerical solution.

The shape of the resonance response curves depends strongly on the values of the parameters of the mathematical model. Let us start with the discussion about the influence of the amplitude of the harmonic force. The following values of the parameters have been assumed

$$\lambda = 0.8, \quad \alpha_1 = 0.1, \quad \alpha_2 = 0.1, \quad \gamma = 0.001, \quad \beta = 0.0001.$$

By implementing the iterative procedure of determining the fixed points and assessing their stability, the detuning parameter was changed from -0.01 to 0.01 with a step of $2 \cdot 10^{-5}$. In Figures 8.20–8.22 there are shown dependencies between: the vibration amplitude and the detuning parameter $(a - \sigma)$, the modified phase and the detuning parameter $(\theta - \sigma)$, and the vibration amplitude and the modified phase $(a - \theta)$, respectively.

When the vibration is excited by the harmonic force with small values of the amplitude, then the curves depicted in Figures 8.20–8.22 are unambiguous. The greater is the amplitude f_0 the nonlinear nature of the response becomes stronger. At $f_0 = 0.0004$, the zone with ambiguous response appears. The fixed points lying on the unstable branches of the resonance response curves are drawn by dots of small size. An increase in the magnitude of the forcing broadens significantly the zone in

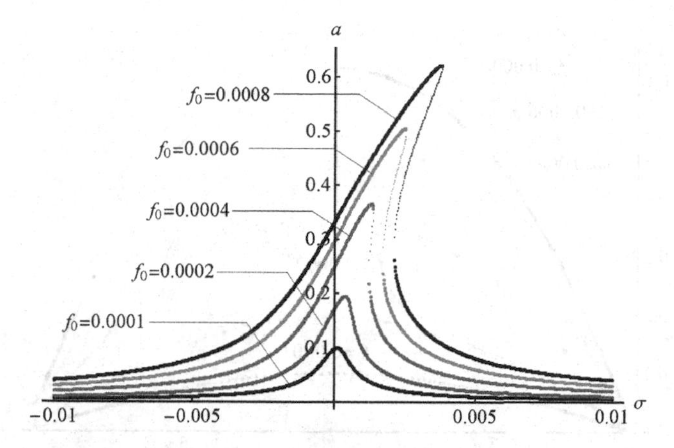

FIGURE 8.20 The amplitude a of stationary vibration versus the detuning parameter σ for various values of the amplitude of the harmonic force f_0.

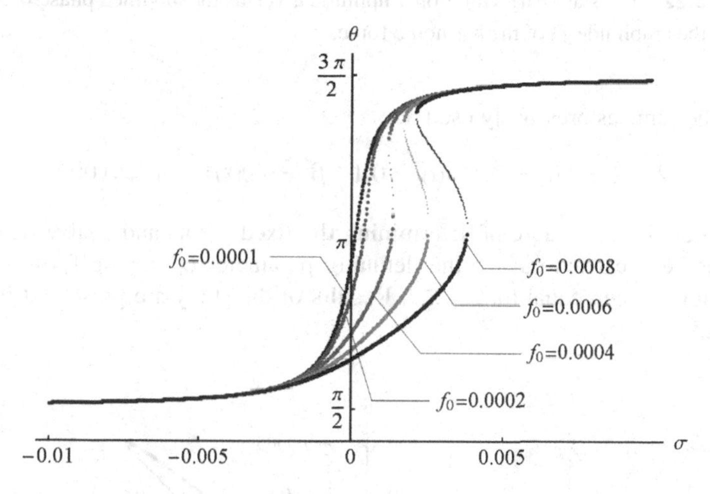

FIGURE 8.21 The modified phase θ versus the detuning parameter σ for various values of the amplitude f_0 of the harmonic force.

which there are stationary vibrations with the phase not synchronized with the phase of the force.

The amplitude of the external harmonic force does not describe the properties of the oscillator, so the curves shown in Figures 8.20–8.22 can be treated as those characterizing one particular system. When examining the influence of subsequent parameters, each graph presented in a given figure concerns a different oscillator.

In order to show the influence of the damping coefficient γ in a wide range, we have chosen the force of large amplitude $f_0 = 0.0008$ for simulation, at which, as noted earlier, the nonlinear properties of the oscillator are revealed. Other parameters

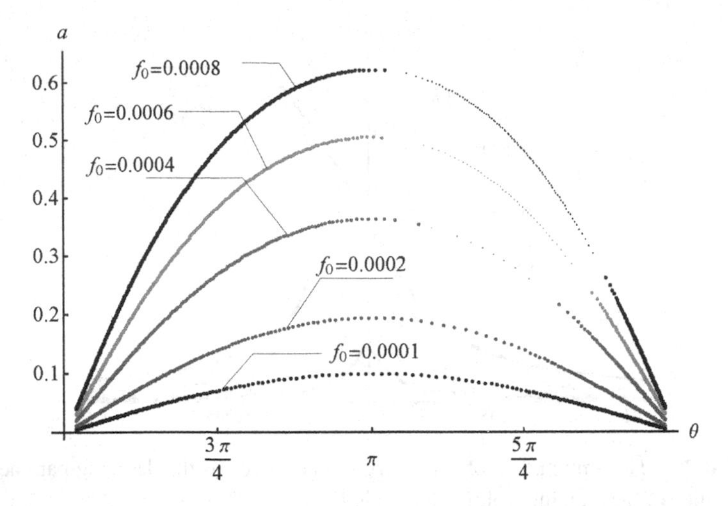

FIGURE 8.22 The stationary vibration amplitude a versus the modified phase θ for various values of the amplitude f_0 of the harmonic force.

remain the same as previously used, i.e.

$$\lambda = 0.8, \quad \alpha_1 = 0.1, \quad \alpha_2 = 0.1, \quad \beta = 0.0001, \quad f_0 = 0.0008.$$

The iterative procedure of determining the fixed points and analyzing their stability has been carried out for the detuning parameter σ varying form -0.008 to 0.008 with the step equal to $2 \cdot 10^{-5}$. Results of the study are presented in Figures 8.23–8.25.

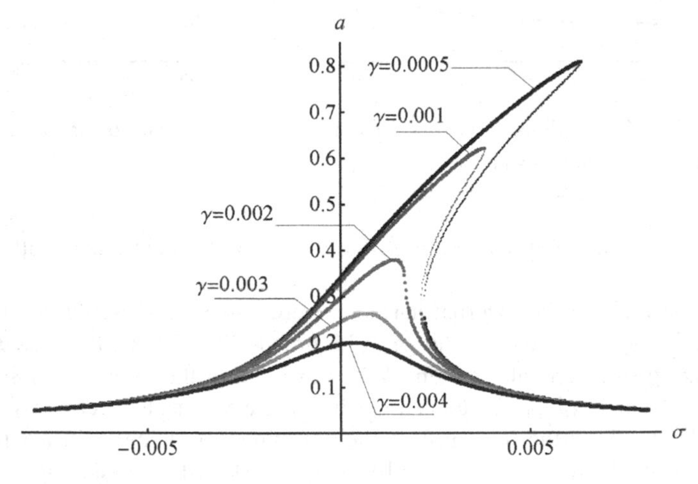

FIGURE 8.23 The amplitude a of stationary vibration versus the detuning parameter σ for oscillators with various values of the damping coefficient γ.

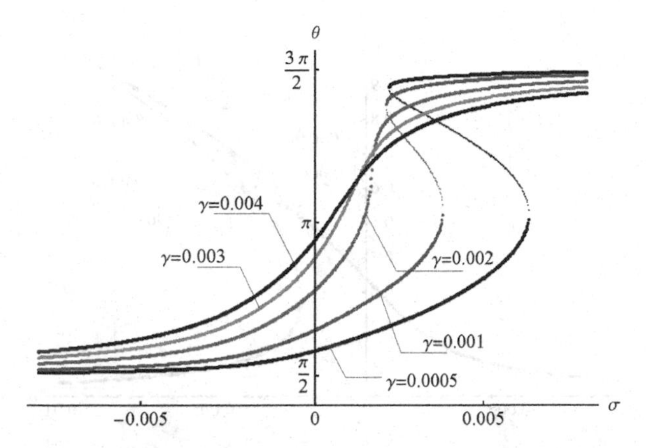

FIGURE 8.24 The modified phase θ versus the detuning parameter σ for oscillators with various values of the damping coefficient γ.

The curves describing the resonance response of oscillators with higher damping are unambiguous. The damping has also a significant qualitative influence on the nature of the phase-parameter resonance dependence. The range of variation of the detuning parameter σ, in which the phase of vibration and the force are out of sync, increases significantly with the increase in the value of the coefficient γ.

Let us pay more attention to the effect of the second damping coefficient in the Rayleigh model on fixed resonance responses. When $\beta > 0$ then the nonlinear effect of the damping model described by the coefficient β increases the dissipation of energy. In order to investigate this case, we have carried out the simulations for the

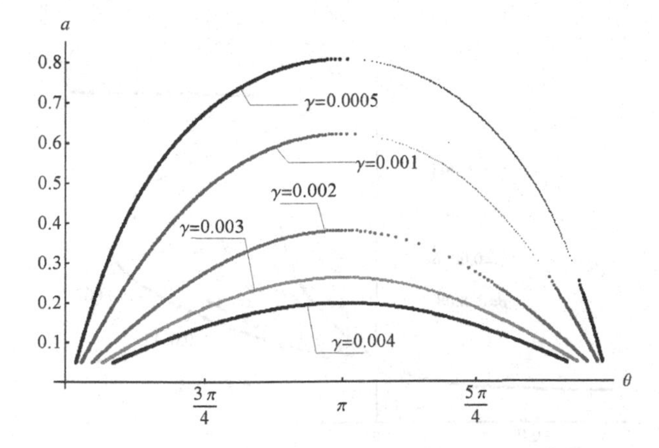

FIGURE 8.25 The stationary vibration amplitude a versus the modified phase θ for oscillators with various values of the damping coefficient γ.

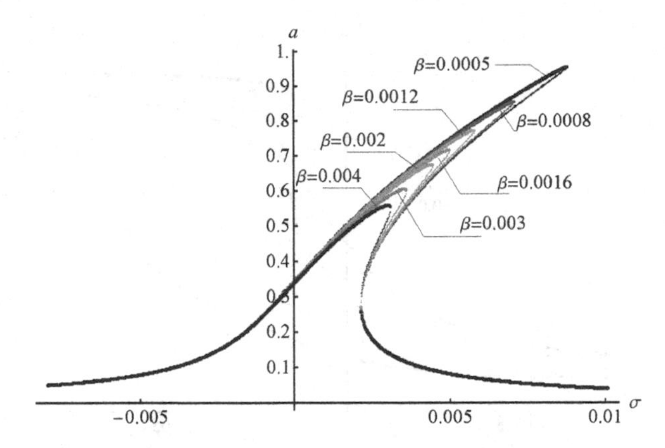

FIGURE 8.26 The amplitude a of stationary vibration versus the detuning parameter σ for oscillators with various positive values of the damping coefficient β.

models with a small value of the damping factor and a relatively large magnitude of the force. The following values have been assumed

$$\lambda = 0.8, \quad \alpha_1 = 0.1, \quad \alpha_2 = 0.1, \quad \gamma = 0.0005, \quad f_0 = 0.0008.$$

The detuning parameter σ was changed from -0.008 to 0.01 with the step equal to $2 \cdot 10^{-5}$.

Each curve of the family presented in Figures 8.26–8.28 has an unstable branch. The influence of this parameter is most significant for the steady states with relatively large amplitudes. In the range of small amplitudes, the differences between particular curves of the family are small. This behavior is a consequence of modulation equations in which the coefficient β occurs only in the expressions with high powers

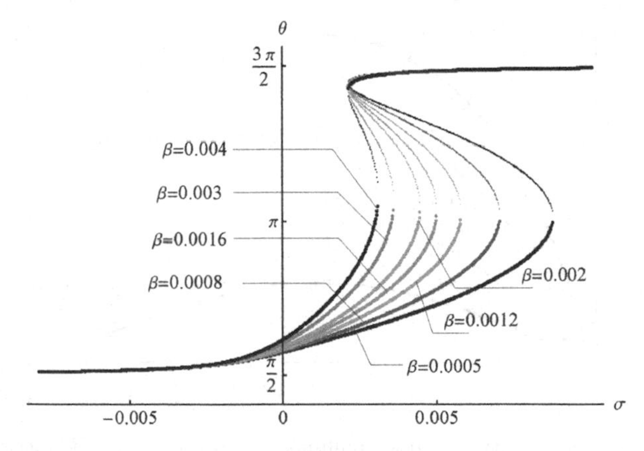

FIGURE 8.27 The modified phase θ versus the detuning parameter σ for oscillators with various positive values of the damping coefficient β.

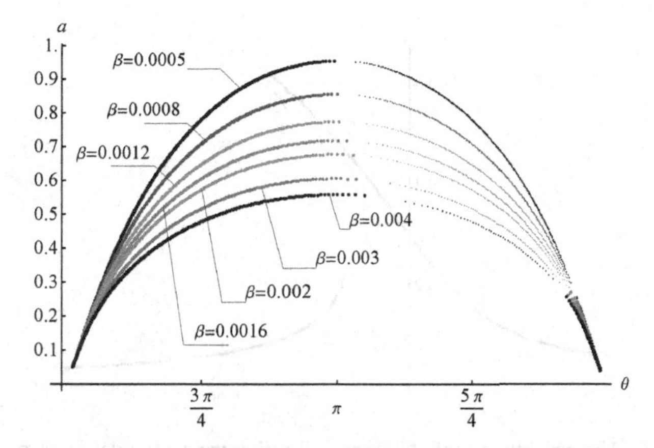

FIGURE 8.28 The stationary vibration amplitude a versus the modified phase θ for oscillators with various positive values of the damping coefficient β.

of the vibration amplitude. Note that the localization of the starting point of the lower stable branch is practically not affected by the parameter β.

For $\beta < 0$ nonlinear relationship (8.2) becomes the model of self-excited-oscillations. WarmiÅĎski [227] studied, employing the MSM, dynamical behaviors of a one-dimensional nonlinear oscillator where dependence (8.2) with negative β plays the role of one of the mechanisms of excitation.

By expecting high-amplitude oscillations due to self-excitation, we set a lower amplitude of the external force as in the previous simulation, to mainly display the effects of the negative beta. The values of the parameters that remained unaltered during the simulation are as follow

$$\alpha_1 = 0.1, \quad \alpha_2 = 0.1, \quad \gamma = 0.001, \quad f_0 = 0.0006.$$

The step assumed for the iterative procedure for determining the fixed points and assessing their stability is the same as previously used and equals $2 \cdot 10^{-5}$. The start value of the detuning parameter is $\sigma = -0.005$. Due to significant differences in the amplitude values achieved in the steady-state, the presentation method of the amplitude-detuning parameter dependency has been changed. Instead of a collective drawing presenting all curves of the family, each curve is shown in a separate figure.

The stable fixed points are drawn with dots of a larger radius in Figures 8.29–8.33. In Figures 8.31–8.33, it can be noticed that the points lying on the upper branch in the zone of ambiguous responses lose their stability. As the absolute value of β increases, this loss of stability includes the stationary oscillations of lower and lower amplitude.

The results of the analysis concerning the dependence between the modified phase and the detuning parameter are depicted in Figure 8.34. The fixed points corresponding to the stable periodic resonance vibration are marked by larger dots. In addition to the curves of a typical shape and sequence: stable branch - unstable branch - stable branch, there are also curves consisting of disjoint branches. While in all other

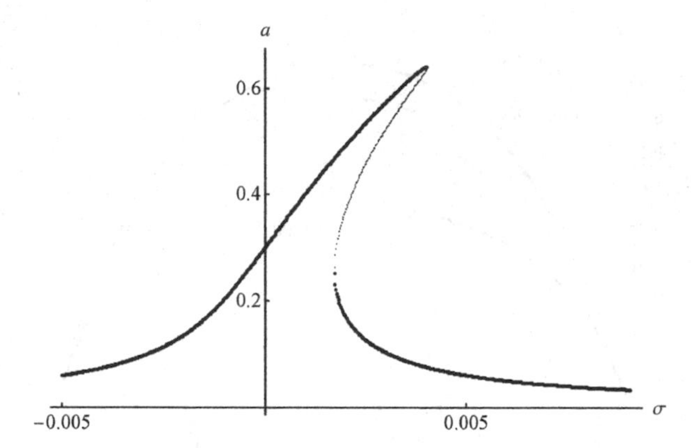

FIGURE 8.29 The amplitude a of stationary vibration versus the detuning parameter σ for $\beta = -0.0002$.

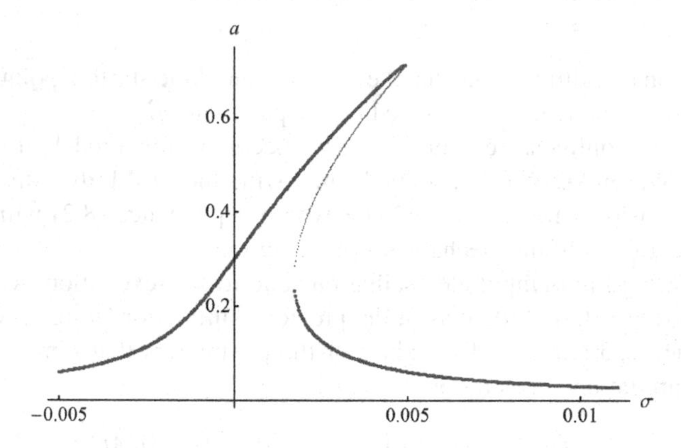

FIGURE 8.30 The amplitude a of stationary vibration versus the detuning parameter σ for $\beta = -0.0004$.

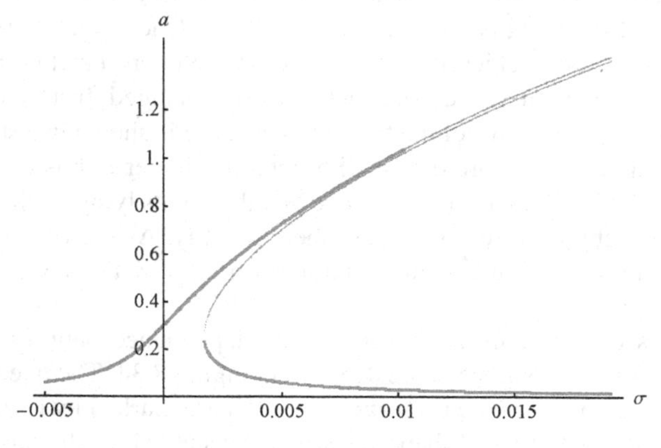

FIGURE 8.31 The amplitude a of stationary vibration versus the detuning parameter σ for $\beta = -0.0006$.

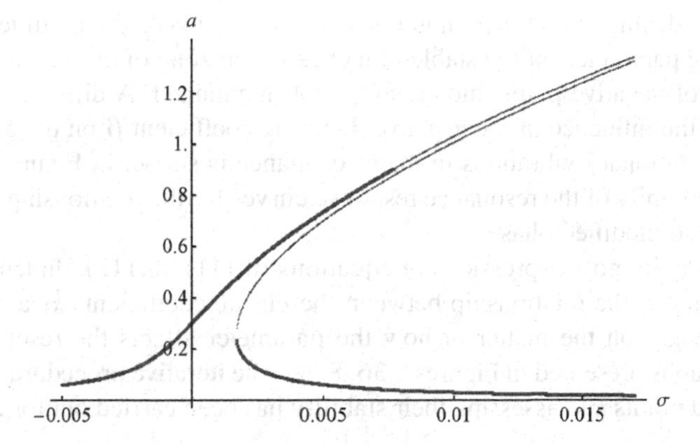

FIGURE 8.32 The amplitude a of stationary vibration versus the detuning parameter σ for $\beta = -0.0008$.

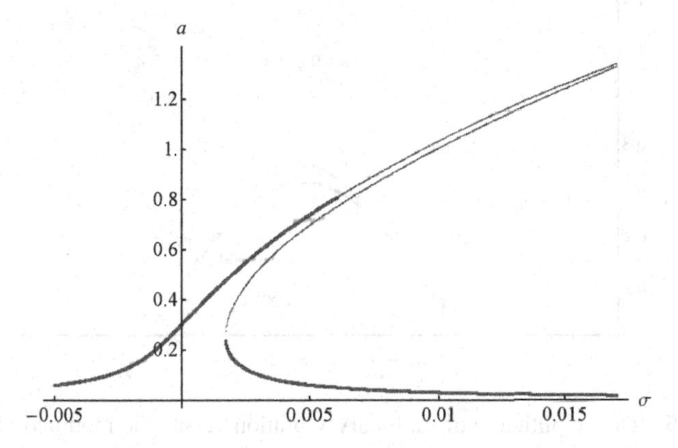

FIGURE 8.33 The amplitude a of stationary vibration versus the detuning parameter σ for $\beta = -0.001$.

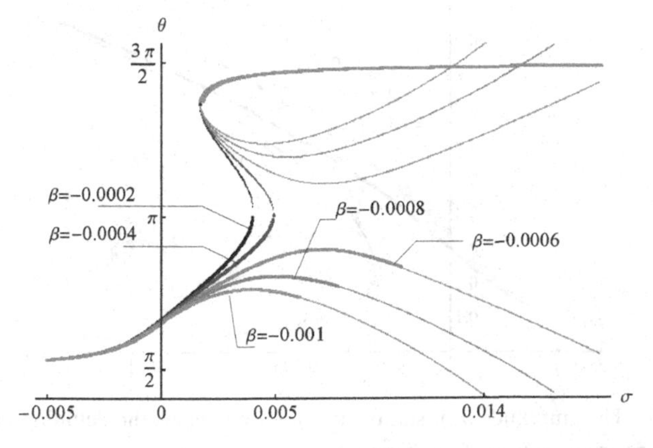

FIGURE 8.34 The modified phase θ versus the detuning parameter σ for oscillators with various negative values of the damping coefficient β.

cases we are dealing with the monotonous relations between the modified phase and the detuning parameter on the stable branches in the zone of ambiguous responses, in the case of negative β this monotony is not maintained. A different view on the question of the influence of the negative damping coefficient β on the existence and stability of stationary vibrations in main resonance is shown in Figure 8.35, which presents the family of the resonance response curves for the relationship between the amplitude and modified phase.

Appearing in most expressions of equations (8.111)–(8.112), dimensionless parameter λ depict the relationship between the elastic coefficients k_1 and k_2 of both springs. A sight on the matter of how the parameter affects the resonance curves gives the graphs presented in Figures 8.36–8.38. The iterative procedure of determining the fixed points and assessing their stability has been carried out for the detuning

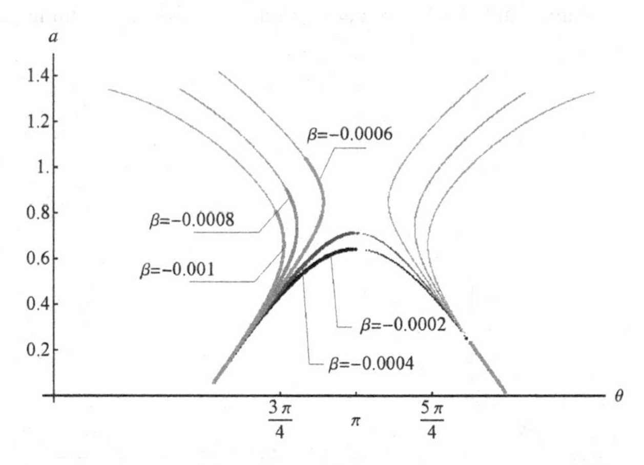

FIGURE 8.35 The amplitude a of stationary vibration versus the modified phase θ for oscillators with various negative values of the damping coefficient β.

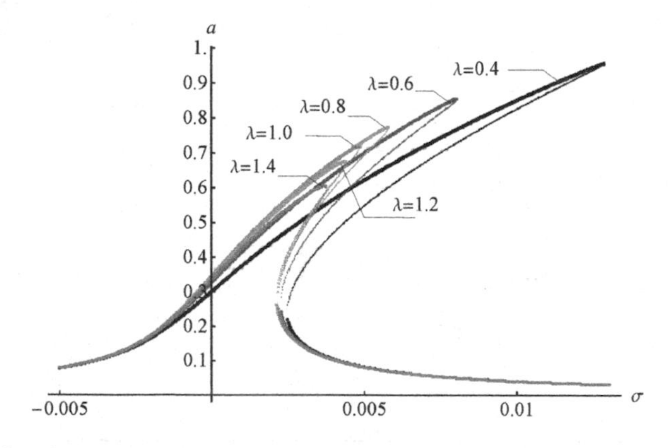

FIGURE 8.36 The amplitude a of stationary vibration versus the detuning parameter σ for oscillators with various values of the parameter λ.

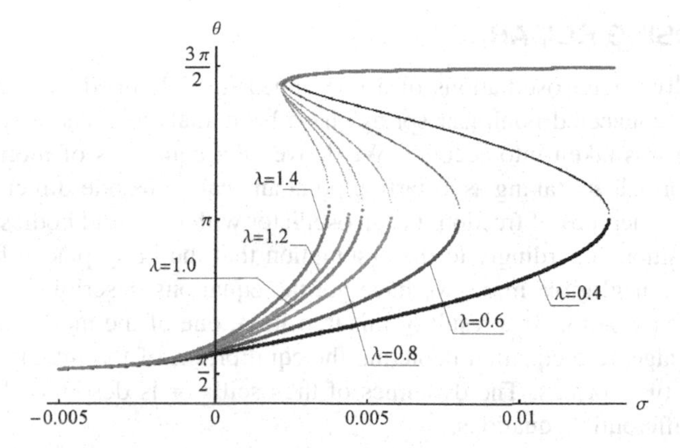

FIGURE 8.37 The modified phase versus the detuning parameter σ for oscillators with various values of the parameter λ.

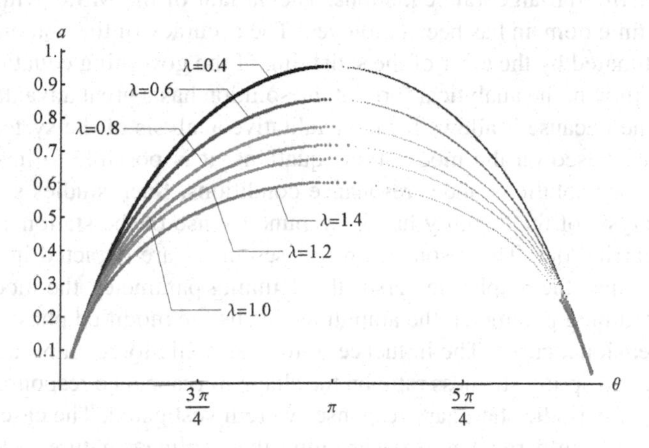

FIGURE 8.38 The amplitude a of stationary vibration versus modified phase θ for oscillators with various values of the parameter λ.

parameter σ varying from -0.005 to 0.014 with the step equal to $2 \cdot 10^{-5}$. The remaining parameters of the model are assumed to be the same for all oscillators and take the following values

$$\alpha_1 = 0.1, \quad \alpha_2 = 0.1, \quad \gamma = 0.0005, \quad \beta = 0.0005.$$

The external harmonic force exciting the oscillators has the dimensionless amplitude $f_0 = 0.0008$.

As the spring rate decreases, the range of amplitude values associated with the steady-state response increases, and the backbone curves become softer. The range of the resonant frequencies, expressed in terms of detuning parameter, is also extended, in which the desynchronization of the vibration phase with the input phase is observed.

8.8 CLOSING REMARKS

Harmonically forced oscillations of a one-dimensional lumped system containing two serially connected nonlinear springs have been analyzed. The Rayleigh model of damping was taken into account. We derived the equations of motion from the Lagrange formalism, taking as a starting point an analogous one-dimensional oscillator with two degrees of freedom i.e. an oscillator with two rigid bodies. By making a limit transition accordingly to the assumption that the body placed between two springs has a negligible mass, we have got the equations describing the dynamics of the studied system. As a result of this transition, one of the model equations has become an algebraic equation depicting the equilibrium of the massless node connecting the two springs. The dynamics of the oscillator is described by the set of algebraic-differential equations.

The mathematical model equations concerning both resonant and non-resonant vibration cases are solved using the MSM which has been appropriately modified to solve the differential-algebraic equations. The variant of the MSM with three variables in the time domain has been employed. The accuracy of the obtained solutions has been estimated by the error of the satisfying of the governing equations.

As well known, the analytical form of the solution has a great advantage over the numerical one because it allows for the qualitative analysis of the system behavior. Among other, based on the modulation equations, it is possible to investigate periodic stationary solutions under resonance conditions. Such studies supplemented with the analysis of the stability in the Lyapunov sense of the stationary responses have been carried out. The resonance response curves are depicted in the form of the dependencies: the amplitude versus the detuning parameter, the modified phase versus the detuning parameter, the amplitude versus the modified phase, and also as a three-dimensional curve. The influence of the harmonic force amplitude, damping coefficients, and spring stiffness ratio on the shape of resonance response curves and the stability of periodic stationary responses were investigated. The case concerning the influence of the parameter β determining the nonlinear nature of the relationship between the damping force and the magnitude of the velocity turned out to be particularly interesting. Positive values of the coefficient β affect in a monotonic way an increase of the energy dissipation effect and are manifested in decreasing the amplitude and shifting the jump of the modified phase towards higher values of the detuning parameter. The trend becomes reverse with negative values of the parameter β. Self-excited-oscillations appear for negative β with sufficiently large absolute value. The division of resonance response curves into stable and unstable branches differs from the standard case. Finally, the interval of the detuning parameter appears on which there are no stable periodic solutions.

9 Periodic Vibrations of Nano/Micro Plates

In this chapter, a study of geometrical nonlinear vibrations of rectangular micro/nanoplates with an account of small-scale effects in a framework of the nonlocal elasticity theory is carried out. Hamilton's principle yields the governing system of nonlinear partial differential equations (PDEs) based on the Kirchhoff-Love hypotheses and geometric nonlinearity introduced through the von Kármán theory. Next, we employed a strategy to get coupled nonlinear and non-autonomous system of ordinary differential equations (ODEs) due to the concept of both reduced order modeling (ROM) truncated to the double-mode approximation of the plate deflection and the Bubnov-Galerkin procedure. The latter one allowed for separation of the initial problem to that of time and position functions, and the used approach is validated. Then, we investigate the reduced model of second-order nonlinear ODEs with coupled geometric nonlinearities through the multiple scales method (MSM) in time domain. Both resonant and non-resonant vibrations have been considered with emphasis on the small-scale effects and the approximation accuracy. In spite of numerous novel results regarding parametric analysis carried out by combined analytical-numerical approach, we have detected and unambiguous back-bone curves which allow one to predict and possibly avoid the pull-in phenomena related to the original problem describing nonlinear vibrations of micro/nanoplates.

9.1 INTRODUCTION

The rapid development of high-tech technologies has given impetus to studies of the behavior of micro- and nano-structural members under various influences since such objects have become part of many modern structures, devices, and so on. When listing the possible applications of micro and nano-objects studied in the current work, we took note of NEMS (nanoelectromechanical systems), MEMS (microelectromechanical systems), energy storage systems, resonators, sensors, DNA detectors, as well as drug delivery system ([41], [46], [83], [110], [243]). However, as shown by theoretical and experimental studies [6–8] the results obtained based on the classical theory in the case of plates with dimensions in the micro or nano-scale may be incorrect (the so-called small-scale effects appear). That is why the nonc-lassical continuum theories received their development ([55], [145], [213], [100], [66], [116]). In the presented work for small-scale analysis of plates, we use the nonlocal elasticity theory developed by Eringen [66]. Note that the results obtained using the nonlocal elasticity theory are in good agreement with the molecular dynamics results, including the study of large deflection vibrations of single-layered graphene sheets [188]. According to this theory, the stress at a given point is a function of strains at all other points of the body. Such assumption leads to the governing equations containing the

DOI: 10.1201/9781003270706-9

additional parameter. This parameter depends on the internal characteristic length (distance between C-C bonds, lattice parameter), as well as the constant associated with the material for adjusting the model to molecular dynamics or experimental results ([61], [67]).

A review of the results of applying the nonlocal theory to vibrational problems of plates shows a large number of publications devoted to the study of linear vibrations of micro/nanoplates ([5], [174], [9], [147], [194], [192]), whereas investigation of the nonlinear vibrations is scarce. Such works include the study of the nonlinear vibrations of isotropic nanoplates by Jomehzadeh et al. [92]. Nonlinear analysis of vibrating plate under influence of magnetic and electric excitation is performed in work of Farajpour et al. [68]. Effect of various types of boundary conditions on the nonlinear vibrations of viscoelastic double-layered plates was analyzed by Wang et al. [222]. *Thus, due to a small amount of works devoted to the nonlinear vibration investigation, we can conclude that this question has not been studied enough. Moreover, in the papers listed above, the deflection of an oscillating plate is approximated by one term. However, such one-mode approximation sometimes does not allow one to study the nonlinear effects yielded by modes interaction.*

Our investigation contains double-mode model for micro/nanoplates based on reduced order modeling (ROM). Application of the Bubnov-Galerkin method allows one to reduce the governing system of partial differential equations (PDEs) to the system of ordinary differential equations (ODEs).

In general, a rigorous treatment of the problem of transition from nonlinear PDEs to nonlinear ODEs has recently been addressed in ([25]) based on consideration of the nanoplate dynamics as well as in the monographs ([22], [23]).

It is well recognized that there are numerous difficulties with analysis of the mass distributed nonlinear systems by using classical approaches like finite difference method (FDM), finite element method (FEM), the higher order Faedo-Galerkin approaches, etc. The attempt to reduce the order of the governing equations is required because the archetypical features of behavior of the governing nonlinear PDEs are hidden, and difficult to predict and extract from thousands/millions of nonlinear ODEs obtained via FEM/FDM. The computational costs of such direct analysis using the commercial programs is high, and in many cases the results validation is either difficult or almost impossible. Since the computational costs of the analysis is high, it practically does not allow for controlling online nonlinear dynamics of the structural members (beams, plates, shells). The integration of large amount of nonlinear ODEs governing the dynamics of infinitely many degrees-of-freedom (DoF), truncated to a set of finite but large system of DoF and supplemented by the standard methods devoted to studying ODEs, including FFT combined with shooting and combination/pseudo-arc techniques, is also expensive from a point of the computational time.

The ROMs, in spite of their strong simplification sometimes possess some meaningful advantages: (i) they may reveal hidden archetypical behavior governed by one- or two-coupled oscillators with quadratic and cubic nonlinearities; (ii) based on the recent spatio-temporal discretization technique employed in numerous papers and collected in the book ([24]), it was discovered that in spite of the finite dimension of the problem the analyzed structural members can exhibit standard well known routes to chaotic vibration such as Feigenbaum, Ruelle-Takens-Newhouse, and Pomeau-Manneville scenarios; (iii) in many cases, location of an attractor occupies a finite and low dimensional space of the original infinite manifold and there is no need to treat the problem in infinite manifold; (iv) the modal interaction can be limited only to a few modes involved, and hence the problem can be strongly truncated; (v) in many cases, ROM technique may allow one to understand and predict nonlinear system features and peculiarities based only on strongly truncated system of ODEs, which can easily be solved using both numerical and analytical approaches (for example, pull-in phenomena can be predicted and hence controlled in some way).

It is worthy to mention that the equations of the system contain nonlinear terms, and therefore, its study is rather complicated. In order to simplify the problem and withdraw hidden nonlinear effects of the first step ROM procedure, we go further in reduction process, by reducing the problem to ODEs governing the dynamics of the associated amplitudes and phases. For investigation of the obtained ODEs, multiple scale method (MSM) was employed. This approach belongs to one of the most commonly used analytical methods for solving problems with various kinds of nonlinearities. Its use allows one to give a better interpretation of nonlinear processes and the influence of various factors on the behavior of the studied object, and follows similar approaches in the framework of the so called reduced order models (ROMs). MSM was used in works by Nayfeh et al. [154], and Nayfeh and Nayfeh [153] where for continuous systems, the nonlinear frequency and modes were studied, taking into account nonlinear stiffness and inertia. Mahmoodi et al. [133] studied the continuous systems with nonlinearity in stiffness, damping and inertia. Shoushtari and Khadem [190] investigated nonlinear vibrations of the plates with MSM considering shear deformation and rotary inertia effects.

Luongo [126] investigated the problem of adaptation of MSM to solve "difficult" bifurcation problems using examples of discrete and continuous systems excited parametrically and externally.

Warmiński [227] investigated regular and chaotic vibrations of a nonlinear structure subjected to self-, parametric, and external excitations. The MSM employed in the study allowed one to find the frequency-locking features and the Neimark-Sacker bifurcation. In particular, the similarities and differences between the van der Pol and Rayleigh models were addressed.

The high efficiency of the MSM in predicting the nonlinear phenomena, including various resonances and amplitude jump-effect, exhibited by back-bone curves and associated with pull-in phenomena of the original structural members was discussed and demonstrated in the recent solved problems that deal with nonlinear dynamics of spring-pendula and beyond ([198], [199], [208]).

Thus, the analysis of existing publications in the field of nonlinear dynamics for non size-dependent objects and a limited number of studies of nonlinear processes for small-scale objects became the impetus for the application of the mentioned approaches to arriving at the governing equations which takes into account, the small-scale effects. The novelty of our work is the use of a two-mode model in combination with the Bubnov-Galerkin method to reduce the nonlocal PDEs to nonlinear coupled ODEs with coefficients depending on the small-scale parameter.

Application of MSM to our problem allows one to analyze non-resonant and resonant vibrations of small-scale plate, as well as establish small scale effects on errors of obtained approximation functions. Moreover, analysis of response curves at the main resonance is performed, the influence of damping, force amplitude and nonlocal parameter on vibrating system is studied.

9.2 MATHEMATICAL FORMULATION

The current study is based on the nonlocal elasticity theory and contains geometrically nonlinear vibrations analysis of the thin isotropic micro/nano plates. Application of the nonlocal theory yields the constitutive relation for the nonlocal stress tensor at a point x as follows

$$\sigma = \int_V K\left(\left|X' - X\right|, \tau\right) \sigma'\left(X'\right) dX', \tag{9.1}$$

where σ, σ' stand for the nonlocal and local stress tensors, $K\left(\left|X' - X\right|, \tau\right)$ is the nonlocal modulus, $\tau = e_0\alpha/l, e_0$ is a constant associated with the material, α and l are the internal characteristic length and external characteristics length, respectively. The more often used differential form ([67]) of the nonlocal constitutive relation is presented below

$$\left(1 - \mu\nabla^2\right)\sigma = \sigma'. \tag{9.2}$$

In relation (9.2) $\mu = (e_0\alpha)^2$ defines the nonlocal parameter, and ∇^2 is the Laplacian operator. For isotropic Kirchhoff's plate one can rewrite relation (9.2) into the following form:

$$\begin{bmatrix} \sigma_{xx} \\ \sigma_{yy} \\ \sigma_{xy} \end{bmatrix} - \mu\nabla^2 \begin{bmatrix} \sigma_{xx} \\ \sigma_{yy} \\ \sigma_{xy} \end{bmatrix} = \begin{bmatrix} \frac{E}{1-\nu^2} & \frac{\nu E}{1-\nu^2} & 0 \\ \frac{\nu E}{1-\nu^2} & \frac{E}{1-\nu^2} & 0 \\ 0 & 0 & \frac{E}{2(1+\nu)} \end{bmatrix} \begin{bmatrix} \varepsilon_{xx} \\ \varepsilon_{yy} \\ \varepsilon_{xy} \end{bmatrix}, \tag{9.3}$$

where E is Young's modulus and ν is Poisson's ratio.

Based on the von Kármán nonlinear theory, the components of strain tensor are presented as follows ([219])

$$\varepsilon_{xx} = \varepsilon_{xx}^0 + zk_{xx}, \quad \varepsilon_{yy} = \varepsilon_{yy}^0 + zk_{yy}, \quad \varepsilon_{xy} = \varepsilon_{xy}^0 + z\,k_{xy}, \tag{9.4}$$

$$\varepsilon_{xx}^0 = \frac{\partial u}{\partial x} + \frac{1}{2}\left(\frac{\partial w}{\partial x}\right)^2, \quad \varepsilon_{yy}^0 = \frac{\partial v}{\partial y} + \frac{1}{2}\left(\frac{\partial w}{\partial y}\right)^2, \quad \varepsilon_{xy}^0 = \frac{\partial u}{\partial y} + \frac{\partial v}{\partial x} + \frac{\partial w}{\partial x}\frac{\partial w}{\partial y}, \tag{9.5}$$

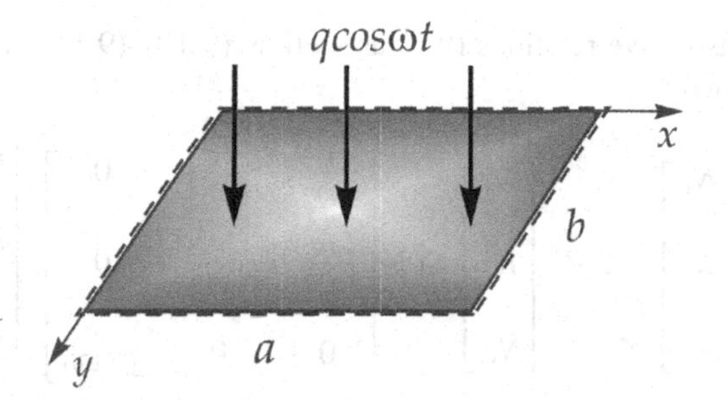

FIGURE 9.1 Simply supported rectangular plate loaded by transverse periodic excitation.

$$k_{xx} = -\frac{\partial^2 w}{\partial x^2}, \quad k_{yy} = -\frac{\partial^2 w}{\partial y^2}, \quad k_{xy} = -2\frac{\partial^2 w}{\partial x \partial y}, \tag{9.6}$$

where u, v are displacements in x- and y-directions and w is the deflection of the plate.

Strains in the middle surface are linked by compatibility equation ([219]) of the following form

$$\frac{\partial^2 \varepsilon_{xx}^0}{\partial y^2} + \frac{\partial^2 \varepsilon_{yy}^0}{\partial x^2} - \frac{\partial^2 \varepsilon_{xy}^0}{\partial x \partial y} = \left(\frac{\partial^2 w}{\partial x \partial y}\right)^2 - \frac{\partial^2 w}{\partial x^2}\frac{\partial^2 w}{\partial y^2}. \tag{9.7}$$

Employing the Hamilton principle yields the system of governing equations as follows

$$\frac{\partial N_x}{\partial x} + \frac{\partial N_{xy}}{\partial y} = \rho h \frac{\partial^2 u}{\partial t^2}, \qquad \frac{\partial N_{xy}}{\partial x} + \frac{\partial N_y}{\partial y} = \rho h \frac{\partial^2 v}{\partial t^2}, \tag{9.8}$$

$$\begin{aligned} &\frac{\partial^2 M_x}{\partial x^2} + 2\frac{\partial^2 M_{xy}}{\partial x \partial y} + \frac{\partial^2 M_y}{\partial y^2} = -\frac{\partial}{\partial x}\left(N_x \frac{\partial w}{\partial x} + N_{xy}\frac{\partial w}{\partial y}\right) - \\ &\frac{\partial}{\partial y}\left(N_{xy}\frac{\partial w}{\partial x} + N_y \frac{\partial w}{\partial y}\right) + \delta_0 \frac{\partial w}{\partial t} + \rho h \frac{\partial^2 w}{\partial t^2} - q\cos\omega t, \end{aligned} \tag{9.9}$$

where h stands for the thickness of the plate, ρ is the density of the plate, and δ_0 denotes the damping coefficient. Equation (9.9) assumes that periodic transverse force of the form $q\cos\omega t$ acts on the plate. On the other hand, the resultant in-plane forces and moments are defined by the following formulas

$$N_x = \int_{-\frac{h}{2}}^{\frac{h}{2}} \sigma_{xx} dz, \quad N_y = \int_{-\frac{h}{2}}^{\frac{h}{2}} \sigma_{yy} dz, \quad N_{xy} = \int_{-\frac{h}{2}}^{\frac{h}{2}} \sigma_{xy} dz, \tag{9.10}$$

$$M_x = \int_{-\frac{h}{2}}^{\frac{h}{2}} \sigma_{xx} z dz, \quad M_y = \int_{-\frac{h}{2}}^{\frac{h}{2}} \sigma_{yy} z dz, \quad M_{xy} = \int_{-\frac{h}{2}}^{\frac{h}{2}} \sigma_{xy} z dz. \tag{9.11}$$

Taking constitutive relations (9.3) as well as (9.10)–(9.11), the following equations are obtained

$$\begin{bmatrix} N_x \\ N_y \\ N_{xy} \end{bmatrix} - \mu\nabla^2 \begin{bmatrix} N_x \\ N_y \\ N_{xy} \end{bmatrix} = h \begin{bmatrix} \frac{E}{1-\nu^2} & \frac{\nu E}{1-\nu^2} & 0 \\ \frac{\nu E}{1-\nu^2} & \frac{E}{1-\nu^2} & 0 \\ 0 & 0 & \frac{E}{2(1+\nu)} \end{bmatrix} \begin{bmatrix} \varepsilon_{xx}^0 \\ \varepsilon_{yy}^0 \\ \varepsilon_{xy}^0 \end{bmatrix}, \tag{9.12}$$

$$\begin{bmatrix} M_x \\ M_y \\ M_{xy} \end{bmatrix} - \mu\nabla^2 \begin{bmatrix} M_x \\ M_y \\ M_{xy} \end{bmatrix} = \frac{h^3}{12} \begin{bmatrix} \frac{E}{1-\nu^2} & \frac{\nu E}{1-\nu^2} & 0 \\ \frac{\nu E}{1-\nu^2} & \frac{E}{1-\nu^2} & 0 \\ 0 & 0 & \frac{E}{2(1+\nu)} \end{bmatrix} \begin{bmatrix} k_{xx} \\ k_{yy} \\ k_{xy} \end{bmatrix}. \tag{9.13}$$

Similarly, taking into account equations (9.4)–(9.5), we get

$$\begin{aligned} M_x - \mu\nabla^2 M_x &= -D\left(\frac{\partial^2 w}{\partial x^2} + \nu\frac{\partial^2 w}{\partial y^2}\right), \\ M_y - \mu\nabla^2 M_y &= -D\left(\nu\frac{\partial^2 w}{\partial x^2} + \frac{\partial^2 w}{\partial y^2}\right), \\ M_{xy} - \mu\nabla^2 M_{xy} &= -D(1-\nu)\frac{\partial^2 w}{\partial x \partial y}, \end{aligned} \tag{9.14}$$

where the flexural plate rigidity $D = \frac{Eh^3}{12(1-\nu^2)}$. We introduce the Airy stress function F (see [238], [219]) through the following relations

$$N_x = \frac{\partial^2 F}{\partial y^2},\ N_y = \frac{\partial^2 F}{\partial x^2},\ N_{xy} = -\frac{\partial^2 F}{\partial x \partial y}, \tag{9.15}$$

and assuming that the in-plane inertia terms are neglected one can transform the governing equation (9.9), taking into account (9.14), to its mixed counterpart form

$$D\Delta^2 w = \left(1 - \mu\nabla^2\right)\left(L(w,F) - \rho h\frac{\partial^2 w}{\partial t^2} - \delta_0\frac{\partial w}{\partial t} + q\cos\omega t\right), \tag{9.16}$$

where operator $L(w,F)$ is defined as

$$L(w,F) = \frac{\partial^2 w}{\partial x^2}\frac{\partial^2 F}{\partial y^2} + \frac{\partial^2 F}{\partial x^2}\frac{\partial^2 w}{\partial y^2} - 2\frac{\partial^2 w}{\partial x \partial y}\frac{\partial^2 F}{\partial x \partial y}, \tag{9.17}$$

and $\Delta^2 = \left(\frac{\partial^2}{\partial x^2} + \frac{\partial^2}{\partial y^2}\right)^2$.

The compatibility equation (9.7) and the relations (9.12) allow one to obtain additional equation of the form

$$\left(1-\mu\nabla^2\right)\frac{1}{E}\Delta^2 F = -\frac{h}{2}L(w,w), \tag{9.18}$$

where

$$L(w,w) = 2\left(\frac{\partial^2 w}{\partial x^2}\frac{\partial^2 w}{\partial y^2} - \left(\frac{\partial^2 w}{\partial x\partial y}\right)^2\right). \tag{9.19}$$

Since the micro/nanoplate with simply supported boundary conditions is considered, the following relations for unknown functions are satisfied

$$\begin{aligned} &w=0, \quad \frac{\partial^2 w}{\partial x^2}+\nu\frac{\partial^2 w}{\partial y^2}=0, \quad \frac{\partial^2 F}{\partial x\partial y}=0, \quad \int_0^b \frac{\partial^2 F}{\partial y^2}dy=0, \quad x=0,a, \\ &w=0, \quad \frac{\partial^2 w}{\partial y^2}+\nu\frac{\partial^2 w}{\partial x^2}=0, \quad \frac{\partial^2 F}{\partial x\partial y}=0, \quad \int_0^a \frac{\partial^2 F}{\partial x^2}dx=0, \quad y=0,b. \end{aligned} \tag{9.20}$$

9.3 THE BUBNOV-GALERKIN METHOD WITH DOUBLE MODE MODEL

Let us present the deflection of the micro/nanoplate $w(x,y,t)$ as follows ([114])

$$w(x,y,t) = w_1(t)\sin\frac{\pi x}{a}\sin\frac{\pi y}{b} + w_2(t)\sin\frac{2\pi x}{a}\sin\frac{2\pi y}{b}, \tag{9.21}$$

where w_1, w_2 are bi-modal amplitudes, $\sin\frac{\pi x}{a}\sin\frac{\pi y}{b}$ and $\sin\frac{2\pi x}{a}\sin\frac{2\pi y}{b}$ are shape functions satisfying simply supported boundary conditions. Substitution of the expression (9.21) into (9.18) yields

$$\begin{aligned} (1-\mu\nabla^2)\Delta F = \frac{hE\pi^4}{2a^2b^2}\Bigg(& w_1^2\cos\frac{2\pi x}{a} + w_1^2\cos\frac{2\pi y}{b} + 16w_2^2\cos\frac{4\pi x}{a} + \\ & 16w_2^2\cos\frac{4\pi y}{b} + 8w_1w_2\left(\cos\frac{3\pi x}{a}\cos\frac{\pi y}{b} + \cos\frac{\pi x}{a}\cos\frac{3\pi y}{b}\right)\Bigg). \end{aligned} \tag{9.22}$$

and hence the expression for the stress function F is as follows

$$\begin{aligned} F = {} & f_1\cos\frac{2\pi x}{a} + f_2\cos\frac{2\pi y}{b} + f_3\cos\frac{4\pi x}{a} + f_4\cos\frac{4\pi y}{b} + \\ & f_5\cos\frac{3\pi x}{a}\cos\frac{\pi y}{b} + f_6\cos\frac{\pi x}{a}\cos\frac{3\pi y}{b} + p_1x^2 + p_2y^2, \end{aligned} \tag{9.23}$$

where coefficients f_1, f_2, f_3, f_4, f_5, f_6 are defined by the following formulas

$$\begin{aligned}
f_1 &= \frac{w_1^2 Eha^4}{32b^2(a^2+4\mu\pi^2)}, \quad f_2 = \frac{w_1^2 Ehb^4}{32a^2(b^2+4\mu\pi^2)}, \\
f_3 &= \frac{w_2^2 Eha^4}{32b^2(a^2+16\mu\pi^2)}, \quad f_4 = \frac{w_2^2 Ehb^4}{32a^2(b^2+4\mu\pi^2)}, \\
f_5 &= \frac{4w_1w_2Eha^4b^4}{(a^2+9b^2)^2(a^2b^2+\mu\pi^2(a^2+9b^2))}, \\
f_6 &= \frac{4w_1w_2Eha^4b^4}{(9a^2+b^2)^2(a^2b^2+\mu\pi^2(9a^2+b^2))}.
\end{aligned} \tag{9.24}$$

It is worthy to mention that for simply supported movable boundary conditions (9.20) the coefficients $p_1 = 0$ and $p_2 = 0$.

The Bubnov-Galerkin method gives

$$\begin{aligned}
&\int_0^a\int_0^b X(x,y)\sin\left(\frac{\pi x}{a}\right)\sin\left(\frac{\pi y}{b}\right)dxdy = 0, \\
&\int_0^a\int_0^b X(x,y)\sin\left(\frac{2\pi x}{a}\right)\sin\left(\frac{2\pi y}{b}\right)dxdy = 0,
\end{aligned} \tag{9.25}$$

where the function $X(x,y)$ is defined as

$$X(x,y) = D\Delta w - \left(1-\mu\nabla^2\right)\left(L(w,F) - \rho h\frac{\partial^2 w}{\partial t^2} - \delta_0\frac{\partial w}{\partial t} + q\cos\omega t\right). \tag{9.26}$$

Performing integration yields the system of the following nonlinear and coupled second-order ODEs:

$$\begin{aligned}
&w_1'' + \bar{\delta}w_1' + \alpha_0 w_1 + \alpha_1 w_1 w_2^2 + \alpha_2 w_1^3 + \bar{q}\cos\omega t = 0, \\
&w_2'' + \bar{\delta}w_2' + \beta_0 w_2 + \beta_1 w_2 w_1^2 + \beta_2 w_2^3 = 0,
\end{aligned} \tag{9.27}$$

with the explicitly defined coefficients

$$\begin{aligned}
\alpha_0 &= \frac{\pi^4 D(a^2+b^2)^2}{\rho ha^2b^2(a^2b^2+\mu\pi^2(a^2+b^2))}, \quad \beta_0 = \frac{16\pi^4 D(a^2+b^2)^2}{\rho ha^2b^2(a^2b^2+\mu\pi^2(a^2+b^2))}, \\
\alpha_1 = \beta_1 &= \frac{16a^2b^2E\pi^4}{\rho}\left(\frac{1}{(9a^2+b^2)^2(a^2b^2+\mu\pi^2(9a^2+b^2))} + \right. \\
&\qquad \left. \frac{1}{(a^2+9b^2)^2(a^2b^2+\mu\pi^2(a^2+9b^2))}\right), \\
\alpha_2 &= \frac{a^2b^2E\pi^4}{16\rho}\left(\frac{1}{a^6(b^2+4\mu\pi^2)} + \frac{1}{b^6(a^2+4\mu\pi^2)}\right),
\end{aligned} \tag{9.28}$$

$$\beta_2 = \frac{a^2b^2E\pi^4}{\rho}\left(\frac{1}{a^6(b^2+16\mu\pi^2)} + \frac{1}{b^6(a^2+16\mu\pi^2)}\right),$$

$$\bar{\delta} = \frac{\delta_0}{\rho h}, \quad \bar{q} = -\frac{16a^2b^2q}{\rho h\pi^2(a^2b^2+\mu\pi^2(a^2+b^2))},$$

Differential equations (9.27) are supplemented with the following initial conditions

$$w_1(0) = \Gamma_1, \quad w_1'(0) = \Gamma_2, \quad w_2(0) = \Gamma_3, \quad w_2'(0) = \Gamma_4, \tag{9.29}$$

where Γ_1, Γ_2, Γ_3, Γ_4 are known.

Assuming the nanoplate thickness h and the characteristic frequency $\omega_1 = \sqrt{\alpha_0}$ as the reference quantities, and introducing dimensionless variables such as

$$y_1 = \frac{w_1}{h}, \quad y_2 = \frac{w_2}{h}, \quad \tau = \omega_1 t, \tag{9.30}$$

one can rewrite equations (9.27) into the following nondimensional form

$$\begin{aligned} &\ddot{y}_1 + \delta\dot{y}_1 + y_1 + \xi_1 y_1 y_2^2 + \xi_2 y_1^3 + \sigma\cos p\tau = 0, \\ &\ddot{y}_2 + \delta\dot{y}_1 + \eta_0 y_2 + \eta_1 y_2 y_1^2 + \eta_2 y_2^3 = 0, \end{aligned} \tag{9.31}$$

where

$$\begin{aligned} &\delta = \frac{\bar{\delta}}{\sqrt{\alpha_0}}, \quad \xi_1 = \frac{\alpha_1 h^2}{\alpha_0}, \quad \xi_2 = \frac{\alpha_2 h^2}{\alpha_0}, \quad \eta_0 = \frac{\beta_0}{\alpha_0}, \\ &\eta_1 = \frac{\beta_1 h^2}{\alpha_0}, \quad \eta_2 = \frac{\beta_2 h^2}{\alpha_0}, \quad \sigma = \frac{\bar{q}}{h\alpha_0}, \quad p = \frac{\omega}{\sqrt{\alpha_0}}. \end{aligned} \tag{9.32}$$

The quantities y_1 and y_2 are functions of the nondimensional time τ while the over dot denotes the differentiation with respect to τ. The dimensionless form of the initial conditions supplementing equations (9.31) is as follows

$$y_1(0) = \gamma_1, \quad \dot{y}_1(0) = \gamma_2, \quad y_2(0) = \gamma_3, \quad \dot{y}_2(0) = \gamma_4, \tag{9.33}$$

where γ_1, γ_2, γ_3, γ_4 are known, and $\gamma_i = \frac{\Gamma_i}{h}$ for $i = 1, 3$, $\gamma_i = \frac{\Gamma_i}{h\omega_1}$ for $i = 2, 4$.

9.4 VALIDATION OF THE PROPOSED APPROACH

Due to the insufficient number of publications devoted to the study of nonlinear oscillations of micro/nano plates, we have performed a comparison with the results reported in [5] when linear vibrations of small-scale plates are studied. The numerical calculations are carried out for small-scale plate with the following parameters

$$E = 3\cdot 10^7\,\frac{\text{N}}{\text{m}^2}, \quad \nu = 0.3, \quad \rho = 1220\,\frac{\text{kg}}{\text{m}^3}, \quad a = b = 10^{-8}\,\text{m}, \quad h = 0.1a.$$

Table 9.1 contains the dimensionless frequency parameters $\Omega_1 = h\sqrt{\frac{\alpha_0\rho}{G}}$, $\Omega_2 = h\sqrt{\frac{\beta_0\rho}{G}}$, ($G = \frac{E}{2(1+\nu)}$) corresponding to considered modes with regard to plate deflection (9.21). As it can be seen, the results presented in the Table 9.1 coincide with the results of work (Aghababaei and Reddy [5]).

TABLE 9.1
Dimensionless frequencies Ω_1, Ω_2 of simply supported isotropic nanoplate

$\mu(nm^2)$	0	1	2	3	4	5
Ω_1	0.09632	0.08802	0.08156	0.07633	0.07200	0.06833
Ω_2	0.38527	0.28799	0.23990	0.20991	0.18893	0.17320

As a validation task, we also considered the nonlinear vibrations of the simply supported isotropic plate. In this study, the governing equations are based on the von Kármán theory and one mode model is employed, while ignoring damping, and load parameter as well as the nonlocal parameter.

TABLE 9.2
Nonlinear frequency ratio ω_n/ω_l for simply supported isotropic rectangular plate $(\nu = 0.3, a/h = 10)$

	$b/a = 1$			$b/a = 0.5$		
A	[218]	[175]	present	[218]	[175]	present
0.2	1.0185	1.0195	1.0195	1.0239	1.0246	1.0241
0.4	1.0717	1.0752	1.0757	1.0918	1.0946	1.0927
0.6	1.1534	1.1601	1.1625	1.1957	1.2011	1.1975
0.8	1.2566	1.2667	1.2733	1.3264	1.3344	1.3293
1	1.3753	1.3881	1.4024	1.4758	1.4872	1.4808

It is worth noting that the comparative study is performed for an immovable plate ([222], [238]) and in this case, p_1, p_2 in stress function (9.23) can be found as in ([139], [222]). The values of the nonlinear frequency ratio ω_n/ω_l are presented in Table 9.2, where $\omega_l = \sqrt{\alpha_0}$ and ω_n are linear and nonlinear frequencies respectively.

9.5 NON-RESONANT VIBRATION

The approximate analytical solution to the initial value problem governed by nonlinear differential equations (9.31) and initial conditions (9.33) are obtained using the multiple scales method (MSM) in time domain. We introduce three time variables τ_0, τ_1, τ_2 describing the system (9.31) evolution in time. They are related to the dimensionless time τ in the following manner:

$$\tau_i = \varepsilon^i \tau, \qquad i = 0, 1, 2. \tag{9.34}$$

The parameter ε which in dimensionless formulation, is a number satisfying the inequalities $0 < \varepsilon \ll 1$. The approach with several time variables that is a key concept of the MSM allows for separating the fast and slow processes from each other in the vibration study. The time variable τ_0 is usually called the fast time scale, and the others variables are regarded as the slow time scales.

The consequence of introducing a few variables is the necessity to substitute the ordinary differential equations with the partial ones. The ordinary derivatives with respect to the time τ are replaced by the partial derivatives. According to the chain rule we can write

$$\frac{d}{d\tau} = \sum_{k=0}^{2} \varepsilon^k \frac{\partial}{\partial \tau_k} = \frac{\partial}{\partial \tau_0} + \varepsilon \frac{\partial}{\partial \tau_1} + \varepsilon^2 \frac{\partial}{\partial \tau_2},$$
$$\frac{d^2}{d\tau^2} = \frac{\partial^2}{\partial \tau_0^2} + 2\varepsilon \frac{\partial^2}{\partial \tau_0 \partial \tau_1} + \varepsilon^2 \left(\frac{\partial^2}{\partial \tau_1^2} + 2 \frac{\partial^2}{\partial \tau_0 \partial \tau_2} \right) + 2\varepsilon^3 \frac{\partial^2}{\partial \tau_1 \partial \tau_2} + O\left(\varepsilon^4\right). \tag{9.35}$$

The functions $y_1(\tau), y_2(\tau)$ are approximated by the following asymptotic expansions

$$y_{1a}(\tau;\varepsilon) = \sum_{k=1}^{3} \varepsilon^k z_{1k}(\tau_0,\ \tau_1,\ \tau_2) + O\left(\varepsilon^4\right),$$
$$y_{2a}(\tau;\varepsilon) = \sum_{k=1}^{3} \varepsilon^k\, z_{2k}(\tau_0,\ \tau_1,\ \tau_2) + O\left(\varepsilon^4\right), \tag{9.36}$$

where functions $z_{1k}(\tau_0, \tau_1, \tau_2)$, $z_{2k}(\tau_0, \tau_1, \tau_2)$, $k = 1, ..., 3$ are to be found.

Assuming the effects caused by the damping and the excitation are weak one can write

$$\delta = \varepsilon^2 \hat{\delta}, \qquad \sigma = \varepsilon^3\, \hat{\sigma}, \tag{9.37}$$

where the values of the parameters $\hat{\delta}$ and $\hat{\sigma}$ are of order $O(1)$ as $\varepsilon \to 0$.

Substituting equations (9.35)–(9.37) into the governing equations (9.31) yields the equations in which the small parameter appears in a few different powers. In accordance with the assumed variant of the MSM, all terms of the order $O\left(\varepsilon^4\right)$ are omitted. Each of the equations should be satisfied for any value of ε. After ordering the terms according to the powers ε, ε^2 and ε^3, this requirement is realized by equating to zero all coefficients standing at these powers. In that way, one can obtain six differential equations that are next organized into three groups as follows:

(i) the equations of the first-order approximation

$$\frac{\partial^2 z_{11}}{\partial \tau_0^2} + z_{11} = 0, \qquad \frac{\partial^2 z_{21}}{\partial \tau_0^2} + \eta_0\, z_{21} = 0; \tag{9.38}$$

(ii) the equations of the second-order approximation

$$\frac{\partial^2 z_{12}}{\partial \tau_0^2} + z_{12} = -2 \frac{\partial^2 z_{11}}{\partial \tau_0 \partial \tau_1}, \qquad \frac{\partial^2 z_{22}}{\partial \tau_0^2} + \eta_0\, z_{22} = -2 \frac{\partial^2 z_{21}}{\partial \tau_0 \partial \tau_1}; \tag{9.39}$$

(iii) the equations of the third-order approximation

$$\frac{\partial^2 z_{13}}{\partial \tau_0^2} + z_{13} = -\hat{\sigma}\cos(p\tau_0) - \hat{\delta}\frac{\partial z_{11}}{\partial \tau_0} - \xi_2 z_{11}^3 - \xi_1 z_{11} z_{21}^2 - \frac{\partial^2 z_{11}}{\partial \tau_1^2} - 2\frac{\partial^2 z_{11}}{\partial \tau_0 \partial \tau_2} - 2\frac{\partial^2 z_{12}}{\partial \tau_0 \partial \tau_1},$$

$$\frac{\partial^2 z_{23}}{\partial \tau_0^2} + \eta_0 z_{23} = -\hat{\delta}\frac{\partial z_{21}}{\partial \tau_0} - \eta_2 z_{21}^3 - \eta_1 z_{11}^2 z_{21} - \frac{\partial^2 z_{21}}{\partial \tau_1^2} - 2\frac{\partial^2 z_{21}}{\partial \tau_0 \partial \tau_2} - 2\frac{\partial^2 z_{22}}{\partial \tau_0 \partial \tau_1}. \tag{9.40}$$

Equations (9.38)–(9.40) are solved recursively starting from the homogenous equations of the first-order approximation whose general solutions are

$$\begin{aligned} z_{11} &= B_1(\tau_1,\tau_2)\, e^{\mathrm{i}\tau_0} + \bar{B}_1(\tau_1,\tau_2)\, e^{-\mathrm{i}\tau_0}, \\ z_{21} &= B_2(\tau_1,\tau_2)\, e^{\mathrm{i}\sqrt{\eta_0}\tau_0} + \bar{B}_2(\tau_1,\tau_2)\, e^{-\mathrm{i}\sqrt{\eta_0}\tau_0}, \end{aligned} \tag{9.41}$$

where symbol i denotes the imaginary unit and complex-valued functions $B_1(\tau_1,\tau_2)$, $B_2(\tau_1,\tau_2)$ and their complex conjugates $\bar{B}_1(\tau_1,\tau_2), \bar{B}_2(\tau_1,\tau_2)$ are unknown.

Substituting solutions (9.41) into equations (9.39) we obtain the following solvability conditions

$$\frac{\partial B_1}{\partial \tau_1} = 0, \qquad \frac{\partial \bar{B}_1}{\partial \tau_1} = 0, \qquad \frac{\partial B_2}{\partial \tau_1} = 0, \qquad \frac{\partial \bar{B}_2}{\partial \tau_1} = 0. \tag{9.42}$$

Taking into account relationships (9.42), one can notice the particular solutions to both equations (9.39) are equal zero.

Substituting solutions (9.41) into equations (9.40) implies occurrence of secular terms again. They must be eliminated from the equations in order to avoid the unlimited oscillations in time. So, the following solvability conditions should be satisfied

$$\begin{aligned} &2\mathrm{i}\,\frac{\partial B_1}{\partial \tau_2} + \left(\mathrm{i}\hat{\delta} + 3\xi_2 B_1 \bar{B}_1 + 2\xi_1 B_2 \bar{B}_2\right) B_1 = 0, \\ &2\mathrm{i}\,\frac{\partial \bar{B}_1}{\partial \tau_2} + \left(\mathrm{i}\hat{\delta} - 3\xi_2 B_1 \bar{B}_1 - 2\xi_1 B_2 \bar{B}_2\right) \bar{B}_1 = 0, \\ &2\mathrm{i}\,\sqrt{\eta_0}\frac{\partial B_2}{\partial \tau_2} + \left(\mathrm{i}\sqrt{\eta_0}\hat{\delta} + 3\eta_2 B_2 \bar{B}_2 + 2\eta_1 B_1 \bar{B}_1\right) B_2 = 0, \\ &2\mathrm{i}\,\sqrt{\eta_0}\frac{\partial \bar{B}_2}{\partial \tau_2} + \left(\mathrm{i}\sqrt{\eta_0}\hat{\delta} - 3\eta_2 B_2 \bar{B}_2 - 2\eta_1 B_1 \bar{B}_1\right) \bar{B}_2 = 0. \end{aligned} \tag{9.43}$$

The particular solutions to the equations of the third-order approximation are

$$z_{13} = \frac{\hat{\sigma}\cos p\tau_0}{1-p^2} + \frac{\xi_1\left(e^{i\left(1+2\sqrt{\eta_0}\right)\tau_0}B_1B_2^2 + e^{-i\left(1+2\sqrt{\eta_0}\right)\tau_0}\bar{B}_1\bar{B}_2^2\right)}{4\left(\eta_0+\sqrt{\eta_0}\right)} +$$
$$\frac{\xi_1\left(e^{i\left(-1+2\sqrt{\eta_0}\right)\tau_0}\bar{B}_1B_2^2 + e^{-i\left(-1+2\sqrt{\eta_0}\right)\tau_0}B_1\bar{B}_2^2\right)}{4\left(\eta_0-\sqrt{\eta_0}\right)} + \frac{\xi_2}{8}\left(e^{3i\tau_0}B_1^3 + e^{-3i\tau_0}\bar{B}_1^3\right),$$
$$z_{23} = \frac{\eta_1\left(e^{i\left(2+\sqrt{\eta_0}\right)\tau_0}B_1B_2^2 + e^{-i\left(2+\sqrt{\eta_0}\right)\tau_0}\bar{B}_1\bar{B}_2^2\right)}{4\left(1+\sqrt{\eta_0}\right)} + \tag{9.44}$$
$$\frac{\eta_1\left(e^{i\left(-2+\sqrt{\eta_0}\right)\tau_0}\bar{B}_1B_2^2 + e^{-i\left(-2+\sqrt{\eta_0}\right)\tau_0}B_1\bar{B}_2^2\right)}{4\left(1-\sqrt{\eta_0}\right)} + \frac{\eta_2}{8\eta_0}\left(e^{3i\sqrt{\eta_0}\tau_0}B_2^3 + e^{-3i\sqrt{\eta_0}\tau_0}\bar{B}_2^3\right).$$

The unknown functions B_1, B_2 and their complex conjugates $\bar{B}_1$, $\bar{B}_2$ are restricted by the solvability conditions. We present all the complex-valued functions in the exponential form

$$B_j = \frac{1}{2}b_je^{i\psi_j}, \quad \bar{B}_j = \frac{1}{2}b_je^{-i\psi_j}, \quad j=1,2. \tag{9.45}$$

Based on equations (9.42), one can state that the unknown real-valued functions $b_1,\ b_2,\ \psi_1,\ \psi_2$ depend only on the variable τ_2. Inserting relationships (9.45) into solvability conditions (9.45), solving the latter ones with respect to the derivatives, and finally, employing relationships (9.34)–(9.35) and (9.37), yields the following system of four ODEs

$$\frac{da_1}{d\tau} = -\frac{1}{2}\delta a_1, \qquad \frac{d\psi_1}{d\tau} = \frac{3}{8}\xi_2a_1^2 + \frac{1}{4}\xi_1a_2^2,$$
$$\frac{da_2}{d\tau} = -\frac{1}{2}\delta a_2, \qquad \frac{d\psi_2}{d\tau} = \frac{\eta_1}{4\sqrt{\eta_0}}a_1^2 + \frac{3\eta_2}{8\sqrt{\eta_0}}a_2^2. \tag{9.46}$$

The reverse employment of the relationships (9.35) transformed the problem from the domain of the slowest time scale τ_2 to the domain of the original nondimensional time τ. It should also be noted that $a_i = \varepsilon\, b_i$ for $i=1,\ 2$. The real-valued functions $a_j,\ \psi_j,\ j=1,2$, stand for the amplitudes and phases of the oscillations, respectively. Therefore, equations (9.46) govern the slow modulation of the amplitudes and phases. The complement to the modulation equations are the initial conditions as follows

$$a_1(0) = a_{10}, \quad \psi_1(0) = \psi_{10}, \quad a_2(0) = a_{20}, \quad \psi_2(0) = \psi_{20}, \tag{9.47}$$

where the quantities $a_{10},\ a_{20},\ \psi_{10},\ \psi_{20}$ can be calculated knowing values the $\gamma_1,\ \gamma_2, \gamma_3,\ \gamma_4$ (see equations (9.33)).

The exact solution to the initial value problem (9.46)–(9.47) has the following form

$$
\begin{aligned}
a_1 &= a_{10}e^{-\frac{\delta}{2}\tau}, \quad \psi_1 = \frac{3\xi_2 a_{10}^2 + 2\xi_1 a_{20}^2}{8\delta}\left(1 - e^{-\delta\,\tau}\right) + \psi_{10},\\
a_2 &= a_{20}e^{-\frac{\delta}{2}\tau}, \quad \psi_2 = \frac{2\eta_1 a_{10}^2 + 3\eta_2 a_{20}^2}{8\delta\sqrt{\eta_0}}\left(1 - e^{-\delta\tau}\right) + \psi_{20}.
\end{aligned} \tag{9.48}
$$

Assembling solutions (9.41) and (9.44) in accordance with (9.36) and omitting all terms of order $O(4)$, one can obtain the approximate solution to the problem described by equations (9.31) and (9.33) which reads

$$
\begin{aligned}
y_{1a} = &-\frac{\sigma}{1-p^2}\cos p\tau + a_1\cos\left(\tau+\psi_1\right) + \frac{\xi_2}{32}a_1^3\cos\left(3\left(\tau+\psi_1\right)\right) + \\
&\frac{\xi_1 a_1 a_2^2}{16\sqrt{\eta_0}\left(\sqrt{\eta_0}-1\right)}\cos\left(\left(1-2\sqrt{\eta_0}\right)\tau+\psi_1-2\psi_2\right) + \\
&\frac{\xi_1 a_1 a_2^2}{16\sqrt{\eta_0}\left(\sqrt{\eta_0}+1\right)}\cos\left(\left(1+2\sqrt{\eta_0}\right)\tau+\psi_1+2\psi_2\right),
\end{aligned} \tag{9.49}
$$

$$
\begin{aligned}
y_{2a} = &\, a_2\cos\left(\sqrt{\eta_0}\tau+\psi_2\right) + \frac{\eta_2 a_2^3}{32\eta_0}\cos\left(3\left(\sqrt{\eta_0}\tau+\psi_2\right)\right) + \\
&\frac{\eta_1 a_1^2 a_2}{16}\left(\frac{\cos\left(\left(\sqrt{\eta_0}+2\right)\tau+2\psi_1+\psi_2\right)}{\sqrt{\eta_0}+1} - \frac{\cos\left(\left(\sqrt{\eta_0}-2\right)\tau-2\psi_1+\psi_2\right)}{\sqrt{\eta_0}-1}\right),
\end{aligned}
$$

where a_1, a_2 and ψ_1, ψ_2 are determined by equations (9.48).

The accuracy of the approximate solutions given by equations (9.49) is estimated using the following measure in the sense of L^2 norm

$$
e_i = \sqrt{\frac{1}{\tau_e-\tau_s}\int_{\tau_s}^{\tau_e}\left(G_i\left(y_{1a}(\tau), y_{2a}(\tau)\right) - 0\right)^2 d\tau}, \qquad i = 1,2, \tag{9.50}
$$

where G_1 and G_2 stand for the differential operators, i.e. the left sides of equations (9.31), while τ_s and τ_e denote the chosen instants. The measure proposed evaluates the error satisfying of the governing equations and, hence, it has the nature of the absolute error. When the exact solution is unknown, any attempts to introduce the relative error are problematic. Assuming the numerical solution is taken as the sufficiently accurate approximation of the exact solution, one can assess the accuracy of the approximate asymptotic solution by means the relative error as follows

$$
\delta_i = \sqrt{\frac{\int_{\tau_s}^{\tau_e}\left(y_{ia}(\tau) - y_{in}(\tau)\right)^2 d\tau}{\int_{\tau_s}^{\tau_e}\left(y_{in}(\tau)\right)^2 d\tau}}, \qquad i = 1,2, \tag{9.51}
$$

where y_{1n} and y_{2n} are the approximate solutions to the considered initial-value problem obtained numerically.

As can be seen from formulas (9.28) and (9.32), the nonlocal parameter μ affects the basic frequency of the function y_{2a} and also the nonlinear terms in equations (9.31), including the elastic coupling. Studying the approximate solutions, one can assess the qualitative impact of the parameter on the problem considered. At a fixed value of the damping coefficient and fixed initial values, the amplitudes of both terms that are conditioned by the nonlinearity of cubic type rise proportionally with the parameters ξ_2 and η_2. The couplings of elastic nature manifest themselves in the form of two terms in each solution. Their amplitudes are proportional to the common value of the parameters ξ_1 and η_1. In turn, the parameter η_0 reduces the magnitude of these amplitudes and affects also the frequency of the terms.

In order to investigate the quantitative influence of the parameter μ, we assume the material properties considered in (Aghababei and Reddy [5])

$$E = 3 \cdot 10^7 \frac{\mathrm{N}}{\mathrm{m}^2}, \qquad \nu = 0.3, \qquad \rho = 1220 \frac{\mathrm{kg}}{\mathrm{m}^3},$$

and the nanoplate dimensions

$$a = b = 10^{-8}\mathrm{m}, \qquad h = 0.1a.$$

Taking into account relationships (9.14), (9.28) and (9.32) one can calculate the values of all dimensionless parameters affecting the approximate solutions (9.49). The parameters values for several values of μ are collected in Table 9.3.

TABLE 9.3
Dimensionless values of the mathematical model parameters depending on the nonlocal parameter μ

μ	ξ_1	ξ_2	η_0	η_1	η_2
0	0.8736	0.34125	16.0	0.8736	5.46
$1 \cdot 10^{-18}$	0.526453	0.292956	10.7055	0.526453	2.53486
$2 \cdot 10^{-18}$	0.409723	0.265969	8.65272	0.409723	1.83141
$3 \cdot 10^{-18}$	0.351166	0.248737	7.5622	0.351166	1.51519
$4 \cdot 10^{-18}$	0.315969	0.236781	6.88581	0.315969	1.33547
$5 \cdot 10^{-18}$	0.29248	0.227999	6.4253	0.29248	1.21956

Generally, one can state values of all dimensionless parameters decrease with the nonlocal parameter μ. Their variabilities are depicted in Figures 9.2–9.3.

In Figures 9.4–9.5 it is shown how the coefficients affecting the system at a fixed value of the damping coefficient and fixed initial values the amplitudes of the terms caused by the couplings of elastic nature depend on the nonlocal parameter μ.

Studying the non-resonant case we focus on the impact of the nonlocal parameter μ on the vibration. The solutions concerning the nanoplate vibration when $\mu = 0$ and $\mu = 5 \cdot 10^{-18}$ have been compared under the same initial conditions and the same

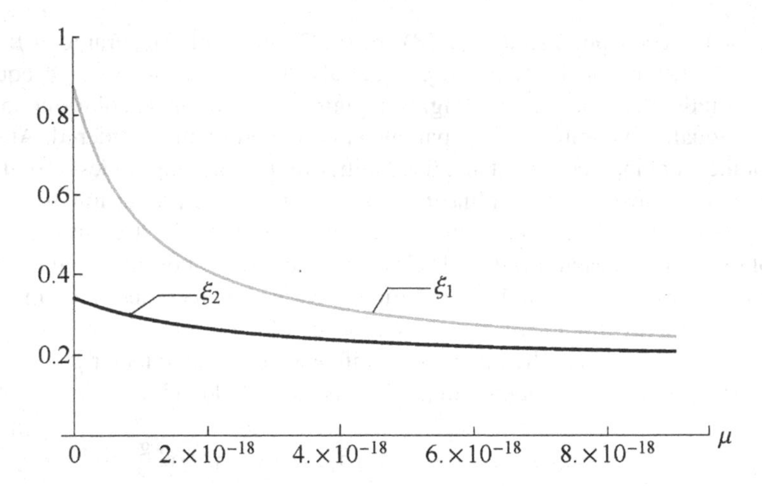

FIGURE 9.2 Parameters $\xi_1 = \eta_1$ and ξ_2 versus μ.

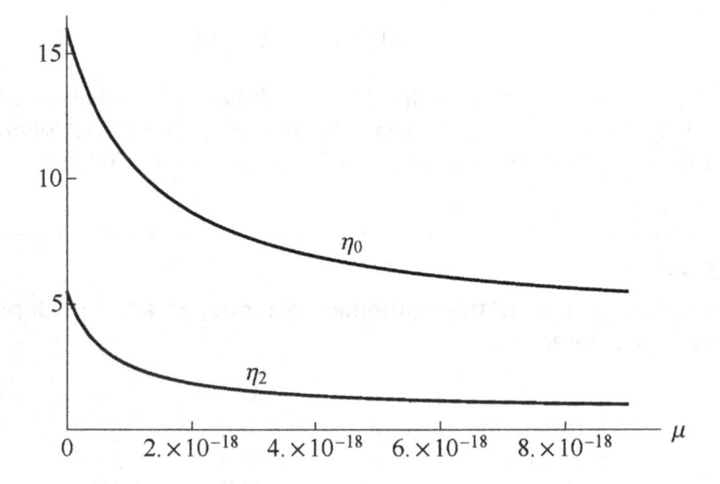

FIGURE 9.3 Parameters η_0 and η_2 versus μ.

harmonic excitation. Also the damping coefficient has the same value for both cases. We set the following nondimensional values of the damping coefficient, the harmonic force amplitude and its frequency

$$\delta = 0.01, \quad \sigma = -0.002, \quad p = 0.341.$$

In order to show the MSM solution efficiency to this problem, we adopted quite large values of the initial nanoplate deflection. So, the initial values fixed in equations (9.33) are assumed as follows

$$\gamma_1 = 0.55, \quad \gamma_3 = 0.2, \quad \gamma_2 = \gamma_4 = 0.$$

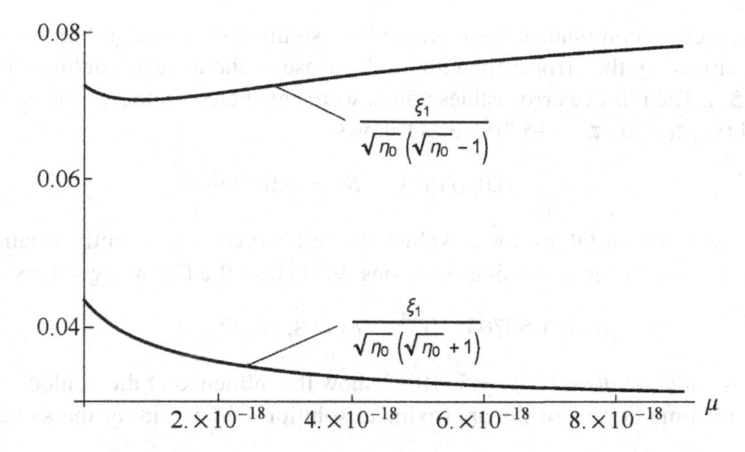

FIGURE 9.4 Coefficients modifying the magnitude of the amplitude of the terms caused by the coupling of elastic nature in of solution y_{1a} versus μ.

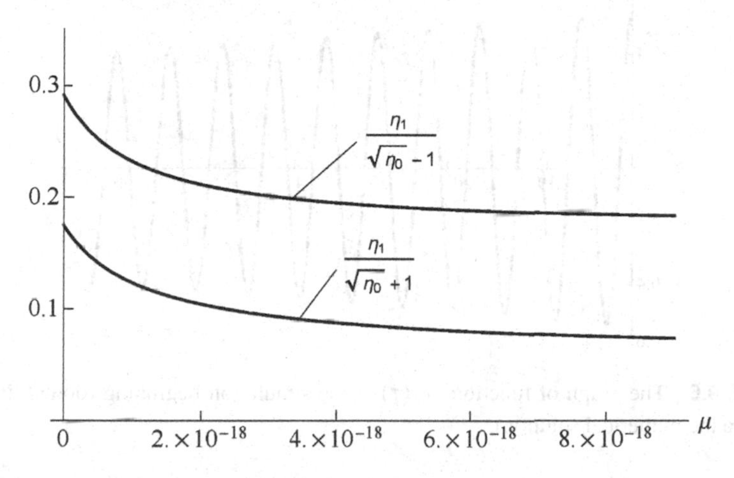

FIGURE 9.5 Coefficients modifying the magnitude of the amplitude of the terms caused by the coupling of elastic nature in solution y_{2a} versus μ.

The following initial values of the amplitudes and phases (see equations (9.47))

$$a_{10} = 0.545849, \quad a_{20} = 0.200356, \quad \psi_{10} = -0.004688, \quad \psi_{20} = -0.0011542$$

remain in compliance with the values $\gamma_j,\ j = 1,4$ when $\mu = 0$.

The time variability of the functions $y_{1a}(\tau)$ and $y_{2a}(\tau)$ concerning the case when $\mu = 0$ is presented in Figures 9.6–9.9. The first two are related to the initial stage of motion. Figures 9.8–9.9 concern the oscillations at the end of the simulation carried out. In each of these group of graphs two curves are depicted. The approximate analytical solution obtained using MSM is drawn using solid gray line, whereas the dotted line in black was used to draw the numerical solution obtained by means of the standard procedure *NDSolve* of the *Mathematica* software.

An excellent compliance between the both solution observed in Figures 9.6–9.9 was confirmed by the error estimation in the sense of the measure defined by equation (9.51). The relative error values which were calculated on the whole simulation interval (i.e. $\tau_s = 0,\ \tau_e = 1500$) are as follows

$$\delta_1 = 0.0203683, \quad \delta_2 = 0.0156907.$$

For the function taking lower values the relative error is slightly smaller. The errors satisfying of the governing equations (9.31) take the following values

$$e_1 = 1.50703 \cdot 10^{-4}, \quad e_2 = 8.47842 \cdot 10^{-5}.$$

The simulation done for $\mu = 5 \cdot 10^{-18}$ show the influence of the nonlocal parameter on the time course of the approximate solutions. Maintaining the same initial

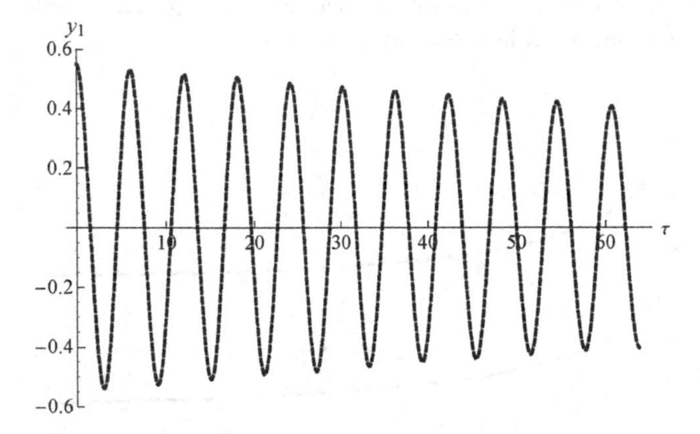

FIGURE 9.6 The graph of function $y_{1a}(\tau)$ at the simulation beginning (dotted line corresponds to the numerical solution).

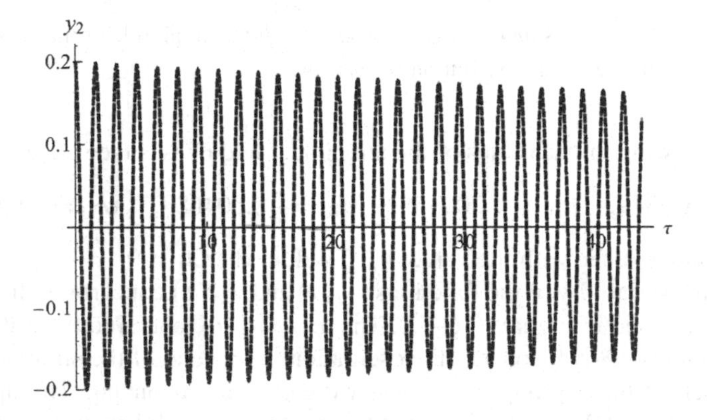

FIGURE 9.7 The graph of function $y_{2a}(\tau)$ at the simulation beginning (dotted line corresponds to the numerical solution).

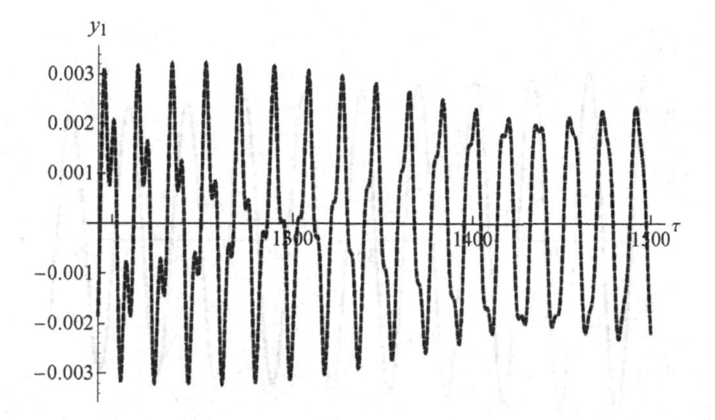

FIGURE 9.8 The graph of function $y_{1a}(\tau)$ at the simulation end (dotted line corresponds to the numerical solution).

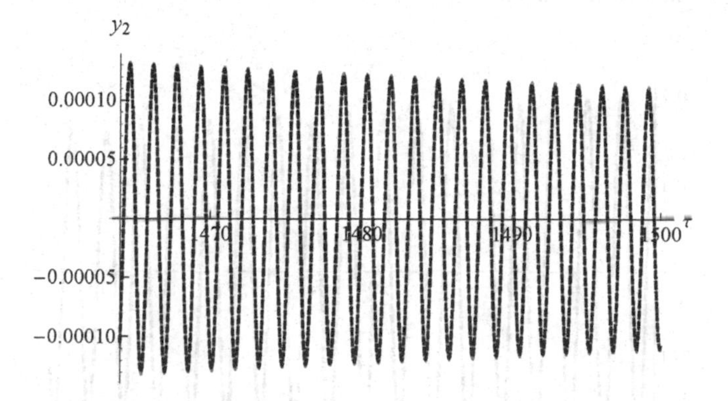

FIGURE 9.9 The graph of function $y_{2a}(\tau)$ at the simulation end (dotted line corresponds to the numerical solution).

values γ_j, $j = 1,..4$, as previous, we calculated the following initial values of the amplitudes and phases

$$a_{10} = 0.546433, \quad a_{20} = 0.200356, \quad \psi_{10} = -0.00480115, \quad \psi_{20} = -0.00190834.$$

The time history of the functions $y_{1a}(\tau)$ and $y_{2a}(\tau)$ on the same ranges as for the case $\mu = 0$ is depicted in Figures 9.10–9.13. The same manner of the results presentation is employed, i.e. the approximate analytical solution obtained using MSM is drawn using solid gray line, whereas the dotted line in black was used to draw the numerical solution obtained by means of the standard procedure *NDSolve*.

The values of the error satisfying the governing equations, and the relative errors are

$$e_1 = 6.73176 \cdot 10^{-5}, \; e_2 = 2.00589 \cdot 10^{-5}, \; \delta_1 = 0.0102189, \; \delta_2 = 0.00449965.$$

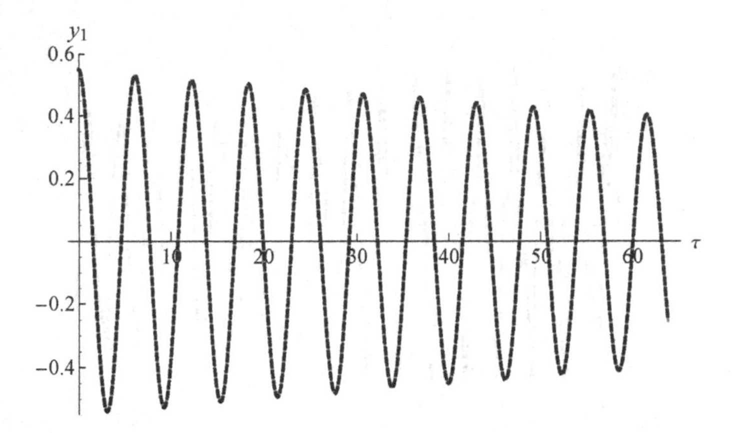

FIGURE 9.10 The graph of function $y_{1a}(\tau)$ at the simulation beginning (dotted line corresponds to the numerical solution).

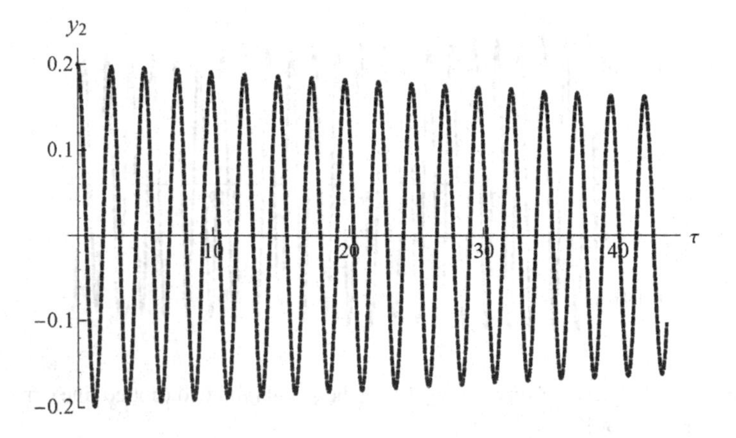

FIGURE 9.11 The graph of function $y_{2a}(\tau)$ at the simulation beginning (dotted line corresponds to the numerical solution).

The graphs presented in Figure 9.14 for $\mu = 0$ and in Figure 9.15 for $\mu = 5 \cdot 10^{-18}$ show how both relative errors vary with the extent of the simulation time. Values of both errors are significantly less for the model employing the nonlocal parameter. In each of the two cases the relative error δ_2, drawn in black, is slightly smaller than δ_1, depicted in gray, on the whole interval considered. Both errors rise rapidly at the vibration beginning. While the error δ_1 stabilizes itself with the simulation duration then the error δ_2 behaves quite chaotic especially when $\mu = 0$. This behavior can be explained by much higher frequency of the function y_{2a}. The numerical integration of highly oscillating functions is ill-conditioned. The situation is made even worse by the fact that there is also a highly oscillating function in the denominator of the formula (9.51).

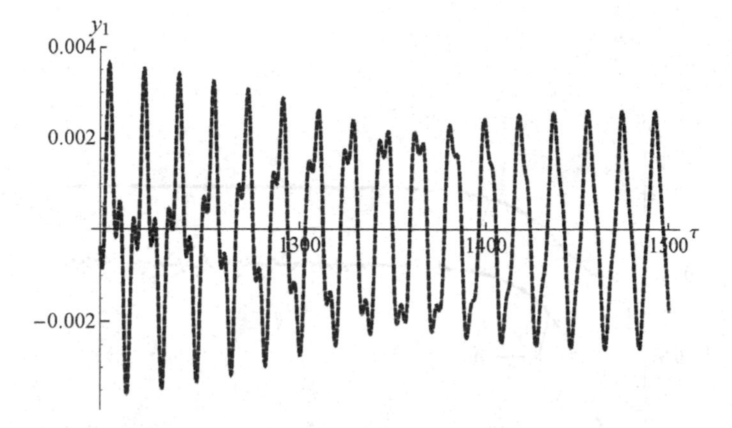

FIGURE 9.12 The graph of function $y_{1a}(\tau)$ at the simulation end (dotted line corresponds to the numerical solution).

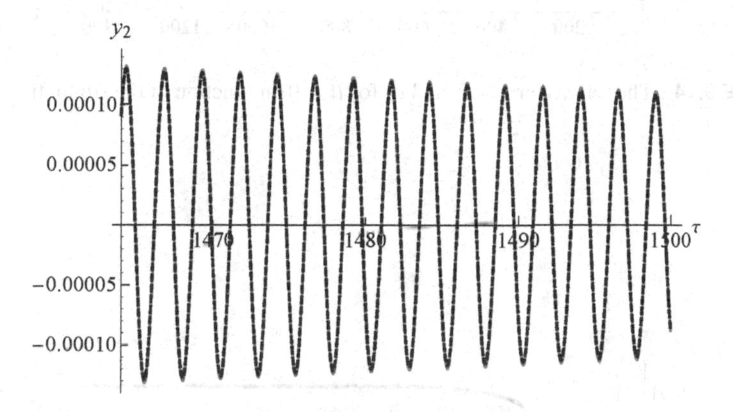

FIGURE 9.13 The graph of function $y_{2a}(\tau)$ at the simulation end (dotted line corresponds to the numerical solution).

The changeability of the absolute errors e_1, presented by gray line, and e_2, drawn in black, with the duration of the simulation time is presented in Figure 9.16, for $\mu = 0$, and in Figure 9.17, for $\mu = 5 \cdot 10^{-18}$. One can notice both errors significantly decrease with time. Random values, i.e., noticeably deviating from the established trend, can be seen mainly in the case of error e_2 related to the second equation from equations (9.31).

In Figures 9.18–9.19 there is shown how the initial deflection $y_1(0) = y_2(0) = y_0$ of the nanoplate impact on the values of the relative errors δ_1 and δ_2. In each of the figures there are six relationships presented for several values of the nonlocal parameter μ. The values of both relative errors grow with the nanoplate initial deflection. The greater the value of the nonlocal parameter μ, the smaller the errors as observed for vibrations initiated by large initial deflections. Some fluctuation observed in Figure 9.19 are the result of the faster oscillating component associated with the function y_{2a}.

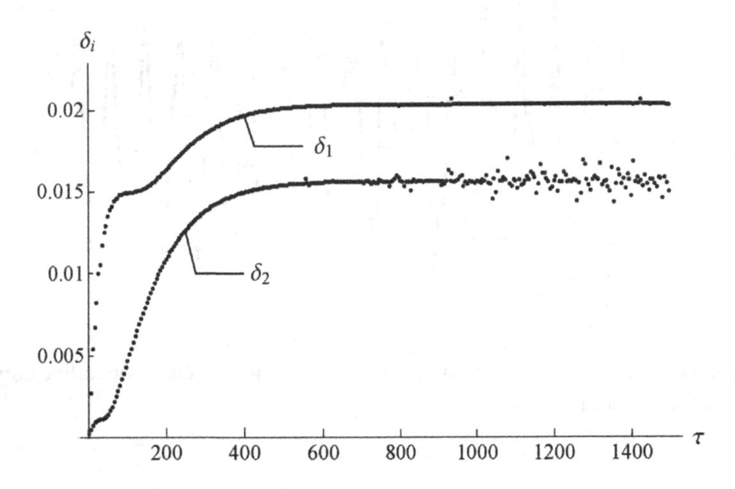

FIGURE 9.14 The relative errors δ_1 and δ_2 for $\mu = 0$ in function of the simulation duration.

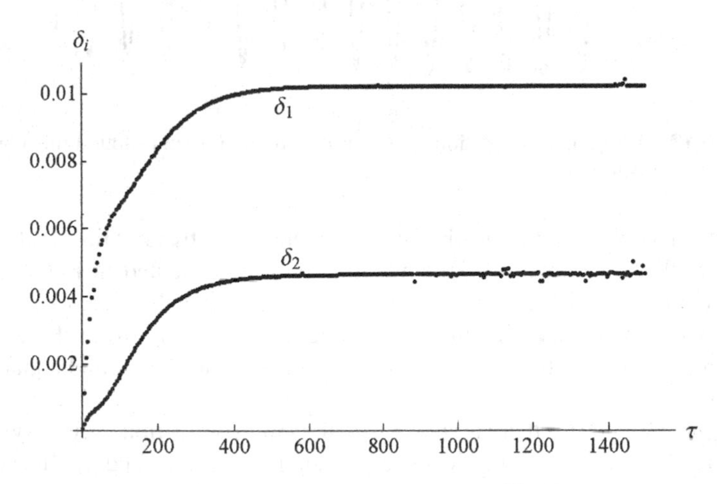

FIGURE 9.15 The relative errors δ_1 and δ_2 for $\mu = 5 \cdot 10^{-18}$ in function of the simulation duration.

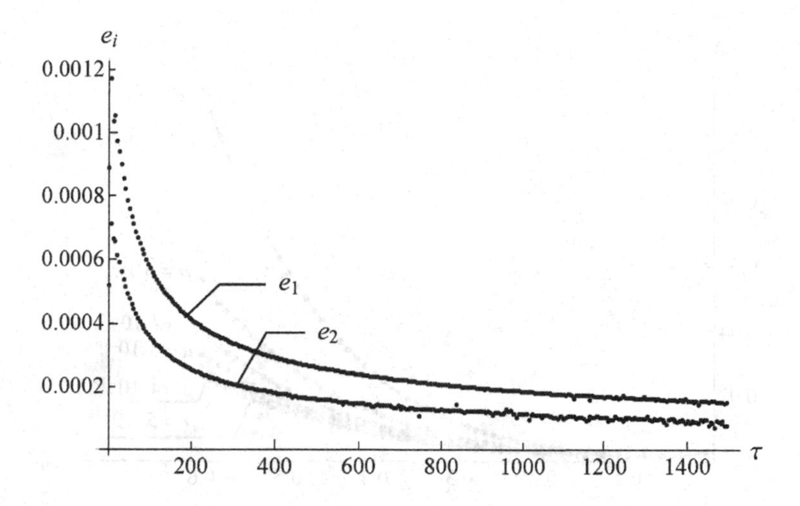

FIGURE 9.16 The errors e_1 and e_2 for $\mu = 0$ in function of the simulation duration.

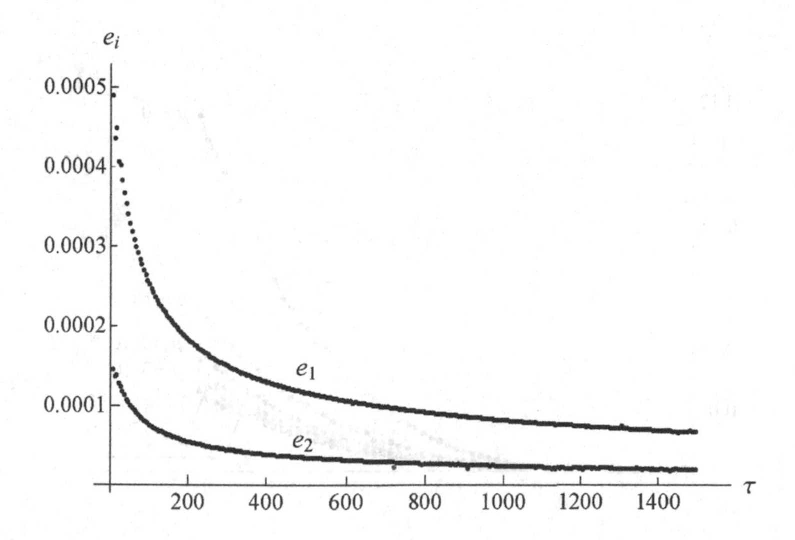

FIGURE 9.17 The errors e_1 and e_2 for $\mu = 5 \cdot 10^{-18}$ in function of the simulation duration.

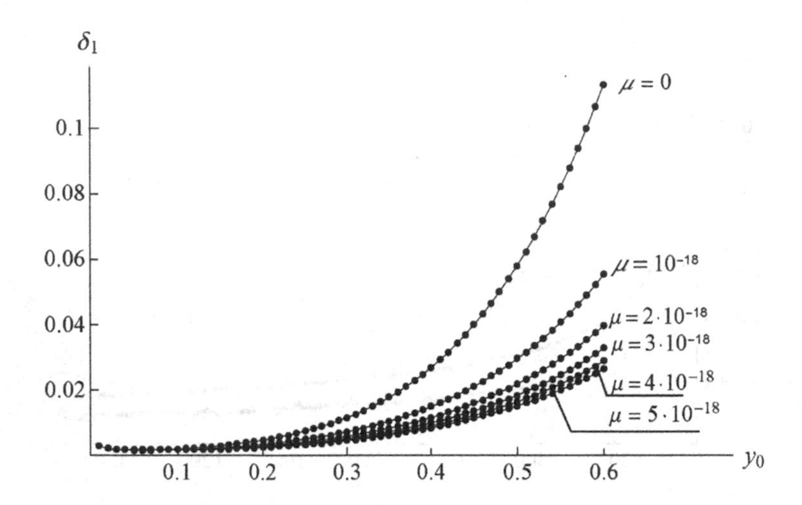

FIGURE 9.18 The error δ_1 versus the initial deflection y_0.

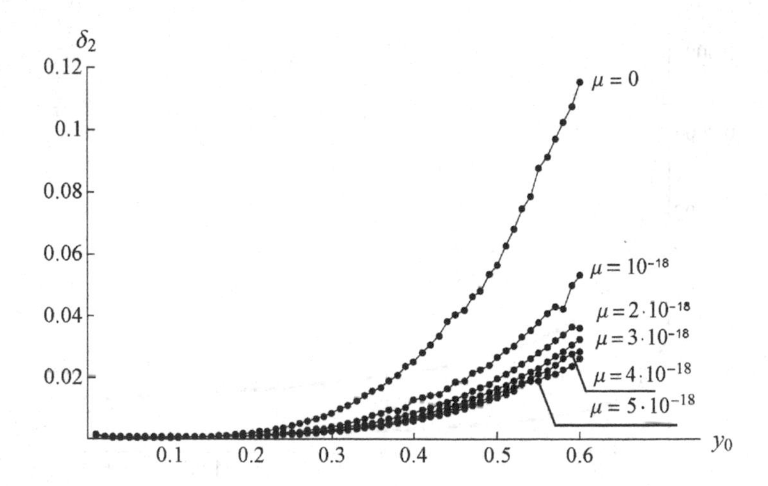

FIGURE 9.19 The error δ_2 versus the initial deflection y_0.

9.6 RESONANT VIBRATION

The parameter η_0 is positive and for the data considered, tends to four as μ grows to one from below. So, of the three theoretical possibilities when the denominators of the terms constituting the approximate solution equal zero, only one can be physically realized. It occurs when the frequency p of the excitation is close to one which means the satisfying conditions of the main resonance. The solutions determined by equations (9.49) then fail. Consequently, it is necessary to solve the problem again basing on the assumption that

$$p = 1 + s, \tag{9.52}$$

where s is any, but rather small, number. Assuming that it is of order $O\left(\varepsilon^2\right)$, we write

$$p = 1 + \varepsilon^2 \hat{s}, \tag{9.53}$$

where $\hat{s}$ is understood as $O(1)$ when $\varepsilon^2 \to 0$.

Inserting equations (9.35)–(9.37) and (9.53) into equations (9.31) yields two equations whose all terms contain different powers of the small parameter ε. Each of the two equations should to be satisfied for any value of ε. After ordering the terms of equations according to the powers ε, ε^2 and ε^3, this requirement is realized by equating to zero all coefficients standing at these powers. In accordance with the assumed variant of the MSM, all terms of the order $O\left(\varepsilon^4\right)$ are omitted. In that manner, one can obtain six differential equations that can be organized into three groups and write as:

(i) the equations of the first-order approximation

$$\frac{\partial^2 z_{11}}{\partial \tau_0^2} + z_{11} = 0, \qquad \frac{\partial^2 z_{21}}{\partial \tau_0^2} + \eta_0\, z_{21} = 0; \tag{9.54}$$

(ii) the equations of the second-order approximation

$$\frac{\partial^2 z_{12}}{\partial \tau_0^2} + z_{12} = -2\frac{\partial^2 z_{11}}{\partial \tau_0 \partial \tau_1}, \qquad \frac{\partial^2 z_{22}}{\partial \tau_0^2} + \eta_0 z_{22} = -2\frac{\partial^2 z_{21}}{\partial \tau_0 \partial \tau_1}; \tag{9.55}$$

(iii) the equations of the third-order approximation

$$\begin{aligned} \frac{\partial^2 z_{13}}{\partial \tau_0^2} + z_{13} = {} & -\hat{\sigma}\cos\left(\left(1+\varepsilon^2\hat{s}\right)\tau_0\right) - \hat{\delta}\frac{\partial z_{11}}{\partial \tau_0} - \xi_2 z_{11}^3 - \xi_1 z_{11} z_{21}^2 - \\ & \frac{\partial^2 z_{11}}{\partial \tau_1^2} - 2\frac{\partial^2 z_{11}}{\partial \tau_0 \partial \tau_2} - 2\frac{\partial^2 z_{12}}{\partial \tau_0 \partial \tau_1}, \end{aligned} \tag{9.56}$$

$$\frac{\partial^2 z_{23}}{\partial \tau_0^2} + \eta_0 z_{23} = -\hat{\delta}\frac{\partial z_{21}}{\partial \tau_0} - \eta_2 z_{21}^3 - \eta_1 z_{11}^2 z_{21} - \frac{\partial^2 z_{21}}{\partial \tau_1^2} - 2\frac{\partial^2 z_{21}}{\partial \tau_0 \partial \tau_2} - 2\frac{\partial^2 z_{22}}{\partial \tau_0 \partial \tau_1}.$$

Since the equations of the first and the second-order approximation are similar in form as in the non-resonant case, the general solutions z_{11}, z_{21} are expressed by (9.41) whereas the particular solutions z_{21}, z_{22} are equal to zero, as previously seen. Moreover, the solvability conditions in the form (9.42) are still valid. A significant difference from the solving concerning the non-resonant regime appears in the formulation of the solvability conditions for the equations of the third-order approximation. Now, they are as follows

$$
\begin{aligned}
&2\mathrm{i}\,\frac{\partial B_1}{\partial \tau_2}+\left(\mathrm{i}\hat{\delta}+3\,\xi_2 B_1\bar{B}_1+2\xi_1 B_2\bar{B}_2\right)B_1+\frac{\hat{\sigma}}{2}e^{\mathrm{i}\varepsilon^2\hat{s}\tau_0}=0,\\
&2\mathrm{i}\,\frac{\partial \bar{B}_1}{\partial \tau_2}+\left(\mathrm{i}\hat{\delta}-3\xi_2 B_1\bar{B}_1-2\xi_1 B_2\bar{B}_2\right)\bar{B}_1-\frac{\hat{\sigma}}{2}e^{-\mathrm{i}\varepsilon^2\hat{s}\tau_0}=0,\\
&2\mathrm{i}\,\sqrt{\eta_0}\frac{\partial B_2}{\partial \tau_2}+\left(\mathrm{i}\sqrt{\eta_0}\hat{\delta}+3\eta_2 B_2\bar{B}_2+2\eta_1 B_1\bar{B}_1\right)B_2=0,\\
&2\mathrm{i}\,\sqrt{\eta_0}\frac{\partial \bar{B}_2}{\partial \tau_2}+\left(\mathrm{i}\sqrt{\eta_0}\hat{\delta}-3\eta_2 B_2\bar{B}_2-2\eta_1 B_1\bar{B}_1\right)\bar{B}_2=0.
\end{aligned}
\tag{9.57}
$$

The particular solutions to the equations of the third-order approximation that contain no secular terms are

$$
\begin{aligned}
z_{13}=&\frac{\xi_1\left(e^{\mathrm{i}\left(1+2\sqrt{\eta_0}\right)\tau_0}B_1B_2^2+e^{-\mathrm{i}\left(1+2\sqrt{\eta_0}\right)\tau_0}\bar{B}_1\bar{B}_2^2\right)}{4\left(\eta_0+\sqrt{\eta_0}\right)}+\\
&\frac{\xi_1\left(e^{\mathrm{i}\left(-1+2\sqrt{\eta_0}\right)\tau_0}\bar{B}_1B_2^2+e^{-\mathrm{i}\left(-1+2\sqrt{\eta_0}\right)\tau_0}B_1\bar{B}_2^2\right)}{4\left(\eta_0-\sqrt{\eta_0}\right)}+\frac{\xi_2}{8}\left(e^{3\mathrm{i}\tau_0}B_1^3+e^{-3\mathrm{i}\tau_0}\bar{B}_1^3\right),\\
z_{23}=&\frac{\eta_1\left(e^{\mathrm{i}\left(2+\sqrt{\eta_0}\right)\tau_0}B_1B_2^2+e^{-\mathrm{i}\left(2+\sqrt{\eta_0}\right)\tau_0}\bar{B}_1\bar{B}_2^2\right)}{4\left(1+\sqrt{\eta_0}\right)}+\\
&\frac{\eta_1\left(e^{\mathrm{i}\left(-2+\sqrt{\eta_0}\right)\tau_0}\bar{B}_1B_2^2+e^{-\mathrm{i}\left(-2+\sqrt{\eta_0}\right)\tau_0}B_1\bar{B}_2^2\right)}{4\left(1-\sqrt{\eta_0}\right)}+\frac{\eta_2}{8\eta_0}\left(e^{3\mathrm{i}\sqrt{\eta_0}\tau_0}B_2^3+e^{-3\mathrm{i}\sqrt{\eta_0}\tau_0}\bar{B}_2^3\right).
\end{aligned}
\tag{9.58}
$$

The solvability conditions in which the exponential representation of the complex-valued functions $B_j,\ j=1,\ 2$, was employed, give four modulation equations restricting the unknown amplitudes and phases as follows

$$
\begin{aligned}
&\frac{da_1}{d\tau}=-\frac{1}{2}\left(\delta a_1+\sigma\sin\left(s\tau-\psi_1\right)\right),\\
&\frac{d\psi_1}{d\tau}=\frac{3}{8}\xi_2a_1^2+\frac{1}{4}\xi_1a_2^2+\frac{\sigma}{2a_1}\cos\left(s\tau-\psi_1\right),\\
&\frac{da_2}{d\tau}=-\frac{1}{2}\delta a_2,\qquad \frac{d\psi_2}{d\tau}=\frac{\eta_1}{4\sqrt{\eta_0}}a_1^2+\frac{3\eta_2}{8\sqrt{\eta_0}}a_2^2.
\end{aligned}
\tag{9.59}
$$

The modulation equations are supplemented by the following initial conditions

$$
a_1(0)=a_{10},\quad \psi_1(0)=\psi_{10},\quad a_2(0)=a_{20},\quad \psi_2(0)=\psi_{20},
\tag{9.60}
$$

where the quantities a_{10}, a_{20}, ψ_{10}, ψ_{20} can be calculated knowing the values γ_1, γ_2, γ_3, γ_4 (see equations (9.33)). *Contrary to the case concerning the non-resonant vibration, the initial-value problem (9.59)–(9.60) cannot be solved analytically.*

Assembling the solutions z_{1k}, z_{2k}, $k = 1,2,3$, in accordance with (9.36) gives the following approximate solution applicable at the main resonance

$$y_{1a} = a_1 \cos(\tau + \psi_1) + \frac{\xi_1 a_1 a_2^2}{16(\eta_0 - \sqrt{\eta_0})} \cos((1 - 2\sqrt{\eta_0})\tau + \psi_1 - 2\psi_2) +$$

$$\frac{\xi_1 a_1 a_2^2}{16(\eta_0 + \sqrt{\eta_0})} \cos((1 + 2\sqrt{\eta_0})\tau + \psi_1 + 2\psi_2) + \frac{\xi_2}{32} a_1^3 \cos(3(\tau + \psi_1)),$$

$$y_{2a} = a_2 \cos(\sqrt{\eta_0}\tau + \psi_2) + \frac{\eta_2 a_2^3}{32\eta_0} \cos(3(\sqrt{\eta_0}\tau + \psi_2)) + \tag{9.61}$$

$$\frac{\eta_1 a_1^2 a_2}{16} \left(\frac{\cos((\sqrt{\eta_0} + 2)\tau + 2\psi_1 + \psi_2)}{\sqrt{\eta_0} + 1} - \frac{\cos((\sqrt{\eta_0} - 2)\tau - 2\psi_1 + \psi_2)}{\sqrt{\eta_0} - 1} \right),$$

where the functions $a_1(\tau)$, $a_2(\tau)$, $\psi_1(\tau)$, $\psi_2(\tau)$ are numerically obtained solutions of the modulation equations (9.59).

Despite the formal similarity of the particular terms in solutions (9.61) and (9.49), a different character of the modulation equations (9.59) and (9.46) yields requirement for new discussions if the influence of the parameters examined in the previous section does not apply to the vibration at main resonance. The crucial significance for the resonant vibration *is the slowly changing modulation of the amplitudes and phases*. The influence on the form of the modulation equations governing these slow changes have all the parameters describing the nanoplate features which are affected by the nonlocal parameter μ and also the excitation parameters. We focus again on the impact of the nonlocal parameter on the resonant vibration by comparison two extreme cases when $\mu = 0$ and $\mu = 5 \cdot 10^{-18}$.

The values of the damping coefficient, the amplitude of the harmonic force and the initial values remain the same as in the examples considered previously and concerning the non-resonant vibration. The common for both simulation value of the detuning parameter s is 0.01. The initial conditions for modulation equations (9.59), applicable when $\mu = 0$, are

$$a_{10} = 0.54809, \quad a_{20} = 0.200353, \quad \psi_{10} = -0.004686, \quad \psi_{20} = -0.0011535.$$

Their counterpart values when $\mu = 5 \cdot 10^{-18}$ differ slightly and they follow

$$a_{10} = 0.54868, \quad a_{20} = 0.200359, \quad \psi_{10} = -0.004800, \quad \psi_{20} = -0.0019079.$$

In order to compare the compliance of the approximate solution with the numerically obtained one, the time courses are shown at the initial stage of vibration, and after a long time when the vibration practically stabilizes. The time intervals chosen are relatively short compared to the time required for achieving the steady state vibration. The time variability of the functions and concerning the case when $\mu = 0$

is presented in Figures 9.20–9.23. The approximate analytical solutions $y_{1a}(\tau)$ and $y_{2a}(\tau)$ given by equations (9.61) are drawn using solid gray line, whereas the dotted line in black was used to depict the numerical solution obtained by means of the standard procedure *NDSolve*. The solid black lines present the amplitudes $a_1(\tau)$ and $a_2(\tau)$. The solutions related to the model when $\mu = 5 \cdot 10^{-18}$ are depicted in Figures 9.24–9.27 using the same method of presentations. A high agreement between the amplitudes graphs and the fast oscillations of the corresponding functions is also worth noting in each of the Figures 9.20–9.27.

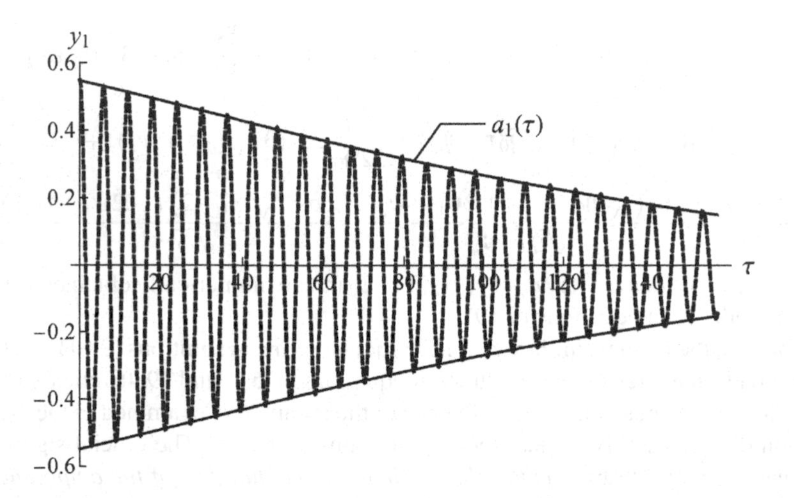

FIGURE 9.20 The graph of function $y_{1a}(\tau)$ at the simulation beginning for $\mu = 0$ (dotted line corresponds to the numerical solution).

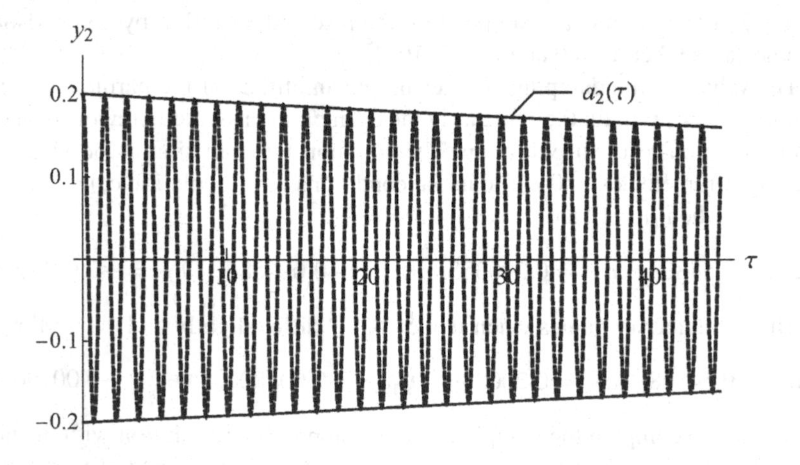

FIGURE 9.21 The graph of function $y_{2a}(\tau)$ at the simulation beginning for $\mu = 0$ (dotted line corresponds to the numerical solution).

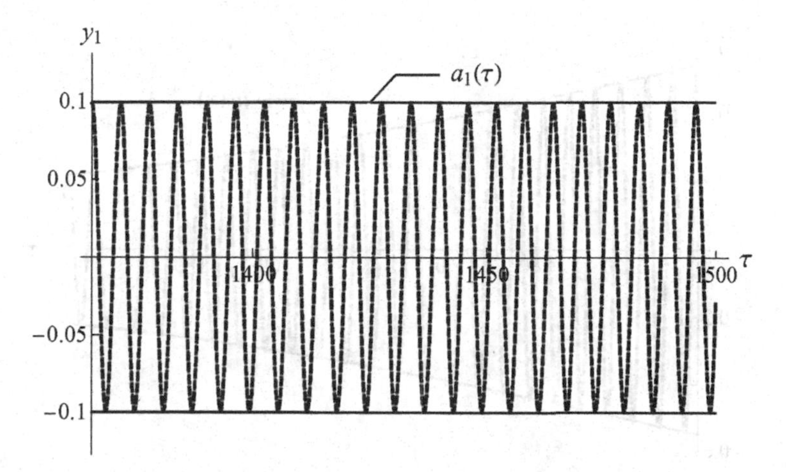

FIGURE 9.22 The graph of function $y_{1a}(\tau)$ at the simulation end for $\mu = 0$ (dotted line corresponds to the numerical solution).

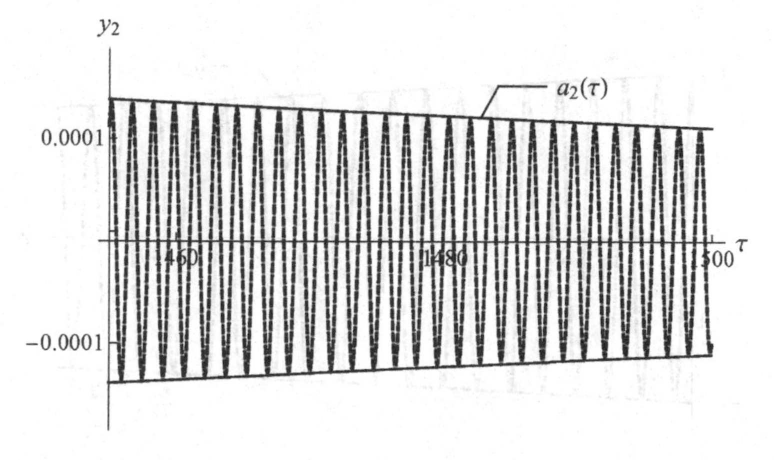

FIGURE 9.23 The graph of function $y_{2a}(\tau)$ at the simulation end for $\mu = 0$ (dotted line corresponds to the numerical solution).

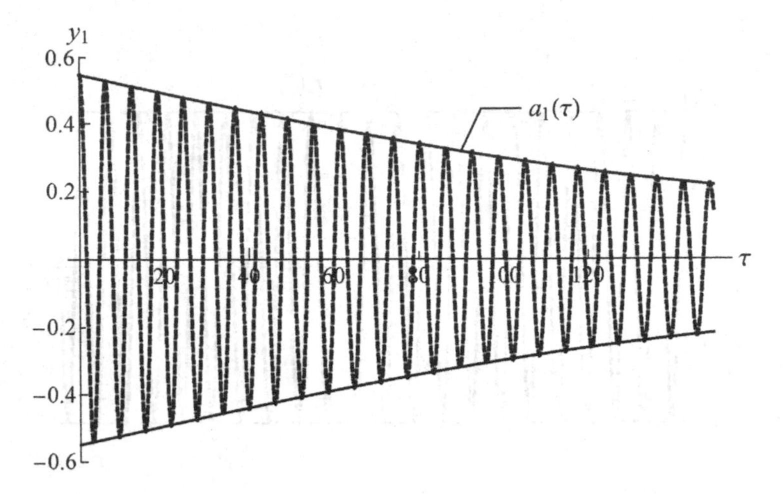

FIGURE 9.24 The graph of function $y_{1a}(\tau)$ at the simulation beginning for $\mu = 5 \cdot 10^{-18}$ (dotted line corresponds to the numerical solution).

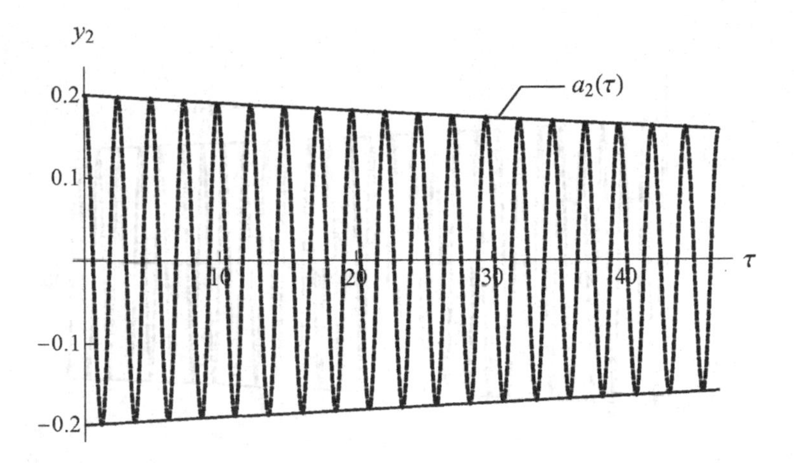

FIGURE 9.25 The graph of function $y_{2a}(\tau)$ at the simulation beginning for $\mu = 5 \cdot 10^{-18}$ (dotted line corresponds to the numerical solution).

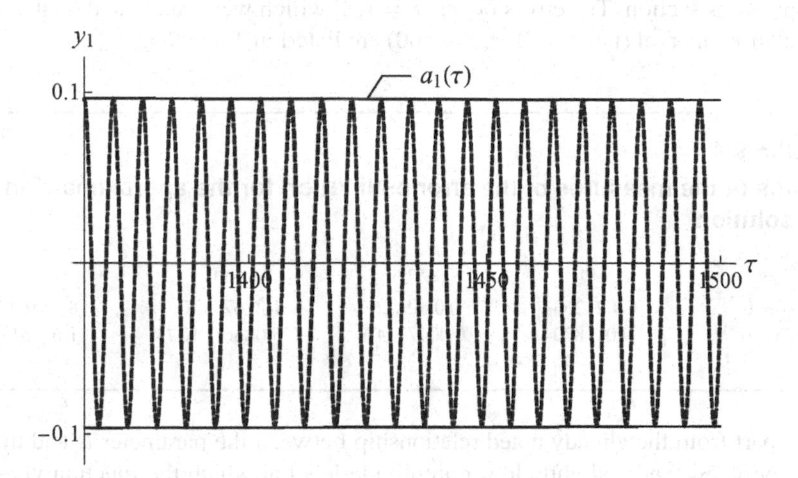

FIGURE 9.26 The graph of function $y_{1a}(\tau)$ at the simulation end for $\mu = 5 \cdot 10^{-18}$ (dotted line corresponds to the numerical solution).

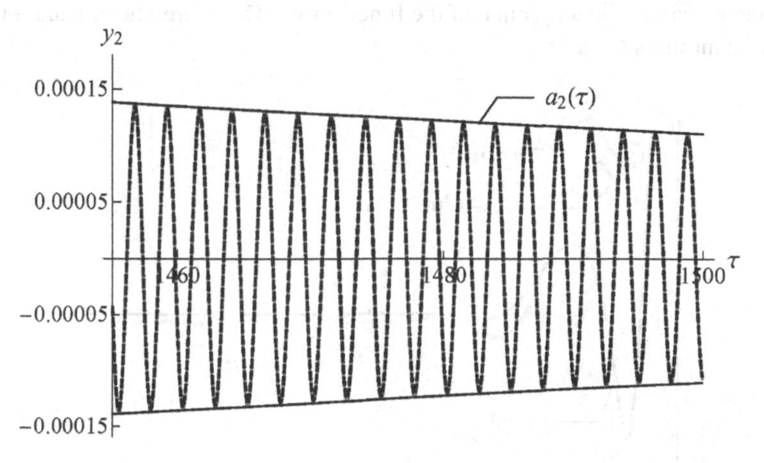

FIGURE 9.27 The graph of function $y_{2a}(\tau)$ at the simulation end for $\mu = 5 \cdot 10^{-18}$ (dotted line corresponds to the numerical solution).

A high compliance between the approximate analytical solutions and the numerical ones was confirmed by the error estimation using both measures introduced in the previous section. The errors δ_k, e_k, $k = 1,2$, which were calculated on the whole simulation interval (i.e. $\tau_s = 0$, $\tau_e = 1500$) are listed in Table 9.4.

TABLE 9.4
Values of the measures of the error estimation for the approximate analytical solution

μ	δ_1	δ_2	e_1	e_2
$\mu = 0$	0.0212664	0.0165126	$1.27679 \cdot 10^{-4}$	$8.28589 \cdot 10^{-5}$
$\mu = 5 \cdot 10^{-18}$	0.0118704	0.00477147	$5.08364 \cdot 10^{-5}$	$1.65351 \cdot 10^{-5}$

Apart from the already noted relationship between the parameter μ and the frequency of y_{2a}, and a slightly lower amplitude level at which the function y_{1a} stabilizes, there are no significant differences in the course of the solutions at the main resonance for both models considered. However, some differences are observable by studying the solutions of modulation equations. The graphs of the amplitude $a_1(\tau)$ for a few values of the nonlocal parameter μ on a sufficiently long interval are presented in Figure 9.28. Three values of μ have been taken into account. The biggest differences concern the transition in vibrations where one can note that higher value of the nonlocal parameter μ, leads to softer transit into the steady state.

The coupling of elastic nature as a weak nonlinear term does not cause any significant growth of the amplitude of the function y_2. Therefore, the attenuation effect is dominant in its behavior.

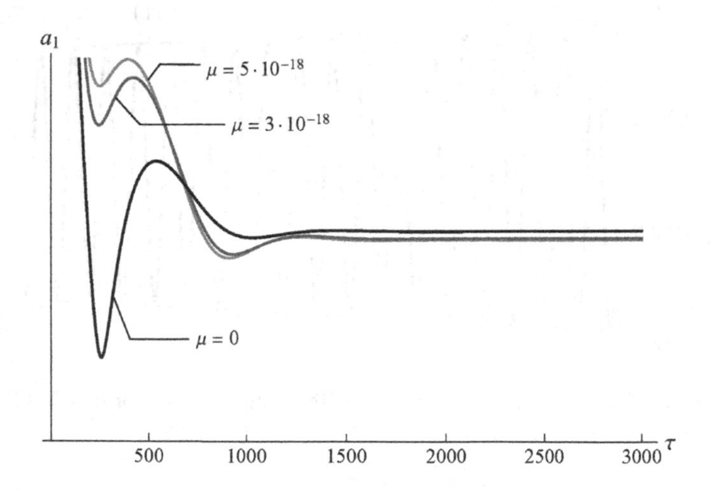

FIGURE 9.28 The slow modulation of the amplitude a_1 for different values of μ.

9.6.1 STEADY STATE RESONANT RESPONSES

Analyzing the modulation equations, i.e. equations (9.59), one can note the amplitude a_2 is affected only by the damping. The general solution for the amplitude a_2 has an exponential form $C_1 \exp(-\frac{\delta}{2}\tau)$, from where it implies that $\lim_{\tau\to\infty} a_2(\tau) = 0$. Therefore, studying the steady states we focus only on first two relations of equations (9.59) and neglect the term $\frac{1}{4}\xi_1 a_2^2$ which allows one to write the modulation equations in the form

$$\begin{aligned}\frac{da_1}{d\tau} &= -\frac{1}{2}\left(\delta a_1 + \sigma \sin\left(s\tau - \psi_1\right)\right), \\ \frac{d\psi_1}{d\tau} &= \frac{3}{8}\xi_2 a_1^2 + \frac{\sigma}{2a_1}\cos\left(s\tau - \psi_1\right),\end{aligned} \tag{9.62}$$

and is adequate for examining the stationary responses. Moreover, it is convenient to introduce into the considerations the modified phase such that

$$\theta_1(\tau) = s\tau - \psi_1(\tau). \tag{9.63}$$

Substituting definition (9.63) into equations (9.62) yields

$$\begin{aligned}\frac{da_1}{d\tau} &= -\frac{1}{2}\left(\delta a_1 + \sigma \sin\theta_1\right), \\ s - \frac{d\theta_1}{d\tau} &= \frac{3}{8}\xi_2 a_1^2 + \frac{\sigma}{2a_1}\cos\theta_1.\end{aligned} \tag{9.64}$$

As expected regarding behavior of the function y_1 at the steady state, we postulate the time derivatives of the amplitude and the modified phase equal to zero which leads to the following equations

$$\begin{aligned}&-\frac{1}{2}\left(\delta a_1 + \sigma \sin\theta_1\right) = 0, \\ &\frac{3}{8}\xi_2 a_1^2 + \frac{\sigma}{2a_1}\cos\theta_1 - s = 0.\end{aligned} \tag{9.65}$$

Thus, the search for periodic steady states reduces to solving algebraic (9.65). The trigonometric identity can be used to eliminating the modified phase from equations (9.65) which allows for expressing the dependence between the amplitude and the force frequency as an algebraic equation of the form

$$\frac{\delta^2 a_1^2}{\sigma^2} + \frac{\left(3\xi_2 a_1^2 - 8s\right)^2}{16\sigma^2} a_1^2 - 1 = 0. \tag{9.66}$$

The polynomial occurring on the left sides of equations (9.66) is of degree six, while with respect to the square of a_1 it is the third degree polynomial. Hence, at most three, significantly different real solutions can be expected.

In order to find the amplitude and phase values corresponding to the steady state vibration for given detuning parameter s we solve equations (9.65) in the numerical

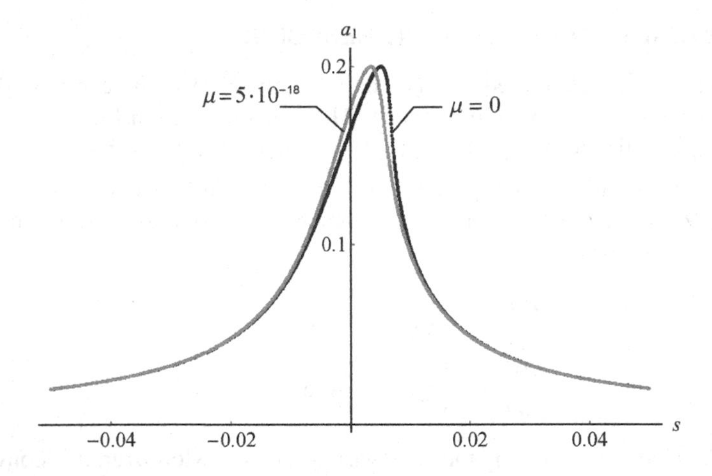

FIGURE 9.29 The amplitude a_1 versus the detuning parameter s.

manner with use of the standard procedure *NSolve* of the *Mathematica*. The steady state amplitude and phase are denoted further using symbols $\tilde{a}_1$ and $\tilde{\theta}_1$, respectively. Stability study of each steady state is performed using the standard approach based on the expansion of the functions on left sides of equations (9.65) in the Taylor series around this fixed point $(\tilde{a}_1, \tilde{\theta}_1)$. Then the eigenvalues of the matrix

$$A = \begin{bmatrix} -\frac{\delta}{2} & -\frac{\sigma}{2}\cos\tilde{\theta}_1 \\ -\frac{3\,\xi_2}{4}\tilde{a}_1 + \frac{\sigma\cos\tilde{\theta}_1}{2\tilde{a}_1^2} & \frac{\sigma\sin\tilde{\theta}_1}{2\tilde{a}_1} \end{bmatrix} \tag{9.67}$$

are analyzed in searching of the ones that are negative.

Continuing the examples described in the previous sections we compare the resonance response curves corresponding to the model when $\mu = 0$ and to the second one when $\mu = 5\cdot 10^{-18}$. The damping coefficient and the external force amplitude remain the same, i.e. $\delta = 0.01$, $\sigma = -0.002$. For these values of the parameters, both resonance response curves are unambiguous. The stability analysis has confirmed that each point is stable. In Figures 9.29–9.30 the resonance responses of the system are presented as the standard dependence between the amplitude and the detuning parameter, and also as the dependence between the modified phase and the detuning parameter. The curves differ slightly, however one can comment on the difference as stiffening the amplitude response curve, and softer changes of the phase in the resonance zone when $\mu > 0$.

How the amplitude σ of the harmonic force affects the resonance response of the system is shown in Figures 9.31–9.32. The graphs concern the nanoplate model with the nonlocal parameter $\mu = 5\cdot 10^{-18}$ and the damping coefficient $\delta = 0.01$. The points representing the stable steady states are drawn using dots of larger size. The nonlinear nature of the curves becomes more visible as the magnitude of the

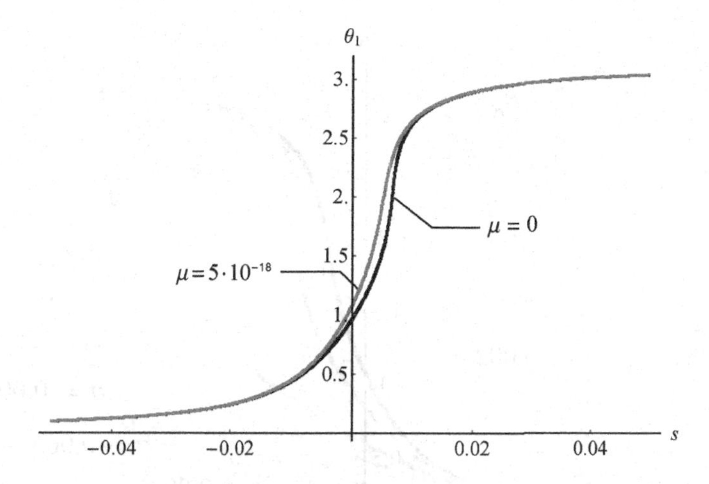

FIGURE 9.30 The modified phase θ_1 versus the detuning parameter s.

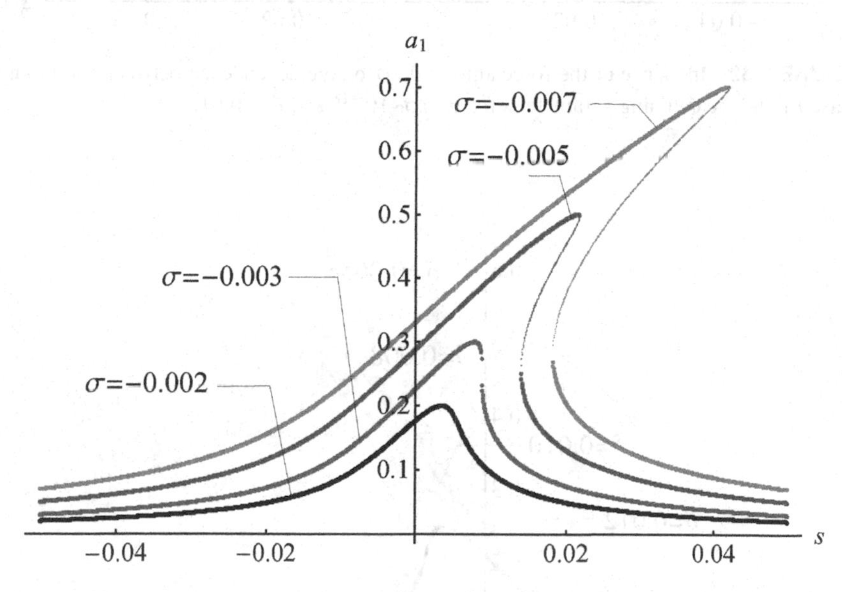

FIGURE 9.31 Influence of the force amplitude σ on the dependence between the amplitude a_1 and the detuning parameter s for $\mu = 5 \cdot 10^{-18}$ and $\delta = 0.01$.

external force σ increases in the negative direction. At bigger values of the excitation magnitude, the curves become ambiguous and the unstable branches occur.

The impact of the damping on the shape and the stability of the resonance response curves is shown in Figures 9.33–9.34. The amplitude of the harmonic force of magnitude $\sigma = -0.003$ is same for each curve which represents the given nanoplate. As in the previous cases, the model with the nonlocal parameter $\mu = 5 \cdot 10^{-18}$ is applied. The nanoplates whose damping features are sufficiently

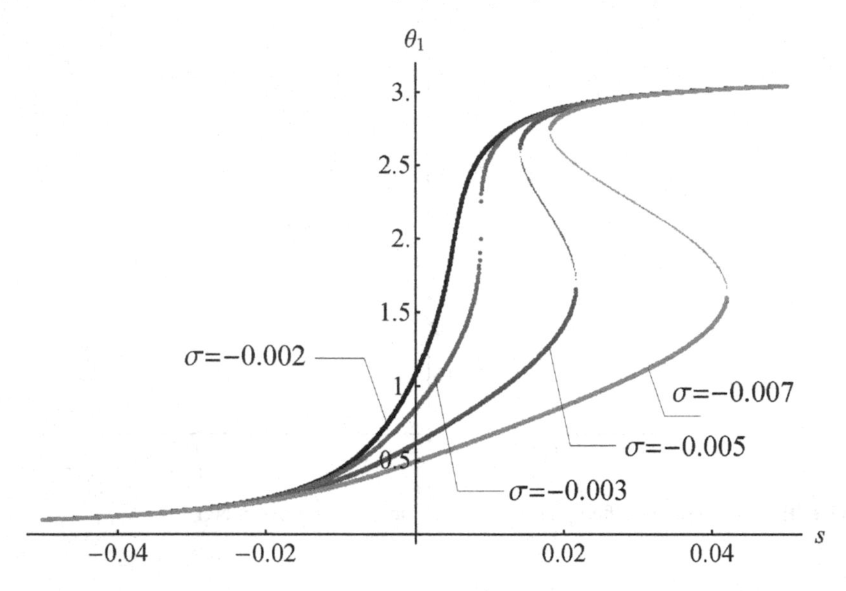

FIGURE 9.32 Influence of the force amplitude σ on the dependence between the modified phase θ_1 and the detuning parameter s for $\mu = 5 \cdot 10^{-18}$ and $\delta = 0.01$.

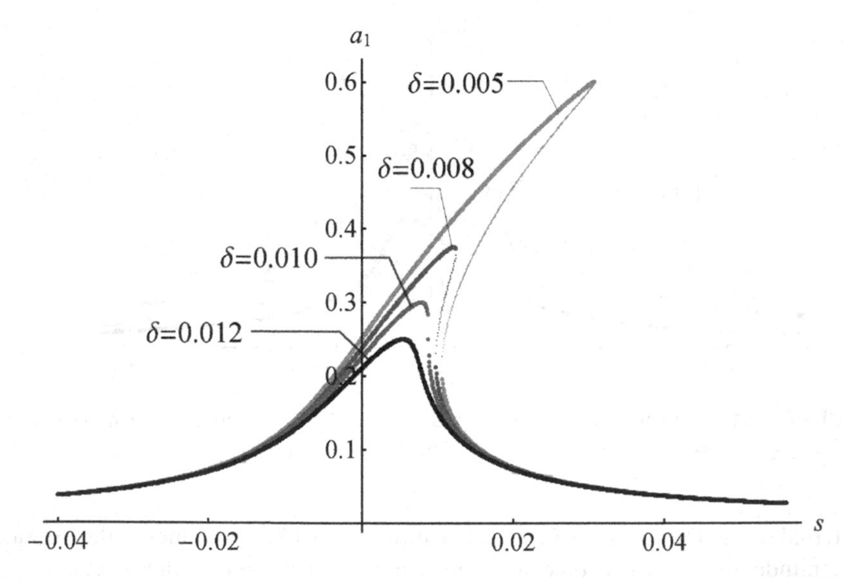

FIGURE 9.33 Influence of the damping coefficient δ on the dependence between the amplitude a_1 and the detuning parameter s under the same load $\sigma = -0.003$ for $\mu = 5 \cdot 10^{-18}$.

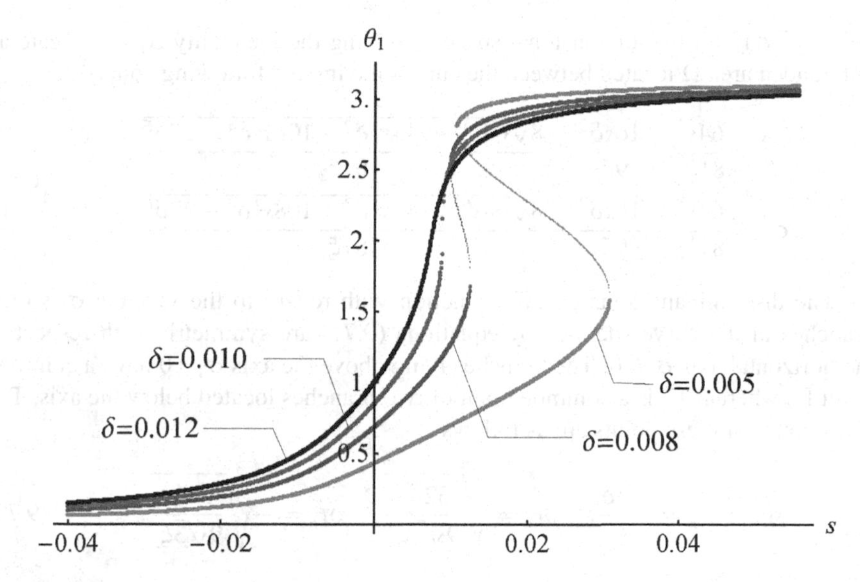

FIGURE 9.34 Influence of the damping coefficient δ on the dependence between the modified phase θ_1 and the detuning parameter s under the same load $\sigma = -0.003$ for $\mu = 5 \cdot 10^{-18}$.

large give unambiguous resonance responses, and achieve significantly smaller amplitudes. More so, the changes in the modified phase are less suddenly in the resonace zone.

9.6.2 AMBIGUOUS RESONANCE AREAS

Substituting a new variable

$$r = a_1^2 \tag{9.68}$$

into equation (9.66) gives the following third degree equation

$$\frac{\delta^2 r}{\sigma^2} + \frac{(3\xi_2 r - 8s)^2}{16\sigma^2} r - 1 = 0. \tag{9.69}$$

The number of real solutions of the cubic equation with real coefficients depends on its discriminant which for equation (9.69) is expressed as

$$\Delta_1 = \frac{64\left(\left(64s^3 + 144\delta^2 s - 81\xi_2\sigma^2\right)^2 + 64\left(3\delta^2 - 4s^2\right)^3\right)}{9^6 \xi_2^2}. \tag{9.70}$$

The ambiguity of the resonance response is expected when three distinct real roots occur which is possible if only if $\Delta_1 < 0$. For a given nanoplate, the quantities ξ_2 and δ have the status of the fixed parameters so the discriminant can be treated as the function of two variables s and σ associated with the external force. The points (s,σ) of a two-dimensional space satisfying the inequality $\Delta_1 < 0$ create an unbounded area Ω located between the curves having the following equations

$$\sigma_u^2 = \frac{64s^3}{81\xi_2} + \frac{16s\delta^2}{9\xi_2} + \frac{8\sqrt{64s^6 - 144s^4\delta^2 + 108s^2\delta^4 - 27\delta^6}}{81\xi_2},$$
$$\sigma_l^2 = \frac{64s^3}{81\xi_2} + \frac{16s\delta^2}{9\xi_2} - \frac{8\sqrt{64s^6 - 144s^4\delta^2 + 108s^2\delta^4 - 27\delta^6}}{81\xi_2}. \tag{9.71}$$

The discriminant Δ_1 is an even function with regard to the variable σ, so the branches of the curves defined by equations (9.71) are symmetric with respect to the horizontal axis $\sigma = 0$. The branches lying above the axis $\sigma = 0$ have a common point Γ_p whereas Γ_n is a common point of two branches located below the axis. The coordinates of these points are as follows

$$s_{\Gamma_p} = s_{\Gamma_n} = \frac{\sqrt{3}\delta}{2}, \quad \sigma_{\Gamma_p} = \sqrt{\frac{32\delta^3}{9\sqrt{3}\xi_2}}, \quad \sigma_{\Gamma_n} = -\sqrt{\frac{32\delta^3}{9\sqrt{3}\xi_2}}. \tag{9.72}$$

At the point Γ_p the branches have a common tangent equation which is

$$\sigma = \sqrt{2\frac{\delta}{\xi_2}}\left(\frac{2}{\sqrt[4]{27}}s + \frac{\delta}{\sqrt[4]{243}}\right). \tag{9.73}$$

The common tangent at the point Γ_n of the branches lying below the horizontal axis is

$$\sigma = -\sqrt{2\frac{\delta}{\xi_2}}\left(\frac{2}{\sqrt[4]{27}}s + \frac{\delta}{\sqrt[4]{243}}\right). \tag{9.74}$$

The region Ω with points which satisfy the inequality $\Delta_1 < 0$ consists of two disjoined subregions Ω_p and Ω_n. The curves σ_u, σ_l and bounded by the subregions Ω_p, Ω_n together with the points Γ_p and Γ_n, and also by the common tangents defined by equations (9.73)–(9.74) as depicted in Figure 9.35 for the case with the dimensionless damping coefficient $\delta = 0.01$ and $\mu = 5 \cdot 10^{-18}$.

Both the curves σ_u, σ_l and the ordinate of the points Γ_p, Γ_n are affected by the damping coefficient δ while the parameter ξ_2 depended in turns on the nonlocal parameter μ. When there is no damping, the points $\Gamma_p = \Gamma_n = \Gamma$ are located at the origin $(0,0)$ while equations (9.71) take the form

$$\sigma_u^2 = \frac{128s^3}{81\xi_2}, \quad \sigma_l^2 = 0. \tag{9.75}$$

The area of ambiguous resonance responses corresponding to the nondamping case is presented in Figure 9.36 for the nonlocal parameter $\mu = 5 \cdot 10^{-18}$. The branch $\sigma_l = 0$ is a common line bounded both parts Ω_p and Ω_n of the whole region Ω. It is also the common tangent of four branches of the curves bounding the subregions. So, the damping occurrence causes four distinct branches of the curves to exist. Two

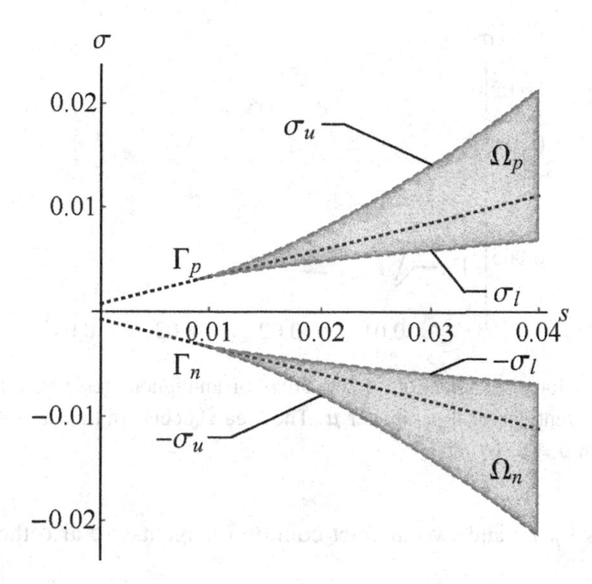

FIGURE 9.35 Area $\Omega = \Omega_p \cup \Omega_n$ in which the resonance response curves are ambiguous for $\delta = 0.01$ and $\mu = 5 \cdot 10^{-18}$.

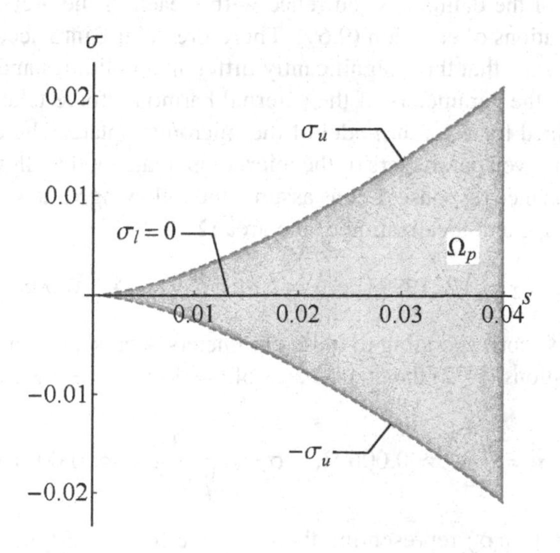

FIGURE 9.36 Area in which the resonance response curves are ambiguous for $\delta = 0$ and $\mu = 5 \cdot 10^{-18}$.

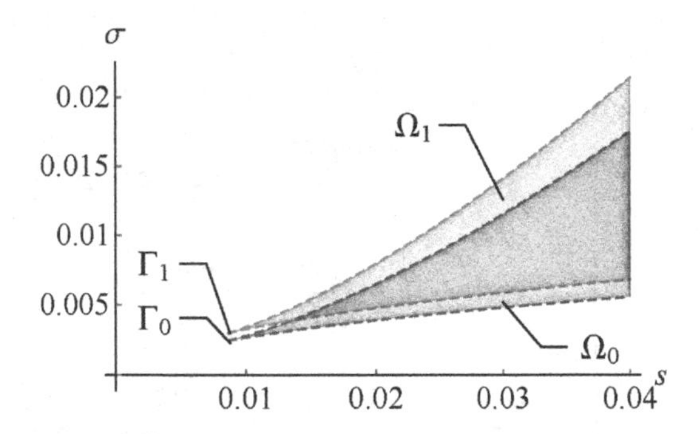

FIGURE 9.37 Mutual localization of two areas of ambiguous resonance responses corresponding to different nonlocal parameter μ. The area Ω_0 concern the case with $\mu = 0$ while Ω_1 the case with $\mu = 5 \cdot 10^{-18}$.

distinct points Γ_p, Γ_n and two distinct common tangents are also the results of the damping.

The nonlocal parameter μ indirectly through the quantity ξ_2, affects the ordinate of the points Γ_p, Γ_n and also the shape af the curves bounding the subregions of the area Ω. The influence is illustrated in Figure 9.37 in which two areas of ambiguous resonance responses are shown together. Due to the symmetry aforementioned, only one part of each of the areas is shown. The subregion denoted by Ω_0 and surrounded by a dark line corresponds to the model with $\mu = 0$. The second one denoted by Ω_1 and surrounded by a gray line relates to the model with $\mu = 5 \cdot 10^{-18}$. The same value 0.01 of the damping coefficient is assumed in both cases.

Regardless of the damping occurrence within each of the areas, there are three distinct real solutions of equation (9.69). Therefore, taking into account relationship (9.68), one can state that three significantly different amplitudes and phases are possible only when the parameters of the external harmonic force take values from the area Ω determined for a given model of the micro/nanoplate. The determination of the region Ω for given parameters of the micro/nanoplate model allows one to predict the kind of resonance response. Let us assume the following values of the parameters affecting the shape and localization of the area Ω

$$\xi_2 = 0.227999 \left(\text{i.e.}\mu = 5 \cdot 10^{-18}\right), \qquad \delta = 0.008.$$

The region Ω corresponding to these parameters is presented in Figure 9.38. According to equations (9.72) the coordinates of the point $\Gamma = \Gamma_p$ are as follows

$$s_\Gamma = \frac{\sqrt{3}\delta}{2} \approx 0.00693, \quad \sigma_\Gamma = \sqrt{\frac{32\delta^3}{9\sqrt{3}\xi_2}} \approx 0.00215.$$

For any point (s, σ) representing the harmonic force and lying outside the area Ω, where the discriminant $\Delta_1 > 0$, equation (9.69) has only one real solution. Any external harmonic force with the amplitude whose value is less than the ordinate of

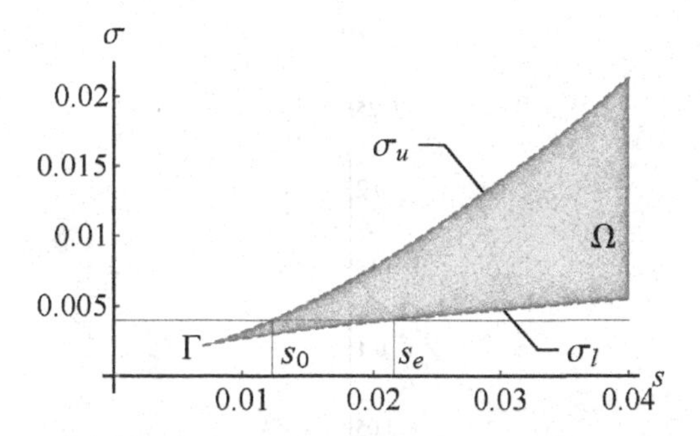

FIGURE 9.38 Application of the graph of the region Ω in prediction of the resonance response curves ambiguity; $\mu = 5 \cdot 10^{-18}$, $\delta = 0.008$.

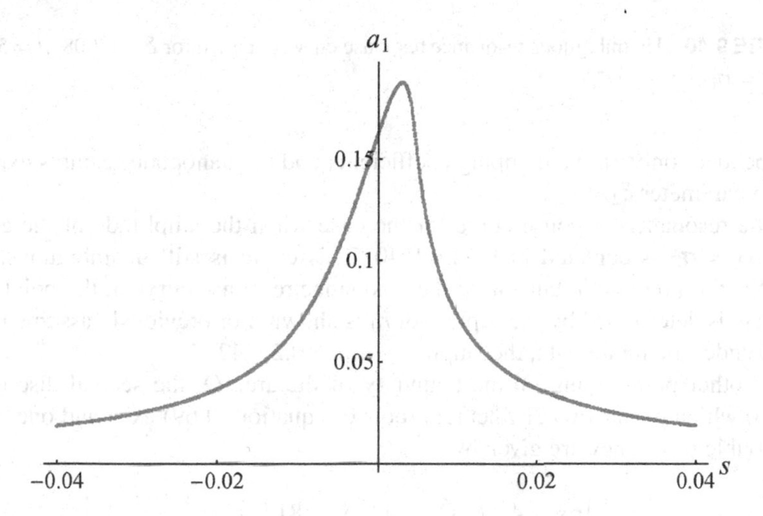

FIGURE 9.39 Unambiguous resonance response curve obtained for $\delta = 0.008$, $\mu = 5 \cdot 10^{-18}$ and $\sigma = 0.0015 < \sigma_\Gamma$.

the point Γ results in the unambiguous resonance response curve which is confirmed in Figure 9.39.

At the boundaries satisfying equations (9.71) and bounding the area Ω, the discriminant $\Delta_1 = 0$ and does not determine unambiguously, the number of the distinct real roots of equation (9.69). The second discriminant of equation (9.69) given by

$$\Delta_2 = \frac{8\left(64s^3 + 144s\delta^2 - 81\xi 2\sigma^2\right)}{729\xi_2^3} \tag{9.76}$$

equals zero at the point Γ, which ensures the existence of a triple real root of equation (9.69). The triple root defined as follows

$$r_0 = \frac{8\delta}{3\sqrt{3}\xi_2} \tag{9.77}$$

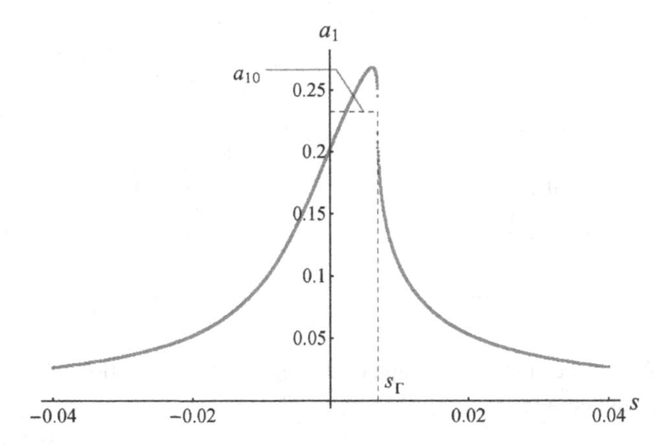

FIGURE 9.40 Unambiguous resonance response curve obtained for $\delta = 0.008$, $\mu = 5 \cdot 10^{-18}$ and $\sigma = \sigma_\Gamma$.

is dependent only on the damping coefficients and the nanoplate features expressed by the parameter ξ_2.

The resonance response curve for the case when the amplitude of the external force $\sigma = \sigma_\Gamma$ is depicted in Figure 9.40. This curve is still unambiguous. Additionally, the line $s = s_\Gamma$ tangent to the resonance response curve at the point whose ordinate is determined by the triple root r_0 is shown. For previously assumed values listed under the figure data, the amplitude $a_{10} \approx 0.23242$.

At other points lying on the boundary of the area Ω, the second discriminant $\Delta_2 \neq 0$ which means two distinct real roots of equation (9.69) exist and one of them is a double root. They are given by

$$
\begin{aligned}
r_1 &= \frac{16s}{9\xi_2} - \frac{4}{9}\left(\frac{64s^3 + 144s\delta^2 - 81\xi 2\sigma^2}{\xi 2^3}\right)^{1/3}, \\
r_2 = r_3 &= \frac{16s}{9\xi_2} + \frac{2}{9}\left(\frac{64s^3 + 144s\delta^2 - 81\xi 2\sigma^2}{\xi 2^3}\right)^{1/3},
\end{aligned}
\tag{9.78}
$$

where s, σ satisfy equations (9.71).

When the amplitude of the harmonic force $\sigma > \sigma_\Gamma$ then the ambiguous branches of the resonance response curves appear. The interval $[s_0,\ s_e]$ of the detuning parameter s on which the ambiguity is observed, and it is determined by the abscissae of the shown Figure 9.38, the points at which the horizontal line representing the value σ crosses the boundries of the region Ω.

Let the amplitude of the external harmonically changing force $\sigma = 0.004$ and the values of ξ_2, μ be the same as previously. Solving equations (9.71) for

$$
\sigma_u = 0.004, \quad \sigma_l = 0.004
$$

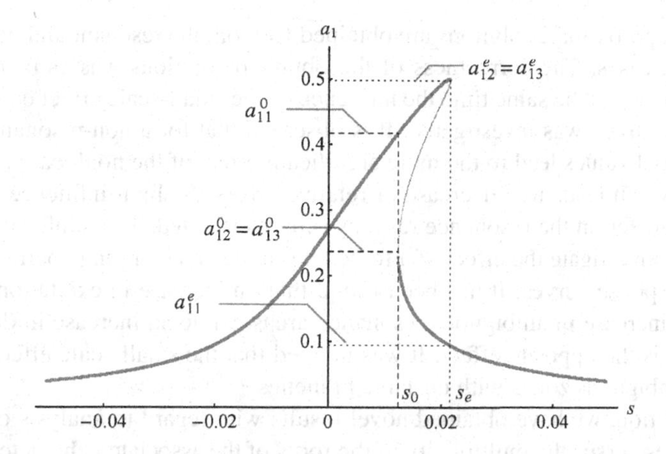

FIGURE 9.41 Ambiguous resonance response curve obtained for $\delta = 0.008$, $\mu = 5 \cdot 10^{-18}$ and $\sigma = 0.004 > \sigma_\Gamma$.

one can find the values $s_0 \approx 0.0122$ and $s_e \approx 0.0216$. The amplitudes calculated employing equations (9.78) and corresponding to these abscissae take values

$$a^0_{11} = \sqrt{r_1} \approx 0.41713, \;\; a^0_{12} = a^0_{13} \approx 0.23681 \;\text{ for }\; s = s_0,$$

$$a^e_{11} = \sqrt{r_1} \approx 0.09442, \;\; a^e_{12} = a^e_{13} \approx 0.49775 \;\text{ for }\; s = s_e.$$

In Figure 9.41, the resonance response curve corresponding to the harmonic force with the amplitude $\sigma = 0.004 > \sigma_\Gamma$ is depicted together with two vertical lines $s = s_0$ and $s = s_e$. The values s_0, s_e determine strictly the interval on which the ambiguous resonance response curves exist. The vertical line $s = s_0$ is tangent to the branch of the resonance response curve at the point representing the double root $a^0_{12} = a^0_{23}$ is indicated, and the line crosses the upper branch at the point the ordinate equals a^0_{11}. The cross point of the line $s = s_e$ and the lower branch corresponds to the root representing the amplitude a^e_{11} . The double root $a^e_{12} = a^e_{23}$ represents the point of tangency.

9.7 CLOSING REMARKS

We have analyzed geometrically nonlinear oscillations of rectangular micro/nano plates subjected to transverse periodic load by employing the von Kármán theory and the Kirchhoff-Love hypotheses. In addition, in order to keep the small-scale-effects, the nonlocal elasticity theory has been adopted. Originality of the proposed method exhibits a combination of a double-mode model, the Bubnov-Galerkin method and the multiple-scale method.

The strong reduction procedure yielded two coupled nonlinear ODEs, which have been further reduced to the system of ODEs with regard to amplitudes and phases based on MSM.

Next, approximate solutions are obtained for both the resonant and non-resonant oscillation cases. The correctness of the obtained solutions was estimated by two types of errors, at the same time the influence of the small-scale effect on the approximation accuracy was investigated. It is observed that for a non-resonant vibration, higher initial values lead to the more significant effect of the nonlocal parameter, an increase which leads to a decrease of relative errors. A slight influence of the nonlocal parameter on the resonance response curves is noted. The nonlocal model was applied to investigate the effect of a force amplitude and damping coefficient on resonance response curves. It has been found that an increase in excitation amplitude yields an increase in ambiguous resonance areas, while an increase in damping coefficient has the opposite effect. It was noticed that the small-scale effect leads to a shift in ambiguous zones with unstable branches.

In addition, we have obtained novel results with regard to analysis of the backbone curves versus the multiplicity of the roots of the associated characteristic equation. For example, three roots of the characteristic equation where two of them are stable allows one to extract the occurrence of the pull-in phenomenon exhibited by the original plate PDEs.

References

1. Abdulle, A., Weinan, E. (2003) Finite difference heterogeneous multi-scale method for homogenization problems. *J. Comput. Phys.* **191**:18–39.

2. Abouhazim, N., Belhaq, M., Lakrad, F. (2005) Three-period quasi-periodic solutions in the self-excited quasi-periodic Mathieu oscillator. *Nonlin. Dyn.* **39**:395–409.

3. Abouhazim, N., Rand, R.H., Belhaq, M. (2006) The damped nonlinear quasiperiodic Mathieu equation near 2:2:1 resonance. *Nonlin. Dyn.* **45**:237–247.

4. Aginsky, Z., Gottlieb, O. (2013) Nonlinear fluid-structure interaction of an elastic panel in an acoustically excited two-dimensional inviscid compressible fluid. *Phys. Fluids* **25**:076104.

5. Aghababaei, R., Reddy, J.N. (2009) Nonlocal third-order shear deformation plate theory with application to bending and vibration of plates. *J. Sound Vib.* **326**(1–2):277–289.

6. Amer, T.S. (2017) The dynamical behaviour of a rigid body relative equilibrium position. *Adv. Math. Phys.* **2017**:8070525.

7. Amer, T., Bek, M. (2009) Chaotic responses of a harmonically excited spring pendulum moving in circular path. *Nonlin. Anal. Real World Appl.* **10**(5):3196-3202.

8. Amer, T.S., Bek, M.A., Hamada, I.S. (2016) On the motion of harmonically excited spring pendulum in elliptic path near resonances. *Adv. Math. Phys.* **2016**:8734360.

9. Analooei, H.R., Azhari, M., Heidarpour, A. (2013) Elastic buckling and vibration analyses of orthotropic nanoplates using nonlocal continuum mechanics and spline finite strip method. *Appl. Math. Model.* **37**(10–11):6703–6717.

10. Andrianov, I.V., Awrejcewicz, J., Danishevskyy, V.V., Ivankov, A.O. (2014) *Asymptotic Methods on the Theory of Plates with Mixed Boundary Conditions*. Wiley: Chinchester.

11. Andrianov, I.V., Awrejcewicz, J., Danishevskyy, J. (2018) *Asymptotical Mechanics of Composites*. Springer: Berlin.

12. Andrzejewski, R., Awrejcewicz, J. (2005) *Nonlinear Dynamics of a Wheeled Vehicle*. Springer: Berlin.

13. Anicin, B.A., Davidovic, D.M., Babovic, V.M. (1993) On the linear theory of the elastic pendulum. *Eur. J. Phys.* **14**:132–135.

14. Anurag, Mondal, B., Bhattacharjee, J.K., Chakraborty, S. (2020) Understanding the order-chaos-order transition in the planar elastic pendulum. *Phys. D* **402**:132256.

15. Anurag, Mondal B., Shah, T., Chakraborty, S. (2021) Chaos and order in liberating quantum planar elastic pendulum. *Nonlin. Dyn.* **103**:2841–2853.

16. Aston, P.J. (1999) Bifurcations of the horizontally forced spherical pendulum. *Comput. Meth. Appl. Mech. Eng.* **170**:343–353.

17. Awrejcewicz, J. (1991) *Bifurcation and Chaos in Coupled Oscillators*. New Jersey: World Scientific.

18. Awrejcewicz, J., Andrianov, I.V., Manevitch, L.I. (1998) *Asymptotic Approach in Nonlinear Dynamics: New Trends and Applications*. Springer-Verlag: Berlin.

19. Awrejcewicz J., Andrianov I.V., Manevitch, L.I. (2004) Asymptotical Mechanics of Thin Walled Structures. Springer-Verlag: Berlin.

20. Awrejcewicz, J., Holicke, M.M. (2007) *Smooth and Non-smooth High Dimensional Chaos and the Melnikov type Methods*. World Scientific: Singapore.

21. Awrejcewicz, J., Krysko, V.A. (2006) *Introduction to Asymptotic Methods*. Chapman and Hall, CRC Press, Taylor and Francis Group: Boca Raton, New York.

22. Awrejcewicz, J., Krysko, V.A., Papkova, I.V., Krysko, A.V. (2016) *Deterministic Chaos in One-Dimensional Continuous Systems*. World Scientific: Singapore.

23. Awrejcewicz, J., Krysko, V.A. (2020) *Elastic and Thermoelastic Problems in Nonlinear Dynamics of Structural Members*. Springer: Berlin.

24. Awrejcewicz, J., Krysko, A.V., Zhigalov, M.V., Krysko, V.A. (2021) *Mathematical Modelling and Numerical Analysis of Size-Dependent Structural Members in Temperature Fields: Regular and Chaotic Dynamics of Micro/Nano Beams, Plates and Shells*. Springer: Berlin.

25. Awrejcewicz, J., Krysko-jr., V.A., Kalutsky, L.A., Zhigalov, M.V., Krysko, V.A. (2021) Review of the methods of transition from partial to ordinary differential equations: from macro- to nano-structural dynamics. *Arch. Comput. Meth. Eng.* (doi.org/10.1007/s11831-021-09550-5).

26. Awrejcewicz, J., Starosta, R. (2010) Resonances in kinematically driven nonlinear system – asymptotic analysis. *Math. Eng Sci. Aero.* **1**(1):1–10.

27. Awrejcewicz, J., Starosta, R. (2012) Internal motion of the complex oscillators near main resonance. *Theor. Appl. Mech. Lett.* **2**:043002.

28. Awrejcewicz, J., Starosta, R., Sypniewska-Kamińska, G. (2013) Asymptotic analysis of resonances in nonlinear vibrations of the 3-dof pendulum. *Diff. Eq. Dyn. Sys.* **21**:123–140.

29. Awrejcewicz, J., Starosta, R. Sypniewska-Kamińska, G. (2014) Decomposition of the equations of motion in the analysis of dynamics of a 3-DOF nonideal system. *Math. Prob. Eng.* **2014**:816840.

30. Awrejcewicz, J., Starosta, R. Sypniewska-Kamińska G. (2015) Decomposition of governing equations in the analysis of resonant response of a nonlinear and non-ideal vibrating system. *Nonlin. Dyn.* **82**:299–309.

31. Awrejcewicz, J., Starosta, R., Sypniewska-Kamińska, G. (2016) Stationary and transient resonant response of a spring pendulum. *IUTAM Symp. Anal. Meth. Nonlin. Dyn.* **19**:201–208.

32. Awrejcewicz, J., Starosta, R., Sypniewska-Kamińska, G. (2019) Complexity of resonances exhibited by a nonlinear micromechanical gyroscope: an analytical study. *Nonlin. Dyn.* **97**:1819–1836.

33. Axinti, A., Axinti, G. (2006) About kinematic excitation induced of the dislevelments bed bearer to the wheel of self-propelled equipments. *Rom. J. Acoust. Vib.* **2**(3):79–82.

34. Bakri, T., Meijer, H.G.E., Verhulst, F. (2009) Emergence and bifurcations of Lyapunov manifolds in nonlinear wave equations. *J. Nonlin. Sci.* **19**:571–596.

35. Bayly, P.V., Virgin, L.N. (1993) An empirical study of the stability of periodic motion in the forced spring-pendulum. *Proc. Math. Phys. Sci.* **443**(1918):391–408.

36. Belhaq, M., Guennoun, K., Houssini, M. (2002) Asymptotic solutions for a damped non-linear quasi-periodic Mathieu equation. *Int. J. Non-Lin. Mech.* **37**:445–460.

37. Belhaq, M., Hamdi, M. (2016) Energy harvesting from quasi-periodic vibrations. *Nonlin. Dyn.* **86**:2193–2205.

38. Belhaq, M., Houssini, M. (1999) Quasi-periodic oscillations, chaos and suppression of chaos in a nonlinear oscillator driven by parametric and external excitations. *Nonlin. Dyn.* **18**:1–24.

39. Belhaq, M., Lakrad, F. (2000) The elliptic multiple scales method for a class of autonomous strongly non-linear oscillators. *J. Sound Vib.* **234**(3):547–553.

40. Benedettini, F., Rega, G. (1987) Non-linear dynamics of an elastic cable under planar excitation. *Int. J, Non-Lin. Mech.* **22**(6):497-509.

41. Bi, L., Rao, Y., Tao, Q., Dong, J., Su, T., Liu, F., Qiam, W. (2013) Fabrication of large-scale gold nanoplate films as highly active SERS substrates for label-free DNA detection. *Biosens. Bioelectron.* **43**(1):193–199.

42. Bishop, R.E.D., Gladwell, G.M.L., Michaelson, S. (2008) *The matrix Analysis of Vibration*. Cambridge University Press.

43. Bouche, D., Molinet, F., Mittra, R. (2011) *Asymptotic Methods in Electromagnetics*. Springer: Berlin.

44. Braghin, F., Resta, F., Leo, E., Spinola, G. (2007) Nonlinear dynamics of vibrating MEMS. *Sens. Actuat. A* **134**:98–108.

45. Breitenberger, E., Muller, R.D. (1981) The elastic pendulum: a nonlinear paradigm. *J. Math. Phys.* **22**:1196.

46. Bu, I.Y.Y., Yang, C.C. (2012) High-performance ZnO nanoflake moisture sensor. *Superlat. Microstruct.* **51**(6):745–753.

47. Bush, A.W. (1994) *Perturbation Methods for Engineers and Scientists*. CRC Press.

48. Cacan, M.R., Leadenham, S., Leamy, M.J. (2014) An enriched multiple scales method for harmonically forced nonlinear systems. *Nonlin. Dyn.* **78**:1205–1220.

49. Caruntu, D.I., Oyervides, R. (2017) Frequency response reduced order model of primary resonance of electrostatically actuated MEMS circular plate resonators. *Comm. Nonlin. Sci. Num. Simul.* **43**:261–270.

50. Cayton, Th.E. (1977) The laboratory spring-mass oscillator: an example of parametric instability. *Am. J. Phys.* **45**:723–732.

51. Chipot, M.M. (2016) *Asymptotic Issues for Some Partial Differential Equations*. Imperial College Press.

52. Chong, A.C.M., Yang, F., Lam, D.C.C., Tong, P. (2001) Torsion and bending of micron-scaled structures. *J. Mater. Res.* **16**(4):1052–1058.

53. Christensen, J. (2004) An improved calculation of the mass for the resonant spring pendulum. *Am. J. Phys.* **72**:818.

54. Clementi, F., Lenci, S., Rega, G. (2020) 1:1 internal resonance in a two d.o.f. complete system: a comprehensive analysis and its possible exploitation for design. *Meccanica* **55**:1309–1332.

55. Cosserat, E., Cosserat, F. (1909) *Theory of Deformable Bodies*. A. Herman Sons: Paris.

56. Cross, R. (2017) Experimental investigation of an elastic pendulum. *Eur. J. Phys.* **38**:065004.

57. Cuerno, R., Ranada, A.F., Ruiz-Lorenzo, J.J. (1992) Deterministic chaos in the elastic pendulum: A simple laboratory for nonlinear dynamics. *Am. J. Phys.* **60**:73.

58. Davidovic, D.M., Anicin, B.A., Babovic, V.M. (1996) The libration limits of the elastic pendulum. *Am. J. Phys.* **64**(3):338–342.

59. De Bruijn, N.G. (1981) *Asymptotic Methods in Analysis*. Dover.

60. Defoort, M., Taheri-Tehrani, P., Nitzan, S., Horsley, D.A. (2016) Synchronization in micromechanical gyroscopes. In: *Solid-State Sensors, Actuators and Microsystems Workshop*, Hilton Head Island, South Carolina June 5–9, pp. 84–88.

61. Duan, W.H., Wang, C.M., Zhang, Y.Y. (2007) Calibration of nonlocal scaling effect parameter for free vibration of carbon nanotubes by molecular dynamics. *J. Appl. Phys.* **101**(2):024305.

62. Duka, B., Duka, R. (2020) On the elastic pendulum, parametric resonance and 'pumping' swings. *Eur. J. Phys.* **40**:025005.

63. El-Dib, Y. (2020) Modified multiple scale technique for the stability of the fractional delayed nonlinear oscillator. *Pramana* **94**:56

64. Elliot, A.J., Cammarano, A., Neild, S.A., Hill, T.L., Wagg, D.J. (2018) Comparing the direct normal form and multiple scales methods through frequency detuning. *Nonlin. Dyn.* **94**:2919–2935.

65. Elliott, S.J., Ghandchi Tehrani, M., Langley, R.S. (2015) Nonlinear damping and quasi-linear modelling. *Phil. Trans. R. Soc. A* **373**: 20140402.

66. Eringen, A.C. (1972) Linear theory of nonlocal elasticity and dispersion of plane waves. *Int. J. Eng. Sci.* **10**(5):425–435.

67. Eringen, A.C. (1983) On differential equations of nonlocal elasticity and solutions of screw dislocation and surface waves. *J. Appl. Phys.* **54**(9):4703–4710.

68. Farajpour, A., Hairi Yazdi, M.R., Rastgoo, A., Loghmani, M., Mohammadi, M. (2016) Nonlocal nonlinear plate model for large amplitude vibration of magneto-electro-elastic nanoplates. *Compos. Struct.* **140**:323–336.

69. Fikioris G., Tastsoglou I., Bakas O.N. (2013) *Selected Asymptotic Methods with Applications to Electromagnetics and Antennas*. Morgan & Claypool.

70. Freundlich, J., Sado, D. (2020) Dynamics of a coupled mechanical system containing a spherical pendulum and a fractional damper. *Meccanica* **55**:2541–2553.

71. Fritzkowski, P., Kamiński, H. (2013) Non-linear dynamics of a hanging rope. *Lat. Am. J. Sol. Struct.* **10**:81–90.

72. Fronk, M.D., Leamy, M.J. (2019) Direction-dependent invariant waveforms and stability in two-dimensional, weakly nonlinear lattices. *J. Sound Vib.* **447**:137–154.

73. Gallacher, B.J., Burdess, J.S., Harish, K.M. (2006) A control scheme for a MEMS electrostatic resonant gyroscope excited using combined parametric excitation and harmonic forcing. *J. Micromech. Microeng.* **16**:320–331.

74. Gendelman, O.V. (2001) Transition of energy to nonlinear localized mode in highly asymmetric system of nonlinear oscillators. *Nonlin. Dyn.* **25**:237–253.

75. Gendelman, O.V. (2004) Bifurcations of nonlinear normal modes of linear oscillator with strongly nonlinear damped attachment. *Nonlin. Dyn.* **37**:115–128.

76. Gendelman, O.V., Vakakis, A.F., Manevitch, L.I., McCloskey, R. (2001) Energy pumping in nonlinear mechanical oscillators. Part I - Dynamics of the underlying hamiltonian system. *ASME J. Appl. Mech.* **68**(1):34–41.

77. Georgiou, J.T. (1999) On the global geometric structure of the dynamics of the elastic pendulum. *Nonlin. Dyn.* **18**:51–68.

78. Ghosh, S., Lee, K., Moorthy, S. (1995) Multiple scale analysis of heterogeneous elastic structures using homogenization theory and Voronoi cell finite element method. *Int. J. Sol. Struct.* **32**(1): 27–62.

79. Gottlieb, O., Cohen, A. (2010) Self-excited oscillations of a string on an elastic foundation subject to a nonlinear feed-forward force. *Int. J. Mech. Sci.* **52**:1535–1545.

80. Guo, T., Rega, G. (2019) Solvability conditions in multi-scale dynamic analysis of one-dimensional structures with non-homogeneous boundaries: a general operator formulation. *Int. J. Non-Lin. Mech.* **115**:68–75.

81. Guo, T., Rega, G. (2020) Direct and discretized perturbations revisited: A new error source interpretation, with application to moving boundary problem. *Eur. J. Mech. Sol.* **81**:103936.

82. Haxton, R.S., Barr, A.D.S. (1972) The autoparametric vibration absorber. *J. Eng. Indust.* 119–225.

83. Hoa, N.D., Van Duy, N., Van Hieu, N. (2013) Crystalline mesoporous tungsten oxide nanoplate monoliths synthesized by directed soft template method for highly sensitive NO2 gas sensor applications. *Mater. Res. Bull.* **48**(2):440–448.

84. Holm, D.D., Lynch, P. (2002) Stepwise precession of the resonant swinging spring. *SIAM J. Appl. Dyn. Sys.* **1**(1):44–64.

85. Hosseini-Pishrobat, M., Keighobadi, J. (2018) Robust output regulation of triaxial MEMS gyroscope via nonlinear active disturbance rejection. *Int. J. Rob. Nonlin. Contr.* **28**:2–38.

86. Hu, Z., Gallacher, B.J. (2019) Effects of nonlinearity on the angular drift error of an electrostatic MEMS rate integrating gyroscope. *IEEE Sens. J.* **19**(22):10271–10280.

87. Hough, M.E. (1988) A new integration theory for conservative two degree-of-freedom systems. *Celest. Mech.* **44**:1–29.

88. Ismail, A.I. (2020) New vertically planed pendulum motion. *Math. Prob. Eng.* **2020**:8861738.

89. Jacobsen, P.K. (2016) Introduction to the method of multiple scales. *arXiv*.

90. Janowicz, M. (2003) Method of multiple scales in quantum optics. *Phys. Rep.* **375**:327–410.

91. Jin, L., Zhang, H., Zhong, Z. (2011) Design of a lc-tuned magnetically suspended rotating gyroscope. *J. Appl. Phys.* **109**:07E525–07E525–3.

92. Jomehzadeh, E., Noori, H.R., Saidi, A.R. (2011) The size-dependent vibration analysis of micro-plates based on a modified couple stress theory. *Phys. E Low-Dimen. Syst. Nanostruct.* **43**(4):877–883.

93. Kacem, N., Hentz, S., Pinto, D., Reig, B., Nguyen, V. (2009) Nonlinear dynamics of nanomechanical beam resonators: improving the performance of NEMS-based sensors. *Nanotechnology* **20**, 275501.

94. Kalashnikov, V., Sergeyev, S.V., Jacobsen, G., Popov, S., Turitsyn, S.K. (2016) Multi-scale polarisation phenomena. *Light: Sci. Appl.* **5**:e16011.

95. Kecik, K., Warmiński, J. (2011) Dynamics of an autoparametric pendulum-like system with a nonlinear semiactive suspension. *Math. Prob. Eng.* **2011**:451047.

96. Kerschen, G., Lee, Y.S., Vakakis, A.F., McFarland, D.M., Bergman, L.A. (2006) Irreversible passive energy transfer in coupled oscillators with essential nonlinearity. *SIAM J. Appl. Math.* **66**:648–679.

97. Kervorkian, J.K., Cole, J.D. (1981) *Perturbation Methods in Applied Mathematics*. Springer: Berlin.

98. Kervorkian, J.K., Cole, J.D. (1996) *Multiple Scale and Singular Perturbation Method, AMS*. Springer: Berlin.

99. Kirillov, O.N., Verhulst, F. (2010) Paradoxes of dissipation-induced destabilization or who opened Whitney's umbrella? *ZAMM* **90**(6):462–488.

100. Koiter, W.T. (1964) Couples-stress in the theory of elasticity. *Proc. K. Ned. Akad. Wet* **67**:17–44.

101. Kovacic, I., Brennan, M.J. (2011) *The Duffing Equation*. Willey.

102. Kovaleva, M., Manevitch, L., Romeo, F. (2019) Stationary and non-stationary oscillatory dynamics of the parametric pendulum. *Comm. Nonlin. Sci. Num. Simul.* **76**:1–11.

103. Kovaleva, M.A., Smirnov, V.V., Manevitch, L.I. (2018) The nonlinear model of the librational dynamics of the paraffin crystal. *Mat. Phys. Mech.* **35**:80–86.

104. Kovaleva, M., Smirnov, V., Manevitch, L. (2019) Nonstationary dynamics of the sine lattice consisting of three pendula (trimer). *Phys. Rev. E* **99**:012209.

105. Kozlowski, J., Parlitz, U., Lauterborn, U. (1995) Bifurcation analysis of two coupled periodically driven Duffing oscillators. *Phys. Rev. E* **51**:1861–1867.

106. Kozmidis-Luburić, U., Marinković, M., Ćirić, D., Tosić, B. (1990) Undamped kinematical excitations in molecular chains. *Phys. A* **2**(163):483–490.

107. Kramer, P., Khan, A., Stathos, P., Lee de Ville, R.E. (2007) Method of multiple scales with three time scales. *Proc. Appl. Math. Mech.*

108. Krasnopolskaya, T.S., Shvets, A.Y. (1992) Chaotic oscillations of a spherical pendulum as an example of interaction with energy source. *Int. Appl. Mech.* **28**:669–674.

109. Král, L., Straka, O. (2017) Nonlinear estimator design for MEMS gyroscope with time-varying angular rate. *IFAC Papers OnLine* **50**(1):3195–3201.

110. Kriven, W.M., Kwak, S.Y., Wallig, M.A., Choy, J.H. (2004) Bio-resorbable nanoceramics for gene and drug delivery. *MRS Bull.* **29**(1):33–37.

111. Kyziol, J., Okniński, A. (2009) Effective equation for two coupled nonlinear oscilators: general case. *10th Conferrence on Dynamical Systems – Theory and Applications*. Łódź, pp. 295–300.

112. Lacarbonara, W., Rega, G., Nayfeh, A.H. (2003) Resonant non-linear normal modes. Part I: analytical treatment for structural one-dimensional systems. *Int. J. Non-Lin. Mech.* **38**:851–872.

113. Lai, H.M. (1984) On the recurrence phenomenon of a resonant spring pendulum. *Am. J. Phys.* **52**(3):219–223.

114. Lai, H.Y., Chen, C.K., Yeh, Y.L. (2002) Double-mode modeling of chaotic and bifurcation dynamics for a simply supported rectangular plate in large deflection. *Int. J. Non. Linear. Mech.* **37**(2):331–343.

115. Lakshmanan, M., Rajaseekar, S. (2003) *Nonlinear Dynamics. Integrability, Chaos and Patterns*. Springer: Berlin.

116. Lam, D.C.C., Yang, F., Chong, A.C.M., Wang, J., Tong, P. (2003) Experiments and theory in strain gradient elasticity. *J. Mech. Phys. Sol.* **51**(8):1477–1508.

117. Leamy, M.J., Gottlieb, O. (2000) Internal resonances in whirling strings involving longitudinal dynamics and material non-linearities. *J. Sound Vib.* **236**(4):683–703.

118. Lee, T., Leok, M., McClamroch, N.H. (2010) Computational dynamics of a 3D elastic string pendulum attached to a rigid body and an inertially fixed reel mechanisms. *Nonlin. Dyn.* **64**:97–115.

119. Lenci, S., Clementi, F., Kloda, L., Warmiński, J., Rega, G. (2021) Longitudinal-transversal internal resonances in Timoshenko beams with an axial elastic boundary condition. *Nonlin. Dyn.* **103**:3489–3513.

120. Lestev, A.M. (2015) Combination resonances in MEMs gyro dynamics. *Gyroscopy Navig.* **6**(1):41–44.

121. Lestev, M.A., Tikhonov, A.A. (2009) Nonlinear phenomena in the dynamics of micromechanical gyroscopes. *Vest. St. Petersburg Univ. Math.* **42**(1):53–57.

122. Leung, A.Y.T., Kuang, J.L. (2006) On the chaotic dynamics of a spherical pendulum with harmonically vibrating suspension. *Nonlin. Dyn.* **43**:213–238.

123. Liang, D.-D., Yang, X.-D., Zhang, W., Ren, Y., Yang, T. (2018) Linear, nonlinear dynamic, and sensitivity analysis of a vibratory ring gyroscope. *Theor. Appl. Mech. Lett.* **8**:393–403.

124. Lifshitz, R., Cross, M.C. (2003) Response of parametrically driven nonlinear coupled oscillators with application to micromechanical and nanomechanical resonator arrays. *Phys. Rev. B* **67**:1343021.

125. Litak, G., Margielewicz, J., Gaska, D., Yurchenko, D., Dabek, K. (2020) Dynamic response of the spherical pendulum subjected to horizontal Lissajous excitation. *Nonlin. Dyn.* **102**:2125–2142.

126. Luongo, A. (2017) On the use of the multiple scale method in solving 'difficult' bifurcation problems. *Math. Mech. Sol.* **22**(5):988–1004.

127. Luongo, A., D'Annibale (2013) Double zero bifurcation of non-linear viscoelastic beams under conservative and non-conservative loads. *Int. J. Non-Lin. Mech.* **55**:128–139.

128. Luongo, A., Di Egidio, A. (2006) Divergence, Hopf and double-zero bifurcations of a nonlinear planar beam. *Comp. Struct.* **84**:1596–1605.

129. Luongo, A., Di Egidio, A., Paolone, A. (2004) Multiscale analysis of defective multiple-Hopf bifurcations. *Comp. Struct.* **82**:2705–2722.

130. Luongo, A., Polone, A. (1998) Multiple scale analysis for divergence-Hopf bifurcation of imperfect symmetric systems. *J. Sound Vib.* **218**(3):527–539.

131. Luongo, A., Polone, A. (1998) On the reconstitution problem in the multiple time-scale method. *Nonlin. Dyn.* **19**(2):135-158.

132. Lynch, P. (2002) Resonant motions of the three-dimensional elastic pendulum. *Int. J. Non-Lin. Mech.* **37**:345–367.

133. Mahmoodi, S.N., Khadem, S.E., Rezaee, M. (2004) Analysis of non-linear mode shapes and natural frequencies of continuous damped systems. *J. Sound Vib.* **275**(1–2):283–298.

134. Manevitch, L.I., Gourdon, E., Lamarque, C.H. (2007) Towards the design of an optimal energetic sink in a strongly inhomogeneous two-degree-of-freedom system. *ASME J. Appl. Mech.* **74**(6):1078–1086.

135. Manevitch, E.L., Manevitch, L.I. (2009) Limiting phase trajectories (LPT) in 1 dof asymmetric system with damping and 1:1 resonance. *DSTA 10th Conference*, Łódź 2009, pp. 559–568.

136. Manevitch, L.I., Musienko, A.I. (2009) Limiting phase trajectories and energy exchange between anharmonic oscillator and external force. *Nonlin. Dyn.* **58**:633–642.

137. Martynenko, Yu.G., Mekuriev, I.V., Podalkov, V.V. (2010) Dynamics of a ring micromechanical gyroscope in the forced-oscillation mode. *Gyroscopy Navig.* **1**(1):43–51.

138. Matheny, M.H., Villanueva, L.G., Karabalin, R.B., Sader, J.E., Roukes, M.L. (2013) Nonlinear mode-coupling in nanomechanical systems. *Nano Lett.* **13**:1622–1626.

139. Mazur, O., Awrejcewicz, J. (2020) Nonlinear vibrations of embedded nanoplates under in-plane magnetic field based on nonlocal elasticity theory. *J. Comput. Nonlin. Dyn.* **15**(12):121001.

140. Mekhtiev, M.F. (2021) *Asymptotic Analysis of Spatial Problems in Elasticity*. Springer: Singapore.

141. Mi, L., Gottlieb, O. (2015) Asymptotic model-based estimation of a wake oscillator for a tethered sphere in uniform flow. *J. Fluids Struct.* **54**:361–389.

142. Miles, J. (1962) Stability of forced oscillations of a spherical pendulum. *Quart. Appl. Math.* **20**:21–32.

143. Miles, J. (1984) Resonant motion of a spherical pendulum. *Phys. D* **11**:309–323.

144. Miller, P.D. (2006) *Applied Asymptotic Analysis*. American Mathematical Society: Michigan.

145. Mindlin, R.D., Tiersten, H.F. (1962) Effects of couple-stresses in linear elasticity. *Arch. Ration. Mech. Anal.* **11**(1):415–448.

146. Mora, K., Gottlieb, O. (2017) Parametric excitation of a microbeam-string with asymmetric electrodes: multimode dynamics and the effect of nonlinear damping. *J. Vib. Acoust.* **139**:040903.

147. Murmu, T., McCarthy, M.A., Adhikari, S. (2013) In-plane magnetic field affected transverse vibration of embedded single-layer graphene sheets using equivalent non-local elasticity approach. *Compos. Struct.* **96**:57–63.

148. Murray, J.D. (1984) *Asymptotic Analysis*. Springer: Berlin.

149. Nabholz, U., Curcic, M., Mehner, J.E., Degenfeld-Schonburg, P. (2019) Nonlinear dynamical system model for drive mode amplitude instabilities in MEMS gyroscopes. *IEEE International Symposium on Inertial Sensors and Systems (INERTIAL)*, 1-5 April, Naples, FL, USA.

150. Nayfeh, A.H. (1973) *Perturbation Methods*. John Wiley & Sons.

151. Nayfeh, A.H. (1981) *Introduction to Perturbation Techniques*. John Wiley & Sons.

152. Nayfeh, A.H. (1985) Topical course on nonlinear dynamics. In: *Perturbation Methods in Nonlinear Dynamics*. Societa Italiana di Fisica, Santa Margherita di Pula, Sardinia.

153. Nayfeh, A.H., Nayfeh, S.A. (1994) On nonlinear modes of continuous systems. *J. Vib. Acoust. Trans. ASME* **116**(1):129–136.

154. Nayfeh, A.H., Nayfeh, J.F., Mook, D.T. (1992) On methods for continuous systems with quadratic and cubic nonlinearities. *Nonlin. Dyn.* **3**(2):145–162.

155. Nayfeh, A.H., Mook, D.T. (1995) *Nonlinear Oscillations*. John Wiley & Sons.

156. Nayfeh, A.H., Younis, M.I. (2005) Dynamics of MEMS resonators under superharmonic and subharmonic excitations. *J. Micromech. Microeng.* **15**:1840–1847.

157. Nitzan, S.H., Zega, V., Li, M., Ahn, C.H., Corigliano, A., Kenny, T.W., Horsley, D.A. (2015) Self-induced parametric amplification arising from nonlinear elastic coupling in a micromechanical resonating disk gyroscope. *Sci. Rep.* **5**:9036.

158. Nitzan, S.H., Taheri-Tehrani, P., Defoort, M., Sonmezoglu, S. (2016) Countering the effects of nonlinearity in rate-integrating gyroscopes. *IEEE Sens. J.* **16**(10):3556–3558.

159. Okniński A., Kyziol, J. (2005) Analysis of two coupled periodically driven oscillators via effective equation of motion. *Mach. Dyn. Prob.* **29**(2):107–114.

160. Okniński, A., Kyziol, J. (2006) Perturbation analysis of the effective equation for two coupled periodically driven oscillators. *Int. J. Diff. Eq.* **2006**:56146.

161. Olsson, M.G. (1976) Why does a mass on a spring sometimes Misbehave? *Am. J. Phys.* **44**(12):1211–1212.

162. Oueini, S.S., Nayfeh, A.H., Pratt, J.R. (1999) A review of development and implementation of an active nonlinear vibration absorber. *Arch. Appl. Mech.* **69**:585–620.

163. Ovchinnikova, N., Panferov, A., Ponomarev, V., Severov, L. (2014) Control of vibrations in a micromechanical gyroscope using inertia properties of standing elastic waves. In: *Proceedings of the 19th World Congress, The International Federation of Automatic Control*, Cape Town, South Africa, August 24–29, pp. 2679–2682.

164. Park, H.S., Liu, W.K. (2004) An introduction and tutorial on multiple-scale analysis in solids. *Comput. Meth. Appl. Mech. Eng.* **193**:1733–1772.

165. Passaro, V.M.N., Cuccovillo, A., Vaiani, L., De Carlo, M., Campanella, C.E. (2017) Gyroscope technology and applications: a review in the industrial perspective. *Sensors* **17**:2284–2288.

166. Paulsen, W. (2013) *Asymptotic Analysis and Perturbation Theory*. Chapman and Hall/CRC.

167. Pavliotis, G.A., Stuart, A.M. (2007) *Multiscale Methods: Averaging and Homogenization*. Springer, Berlin.

168. Pender, J., Rand, R.H., Wesson, E. (2018) An analysis of queues with delayed information and time-varying arrival rates. *Nonlin. Dyn.* **91**:2411–2427.

169. Perig, A.V., Stadnik, A.N., Deriglazov, A.I. (2014) Spherical pendulum small oscillations for slewing crane motion. *Sci. World J.* **2014**:451804.

170. Pielorz, A. (2001) Vibrations of discrete-continuous models of low structures with a nonlinear soft spring. *J. Theor. Appl. Mech.* **39**:153–174.

171. Pokorny, P. (2008) Stability condition for vertical oscillation of 3-dim heavy spring elastic pendulum. *Reg. Chaot. Dyn.* **13**(3):155–165.

172. Polach, P., Hajzman, M., Sika, Z., Mrstik, J., Svatos, P. (2012) Effects of fibre mass on the dynamic response of an inverted pendulum driven by fibres. *Eng. Mech.* **19**(5):341–350.

173. Porter, B. (1965) Non-linear torsional vibration of a two-degree-of-freedom system having variable inertia. *J. Mech. Eng. Sci.* **7**(1):101–113.

174. Pradhan, S.C., Phadikar, J.K. (2009) Nonlocal elasticity theory for vibration of nanoplates. *J. Sound Vib.* **325**(1–2):206–223.

175. Raju, K.K., Hinton, E. (1980) Nonlinear vibrations of thick plates using mindlin plate elements. *Int. J. Numer. Methods Eng.* **15**(2):249–257.

176. Ramnath, R.V. (2012) *Multiple Scales Theory and Aerospace Applications*. ARC.

177. Rand, R.H., Holmes, P.J. (1980) Bifurcation of periodic motions in two weakly coupled van der Pol oscillators. *Int. J. Non-Lin. Mech.* **15**:387–399.

178. Rand, R.H., Zehnder, A.T., Shayak, B., Bhaskar, A. (2020) Simplified model and analysis of a pair of coupled thermo-optical MEMS oscillators. *Nonlin. Dyn.* **99**:73–83.

179. Rayleigh, J.W.S. (1945) *The Theory of Sound*. Section 68a, 2nd revised edn Mineola, NY: Dover Publications Inc.

180. Rega, G., Lacarbonara, W., Nayfeh, A.H., Chin, C.M. (1999) Multiple resonances in suspended cables: direct versus reduced-order models. *Int. J. Non-Lin. Mech.* **34**:901–924.

181. Rhoads, J.F., Shaw, S.W., Turner, K.L. (2008) Nonlinear dynamics and its applications in micro- and nanoresonators. *Proceedings of DSCC2008 ASME Dynamic Systems and Control Conference*, October 20-22, Ann Arbor, Michigan, USA, 2406.

182. Roseau, M. (1976) *Asymptotic Wave Theory*. North Holland/American Elsevier.

183. Rusinek, R., Warmiński, J., Waremczuk, A., Szymański, M. (2018) Analytical solutions of a nonlinear two degrees of freedom model of a human middle ear with SMA prosthesis. *Int. J. Non-Lin. Mech.* **98**:163–172.

184. Sado, D. (1997) *Energy Transfer in the Nonlinear Coupled Systems of Two-Degree of Freedom*. OWPW: Warszawa(in Polish).

185. Sado, D., Gajos, K. (2008) Analysis of vibrations of three-degree-of-freedom dynamical system with double pendulum. *J. Theor. Appl. Mech.* **46**(1):141–156.

186. Sanchez, N.E. (1996) The method of multiple scales: asymptotic solutions and normal forms for nonlinear oscillatory problems. *J. Symbol. Comput.* **21**:245–252.

187. Settimi, V., Gottlieb, O., Rega, G. (2015) Asymptotic analysis of a noncontact AFM microcantilever sensor with external feedback control. *Nonlin. Dyn.* **79**:2675–2698.

188. Shen, L., Shen, H.S., Zhang, C.L. (2010) Nonlocal plate model for nonlinear vibration of single layer graphene sheets in thermal environments. *Comput. Mater. Sci.* **48**(3):680–685.

189. Shivamoggi, B.K. (2002) *Perturbation Methods for Differential Equations*. Birkhauser: Boston.

190. Shoushtari, A., Khadem, S.E. (2007) A multiple scale method solution for the nonlinear vibration of rectangular plates. *Sci. Iranica* **14**(1):64–71.

191. Shu, X., Han, Q., Yang, G. (1999) The double mode model of the chaotic motion for a large deflection plate. *Appl. Math. Mech.* **20**(4):360–364.

192. Singh, P.P., Azam, M.S., Ranjan, V. (2018) Analysis of free vibration of nano plate resting on Winkler foundation. *Vibroeng. Proc.* **21**:65–70.

193. Smirnov, V.V., Kovaleva, M.A., Manevitch, L.I. (2018) Nonlinear dynamics of torsion lattices. Russ. *J. Nonlin. Dyn.* **14**(2):179–193.

194. Sobhy, M. (2014) Natural frequency and buckling of orthotropic nanoplates resting on two-parameter elastic foundations with various boundary conditions. *J. Mech.* **30**(5):443–453.

195. Srinil, N., Rega, G., Chucheepsakul, S. (2007) Two-to-one resonant multi-modal dynamics of horizontal/inclined cables. Part I: theoretical formulation and model validation. *Nonlin. Dyn.* **48**:23–252.

196. Starosta, R. (2011) *Nonlinear Dynamics of Discrete Systems – Asymptotic Approach*. Dissertation No 457, WPP, Poznań (in Polish).

197. Starosta, R., Awrejcewicz, J. (2009) Asymptotic analysis of parametrically excited spring pendulum. In: I. Visa (Ed.) *Proc. of the 10th IFToMM Symposium - SYROM 2009*, Brasov, Romania, Springer.

198. Starosta, R., Awrejcewicz, J., Sypniewska-Kamińska, G. (2017) Quantifying nonlinear dynamics of mass-springs in series oscillators via asymptotic approach. *Mech. Sys. Sig. Pr.* **89**: 149–158.

199. Starosta, R., Sypniewska-Kamińska, G., Awrejcewicz, J. (2011) Parametric and external resonances in kinematically and externally excited nonlinear spring pendulum. *Int. J. Bif. Chaos* **21**(10):3013–3021.

200. Starosta, R., Sypniewska-Kamińska, G., Awrejcewicz, J. (2011) Resonances in kinematically driven nonlinear spring pendulum. *DSTA 11th Conference*, Lodz, 2011, pp. 103–108.

201. Starosta, R., Sypniewska-Kamińska, G., Awrejcewicz, J. (2012) Asymptotic analysis of kinematically excited dynamical systems near resonances. *Nonlin. Dyn.* **68**(4):459–469.

202. Starosta, R., Sypniewska-Kamińska G., Awrejcewicz, J. (2017) Nonlinear effects in dynamics of micromechanical gyroscope. In: Awrejcewicz J., Kaźmierczak M., Mrozowski J., Olejnik P. (Eds.) *Engineering Dynamics and Life Sciences*. Department of Automation, Biomechanics and Mechatronics, Lodz, pp. 511–520.

203. Starosvetsky, Y., Gendelman, O.V. (2008) Dynamics of a strongly nonlinear vibration absorber coupled to a harmonically excited two-degree-of-freedom system. *J. Sound Vib.* **312**:234–256.

204. Stölken, J.S., Evans, A.G. (1998) A microbend test method for measuring the plasticity length scale. *Acta Mater.* **46**(14):5109–5115.

205. Suchorsky, M.K., Sah, S.M., Rand, R.H. (2010) Using delay to quench undesirable vibrations. *Nonlin. Dyn.* **62**:407–416.

206. Suciu, C.V., Tobiishi, T., Mouri, R. (2012) Modeling and simulation of a vehicle suspension with variable damping versus the excitation frequency. *J. Telecom. Inf. Technol.* **1/2012**: 83–89.

207. Sung, S.H., Culver, D.R., Nefske, D.J., Dowell, E.H. (2020) *Asymptotic Modal Analysis of Structural and Acoustical Systems*. Morgan & Claypool.

208. Sypniewska-Kamińska, G., Awrejcewicz J., Kamiński, H., Salomon, R. (2021) Resonance study of spring pendulum based on asymptotic solutions with polynomial approximation in quadratic means. *Meccanica* **56**:963–980.

209. Sypniewska-Kamińska, G., Starosta, R., Awrejcewicz, J. (2018) Two approaches in the analytical investigation of the spring pendulum. *Vib. Phys. Sys.* **29**:2018005.

210. Szolc, T. (2001) Simulation of dynamic interaction between the railway bogie and the track in the medium frequency range. *Multibod. Sys. Dyn.* **6**:99–122.

211. Telli, S., Kopmaz, O. (2006) Free vibrations of a mass grounded by linear and nonlinear springs in series. *J. Sound Vib.* **289**: 689–710.

212. Tondl, A., Nabergoj, R. (2000) Dynamic absorbers for an externally excited pendulum. *J. Sound Vib.* **234**(4):611–624.

213. Toupin, R.A. (1962) Elastic materials with couple-stresses. *Arch. Ration. Mech. Anal.* **11**(1):385–414.

214. Turner, K.L., Miller, S.A., Hartwell, P.G., MacDonald, N.C., Strogatz, S.H., Adams, S.G. (1998) Five parametric resonances in a micromechanical system. *Nature* **396**:149–152.

215. Tuwankotta, J.M., Verhulst, F. (2000) Symmetry and resonance in Hamiltonian systems. *SIAM J. Appl. Math.* **61**(4):1369–1385.

216. Vakakis, A., Gendelman, O. (2001) Energy pumping in nonlinear mechanical oscillators. Part II – Resonance Capture. *J. Appl. Mech.* **68**(1):42–48.

217. Vakakis, A.F., Manevitch, L.I., Mikhlin, Y.V., Pilipchuk, V.N., Zevin, A.A. (1996) *Normal Modes and Localization in Nonlinear Systems*. Wiley-Interscience: New York.

218. Venkateswara, R.G., Raju, I.S., Kanaka, R.K. (1976) A finite element formulation for large amplitude flexural vibrations of thin rectangular plates. *Comput. Struct.* **6**(3):163–167.

219. Volmir, A.S. (1972) *Nonlinear Dynamics of Plates and Shells*. Nauka: Moscow.

220. Vitt, A., Gorelik, G. (1933) Oscillations of an elastic pendulum as an example of the oscillations of two parametrically coupled linear systems. *J. Tech. Phys.* **3**:294–307.

221. Wang, L.S., Cheng, S.F. (1996) Dynamics of two spring-connected masses in orbit. *Celest Mech. Dyn. Astr.* **63**: 288–312.

222. Wang, Y., Li, F., Jing, X., Wang, Y. (2015) Nonlinear vibration analysis of double-layered nanoplates with different boundary conditions. *Phys. Lett. Sect. A Gen. At. Solid State Phys.* **379**(24-25):1532–1537.

223. Warmiński, J. (2001) Synchronisation effects and chaos in the van der Pol-Mathieu oscillator. *J. Theor. Appl. Mech.* **39**(4):861–884.

224. Warmiński, J. (2005) Regular and chaotic vibrations of a parametrically and self-excited system under internal resonance condition. *Meccanica* **40**:181–202.

225. Warmiński, J. (2010) Nonlinear normal modes of a self-excited system driven by parametric and external excitations. *Nonlin. Dyn.* **61**:677–689.

226. Warmiński, J. (2015) Frequency locking in a nonlinear MEMS oscillator driven by harmonic force and time delay. *Int. J. Dyn. Contr.* **3**:122–136.

227. Warmiński, J. (2020) Nonlinear dynamics of self-, parametric, and externally excited oscillator with time delay: van der Pol versus Rayleigh models. *Nonlin. Dyn.* **99**:35–56.

228. Warmiński, J., Kloda, L., Lenci, S. (2020) Nonlinear vibrations of an extensional beam with tip mass in slewing motion. *Meccanica* **55**:2311–2335.

229. Warmiński, J., Zulli, D., Rega, G., Latalski, J. (2016) Revisited modelling and multimodal nonlinear oscillations of a sagged cable under support motion. *Meccanica* **51**:2541–2575.

230. Wasow, W. (2018) *Asymptotic Expansions for Ordinary Differential Equations*. Dover Publications.

231. Weggel, D.C., Boyajian, D.M., Chen, S.E. (2007) Modelling structures as systems of springs. *World Trans. Eng. Technol. Edu.* **6**(1):169–172.

232. Weibel, S., Kaper, T., Baillieul, J. (1997) Global dynamics of a rapidly forced cart and pendulum. *Nonlin. Dyn.* **13**:131–170.

233. Wenming, Z., Bohua, W., Shuangshuang, Z., Shuang, L. (2013) *Int. J. Com. Sci.* **10**(3):

234. Wilbanks, J.J., Adams, Ch.J., Leamy, M.J. (2018) Two-scale command shaping for feedforward control of nonlinear systems. *Nonlin. Dyn.* **92**:885–903.

235. Williams, C.B., Shearwood, C., Mellor, P.H., Mattingley, A.D., Gibbs, M.R., Yates, R.B. (1996) Initial fabrication of a micro-induction gyroscope. *Microelectron. Eng.* **30**:531–534.

236. Wirkus, S., Rand, R. (2002) The dynamics of two coupled van der Pol oscillators with delay coupling. *Nonlin. Dyn.* **30**:205–221.

237. Xuefeng, S., Qiang, H., Guitong, Y. (1999) The double mode model of the chaotic motion for a large deflection plate. *Appl. Math. Mech.* **20**:360–364.

238. Yamaki, N. (1961) Influence of large amplitudes on flexural vibrations of elastic plates. *ZAMM* **41**(12):501–510.

239. Ying, Y.H., Yang, S., Zhang, F., Zhao, C., Ling, Q., Wang, H. (2012) *Chinese J. Mech. Eng.* **25**(4)

240. Yoon, S.W., Lee, S., Najafi, K. (2011) Vibration sensitivity analysis of MEMS vibratory ring gyroscopes. *Sens. Actuator A* **171**:163–177.

241. Yoon, S., Park, U., Rhim, J., Yang, S.S. (2015) Tactical grade MEMS vibrating ring gyroscope with high shock reliability. *Microelectron. Eng.* **142**:22–29.

242. Zhang, W.-M., Meng, G., Wei, K.-X. (2010) Dynamics of nonlinear coupled electrostatic micromechanical resonators under two-frequency parametric and external excitations. *Shock Vib.* **17**:759–770.

243. Zhong, Y., Guo, Q., Li, S., Shi, J., Liu, L. (2010) Heat transfer enhancement of paraffin wax using graphite foam for thermal energy storage. *Sol. Energy Mater. Sol. Cells* **94**(6):1011–1014.

244. Zhu, S.J., Zheng, Y.F., Fu, Y.M. (2004) Analysis of non-linear dynamics of a two-degree-of-freedom vibration system with nonlinear damping and non-linear spring. *J. Sound Vib.* **271**:15–24.

245. Zulli, D., Luongo, A. (2012) Bifurcation and stability of a two-tower system under wind-induced parametric, external and self-excitation. *J. Sound Vib.* **331**:365–383.

Index

Printed in the United States
by Baker & Taylor Publisher Services

Printed in the United States
by Baker & Taylor Publisher Services